T0189591

Coordinate Measuring Machines and Systems

Second Edition

MANUFACTURING ENGINEERING AND MATERIALS PROCESSING

A Series of Reference Books and Textbooks

SERIES EDITOR

Geoffrey Boothroyd
Boothroyd Dewhurst, Inc.
Wakefield, Rhode Island

Coordinate Measuring Machines and Systems

Second Edition

Edited by

Robert J. Hocken
& Paulo H. Pereira

CRC Press
Taylor & Francis Group
Boca Raton London New York

CRC Press is an imprint of the
Taylor & Francis Group, an **informa** business

CRC Press
Taylor & Francis Group
6000 Broken Sound Parkway NW, Suite 300
Boca Raton, FL 33487-2742

First issued in paperback 2017

© 2012 by Taylor and Francis Group, LLC
CRC Press is an imprint of Taylor & Francis Group, an Informa business

No claim to original U.S. Government works

ISBN 13: 978-1-138-07689-1 (pbk)
ISBN 13: 978-1-57444-652-4 (hbk)

This book contains information obtained from authentic and highly regarded sources. Reasonable efforts have been made to publish reliable data and information, but the author and publisher cannot assume responsibility for the validity of all materials or the consequences of their use. The authors and publishers have attempted to trace the copyright holders of all material reproduced in this publication and apologize to copyright holders if permission to publish in this form has not been obtained. If any copyright material has not been acknowledged please write and let us know so we may rectify in any future reprint.

Except as permitted under U.S. Copyright Law, no part of this book may be reprinted, reproduced, transmitted, or utilized in any form by any electronic, mechanical, or other means, now known or hereafter invented, including photocopying, microfilming, and recording, or in any information storage or retrieval system, without written permission from the publishers.

For permission to photocopy or use material electronically from this work, please access www.copyright.com (http://www.copyright.com/) or contact the Copyright Clearance Center, Inc. (CCC), 222 Rosewood Drive, Danvers, MA 01923, 978-750-8400. CCC is a not-for-profit organization that provides licenses and registration for a variety of users. For organizations that have been granted a photocopy license by the CCC, a separate system of payment has been arranged.

Trademark Notice: Product or corporate names may be trademarks or registered trademarks, and are used only for identification and explanation without intent to infringe.

Library of Congress Cataloging-in-Publication Data

Coordinate measuring machines and systems / editors, Robert J. Hocken and Paulo H. Pereira. -- 2nd ed.
    p. cm. -- (Manufacturing engineering and materials processing ; 76)
    Includes bibliographical references and index.
    ISBN 978-1-57444-652-4 (hardback)
    1. Coordinate measuring machines. 2. Dimensional analysis. I. Hocken, Robert J. II. Pereira, Paulo H. III. Title. IV. Series.

TA165.5.C66 2011
670.42'5--dc22
                                                  2011005327

Visit the Taylor & Francis Web site at
http://www.taylorandfrancis.com

and the CRC Press Web site at
http://www.crcpress.com

Dedication

To our beloved wives, Dene and Roseli,
for their unconditional support and
encouragement. Bob and Paulo

Contents

Preface

Since John Bosch edited and published the first version of this book in 1995, the world of manufacturing and coordinate measuring machines (CMMs) and coordinate measuring systems (CMSs) has changed considerably. Perhaps most significantly, we have seen a large volume of manufacturing move to Asia, particularly the People's Republic of China. We have also seen enormous growth in the capability of what were once called microcomputers and the incredible strides in communication through the multifaceted ever-changing marvel, the World Wide Web. In addition to that the proliferation and miniaturization of the cell phone, personal digital assistants of all sorts, digital books, etc. and much of the work we do and the tools we do it with would have seemed quite alien a decade and a half ago.

So, what has changed in *Coordinate Measuring Machines and Systems*? First and foremost, it is helpful to remind ourselves that the basic physics of the machines has not changed at all. A good deal of this book, therefore, deals with topics that have not changed in essence but have just become more deeply understood. In other areas, software as an example, the expectations of the user for operator interfaces, ease of use, algorithms, speed, communications, and computational capabilities have expanded remarkably. Further, some types of machines, particularly the non-Cartesian CMMs, have expanded in market share and increased in accuracy and utility. We have also seen big changes in probing systems, called accessory elements in this text, and the number of points they can deliver to ever more sophisticated software. New applications have multiplied and pressures to improve machine performance have continued to increase. The concept of uncertainty has been better defined and is now widely used. It has been an interesting and exciting 16 years.

In addition to two new editors, one from academia and one from a metrology-intensive user industry, this book has many new authors and a known cadre of experts who have grown with the field since the last version. Many of them the reader will recognize from the literature on metrology, machine, and software standards development, and their activities in technical societies. We, the editors, are confident that we have assembled a first-rate team and believe that this book will be a valuable resource for students, practitioners, and researchers. Our authors come from around the world, and we intend that this book will play an important part in the global economy of manufacturing.

Contributors

Dean E. Beutel has responsibility for global manufacturing process execution for Caterpillar Inc. His organization supports development, maintenance, and improvement of manufacturing processes internationally. Dean joined Caterpillar in 1978 as a sweeper, and has performed a variety of manufacturing and quality engineering functions for over 33 years. He obtained his certification as a journeyman machinist in 1984, as well as his bachelor's degree from Bradley University in production operations the same year. He was certified as a quality engineer by the ASQ in 1987 and maintains this certification.

John A. Bosch is the chairman and CEO of Commander Aero, Inc. Mr. Bosch joined the U.S. Air Force as an aircraft maintenance officer immediately after graduating from Penn State with an engineering degree in 1951 and then spent 28 years at General Electric with assignments in engineering and general management before joining Sheffield as president, a position he held from 1982 to 1993. Mr. Bosch was a research associate at the National Institute of Standards and Technology (NIST) and consultant, advance measurement technology for Giddings & Lewis. Mr. Bosch has authored a number of technical and business publications and served on the board of directors of Leland Electrosystems and Shaw Aero Devices, Inc.

James B. Bryan is an independent consultant in precision engineering. Mr. Bryan retired in 1985 from the Lawrence Livermore National Laboratory after 30 years of service in metrology. He is the recipient of the 1977 Society of Manufacturing Engineers International Medal for Research in Manufacturing, an honorary member of the American Society of Precision Engineering, a member of CIRP since 1964, a charter member of ANSI/ASME Committee B89 on Dimensional Metrology, and the first chairman of B89.6.2 Temperature and Humidity Environment for Dimensional Measurement. Mr. Bryan was selected by *Fortune* magazine as one of its heroes of the year in 2000, received a Lifetime Achievement Award also in 2000 from EUSPEN, was selected by SME's *Manufacturing Engineering* magazine as a 2007 Master of Manufacturing, and received the M. Eugene Merchant Manufacturing Medal of ASME/SME in 2008.

Ralf Christoph studied precision mechanics at the Friedrich Schiller University in Jena, Germany. In 1985, he received his PhD in the field of application of image sensors in optical coordinate measuring machines. He earned his postdoctoral thesis qualification in the field of optical sensors for geometric measurement in 1989. Since 1990, Dr. Christoph has been employed at Werth Messtechnik GmbH in Giessen, Germany, first as engineering and development manager and since 1993 as owner and managing director. For more than 10 years, he has participated in the preparation of guidelines and standards for VDI, DIN, and ISO in the field of coordinate measuring technology.

Ted Doiron is the leader of and a physicist in the Engineering Metrology Group of the Precision Engineering Division, one of five divisions of the Manufacturing Engineering Laboratory at NIST. He is the author or the coauthor of many technical papers and is considered one of the U.S. experts on gage blocks and gage block metrology. He is also responsible for complex dimensional standards at NIST.

Marion B. (Bill) Grant is a technical steward for metrology and advanced manufacturing for Caterpillar Inc. in Peoria, Illinois. He received his PhD in physics from the University of Illinois in 1985. Dr. Grant is active in ASME serving on the B46 and B89 committees, is the U.S. representative on the ISO TC213 Committee on coordinate metrology, and has several publications to his credit.

Robert J. Hocken is the Norvin Kennedy Dickerson, Jr., Distinguished Professor of Precision Engineering and Director of the Center for Precision Metrology at the University of North Carolina, Charlotte. Before 1988, Dr. Hocken worked at the National Bureau of Standard—NBS (now NIST) as chief, Precision Engineering Division. Dr. Hocken received his PhD in physics from the State University of New York at Stony Brook in 1969 and is an author or a coauthor of 60 articles and research reports. Dr. Hocken is a key contributor to many national and international standards organizations. He has received many awards including the Presidential Executive Award.

Jörg Hoffmann is a research assistant at the Chair Quality Management and Manufacturing Metrology of University Erlangen-Nuremberg (Germany), active member of VDI-GMA Section 3.44 "Dimensional Measurands" and is nominated for CIRP Research Affiliates. Dipl.-Ing. Hoffmann is doing research in the field of probing systems for the measurement of microparts and multisensor coordinate metrology. He is the author or the coauthor of 21 scientific papers, lecturer at the VDI seminar "Multisensor Coordinate Metrology" and developed an STM-based probing system for nanometer resolving CMMs (patent pending).

Wolfgang Knapp is the head of metrology at the Institute of Machine Tools and Manufacturing (IWF) at the Swiss Federal Institute of Technology (ETH) and a consultant in precision manufacturing, with his office located in Schleitheim, Switzerland. Dr. Knapp received his PhD from the Swiss Federal Institute of Technology in 1984. The theme of Dr. Knapp's thesis was a proposed method for testing CMMs. He has authored several technical papers and is currently the Swiss expert on international standards committees pertaining to machine tools.

Edward Morse has more than 20 years of experience with both coordinate measuring machines and geometric tolerancing. He was first interested in tolerancing and metrology while in the master of engineering program at Cornell University in the late 1980s. After earning his ME, he worked at the Brown & Sharpe Manufacturing Company—first as an applications engineer and then as a project leader in the advanced systems group, linking industrial shop floor controls to measuring equipment. He returned to Cornell to earn his MS and PhD in mechanical engineering.

The focus of his doctoral dissertation was a theoretical investigation of the "Physics of Mechanical Assembly," namely, how tolerances can be analyzed to determine if the assembly of intolerance components can be guaranteed. Since completing his graduate studies in 1999, Dr. Morse has been a member of the faculty in the Mechanical Engineering and Engineering Science Department at the University of North Carolina, Charlotte, home of a world-renown graduate program in dimensional metrology. His research interests include tolerancing for assembly, CMM testing and standards, estimation and evaluation of task-specific measuring uncertainty, and large-scale metrology systems and standardization. Dr. Morse holds Senior Level Certification as an ASME Geometric Dimensioning and Tolerancing Professional. He is a member of the ASME Y14 subcommittee 5.1 (Mathematical definition of Y14.5 dimensioning and tolerancing principles). He is also a member of ASME B89 Committee (Dimensional Metrology), B89.4 (Coordinate Measuring Technology), and B89.7 (Measurement Uncertainty), in addition to several project teams within the B89.4 group. In the area of international standards, Dr. Morse serves as a subject matter expert for the United States in ISO Technical Committee 213 for Working Group 10 (Coordinate Measuring Machines), Working Group 4 (Uncertainty), and Advisory Group 12 (Mathematical support group for GPS).

Hans Joachim Neumann studied radio engineering at the Mittweida High School of Engineering in Saxony, Germany. After two years of development in optoelectronic engineering at the Carl Zeiss Company in Jena, he transferred to Carl Zeiss in Oberkochen, Germany, in 1957. There, he first worked in managerial roles in the fields of electronic engineering for telescopes and precision measuring equipment, then in software and applications engineering, and finally as manager of marketing communication in the industrial measuring technology division. Until 2001, he was in charge of standardization and technical information as a consultant for the corporation and a member of the ISO committee for coordinate measuring technology. For 11 years, he was the chairman of the VDI/DIN joint committee for coordinate measuring technology, for which he was awarded an honorary badge by VDI. He currently works as a technical author and instructor.

Jun Ni is the Shien-Ming (Sam) Wu Collegiate Professor of Manufacturing Science and professor of mechanical engineering at the University of Michigan, U.S. He is as the founding Dean of the University of Michigan–Shanghai Jiao Tong University Joint Institute located in Shanghai, China (2006–2014). He also serves as the director of the S. M. Wu Manufacturing Research Center and as the co-director of a National Science Foundation sponsored Industry/University Cooperative Research Center for Intelligent Maintenance Systems. Professor Ni's research and teaching interests are in the areas of manufacturing science and engineering, the design of optimal maintenance operations, and statistical quality control and improvement. He has received many honors and awards, including ASME's William T. Ennor Manufacturing Technology Award, the Presidential Faculty Fellows Award from the National Science Foundation, and the elected Fellow of ASME and SME.

Paulo H. Pereira has over 28 years of experience in manufacturing and is currently part of the Global Quality Processes team at Caterpillar Inc. in Peoria, IL responsible for the corporate quality management system. Before that, he was the chief metrologist for five years at the Integrated Manufacturing Operations Division—East Peoria facility of Caterpillar Inc., where he was in charge of metrology planning and implementation. Before that, he was an internal consultant in metrology for Caterpillar for about five years. Dr. Pereira received his PhD in mechanical engineering (metrology) from the University of North Carolina, Charlotte, in 2001. Dr. Pereira has been a certified quality engineer by ASQ since 2006 and holds both his bachelor's and master's degrees in mechanical engineering from the University of São Paulo–São Carlos, Brazil. Dr. Pereira is a member of the ASME B89 committee (Dimensional Metrology) and also serves as a subject matter expert for the United States in the ISO Technical Committee 213 for Working Group 10 (Coordinate Measuring Machines).

Steven D. Phillips is the group leader of the Large Scale Coordinate Metrology Group at NIST and serves as the vice chair of the ASME B89 committee for dimensional metrology. He holds three U.S. patents and received the Department of Commerce's Gold and Silver Medals for work in coordinate metrology. Dr. Phillips is the author of 25 archival research publications in diverse fields such as chemistry, physics, applied optics, and precision engineering. Dr. Phillips holds a MS and PhD in physics from the University of California at Santa Barbara, a BS in mathematics and an MBA. He is also the SME-1 U.S. representative to TC213WG10 and TC213WG4 responsible for developing international standards for coordinate metrology applications and dimensional measurement uncertainty.

Craig M. Shakarji heads NIST's Algorithm Testing and Evaluation Program for Coordinate Measuring Systems. Dr. Shakarji chairs the ASME B89 project team on CMM software and serves as a subject matter expert and editor of several standards in the ISO 213 standards committee on CMMs. He was awarded the Department of Commerce Gold Medal for his achievements in CMM standards harmonization. Dr. Shakarji, a mathematician, received his PhD from the University of California, Los Angeles and his master's degree from Caltech, and in 1996 joined NIST, where he has done extensive research in the computational metrology field.

Dennis A. Swyt received his PhD in physics from Case Western Reserve University in 1971. After joining NIST in 1972, he had a number of assignments with increasing responsibilities. Dr. Swyt is the author of 50 technical papers, has a U.S. patent, and serves on several advisory panels. In 1980, he received the NIST Silver Medal Award for his development of a photomask linewidth standard. Dr. Swyt retired from NIST as chief of the Precision Engineering Division.

Albert Weckenmann is professor at and head of the Chair Quality Management and Manufacturing Metrology of University of Erlangen-Nuremberg (Germany), fellow of CIRP and vice-chairman of CIRP STC P. Dr.-Ing. Weckenmann has been doing research in coordinate metrology for more than 32 years and is the coauthor of a number of VDI/VDE and DIN standards, as well as books on coordinate metrology and geometrical product specification. He is chairman of IMEKO TC 14 (Measurement of Geometrical Quantities), chairman of the German VDI-GMA Section 3.40 "Micro- and Nanometrology" and member of the advisory board of DIN-Section NATG (Fundamentals in Engineering). Dr.-Ing. Weckenmann is the author or coauthor of more than 300 scientific papers and 5 books and editor or coeditor of 19 books. He is the inventor or coinventor of 17 patented inventions.

Guoxiong Zhang was conferred a degree of Honorary Doctor by Moscow State University of Technology (Stankin) in 1996. He has been honored with the titles of All-China Model Teacher, Model Worker of Tianjin City, Honorary Expert in Measuring and Testing Technologies and Instruments of Tianjin City for his outstanding contributions in teaching and research work. He was the chairman of the Department of Precision Instrument Engineering, Tianjin University, China, 1986–1995; and dean of the College of Precision Instrument and Opto-Electronics Engineering, Tianjin University, 1995–1997. He was elected as the chairman of the Scientific and Technical Committee on Precision Engineering of the International Academy for Production Engineering (CIRP), 1991–1994; president of the Chinese Production Engineering Institution, 1995–1999; chairman of the All-China Teaching Guiding Committee on Instruments and Gauges, 1996–2001. He worked as a visiting scholar at the U.S. National Bureau of Standards, 1981–1984; visiting professor at the University of North Carolina at Charlotte in 1991, 1998, 2003, and 2007. He has completed more than 60 projects including 12 international cooperative programs and three United Nation Development Programs. He has published 13 books and more than 500 academic papers. Among them, more than 50 papers were indexed by the Science Citation Index (SCI) and 180 by the Engineering Index (EI). He received a China National Invention Award and six awards for Advancement for Science and Technology conferred by The Ministry of Education of China, Tianjin City and other provinces of China. One of the projects completed by him and his American colleagues when he worked at the U.S. National Bureau of Standards received a U.S. government award.

1 Evolution of Measurement

Robert J. Hocken and John A. Bosch

CONTENTS

Measurement is an integral part of our everyday lives. It is something that most people take for granted. In looking back at the evolution of measurement, one finds that it relates directly to the progress of mankind. This chapter provides a brief historical summary of this evolution from the perspective of industrial metrology.

Measurement standards and devices were required to build the pyramids and other ancient structures. With land ownership and the beginning of farming, a means of measuring distances was required. To explore the world, navigation techniques requiring great accuracy needed to be developed. To overcome the poor reliability and high maintenance associated with the military rifle, the concept of interchangeable parts gained increased recognition. The manufacturing needs of producing interchangeable parts gave rise to gage blocks and functional gaging. As the automobile industry flourished with mass production, it was necessary to have parts made to exacting standards. From this need, the comparator became extensively used in the factory and new gaging techniques were developed. The automation of machine tools created the need for a faster and more flexible means of measuring. This requirement resulted in a new industry manufacturing three-dimensional measuring machines. In more recent times, the emphasis on quality improvement and international competition has accelerated the demand for faster and more accurate measurements. Coordinate measuring machines (CMMs) and systems of many types have evolved to fulfill these growing requirements.

1.1 PYRAMIDS PROVIDE EVIDENCE OF EARLY MEASURING SKILLS

Many ancient civilizations left behind great stone structures that leave one wondering how they could have been created with the tools and measuring equipment then available. Examples include the Great Wall of China, the monoliths on Easter Island, Mayan temples in South America, and Stonehenge in England. The pyramids of Egypt are among the most impressive. The Great Pyramid of Cheops (Khufu), built about 4,500 years ago, covers 52,000 square meters and contains approximately 2,300,000 stone blocks having an average mass of 2,270 kg each. It has been estimated that it took 100,000 men from 20 to 30 years to complete the pyramid. This is about the same effort in man-years as it took to put a man on the moon. The only heavy construction tools available to the Egyptians were levers, rollers, and immense earthen ramps. The Egyptians' measuring capability may have been highly refined because it has been estimated that the difference in height of opposite corners of the pyramid at its base is only 13 mm (Morse and Babcock 2009).

An early recording of measuring was found in the tomb of Rekhmire at Thebes (Figure 1.1), which dates back to the fifteenth century BC. To those involved in industrial metrology, it is interesting to note that the measuring task is occurring concurrently with the work process.

The trend today is to place measuring machines close to the machining process. Modern manufacturers are simply striving to return to the fundamentals as shown in this first recording of measuring. A possible interpretation of the illustration suggests that the surfaces around the edges were cleared off by using a square and a cord. It was much harder to get the entire surface flat. To achieve this, three bronze pegs of equal heights were used. The top of two of them were mounted on the cleared edges and the string pulled taut. The third peg was then passed under it to determine where the surface was too high and to clear the surface off with a chisel. When this was done in all directions, a flat face was attained. The picture clarifies the procedure.

1.1.1 THE CUBIT—ONE OF THE EARLIEST UNITS OF MEASURE

Man followed his natural instinct and ego by selecting his own body as a basis for the first units of measure—the length of his forearm, foot, and width of his finger. Such units were always available and easily understood. Of these, the cubit became the most widely used throughout the ancient world. The cubit is defined as the distance from the elbow to the end of the outstretched middle finger.

Figure 1.2 shows a replica of the master standard of the royal Egyptian cubit. The standard was based on the length of the forearm of the Pharaoh Amenhotep I (ca. 1550 BC). It was about 524 mm long and was known as the royal cubit. As shown in this replica, the standard was subdivided by scribed lines that divided the cubit into 2 spans, 6 palms, and 24 digits. The digits, in turn, were divided into halves, thirds, quarters, and down to 16 parts. The royal cubit master was made of black granite and placed in the custody of the royal architect. "Working" cubits made of wood were duplicated from the royal cubit and used by artisans in the great pyramid, tombs, and temple.

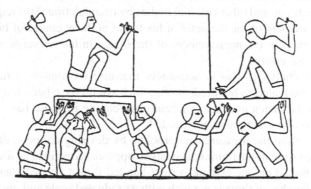

FIGURE 1.1 One of the earliest records of precise measurement is the Egyptian wall painting in the tomb of Rekhmire at Thebes built in ca. 1440 BC.

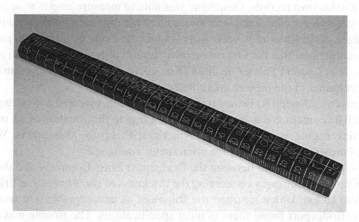

FIGURE 1.2 A replica of the royal Egyptian cubit.

The Greeks and Romans inherited the cubit from the Egyptians. From archaeological evidence, it has been found that none of these cubits agreed in length. The royal Egyptian cubit (Figure 1.2) is equal to 524 mm. The Greek cubit was 60.5 mm shorter. The area of a square with sides of one Greek Olympic cubit is very close to that of a circle having a diameter of one royal Egyptian cubit. If this is a coincidence, it is a very interesting one.

As far as it is known, the Greeks subdivided the cubit into 2 spans, 6 palms, and 24 digits. The Romans, in turn, took the Greek Olympic cubit but subdivided it according to their own ideas into 12 thumbnail breadths.

1.2 ACCURACY IN NAVIGATION IS BASIS FOR THE MICROMETER

Navigation using the position of the sun, moon, and stars requires very accurate measurement of angles. William Gascoigne, an astronomer, killed at the age of 24 in one of the battles of the English Civil War of 1642, developed an astronomical instrument by using a differential screw thread (Towneley 1666). Gascoigne measured the diameter of the sun, moon, and other celestial bodies by triangulation. This required a very accurate measurement of the diameter of his target at the eyepiece of his telescope. He knew any error in the measurement of the image in his telescope would cause large navigational errors.

Gascoigne's challenge was to accurately measure the image in his telescope. Using a scale graduated to hundredths of an inch would have been too crude. Even if a finer scale had been available, difficulty in reading it would have ruled it out. Another means of measurement was needed.

Gascoigne solved the measurement problem by devising calipers, the indicating fingers of which were moved simultaneously in opposite directions by a screw having a left-handed thread on one end and a right-handed thread on the other. Gascoigne measured the number of threads per inch with a graduated scale and, thus, computed the pitch of the screw. Likewise, he computed the advance for any fractional turn of the screw. This is the fundamental idea on which the modern micrometer is based. According to his own records, Gascoigne was able to measure angles to seconds.

Gascoigne was not concerned with mechanical measurements and could not foresee the future possibilities of his discovery. In fact, he made no attempt to patent his device. Records show that the first patent on a "screw caliper" was issued to a French mechanic, Jean Laurent Palmer, in 1848 (Roe 1916). This was a pocket instrument and the forerunner of the present micrometer.

A biography of Joseph R. Brown states that in 1852 he invented a dividing engine from which he constructed a vernier caliper reading to thousandths of an inch. His first dividing engine is located at the American Precision Museum in Windsor, Vermont. Further development of the micrometer continued.

In 1867, trouble occurred between the Bridgeport Brass Company and the Union Metallic Cartridge Company concerning the thickness of sheet brass that Bridgeport furnished to Union. Union returned the shipment as unacceptable. On rechecking the sheets, Bridgeport found them to meet specifications. The trouble was that the

Union Company's gage differed from Bridgeport's gage. Both differed from a third gage brought in to settle the dispute. All of the gages were supposed to be based on the U.S. Standard for wire gages adopted in 1857. This situation called for a remedy.

The superintendent of Bridgeport, S. R. Wilmot, designed a micrometer that measured to thousandths of an inch. It was read by a pointer that moved across an engraved spiral having the same pitch as the micrometer screw. Axial lines on the engraved spiral indicated the size of the micrometer opening. Wilmot's attempts to market his micrometer were unsuccessful because the instrument was too complex for practical use. The closeness of the graduations made reading them difficult and allowed no room for figures.

About the same time, J. R. Brown and Lucian Sharpe, while visiting the Paris Exposition, saw a Palmer micrometer. Using what they considered the best features of both the Palmer and Wilmot designs, they introduced the Brown & Sharpe micrometer in 1867. This was the first practical mechanic's micrometer marketed in the United States (Figure 1.3). By 1877, these micrometers were well established in the metalworking industry.

1.2.1 GAGE BLOCKS SATISFY NEED FOR MEASURING REFERENCES

Carl Edvard Johansson did more than any of his predecessors to bring accurate measurement directly into the machine shop. In 1887, at the age of 23, Johansson started his apprenticeship in Carl Gustafs Rifle Factory at Eskilstuna, Sweden. This company supplied the Swedish army with rifles. Because the very purpose of rifles made reliability imperative, accurate manufacturing was required (Althin 1948).

The accuracy of that day left much to be desired. The measuring instruments in the rifle factory consisted of snap gages, sliding calipers, and one micrometer. The manufacturing gages were steel blocks, one for each dimension. Although the idea of tolerances had been recognized, none were shown on working drawings.

When the Swedish government selected a rifle with a magazine, the manufacturing problem was compounded. An order for magazine rifles was placed with Mauser-Werke, a German firm. The contract provided that a commission from the Eskilstuna

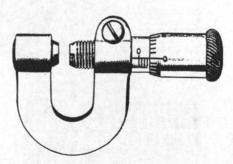

FIGURE 1.3 The first practical micrometer, 1867. (Courtesy of Brown & Sharpe [now a part of Hexagon Metrology, Inc.].)

plant inspect the new rifles and be given sufficient information to permit further production at Eskilstuna. Johansson was a member of this commission that went to Germany in 1894.

The task of producing individual gage blocks in the customary way for the many critical dimensions of the new Mauser rifle appalled Johansson. He believed that there must be a more effective way to control dimensions. Johansson then had his great idea to create a set of blocks increasing in uniform size, which could be used singly or in combination to equal the many manufacturing dimensions encountered. On his return to Sweden in 1896, Johansson had completed his computation of block sizes. His patent application (Althin 1948) specified the dimensions of 111 blocks in four series. From these blocks, any dimension from 2 to 202 mm could be set up in steps of 0.001 mm for a total of 200,000 different measurement combinations. An early typical set is shown in Figure 1.4.

Johansson had the utmost faith in his basic idea, but lack of funds to launch its full-scale development as an independent business caused gage block manufacturing to be done on a part-time basis. Preliminary work on the blocks was done under contract by the rifle factory. All finishing was done in Johansson's home workshop with improvised equipment during his free time.

The first set of blocks, with an accuracy of 0.001 mm, was sent to the Mauser rifle factory in 1896 and immediately put into service. Johansson proceeded with this part-time business development until 1914 when he resigned his position as armorer at the factory and, thereafter, devoted all his time to gage blocks.

Obtaining a Swedish patent on gage blocks proved difficult, but the patent was finally issued in January, 1904 with priority allowed to 1901. The British patent was issued in 1902 (Althin 1948).

Another idea developed and patented by Johansson in 1907 was termed *progressive tolerances*. Prior to that, it had been customary to use a bilateral constant tolerance in making fixed-size gages, regardless of their nominal size. In other words, the same tolerance for a 10-mm gage would apply as that for a meter gage.

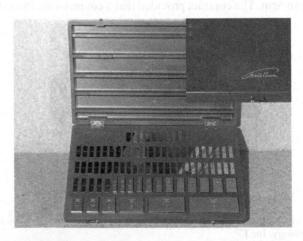

FIGURE 1.4   Set of gage blocks. (Courtesy of the American Precision Museum.)

As long as end measures such as gage blocks are used singly and temperature is ignored, the bilateral constant tolerance concept is tenable. But when used in combination, there can be a considerable discrepancy in the aggregate length, depending on the number of blocks and whether the individual tolerances are plus or minus. Johansson advocated the principle of using graduated tolerances based on the lengths of the blocks. This means that the smaller the block, the smaller the tolerance. This length-dependent tolerancing practice is commonly used today in some standards for specifying the accuracies of CMMs (also refer to Chapter 9, Performance Evaluation).

By 1907, American industry recognized the importance of Johansson's gage block sets. Henry M. Leland of Cadillac was the first automobile manufacturer to have a set of gage blocks. When the United States began to tool up for World War I, gage blocks "were not to be had at any price." The Bureau of Standards, persuaded by William E. Hoke, obtained the sum of $375,000 from the Ordnance Department to make them, and after some effort produced 50 sets for the war production effort (Cochrane 1966). "Jo Blocks" remain in common use today. Over the years, millions of factory personnel have been trained on the proper use and handling of gage blocks.

1.2.2 EARLY COMPARATORS SET NEW STANDARDS FOR ACCURACY

Without detracting from Johansson's important achievement, it must be stated that the indicating shop comparator was necessary to give the gage block concept its full effectiveness and flexibility. The shop comparator is the only practical means of bridging block increments.

Early length standards, such as "Iron Ulna," dating from the reign of Edward I, and the brass yard standards made during the reign of Henry VII (1496) and Queen Elizabeth I (1588) were end standards. The earliest scientific device used as a comparator was a form of calipers or beam compass. Incidentally, an early means of disseminating "true" measures was to embed inside end standards (i.e., calipering) in masonry walls at some convenient central location, such as the market square (Evans 1989).

In the eighteenth century, pressure for improved accuracy increased. In England in 1742, George Graham compared standards of length by using a device where the jaws were moved by micrometer screws with divided heads. About the same period, the clock makers were expressing an interest in thermal effects on materials. This led to the development by Graham and others of "precision dilatometers," which use either a micrometer screw to measure thermal expansion of the test price or a mechanical lever to amplify the displacement and a pulley system to connect the lever movement to a dial indicator or optical system. In later years, special alloys were developed to minimize thermal effects around room temperature for the construction of length standards, leading to a Nobel Prize in physics for Charles Edouard Guillaume in 1920.

Microscope-based comparators are clearly more applicable to line standards than to end standards. Their emergence, perhaps, reflects the growth of de facto standardization on scales, a situation clarified in England in 1824 when use of John Bird's 1760 standard was legalized. End standards have long remained the working standards of industry; however, means of comparing end and line standards has been an underlying theme in the history of engineering metrology (Evans 1989).

FIGURE 1.5   A comparator commonly used in factories.

One of the important practical developments of a precision comparator was made by Joseph Saxton when he built his reflecting comparator. In recognition of its value to science, Saxton received the Johan Scott Legacy Medal in 1837. Saxton's comparator used a light beam to amplify very small displacements of the lever arm as a means of measurement. The movement on the scale was 97.536 mm for 0.0254 mm displacement of the anvil, providing an amplification factor of 3840.

The Rogers-Bond Universal Comparator was constructed in 1879 by the Pratt & Whitney Company from plans proposed by Professor W. A. Rogers and Mr. George M. Bond. This comparator was more elaborate, had more refinements, and was more versatile than its predecessors. It was adaptable for measuring both line and end standards and offered a choice of five separate comparison methods. A more modern comparator is shown in Figure 1.5.

1.3 INTERCHANGEABLE PARTS GAIN INTERNATIONAL RECOGNITION

Until the end of the eighteenth century, it was the general practice for the individual craftsman to make a complete product (Battison 1976). This involved fitting each component as it was being constructed. Depending on the craftsman's training and ability, the products were more or less acceptable, but no two similar products or components were functionally the same.

The need for interchangeable parts began to be recognized as manufacturing volumes increased. Examples of early products made with interchangeable parts are clocks and pulley blocks for naval vessels. The military rifle, however, brought international recognition to the system of interchangeable parts.

One of the first artisans to recognize the benefits of interchangeable parts for muskets was a French gunsmith named Honoré LeBlanc. In 1785, Thomas Jefferson, as the U.S. minister to France, heard about LeBlanc, who had developed a plan for manufacture of interchangeable musket parts 15 years previously. Jefferson was keenly interested and tried unsuccessfully to bring LeBlanc to the United States, where he would have been encouraged to implement his plan for the manufacture of muskets with interchangeable parts.

The period of the Napoleonic Wars emphasized the imperative need for interchangeable parts. For the first time, very large armies with muskets were put into the field. Maintenance of the muskets soon became a huge problem because each spare part had to be made and fitted individually.

Making and fitting individual parts was a very slow and expensive operation. At one time, the British government had more than 200,000 muskets waiting for repair. This was reported to be more than were in serviceable condition. It was a problem that had to be solved.

The two essentials for interchangeability are reliable measurements and good machine tools. At this time, no uniformity in measurement standards existed, and machine tools were in their infancy.

The developments in machinery and gaging methods over the preceding 50 years came together in the small town of Windsor, Vermont, in 1845. It was in Windsor that the American system was brought to the highest level then known, and from there, it began to extend its influence to the rest of the world.

Richard Lawrence was a man with an innate technical sense and mechanical skills. He first used his talents at N. Kendall & Company, a Windsor gunsmithing firm. In 1843, Lawrence and Kendall became partners and started a new custom gun shop. One year later, S. E. Robbins, a local retired businessman, approached the two gun makers with the suggestion to bid on a government contract for 10,000 rifles. They quickly formed a new partnership named Robbins, Kendall, and Lawrence, submitted a bid of $10.90 per rifle, and won the contract by underbidding established contractors. The contract was to be completed in three years.

By 1846, the partners had built a new plant. They recruited top talent from government and private armories and selected the best manufacturing methods, especially those used at the well-established Springfield Armory in Massachusetts. The contract for the 10,000 model 1841 rifles, was so successfully completed that the partners immediately received another contract for 15,000 more in 1848. At this point, Kendall sold his interest in the company.

With a second government contract, Robbins & Lawrence began designing and building improved machinery. The most important machines developed from their efforts included an improved standard milling machine, a profile miller, and a vertical-axis turret lathe. They also began producing the new Jennings breech-loading rifle that used special ammunition patented in 1849. The Jennings' rifle evolved into the Volcanic, Henry, and eventually the well-known Winchester.

Robbins & Lawrence recognized the commercial potential of their work and exhibited their guns at the "Crystal Palace" industrial exposition in London in 1851, which turned the attention of the industrialized world to America (Battison 1976). Success must have produced excitement for them with their new manufacturing concepts, new machine tools, new gun technology, and refined gaging practices.

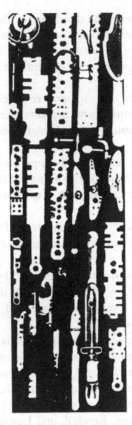

FIGURE 1.6 U.S. government inspection gages used to assure parts interchangeability for rifles made at Robbins & Lawrence. (Courtesy of Smithsonian Institution.)

Robbins & Lawrence rifles, along with other American light goods, took the European exhibition at the Crystal Palace by storm. A group of England's best engineers attended the exhibition to study manufacturing methods. The engineers' report coined the phrase, "the American system of manufacture," and created a major stimulus to develop interchangeable parts manufacturing (Evans 1989).

The Robbins & Lawrence rifles won an award and attracted a great deal of attention, inducing Parliament to send a commission to the United States. Its mission was to study the "American System" of interchangeable manufacturing and to secure the machinery necessary to introduce the system at the Enfield Armory near London. Some typical gages used at the Robbins & Lawrence factory are shown in Figure 1.6.

The Robbins & Lawrence firm received an immediate order for 25,000 Enfield rifles with interchangeable parts and for 141 metalworking machines to equip the Enfield Armory. The Crimean War created an urgent need for the rifles.

The financial risks associated with early efforts toward mass production are illustrated by the fact that only four years after the Crystal Palace Exposition, Robbins & Lawrence declared bankruptcy, the victim of poor management and some bad luck. A venture into railroad car building had left the company financially weakened. Further heavy expenditures for new plants and equipment were made on the promise of a large additional contract for Enfield rifles, which did not materialize. Although Robbins & Lawrence's success was short-lived, it does not detract from the ingenuity and enterprise displayed at their factory in Windsor, Vermont. The machine tool industry and the role of measuring in manufacturing had taken shape.

The Robbins & Lawrence factory in Windsor that produced rifles for the Crystal Palace Exposition is still standing and is the home of the American Precision Museum. After Robbins & Lawrence's bankruptcy, the factory became an armory and was used to produce 50,000 special model Springfields for the Union army during the Civil War. The building was then used to produce basic machine tools and products, such as sewing machines, for the commercial market. In 1870, the building became a cotton mill and served as a hydroelectric power station from 1898 until it was donated to the American Precision Museum in 1966.

1.4 DIAL INDICATOR SIMPLIFIES MEASURING

The familiar dial indicator is a legacy from European watchmakers. In the nineteenth century, New England watchmakers made further developments on the dial indicator. It was developed primarily as a gage for the production of accurate watches. Its broad usefulness to industry was recognized later.

The records of the U.S. Patent Office show that on May 15, 1883, John Logan of Waltham, Massachusetts filed a patent application for a dial indicator to which he referred as "an improvement in gages." In appearance, the instrument did not differ much from present dial indicators, but its movement was quite different. Instead of using the familiar rack and pinion, Logan used a fine Swiss watch chain to transmit the motion of the gaging spindle to the indicating pointer. The practical limitation on the amount of amplification by this method was probably the reason that Logan later switched from the chain movement to a rack and pinion mechanism.

Recognizing the broad market for the dial indicator, Frank E. Randall, an inventor, bought the Logan patents in 1896. In partnership with Francis G. Stickney, Logan undertook the manufacture and marketing of dial indicators. Afterwards, B. C. Ames entered the field with a gear-type indicator.

Felix Auerbach, writing of the Zeiss Works in Jena, indicated that German industry was searching for new measuring instruments of greater accuracy. In 1890, Professor Ernst Abbe established the measuring instrument department of the Zeiss Works. By 1904, Zeiss had developed a number of instruments for the general market among which was the dial indicator (Auerbach 1904). In the United States, one of the major producers of dial indicators was Federal Products. An example of their dial indicators is shown in Figure 1.7.

FIGURE 1.7 Early dial indicator, ca. 1924. (Courtesy of Federal Products [now a part of Mahr-Federal, Inc.].)

1.5 AUTOMOBILE ACCELERATES DEVELOPMENTS
IN METROLOGY

No mechanical product has had a more profound effect on our contemporary society and economy than the automobile. Dimensional metrology has greatly benefited by the production demands of the automobile. The industry had its start in Germany when Karl Benz began building his first gasoline engine in 1878. Together with Gottlieb Daimler, they produced the first motor car in 1885. Their three-wheeled car had an electric ignition, a water-cooled engine, and a differential gear. Karl Benz later developed a float-type carburetor and a transmission system.

In 1893, in Springfield, Massachusetts, Frank Duryea built the first successful American gasoline-powered car. The American car was a single cylinder, 4-hp, buggy-type vehicle. The Detroit Automobile Company, which later became Cadillac, began in 1899. The Henry Ford Automobile Company was formed in Detroit in 1901 and changed its name to the Ford Motor Company in 1903. A dentist, Dr. E. Pfennig, from Chicago, bought the first Ford Motor Company car for $850 in July 1903. By March 1904, Ford had sold 658 cars. Henry Ford is widely considered to be the developer of the moving assembly line (Figure 1.8), which revolutionized manufacturing.

Another event that would eventually propel America into the forefront of automobile production occurred in 1908. Three Cadillac cars that had been shipped to England were completely disassembled, new parts intermixed, and then reassembled (Wren 1991). The impact on the industrial observers was as profound as it had been in 1851 at the Crystal Palace Exposition in London where the Robbins & Lawrence rifles were on display. Conceived by Henry Leland, the demonstration

FIGURE 1.8 Early moving assembly line. Flywheels and magnetos being assembled at Ford's Highland Park, Michigan plant in 1913. (Courtesy of Ford Motor Company.)

of interchangeable parts won the Dewar Trophy for Cadillac. The Dewar trophy was awarded each year by the Royal Automobile Club (RAC) of England "to the motor car which should successfully complete the most meritorious performance or test furthering the interests and advancement of the (automobile) industry." Of more importance, the use of interchangeable parts revolutionized automobile manufacturing. To achieve this, considerable effort was being made in the area of gaging. Equally important was the growing ability to establish manufacturing practices needed to control dimensions traceable to the accepted standards at that time.

1.5.1 Reed Mechanism Provides Greater Shop Floor Precision

In the 1920s, industrial leaders, especially those in the young automotive industry, began to recognize the benefits of tolerances closer than those for which fixed-size gages were suitable. William Bagley, Chief Inspector for Studebaker, was among the first to advocate replacing fixed-size gages with instrument gages for mass production operations.

Without precision comparators, the full potential of the gage block standards introduced by Carl Johansson could not be realized. Also, parts could not be classified for selective assembly. Because of the lack of accuracy of many machine tools of that day, selective assembly was the only way to achieve accurate fits.

The chief obstacle to such progressive concepts was that no practical precision comparator adaptable to mass production inspection was available at that time. Although precision comparators existed, they were laboratory instruments requiring a high degree of skill to operate and were too slow for inspecting automotive parts.

The idea chiefly responsible for circumventing this impasse and opening the way for widespread instrument gaging in production was the principle of the reed mechanism. It consisted of an ingeniously simple mechanical device for amplifying small displacements, such as the displacement of the gaging spindle. Essentially, the reed mechanism consists of two steel blocks and four reeds, the reeds being made of flexible strips of spring steel. The slightly separated blocks are connected by two horizontal reeds. One block is anchored to the gage head; the other, which carries the gaging spindle, is free to move vertically. The two remaining reeds are solidly fastened to the top inside faces of the blocks, one to each block. Displacement of the floating block causes the horizontal reeds to flex and the connected ends of the vertical reeds to sweep through an arc, amplified by a pointer and further by an optical system (Figure 1.9). A good description of the mechanics of these reed mechanisms is provided in the literature (Dotson 2006).

The idea of the reed mechanism was conceived by two metrologists, independently, but not concurrently. The first, E. Mark Eden, developed the idea in 1918, while he was on the staff at the Metrology Division of the National Physical Laboratory of England. The second, Arthur Schoof, an American from Western Electric, was apparently unaware of Eden's work. Those attending a gage manufacturers' meeting in Chicago in 1927 were invited to visit the Western Electric plant and see the company's inspection facilities. Although no special attention was drawn to it, the display included an instrument incorporating the reed principle.

FIGURE 1.9 This visual gage uses a reed mechanism to provide a 10,000 to 1 magnification. (Courtesy of Sheffield Measurement [now part of Hexagon Metrology, Inc.].)

Among the visitors that day and apparently the only ones to grasp the significance of the reed principle, were C. H. Reynolds and Charles E. Watterson, representing the Sheffield Gage Corporation. Reynolds and Watterson immediately recognized the great potential of this sensitive, frictionless, amplifying device. They recognized this device as the basis for a new line of comparators that could be used in the shop for fast, accurate inspection by workers who had no special skill in that area.

Sheffield started negotiating with Western Electric for a license and received one in 1929. The following year, Sheffield introduced the "Visual Gage," which was an indicating comparator using the frictionless reed mechanism together with a weightless light beam lever arm for amplification. Ford was the first automobile manufacturer to use the "Visual Gage," which proved to be ideal for measuring close tolerances in production. Tens of thousands of these comparators were produced and are still used today, which confirms their value as a dimensional inspection tool.

1.5.2 Air Gaging Proves Effective for Checking Tight Tolerance Parts

Another development that influenced industrial metrology was pneumatic gaging (Curtis and Farago 2007). It provided the capability to measure closer tolerances at higher speeds demanded for mass production.

According to historical records, the idea of using a fluid as a gaging medium stems from Cruikshank and Fairweather. As early as 1917, they suggested the relationship of fluid pressure and nozzle area as the basis of measurement in the field of paper manufacturing.

Interest in fluids as a measurement means arose spontaneously on both sides of the Atlantic. A U.S. patent on an air gage was issued to N.T. Harrington in 1922 (Harrington 1922).

Although the early work in using fluid as a gaging medium was in the investigation of such characteristics as flatness and area, the experience gained prepared the way for the dimensional air gage.

The air gage is essentially a comparator that uses the effect of small dimensional changes on metered air in the pneumatic gaging circuit. It is based on the fact that the free flow of compressed air through an open orifice is restricted when an obstruction is brought close to it. Such an obstruction reduces the velocity of flow while raising the pressure in the circuit behind the orifice. Within certain limits, the closer the obstruction is to the orifice, the more pronounced the effects.

In the case of the air gage, the orifice is a jet (or jets) in the air gage tooling. The obstruction is the surface of the workpiece being gaged. By metering either the change in back pressure in the pneumatic circuit or the change in its velocity by flow, the clearance between jets in the air gage tooling and the adjacent workpiece surface is determined. A typical air gage is shown in Figure 1.10.

The development of the Plunjet in the 1950s expanded the capability of air gaging. First developed for measuring the dimensions of turbine blades for General Electric, the Plunjet extended the range of useful measurement variations from 0.13 to approximately 2.54 mm. The principle of the Plunjet is based on limiting the flow of air by using a conical-shaped valve instead of simply obstructing the flow from an open jet by the master or part being measured. The sliding valve contracts the

FIGURE 1.10 Air gaging for checking valve concentricity in an automobile engine. (Courtesy of Sheffield Measurement [now a part of Hexagon Metrology, Inc.].)

master or workpiece at an accurate location, and the metered air is exhausted to the atmosphere from side ports on the Plunjet.

Whether the designers of the first commercial dimensional air gage realized their purpose could be accomplished by metering either back pressure or flow velocity is unknown. No practical velocity-metering device had yet been developed. The first air gage to be marketed was a back-pressure gage, which Solex, of Germany, introduced in 1926.

The first air gage widely marketed in the United States was introduced in 1935 by the Sheffield Gage Corporation to explore the form of flat and cylindrical surfaces in three dimensions simultaneously. The measurements were made by metering the pressure loss between the workpiece surface and the master.

As the possibilities of air gaging became apparent, more gage makers became interested and variations in circuitry began to appear. One of these is the Venturi-type circuit, which combines characteristics of the back-pressure gages and the flow-type gage.

In the 1940s, air/electronic gages were introduced. The dimensions were displayed and recorded electronically, while the sensors remained conventional air gages. Transducers were used to convert the air pressure signals to electronic signals. These systems made it possible to directly measure high-volume production quantities in the production line without increasing the required labor. Automatic gaging systems of this type were used for classifying parts for selective assembly (Figure 1.11), as well as for identifying and sorting out failed parts. These systems

FIGURE 1.11 Automatic air/electronic gage for measuring cylinder bores in a V8 engine block. Pistons are matched with the cylinder bores for selective assembly. (Courtesy of Sheffield Measurement [now a part of Hexagon Metrology, Inc.].)

proved to be most effective for measuring pistons, wrist pins, crankshafts, engine blocks, axles, and many other parts that require accurate dimensions.

1.5.3 ELECTRONIC GAGING EXPANDS CAPABILITY FOR PROCESS CONTROL

The practical application of both mechanical and electrical methods of dimensional gaging appeared in the 1920s. Today, the term *electronic gaging* comprises a category of measuring instruments capable of detecting and displaying extremely small dimensional variations. Any dimensional change causes an electrical signal that is amplified and displayed. The sensing element is frequently a linear variable differential transformer (LVDT). Because variations are being sensed, the electronic gage requires the use of a master to establish data or references from which the variations are to be measured. Some manufacturers supply both "minimum" and "maximum" masters. This method provides the highest degree of accuracy in setting the gages to determine whether the production parts are within the allowable tolerances. Examples of electronic gages are shown in Figure 1.11.

In recent years, electronic gaging has become more effective by adding computer-controlled systems for automatic mastering, statistical data presentations, and improved flexibility to handle a wide range of parts. Their accuracies in shop environments are well proven because the gage is simply a sophisticated comparator. Dimensional integrity is contained within the master that is checked by standards traceable to the international standard of length (refer to Chapter 2, The International Standard of Length).

1.5.4 MACHINE TOOLS EVOLVE INTO EARLY COORDINATE MEASURING MACHINES

Several measuring machines, which were more than comparators in the commonly accepted sense, have been manufactured (Hume 1953). Although different types of early measuring machines varied in their designs and principles of operation, they had the common feature of containing their own standards of measurement in the form of a scale or scales, micrometer, or other device.

The Matrix machine was designed as an intermediate between the short-range comparator and the self-contained measuring machine. It relied on slip gages along with a micrometer used in conjunction with a fiducial indicator. The Matrix machine was suitable for measuring both external and internal diameters on plain, tapered, and threaded work. The principle of operation is rather unusual because a diameter is determined by measuring two radii at exactly 180°.

The Newall measuring machine was made of a rigid bed on which a headstock and a tailstock that carry the measuring faces were mounted. A number of one-inch rollers in a slot were placed along the center of the bed. The exact sizes of the rollers were selected so the cumulative deviation from the nominal distance over a number of rollers was very small. A micrometer was fitted into the headstock and a bubble indicator mounted on the tailstock. The machine was particularly suitable for the measurement of long gages.

The Microptic measuring machines made by Messrs. Hilger & Watts, Ltd., of the United Kingdom, incorporated their own standard in the form of a glass scale. Carl Zeiss of Germany first introduced this type of machine in the late 1920s, and it proved very popular in many parts of the world. The machine was quite useful for the direct calibration of plug gages, and it had an almost limitless range of applications in other types of measurements.

The Société Genevoise of Switzerland has played an active part in the development of metrology since 1865. They are the originators of jig-boring machines in the modern sense. The well-known SIP, Société Genevoise d'Instruments de Physique, machines are of the universal type and their basic designs are used in both jig-boring machines and CMMs. As an example, the SIP "Trioptic," which was introduced in 1961, was later transformed into the SIP 560M CMM.

The Universal measuring machine, introduced by Moore Special Tool of Bridgeport, Connecticut was based on the jig-boring machine just as the earlier-mentioned SIP machines (Moore 1970). A Moore No. 3 machine is shown in Figure 1.12.

Both SIP and Moore Special Tool Company continued producing measuring machines based on jig bore technology that was used for very high accuracy applications. Moore introduced the Moore 48 measuring machines in about 1968, and a modified version of this machine, called the Moore 5Z, was used at the National Bureau of Standards (NBS; now National Institute of Standards and Technology [NIST]) for many years. This machine, shown in Figure 1.13, was the machine on which software correction was developed (Hocken et al. 1977). A slightly modified Moore machine, called the M60, is probably still, for its size, the most accurate machine in the world.

FIGURE 1.12 Moore No. 3 universal measuring machine introduced in 1957. Richard F. Moore (right) with customer. (Courtesy of Moore Tool Company.)

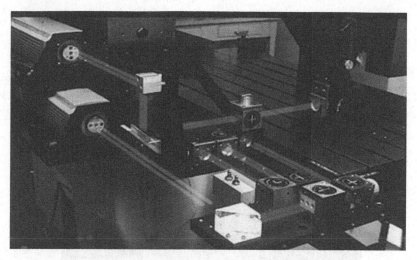

FIGURE 1.13   The fixed-bridge measuring machine from Moore Special Tool (a modified M48), which was used to develop software correction of CMMs. (Courtesy of NIST.)

1.6   FIRST COORDINATE MEASURING MACHINE DEVELOPED AS AID TO AUTOMATED MACHINING

The first measuring machine that falls into the category most commonly called CMMs was developed by Ferranti, Ltd., of Dalkeith, Scotland (Ogden 1970). An early model is shown in Figure 1.14. This CMM was developed as a companion product to their growing family of numerical-controlled machine tools. It is interesting to note that Ferranti had not been in the measuring equipment business. Ferranti developed the CMM in response to the need for faster and more flexible measuring when machining became more automated. As is often the case, the existing measuring equipment manufacturers had not recognized the potential of the CMM market that was to emerge.

In 1956, just two years after Harry Ogden joined the Numerical Control (NC) Division of Ferranti, Ltd. as chief mechanical engineer, he invented the Ferranti inspection machine. It became apparent to Ogden and others at Ferranti that parts made in a matter of minutes on their new NC machines took hours to inspect. Ogden then conceived that a freely moving, mechanical measuring machine with electronic numerical display would facilitate inspection of machined components. This, in effect, changed the whole economic aspect of conventional inspection methods by reducing inspection time and the skill required for inspection.

The key element that made the development of the CMM possible was the availability of an accurate, long-range, electronically compatible digital measuring system. The optical grating and the moiré fringe sensing system were ideal. They were accurate, relatively inexpensive to manufacture, and easily reset to zero. Their accuracy was based upon averaging a number of lines and was not dependent upon the accuracy of any one line. Credit for the optical grating system should properly go to David T. N. Williamson of Ferranti, who is remembered as being very bright and inventive.

FIGURE 1.14 Ferranti coordinate measuring machine. (Courtesy of Ferranti, Ltd. [now International Metrology Systems, Ltd.].)

The initial Ferranti development was an inspection machine with X and Y movements of 610 mm and 381 mm, respectively, and a Z "daylight" of 254 mm designed for production inspection accuracy of 0.025 mm and a resolution of 0.012 mm. The machine was equipped with tapered probe tips that established the constraint type of inspection process and revolutionized the approach to production inspection. The machine was also fitted with locks and fine adjustments on the x and y axes so that measurements could be made by means of a projector microscope on components, which were flexible or delicate. Initially, the z-axis measurement was made by adding a support bracket to the front of the cantilever arm bearing on a steel straightedge mounted on the front edge of the table. Full Z measurement was introduced in 1962 (Ogden 1970).

The Ferranti CMM was a classic kinematic design rather than a conventional machine tool design. The underlying principle of kinematics involves minimum constraint with provision for alignment of the moving elements (Maxwell 1890). The design was radically different and caused machine tool designers to rethink their design principles.

This original machine created a large market throughout the industrial world and led to the development of similar machines with larger capacities and improved accuracies and resolutions. Competitors entered the market at an average of two each year for the next 25 years. Such a proliferation of competitors has led to a series of consolidations. In the case of Ferranti, they ceased production of CMMs in 1992 and sold their remaining assets to their previous Korean dealer, which operated the business under the name "International Metrology Systems." A search of the World Wide Web* reveals that a company called International Metrology Systems still markets CMMs, though the company ownership is unclear.

---

* Accessed February 23, 2011.

1.6.1 SHEFFIELD INTRODUCES COORDINATE MEASURING MACHINES TO THE NORTH AMERICAN MARKET

In the summer of 1959, George Knopf, general manager of the Industrial Controls section of the Bendix Corporation, attended the International Machine Tool Show in Paris. While he was examining a NC positioning system in the Ferranti display, a two-axis CMM, also made by Ferranti, captured his interest. Knopf immediately recognized its sales potential for the Bendix Controls business and the Sheffield Corporation, which had been acquired by Bendix in 1956.

Knopf decided to take immediate action and flew directly from the Paris show to the Ferranti facilities in Edinburgh, Scotland. Knopf learned more about details of the machine and established contacts for negotiations. Later, a contract was signed providing Bendix with exclusive rights to sell Ferranti CMMs in the North American market.

The first Ferranti machines arrived at the Industrial Controls section in early 1960. One machine was shipped to the Sheffield subsidiary of Bendix for evaluation. The Sheffield engineers were dubious about the accuracy of the machine, but it proved to be better than they had expected. Not wasting any time, Bendix displayed the first CMMs in the North American market at the National Machine Tool Show in Chicago in 1960.

In 1959, even before the first machines were shipped to Bendix, one was delivered to the Western Electric plant in Winston-Salem, North Carolina, for John Haney, who was the supervisor of the tool and gage laboratory. Having read about the Ferranti machine in *American Machinist* magazine and having recognized its time-saving potential, Sheffield representatives Marcus Crotts and Charles Saunders became interested. Their customers were measuring with a granite surface plate, height gages, and gage blocks. Although the height gage technique was accurate, it was extremely time-consuming and cumbersome. When John Haney saw the article, his interest began to grow. Marcus Crotts contacted Ferranti and requested distribution rights in his territory but learned that Bendix would handle sales in the United States. Bendix received sales credit for the first machine.

Western Electric had just initiated the return-on-investment method for evaluating capital equipment purchases. Although John Haney was unfamiliar with the technique, he projected that the $12,000 machine would pay for itself in a reasonable time. Marcus and John spent many hours estimating the payback period and felt that they may have been exaggerating the numbers to justify a payback period of three years. John submitted the requisition for approval and, after a series of meetings with management, received authorization to purchase the machine. The actual payback period was only nine months. John kept meticulous records comparing inspection times and accuracies between the CMM and the conventional surface plate, height stand, and gage block technique. A single center distance measurement with the conventional technique would often take 20 minutes, whereas the same measurement could be made with the CMM in less than 1 minute.

The senior managers at Western Electric were elated and within a few months approved the purchase of a second machine. The first machine was known as serial number 1 and remained in service until 1976 when the plant closed. The experience at Western Electric was repeated worldwide, which created the demand for CMMs.

In the spring of 1961, Bendix management decided that Sheffield, with its well-established measurement business, should market the Ferranti CMMs. This decision was not unanimous, as many within Sheffield believed it was a major mistake. Sheffield sold the first machine in August 1961. After several more machines were sold, the attitude within Sheffield began to change. The machines were proving themselves, meeting the high standards of the Sheffield engineers. It also was a case of hitting the market with the right product at the right time. The name "CORDAX," which stood for coordinate axes, was applied. The name became a generic term for CMMs and is still used today. An early CORDAX CMM is shown in Figure 1.15.

More than 250 CMMs were sold from 1961 to 1964. Although the machine's overall performance was good, changes needed to be made. Competition was entering the market and faster deliveries were required. With this in mind, Sheffield negotiated to produce the CMMs under a license agreement from Ferranti. Immediately following the agreement, a design team was established to improve the machine's accuracy and resolution while expanding the number of models to cover a broader range of applications. Professional product design stylists were brought in to create a design to promote the accuracy capabilities of the machine. The stylists did an excellent job, which set the standard of excellence for the industry. Today Sheffield is a division of Hexagon Metrology, Inc., and has delivered more than 10,000 CMMs throughout the industrial world.

FIGURE 1.15 Early CORDAX CMM, 1966. (Courtesy of Sheffield Measurement [now a part of Hexagon Metrology, Inc.].)

1.6.2 Digital Electronic Automation Is First Company Formed to Produce Coordinate Measuring Machines

As a Quality Control Engineer for Fiat in Italy, Franco Sartorio was frustrated by the lack of advancements made in measuring technology. The problem was compounded because technology associated with machining was rapidly advancing and the inability to make timely measurements impeded manufacturing operations. Although others made the same observation, Sartorio decided to do something different. He first assessed the state of measuring capability by visiting major companies throughout Europe and the United States. To justify his trip, Sartorio argued that it was costly and ineffective to have 400 people at Fiat making measurements from surface plates. In the United States, Sartorio visited one automobile company with 1200 employees in the same job classification. Other companies had the same problem, but they did not recognize it.

Although he saw the Ferranti machine, Sartorio concluded from his trip that Italy was not behind other industrial countries. He further recognized that a large opportunity existed for a company devoted to meeting the growing needs for fast and flexible measuring machines.

Aware of this opportunity, Franco Sartorio, Luigi Lazzaroni, and Giorgio Minucciani formed Digital Electronic Automation (DEA) at Torino, Italy in November 1962. It may be of interest to note that they selected the name because of the appeal of digital technology to financial investors at that time. Sartorio was the machine inventor and promoter for the business, Lazzaroni brought financial expertise, and Minucciani was the electronic specialist. After applying for a patent for a servo-driven gantry style machine and showing a prototype at the European Machine Tool Show in Milan (Figure 1.16), DEA delivered its first machine in November 1965.

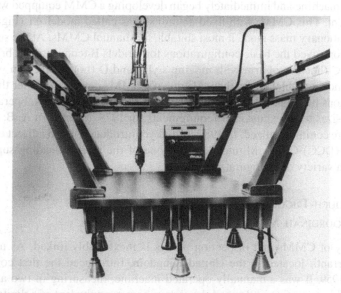

FIGURE 1.16 First prototype of a 3D CMM displayed by DEA at the European Machine Show in Milan in October 1963. The measuring range was 2500 × 1600 × 600 mm. (Courtesy of Franco Sartorio.)

The machine was a 3D gantry CMM built for Läpple-Heilbronn in Germany. At the October 1965 Machine Tool Show in Brussels, DEA showed its first manually driven 3D CMM for piece part inspection. At the same show in Hanover in 1967, DEA demonstrated its newly developed touch probe on a new high-accuracy manual CMM. In 1973, DEA delivered its first fully automatic CMM with probe changing to the Caterpillar Tractor Company (now Caterpillar Inc.) in Peoria, Illinois.

DEA also contributed to the advancement of automobile car body development and to "body in white" dimensional inspection. In 1966, DEA equipped the design center of Pininfarina with a very large CMM that was capable of milling wood models, and another machine for 2D drafting and scanning measurements that was connected in real time to the larger machine. DEA flourished to become one of the largest worldwide suppliers of CMMs. In 1994, DEA became part of the Brown & Sharpe Corporation. Brown & Sharpe was purchased by Hexagon Metrology, Inc. in 2001. This change was one among many consolidations taking place in the CMM industry.

1.6.3 COORDINATE MEASURING MACHINE DEVELOPMENTS INITIATED IN JAPAN BY MITUTOYO

The development of CMMs started at Mitutoyo in 1968 with the design of an X–Y measuring instrument for measuring the cases of household electric appliances. The instrument was developed at the request of appliance manufacturers and was used to measure the pitch between holes by using either a microscope or tapered probe and scribing. The machine proved to be several times more effective than traditional measuring techniques.

From this experience, Mitutoyo recognized the advantages of a freely moving measuring machine and immediately began developing a CMM equipped with z-axis measurement. This CMM, with fixed columns, was called model A1 (Figure 1.17). Its more stationary mass made it most suitable for manual CMMs. At the same time, Mitutoyo conceived the basic configurations for models B (conventional bridge construction), C (horizontal arm with moving styli), and D (horizontal arm with fixed stage and moving table). Since then, Mitutoyo's CMMs developed under these basic configurations have been designed with a broad range of accuracies, operator interfaces, and sizes to meet customer requirements. In 1978, Mitutoyo's A, B, C, and D models were commercialized. In 1980, Mitutoyo introduced its first direct computer-controlled (DCC) CMM. Mitutoyo is now one of the world's leading suppliers of CMMs in a variety of configurations.

1.6.4 TOUCH-TRIGGER PROBES* EXPAND VERSATILITY OF COORDINATE MEASURING MACHINES

The history of CMMs and inspection probes is inextricably linked. As mentioned earlier, Ferranti, located in the United Kingdom, introduced the first commercial CMM in 1959. It was a manually operated machine, measuring in two axes using Ferranti's fringe grating scales and the contemporary equivalent of a digital readout

---

* Refer also to Chapter 6, Probing Systems for CMMs.

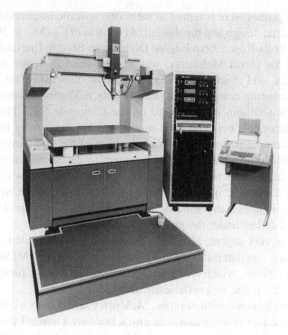

FIGURE 1.17 An early CMM developed by Mitutoyo Corporation. (Courtesy of Mitutoyo Corporation.)

that was operated by a foot switch when the hard probe was in the measuring position. To measure the position of the centers of holes, the CMM used a tapered probe, which was lowered into the hole until it fit centrally in the bore. Several different-sized plugs were needed to cover the full range of hole sizes. Positions of edges, hole diameters, and other features were measured using a rigid ball-ended probe that was manually moved to contact the surface. It was necessary to allow any vibration to disappear before reading the machine's scales. The measuring task was further complicated because there was no automatic probe radius compensation. A half-flat cylindrical rotating probe was developed to measure edges without the need to compensate for offset. There was no way to avoid offsets when measuring inside diameters. The task of performing arithmetic calculations for probe offsets was very difficult and prone to errors.

Solid probes had limitations. To allow the operator access to the whole workpiece, CMMs with solid probes could be conveniently used only on a small table. A larger machine was difficult to move manually. The increased inertia made it very cumbersome to find the correct location for measuring. Another problem with hard probes was that the measuring force could not be controlled. Consequently, relatively large errors could be caused by deflection of the machine/probe system. This was less true when checking center distances between holes with a tapered probe. Also, the operation could not be assisted by putting motor drives on the axes because the hard probes could not be positioned with any great accuracy. Probes were likely to break or the machine itself could become damaged. Further development of DCC machines had to wait for the introduction of nonrigid probes.

Touch-trigger probes were invented to solve one specific inspection problem, but later transformed the design and revolutionized the use of CMMs. In 1972, the problem arose at the Rolls-Royce Aero Engine Division in Bristol, England where David McMurtry (now Sir David McMurtry) was deputy chief designer. McMurtry, the founder, chairman, and chief executive of Renishaw, was asked for his advice on a problem of measuring complex pipe runs of only 6.35 mm diameter. These fuel lines were for the Olympus engines used on the Concorde SST. The pipes had to be made to fit accurately between solid mountings but measuring them after they had been manufactured was difficult. With the help of the company's mathematics department, McMurtry's team developed a system to define and measure points in space using v-shaped probes that fit over the pipe. The CMM used was made of granite and steel and difficult to operate due to its weight. Using this CMM to measure the small diameter pipes was virtually impossible with a solid probe because the work itself would deflect under the pressure of the probe head.

As the job was very urgent, McMurtry looked at it and decided that an optical probe or something else that did not deflect the pipe was needed. McMurtry built the first touch-trigger probe, which was primitive. Because it used fundamental kinematic principles, the probe was sufficiently accurate for the task.

Describing how he made the first probe, McMurtry said, I drew a circle and divided it into three with a pair of compasses, as any schoolboy knows. I put six ball bearings on the table at the three radii and embedded them in plastic padding [a molding compound used in model making]. I soldered up the connections between the balls and built up the plastic padding to form a body. Then I made a crucifix and stylus on a micro lathe. Next day I took it into the factory and it worked. (Waurzyniak 2008). When the probe, which was a switch, touched the work, electrical continuity in the circuit was broken. The broken contact froze the reading of the digital display on the CMM. This was done quite simply with a battery in circuit and a solenoid that took the place of the machine's normal foot switch.

Although this relatively simple device greatly improved the versatility and usefulness of CMMs, it had a more profound effect on measurement than was realized. This device, an example of which is shown in Figure 1.18, changed the measuring function from a static to a dynamic operation. Measurements were previously taken with the measuring device and part at rest, including measurements taken with CMMs. With the new touch-trigger probe acting as a switch and operating on the basis of motion in space, dynamics became a consideration. Such factors as vibration, acceleration and deceleration forces, settling time, and static versus dynamic lift of air bearings became important when assessing accuracy. Years after the development of the probe, the importance of system accuracy was understood and the error sources were clarified.

1.6.5   Software* Becomes Essential to Coordinate Metrology

In reviewing the evolution of measurement, it is necessary to include the contributions made by computers. The first CMMs simply had a digital readout to provide the center location of a hard probe. Using the readouts, length measurement could be manually

---

* Refer also to Chapter 8, Coordinate Measuring Systems Algorithms and Filters.

FIGURE 1.18 An early touch-trigger probe, 1973. (Courtesy of Renishaw.)

calculated. From these very basic machines, computational capability has permeated every function of coordinate metrology. These functions include geometric feature development, CMM automation and part programming, temperature compensation, geometric error correction, data analysis, and networking for system integration.

The first application of computers to CMMs was to compensate for probe offsets and to calculate geometric features from individual data points. It is hard to conceive a CMM without this fundamental capability. Today, probe tip calibration is well-known to every CMM user, and the range of geometric features whose parameters can be calculated automatically is extensive.

With the development of DCC CMMs, software was needed for part programming and machine control. The original part programs were written in machine codes. Subsequently, more advanced computer languages were adopted for the development of macros or subroutines required for feature calculations. As every CMM manufacturer developed its own part programming macros, providing common interfaces with computer-aided design (CAD) systems was impossible. An industry-wide effort resulted in the development of a dimensional measurement interface specification (DMIS; ISO 2003c), which forms the basis of several commercially available CMM software languages in use today (PC-DMIS, Open-DMIS, Virtual-DMIS, Calypso, etc.). This specification provides a neutral interchange format between CAD systems and dimensional measuring equipment. Today, part programming is a highly developed process with different levels of automation for various users. The range of capability extends from basic user-interactive programming to the fully automated inspection routine generation, including verification through simulation.

The first implementation of software-based geometric error correction was for scale length, a simple linear compensation. For measuring machines used in national standards laboratories, a mapping process to correct for geometric errors throughout the measuring volume was developed (Hocken et al. 1977). This process was very time-consuming and required considerable computer memory capacity. In the mid-1980s, a more effective parametric compensation procedure that is widely used today was developed (Zhang et al. 1985). Chapter 12, Error Compensation of Coordinate Measuring Systems discusses this technique and others that have been developed. The concept represents a major advancement in coordinate metrology. It also increases the sophistication of CMMs and makes increasingly important the use of traceability, as described in Chapters 2, The International Standard of Length, and 14, Measurement Uncertainty for Coordinate Measuring Systems to assure that the measurement uncertainties are within acceptable limits.

Temperature compensation followed a similar evolutionary pattern as geometric error correction. It was first applied by correcting the change of scale length through computers. From this early beginning, temperature compensation has become very sophisticated with many temperature sensors applied throughout the measuring machine and on the part to be measured. Most manufacturers have developed empirical formulations on the performance of their machines under varying temperature conditions and have also developed the computational algorithms for compensation. These algorithms are, of course, subject to the uncertainties that are described in both B89 and ISO Standards (see Chapter 9, Performance Evaluation).

The computer performs sophisticated analysis tasks such as applying statistics to process measurement data. Today's three-dimensional data analysis routines make it possible for a CMM to efficiently measure compound formed surfaces such as gears and airfoils and analyze the resulting measurement data.

The integration of measuring machines with other factory systems involves networking to provide interface with part-handling systems, machine tools, quality-monitoring systems, and engineering databases. Numerous application cases involve networking. An example is gage calibration, where inventory control and monitoring systems for the gages are included.

Software has become as important as hardware and still provides a great opportunity for future advancements in dimensional metrology.

1.6.6 CARL ZEISS CONTRIBUTIONS TO COORDINATE METROLOGY

In 1973, Carl Zeiss of Germany launched the first universal CMM with a three-dimensional measuring probe head. The new Zeiss probe head made continuous probing (scanning) possible. Scanning of a surface with continuous-time or distance-dependent data transfer to the computer is an upgraded form of single-point probing. Zeiss' first CMM was the UMM 500 (Figure 1.19), which continued in production into the 1990s. In 1978, Carl Zeiss introduced a CMM with a touch-trigger probe, WMM 850, which contained a piezoelectric sensor. This probe made it possible to attain higher accuracies with less expensive CMMs.

In 1982, Carl Zeiss introduced the first CMM specifically designed to meet the requirements of gear and transmission production. This machine, ZMC 550, greatly influenced the development of traditional gear testers.

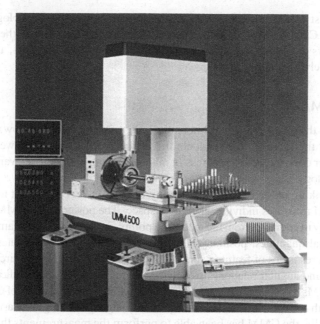

FIGURE 1.19   The Zeiss UMM 500, which was introduced in 1973. (Courtesy of Carl Zeiss IMT.)

In 1985, Carl Zeiss and Sheffield became the first manufacturers to produce CMMs with full software error correction. The Zeiss process is called computer-aided accuracy (CAA), while Sheffield called its technique microprocessor-enhanced accuracy (MEA). This technique represented a major leap forward in the performance of production CMMs.

In 1989, Carl Zeiss launched coated aging resistant aluminum technology (CARAT) to improve guideway stability under varying thermal conditions.

1.6.7   Coordinate Measuring Machine Industry Follows Traditional Business Patterns

When a new product that offers significant value is developed, the business opportunities become obvious and many people enter the field. The most prominent examples are the automobile, airplane, and more recently, the computer. Large manufacturers, small technology firms, and entrepreneurs from the major industrial countries entered the CMM market. Many of their names have been dropped or have been acquired by larger companies. Among these were Boice, Portage, Sheldon Metrology, Numerex, Renault, C. E. Johansson Company, Olivetti, Starrett, Federal Products, LK (now Metris, acquired by Nikon in 2009), Mauser, and Hansford. Other companies' names have been retained, but their ownership has changed. Examples are Leitz, Brown & Sharpe, Leica, Romer, Sheffield, DEA, etc. Since Sheffield was sold to Bendix in 1956, the business was aligned with machine tool divisions of major corporations for a long time. Sheffield's ownership changed from Bendix to Allied in 1982, to Cross & Trecker in 1984, to Giddings & Lewis in 1991, to ThyssenKrupp AG in 1997, and to Hexagon Metrology in 2004.

The largest CMM manufacturers are Mitutoyo, Hexagon Metrology, Metris, Wenzel, and Carl Zeiss, with about 50 other companies also listing themselves as manufacturers of CMMs. As in other industries, the shakeout, mergers, and realignments are likely to continue.

1.7 SUMMARY

As shown in this chapter, advancements in industrial metrology follow the axiom, "necessity is the mother of invention"; the developments described were selected based on their contribution to progress in manufacturing and their relevance to coordinate metrology.

While thinking of the enormous capability of CMMs, it is important to remember that the CMM measures individual points in space. The power of the CMM is derived from its ability to compute, from the measured points, any one of a whole family of types of dimensional quantities: position of features relative to part coordinates; distances between features; sizes of features; forms of features, such as flatness, circularity, and cylindricity; and angular relationships between features, such as perpendicularity. The power of CMMs to go from measured points to a whole family of types of dimensional quantities both caused and made necessary a new language of length measurements.

Additionally, the CMM has been able to perform the measurements that had been previously performed by a number of individual measuring devices such as height gages, calipers, roundness-measuring equipment, micrometers, and surface plates. In doing so, the CMM has revealed the systematic differences between the many length-measuring devices. More importantly, computer-based CMMs and systems are transforming industrial length metrology from a handicraft to a science.

Advancements in CMMs in the last few years have been unprecedented. Little resemblance exists between today's CMMs and the original CMMs. Since 1985, not only have CMMs' accuracies improved greatly, but the evaluation methods have become more rigorous; the speed of CMMs has greatly increased; software for part programming, feature definition, and data analyses have expanded; and the environmental capabilities have greatly improved. With the need for more competitive manufacturing capabilities growing from the unrelenting competitiveness of the global market, future advancements will occur at an even greater rate.

ACKNOWLEDGMENTS

Special thanks and appreciation are extended to Ed Battison (now deceased) of the American Precision Museum for his willingness to share his wealth of knowledge regarding early precision manufacturing, Chris Evans of UNC-Charlotte who wrote *Precision Engineering: An Evolutionary View*, and Warren White, who has been a key contributor to a number of classic books on manufacturing. Klaus Herzog of Carl Zeiss, Hiroshi Sasaki of Mitutoyo, Ben Taylor of Renishaw, Harry Ogden formerly of Ferranti, Franco Sartorio formerly of DEA, George Knopf formerly of Bendix, and Marcus Crotts of Crotts & Saunders, all willingly gave of their time by researching their records to help reconstruct the early history of the development of CMMs. Their efforts are appreciated and their fondness for CMMs is contagious.

2 The International Standard of Length

Dennis A. Swyt and Robert J. Hocken

CONTENTS

Whether used in direct inspection or in the monitoring of a manufacturing process, the principal function of a coordinate measuring machine (CMM*) is to check whether the dimensions of manufactured parts conform to design tolerances. This function entails a comparison of measured results to a standard of length.

Interchangeability of manufactured parts, on which both national and international trade in such parts depends, requires that the calibration of the CMM be traceable to the national standard and, ultimately, to the international standard. The international standard of length to which CMM measurements need to be traceable is the meter, which is part of the International System of Units (SI) or the modernized metric system (BIPM 2006; Taylor and Thompson 2008; Thompson and Taylor 2008).

The modern concept of traceability has two requirements: (1) the ability to demonstrate the chain by which measurements are referenced to the SI unit and (2) the ability to express the uncertainty of those measurements relative to the international standard (ASME 2006b; JCGM 2008a).

This chapter discusses the modern concept of traceability as it applies to CMM measurements of manufactured parts. It describes the means by which these dimensional

---

* The abbreviation CMM is used generically in this chapter to include all measuring machines capable of measuring coordinates such as traditional Cartesian models as well as other types such as articulated arm CMMs and laser trackers.

measurements are functionally related to the international standard of length, the various paths by which a specific chain of traceability may be established, and some comments about how to determine uncertainty* in some particular measurements.

2.1 DEFINITION OF THE METER

Throughout the industrialized world, the international standard of length is the meter. All dimensional measurements are to be referenced to this SI unit of length. As established by the International Committee for Weights and Measures (CIPM) and subscribed to by all the industrialized nations, the definition of the meter is " ... the length of the path traveled by light in vacuum during a time interval of 1/299,792,458 of a second" (BIPM 1984, p. 163).

This statement, issued in 1983, represents the second redefinition of the meter since it was defined in terms of the international prototype meter in 1889, following the first Treaty of the Meter in 1875 (Taylor and Thompson 2008). The first redefinition was in 1960 when the artifact standard was replaced by one that was based on the wavelength of light emitted by krypton-86 gas in a high-pressure gas discharge lamp (BIPM 1960).

2.1.1 Realization of the International System of Units: Unit of Length

The definition of the meter can be practically realized in a number of recommended methods by using

- The length, l, of path traveled in vacuum by a plane electromagnetic wave in time, t, using the relation $l = c \times t$ and the defined value of the speed of light in vacuum, $c = 299{,}792{,}458$ m/s
- The wavelength, λ, of a plane electromagnetic wave of frequency, f, using the relation $\lambda = c/f$ and the value of the speed of light given in the previous list item
- One of the radiations listed by CIPM (Quinn 2003), whose stated wavelength in vacuum, or whose stated frequency, can be used with the uncertainty shown by CIPM, provided that given specifications and accepted good practice are followed

According to the CIPM definition, "in all cases any necessary corrections must be applied to take into account actual conditions such as diffraction, gravitation, or imperfections in the vacuum" (p. 165). The last of these refers to the index of refraction of the medium.

With the definition in terms of time and the speed of light, the theoretical lowest uncertainty in length is determined by uncertainty in measurements of intervals of time. The best uncertainty currently attained in time measurements is that by a cesium fountain atomic clock, NIST-F1, at the U.S. National Institute of Standards

---

* Refer also to Chapter 14, Measurement Uncertainty for Coordinate Measuring Systems.

and Technology (NIST) with a standard uncertainty of 5 parts in 10^{16} (http://www.nist
.gov/pml/div688/grp50/primary-frequency-standards.cfm, accessed on March 2011).
In terms of time this uncertainty amounts to 1 second in more than 63 million years,
and in terms of length it corresponds to about 20 nm at a distance equal to the circumference of Earth.

There is a list of recommended radiations for realizing the SI unit of length
(Quinn 2003; Riehle et al. to be published). Among them the radiation to which measurements of lengths of material objects are customarily referred is the wavelength
of the red–orange light of the helium–neon laser. When stabilized by a particular
atomic transition of a saturated vapor of the iodine isotope $^{127}I_2$, the wavelength is
632.99121258 nm in vacuum.

For the decade starting with the adoption of the new definition of the meter in
1983 until 1993, the relative standard deviation of a properly stabilized HeNe laser
containing an iodine cell was estimated by CIPM to be 3.4 parts in 10^{10}, corresponding to 0.34 nm at 1 m. Improvements in cell design and related conditions were made
during this period. In 2003, CIPM revised that relative uncertainty and brought it
down to 2.1 parts in 10^{11}, corresponding to 0.021 nm at 1 m (Quinn 2003). Special
measuring machines using the working laser's frequency offset locked to iodine-
stabilized laser have been built (Stroup, Overcash, and Hocken 2008).

The uncertainty of the iodine-stabilized laser, whereas it was once effectively
set by convention, can now be reduced somewhat further if the laser is calibrated
against an optical frequency comb. The comb is arbitrarily good (limited only by
the uncertainty of the frequency standard that sets its pulse repetition rate), and the
uncertainty of a calibrated iodine-stabilized laser is better than the uncertainty of an
uncalibrated iodine-stabilized laser.

Although the wavelength of light in vacuum is known very well as a practical matter,
it is not currently possible to achieve an uncertainty better than a few parts in 10^8 when
operating in air. This is due to uncertainty in the refractive index of air as measured
either by a refractometer or via the Edlen equation (J. A. Stone, pers. comm. 2009).

2.1.2 LASER DISPLACEMENT INTERFEROMETRY

The means by which the known wavelength of light becomes the practical standard
for measurements of length is known as "laser displacement interferometry." This
technique is commonly used in instruments for measuring the dimensions of objects.
The displacement, D, of a moveable plane mirror in the interferometer is measured
in terms of wavelengths of light. This displacement measurement is defined by the
following equation:

$$D = \frac{M\lambda_0}{2nmK \times \cos\theta}$$

where

M: The combined counted integer order and measured fringe fraction of the
     interference

λ_0: Vacuum wavelength of laser light

n: Index of refraction of the medium
m: The number of passes of light through the interferometer
K: The number of electronic subdivisions of the interference fringes (whose value
    is typically a power of 2 ranging from 16 to 1024 depending on the system)
θ: Angle between the direction of propagation of laser light and the normals to
    the interferometer mirrors

For a high-resolution commercial heterodyne interferometer, the limit of resolution, or the least count, is about $\lambda/6000$ or 0.1 nm. Lawall and Kessler (2000)
describe a heterodyne Michelson interferometer for displacement measurements
capable of fringe interpolation resolution of 1 part in 36,000. Halverson et al.
(2004) demonstrated the progress achieved in developing laser metrology for space
interferometry. Over a restricted range of motion, they were able to achieve 20-pm
(pm stands for picometer) root-mean-square (RMS) relative agreement between
independent interferometers arranged in a two-dimensional truss. (Agreement
degraded to 215 pm with larger motions on the order of 1 m). Křen (2007) claims
residual nonlinearity of under 100 pm of a method for linearizing interference
fringes interpolation of quadrature phase detectors used in homodyne counting
laser interferometers.

2.1.3 INTERNATIONAL COMPARISON OF ARTIFACTS

In current practice, the central standards laboratories (National Measurement
Institutes [NMIs]) of various nations determine the degree to which they are capable
of realizing the meter by round-robin comparison of measurements made on material standards. Most often, the standard is the 1-m spacing of two scribe marks on a
meter bar with measurements made by laser displacement interferometry on special
single-axis measuring machines.

In a round-robin comparison on meter-long scribed-mark artifacts among the
International Bureau of Weights and Measures (BIPM) and the NMIs of 11 industrialized nations, BIPM determined that the standard deviation (1σ) of the mean
of the best values, among those of the major CMM-producing nations, is approximately 30 nm (Hamon, Giacomo, and Carré 1987). In a comparison against these
values, measurements made at the U.S. NIST had a standard uncertainty (1σ)
estimated by error budget analysis to be less than 5 parts in 10^8 (50 nm) and it
lay within 2.5 parts in 10^8 (25 nm) of the mean of other laboratories (Beers and
Penzes 1992).

In another round-robin comparison (Bosse et al. 2003), BIPM circulated two
high-quality line scales, one made of Zerodur and the other made of fused silica
(quartz), with main graduation length of 260 mm and additional smaller linescales
of only a few millimeters. Long gage blocks were also produced out of the same
piece of substrate material used for the scales. The gage blocks were used to determine important material parameters such as thermal expansion, compressibility, and
long-term stability. These artifacts were measured by 13 NMIs. The measurement
uncertainty evaluated by the participants over 280 mm showed a variation from
about 300 nm down to 30 nm (for $k = 2$).

2.2 COORDINATE MEASURING MACHINE MEASUREMENT TRACEABILITY

For many years traceability for coordinate measurement has been a matter of much debate, because it is not a simple issue due to the complications associated with estimating uncertainty for CMMs. Yet traceability is arguably one of the most fundamental and important aspects of the proper use of a CMM in a quality management system.

The various aspects of traceability that are of concern are the actual physical chains by which measurements may be related to the SI unit of length, the documentary standards that define the basis for traceability in general, and the approaches for realizing traceability in specific cases taking uncertainty into consideration.

2.2.1 THE PHYSICAL CHAIN OF TRACEABILITY

Measurements on different CMMs can be related to the international standard of length in a variety of ways depending on the design of the CMM and the transfer standards used. Figure 2.1 shows how "traceable" CMM measurements of dimensions of manufactured parts are ultimately related to the international standard of length. In the figure, at the top level there is the frequency of the cesium* fountain atomic clock and its standard uncertainty (1σ) of 5 parts in 10^{15}. Also at the top there is the wavelength of the iodine-stabilized HeNe laser, which is measured relative to the frequency of the atomic clock, and its standard uncertainty of 2.5 parts in 10^{11}. At the second level there is a laser displacement interferometer system and its practical standard uncertainty in the range 0.7–15.0 parts in 10^8 (where this uncertainty does not include the large uncertainty component associated with atmospheric pressure compensation). Below the second level, at the third level there is the measurement of check standards such as the distance between graduations on a meter bar with its standard uncertainty of 2–5 parts in 10^8; or calibrations of transfer standards such as step gages, ball bars, or end standards, which have standard uncertainties on the order of 1 µm at 1 m or 1 part in 10^6. As shown in Figure 2.1, CMM provides the fourth level or the base of the pyramid.

2.2.2 DOCUMENTARY STANDARDS DEFINING TRACEABILITY

Key national and international standards dealing with compatibility of physical measurements introduce the requirement of traceability. The basic notion common to these different standards is that traceability is the demonstration of an unbroken chain of measurements referenced to national and, ultimately, international standards of the quantity being measured.

For comparison, Table 2.1 shows the bases for traceability shared by the ISO/IEC 17025 on "General Requirements for the Competence of Testing and Calibration Laboratories" (ISO/IEC 2005), the U.S. military standard MIL-STD-45662A† on

---

* Work is being done at NIST, with a target completion date of 2019, to replace cesium with new optical standards having better reproducibility (J. A. Stone, pers. comm. 2009).
† Officially canceled in 1995 (Department of Defense 1995) in favor of civilian documents ISO 10012 (ISO 2003a) and ANSI/NCSL Z540.1 (ANSI/NCSL 1994).

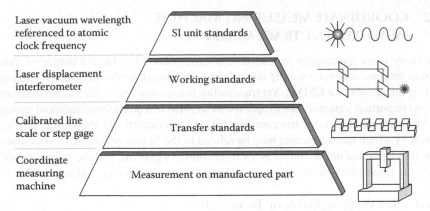

FIGURE 2.1 Pyramid depicting the physical means for relating CMM measurements on manufactured parts to the international standard of length defined in terms of the wavelength of light.

TABLE 2.1
Physical Bases for Traceability Specified by ISO/IEC 17025, MIL-STD-45662A, and Z540.1

Physical Basis	ISO/IEC 17025	MIL-STD-45662A	Z540.1
International standard	International reference materials		International standard
National standard	National standards, reference materials	Standards maintained by NIST or the U.S. Naval Observatory	National standard
Phenomenon of nature		Fundamental or natural constants with values assigned or accepted by NIST	Intrinsic standards
Symmetry-type relations		Ratio types of calibrations	Ratio or reciprocity measurements
Interlaboratory comparison	Correlation by interlaboratory comparison, proficiency testing	Comparison with consensus standard	Interlaboratory comparisons or proficiency testing

"Calibration System Requirements" (Department of Defense 1988), and the U.S. national standard Z540.1* on "Calibration Laboratories and Measuring and Test Equipment—General Requirements" (ANSI/NCSL 1994).

In the table, each row represents a basis for traceability, such as international, national, or intrinsic standards. As indicated in the table, the International Organization

---

* This standard was withdrawn in 2007 in favor of ISO/IEC 17025:2005 according to http://www.ncsli .org/NCSLIORG/Store/Core/Orders/product.aspx?catid=84&prodid=647 (accessed March 18, 2011). It is used in this chapter only for comparison purposes.

for Standardization (ISO), U.S. military, and U.S. national standard provide traceability to national standards. The CMM measurements based on step gages or end standards that have been calibrated by NIST or directly referenced to NIST meet the traceability requirement of each of the aforementioned three specifications.

Note that beyond national standards, for the three documentary standards shown in the table, there are other bases of traceability that may be invoked. For example, the U.S. national standard allowed intrinsic standards and the U.S. military standard allows fundamental or natural physical constants with values assigned or accepted by NIST. As such, an iodine-stabilized HeNe laser in principle would satisfy a traceability requirement for either of these two standards. The U.S. military standard also allows the use of national standards of other countries correlated with U.S. national standards. The U.S. national standard allows the use of suitable reference materials and internationally accepted standards.

Given the different bases whose traceability can be established, there is a wide variety of ways by which the traceability of measurements of particular manufactured parts can be asserted. For example, in the case of CMM measurements of the dimensional form of production gears, one might assert that traceability to national and international standards is based on laser wavelength, general length, or gear form standards.

Some of the alternate chains of traceability, based on different types of length standards and paths to the unit of length, are shown in Figure 2.2. In the case of laser wavelength as the standard, chains of traceability can involve the use of a commercial stabilized laser, the wavelength of which is asserted to be known from first principles based on the physics of the laser transition (Stone et al. 2009); the use of such a laser whose wavelength has been calibrated against an iodine-stabilized laser, the uncertainty of which is known by conformity to the CIPM-specified prescription; or the use of a laser with its wavelength known from calibration by an NMI. When making measurements with a laser in air, the need for proper correction for air refractive index cannot be overemphasized. This requires, at a minimum, calibrated pressure and temperature sensors and, in some cases, humidity and carbon dioxide sensors and knowledge of correct equations* (ASME 1997).

In the case of general length standards, this can involve the use of an end standard, step gage, ball bar, or ball plate, any one of which could be calibrated by means of a laser interferometer, the wavelength of which is known by any of the aforementioned three methods; or by calibration of the artifact standard by an NMI. Correction for the nominal differential expansion, which requires calibrated thermometry, is also required (see Chapter 10, Temperature Fundamentals).

Finally, in the case of gear form standards this can involve the use of gear elements (such as involutes and leads), master gears, or prototype production gears. Again, any of these reference standards could have been calibrated against other calibrated gear standards; an end standard, step gage, ball bar, or ball plate; or a laser displacement interferometer system. For whichever of these the traceability paths apply, the uncertainty at each step needs to be defined and propagated in order to provide the basis for the uncertainty of final measurement results.

---

* See also "Engineering Metrology Toolbox" at http://emtoolbox.nist.gov/Main/Main.asp.

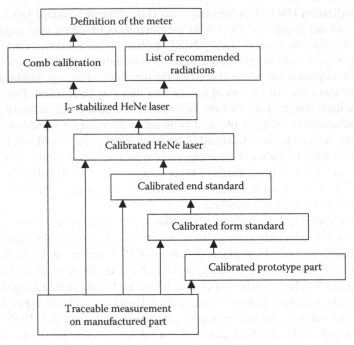

FIGURE 2.2 Representation of some alternate paths for the chain of traceability of dimensional measurements of a manufactured part.

2.2.3 THE OLD AND THE NEW TRACEABILITY

Prior to the 1990s, traceability to manufacturers in the U.S. market was an issue in general only for those dealing with defense procurements. This "old traceability" meant the ability to show a military inspector a paper trail of calibrations leading back to those of NIST. Nowadays, U.S. companies competing in the global marketplace, like their European and Asian counterparts, need traceability to meet the requirements of ISO 9001 (ISO 2008a) or their own internal quality management systems. This "new traceability" includes not only a requirement that measurements be referenced to recognized standards but also a requirement that measurement uncertainties be stated and the bases for that statement shown (ISO 2003a; ASME 2006b; JCGM 2008a).

This new traceability is exceptionally difficult to demonstrate for CMMs. Their ability to measure components with virtually any geometric feature or combination of features defeats attempts to demonstrate traceability in the general sense for all measurements. Although there are alternative bases for asserting traceability in principle, it is apparent that the only practical way to deal with the problem is to provide task-specific statements about traceability when a CMM is used to obtain the measurements. This is because a general uncertainty value for a CMM covering all possible measurements would be too large for the majority of the tasks, reducing the applicability of the system.

2.2.4 TRACEABILITY IN COORDINATE MEASURING MACHINES—
TASK-SPECIFIC MEASUREMENT UNCERTAINTY

From a metrological point of view, the establishment of a sound estimate of uncertainty of a CMM measurement necessary to establish traceability is a complex problem. The complexity arises from the nature of the CMM itself, with its ability to carry out a wide variety of measurements in any position within its working volume based on a number of sampling strategies. In general, the sources of uncertainty in CMM measurements are considered to include those associated with*

- Geometric imperfections resulting from the manufacture of the CMM, elastic deformations, and dynamic effects in the geometry-defining elements of the CMM with its 21 degrees of freedom
- Geometric imperfections of the probing system as applied to each measuring point
- Mechanical distortions of the machine arising from nonuniform, nonconstant, and nonstandard temperature within the measuring machine and part, including uncertainties in nominal temperature expansion coefficients of scales and parts
- Variabilities associated with the combination of the sampling strategy for selecting points to be probed, the mathematical algorithm for computing a substitute feature from these sampled points, and the particular software implementation
- Part properties, including form errors, surface roughness, and various interactions with the probe such as elastic deformations under force or deflections of thin-walled elements

2.3 A NOTE ON CUSTOMARY UNITS OF
LENGTH IN THE UNITED STATES

The United States is unique among the industrialized nations in its use of the metric system. Although by federal statute the use of the metric system is legal in contracts, it has not been adopted as mandatory for use in trade. Progress is being made, however, in the conversion process. NIST uses the metric system in its scientific work and for international intercomparisons. By a 1988 congressional directive, NIST and all other federal agencies started using the modernized metric system in their business operations.

For historical reasons, the actual system of measure in use in the United States is a complex mixture of metric and customary units. In length measurements, three units of length are in use: (1) There are measurements made directly in terms of the meter, such use being made permissible in law and legal in contracts by an act of Congress in 2007[†] (U.S. Congress 2007). These are called the SI units of measure. (2) There are measurements made in terms of generally used customary units of

---

* For more details, refer to Chapter 14, Measurement Uncertainty for Coordinate Measuring Systems.
[†] The initial Metric Act dates from 1866.

length related to the meter through the following conversion: 1 inch = 25.4 mm exactly. These are officially referred to as the "U.S. international foot," which has been related to SI units by U.S. industry standard since 1933 (ASA 1933) and by agreement among the NMIs of the United States and the British Commonwealth nations since 1959 (Astin 1959). (3) Finally, there are measurements made in terms of the U.S. surveyor's units of length related to the meter through the following conversion: 1 m = 39.37 inches exactly. This yields the relation 1 inch = 25.40005 mm. These are officially referred to as being based on the "U.S. survey foot," which has been in use since 1893 (Mendenhall 1893).

The difference between the U.S. survey foot, used for geodetic scale measurements, and the U.S. international foot, used for industrial scale measurements, is 2 ppm, corresponding to a difference of approximately 2 mm/km. In the United States, standard factors for converting among customary and metric units and practices for rounding converted fractions and tolerances have been established by American National Standards Institute/Institute of Electrical and Electronics Engineers (ANSI/IEEE; ANSI/IEEE 1992).

2.4 SUMMARY

This chapter discusses how CMM measurements of the dimensions of manufactured parts need to be related to the international standard of length to satisfy the modern concept of traceability. For measurements to meet this concept of traceability, they must first relate to the SI unit of length through an unbroken chain of physical comparisons. This chain links the CMM through calibrated artifacts, laser displacement interferometry, and the wavelength of light to the definition of the meter. For measurements to meet the new concept of traceability, they must also relate to the SI unit through a complete analysis of measurement uncertainty. This analysis relates the uncertainty in the result of a specific measurement task to that of the SI unit through the uncertainties of the calibration standards on which the CMM measurement is based. As discussed earlier (Section 2.2), there are alternative physical paths from CMM measurements to the SI unit of length and there are alternative approaches to the assessment of CMM measurement uncertainty.

From a practical point of view, establishment of a sound estimate of uncertainty necessary to establish traceability is a complex problem. Although there are certain documentary standards that provide guidance on how to establish traceability in CMM measurements, there is still the question of who is the final arbiter in determining whether particular CMM measurements meet the definition of traceability.

Given that the purpose of establishing traceability is for a supplier of measurements to be able to demonstrate a firm basis for a customer's confidence in the stated accuracy of a measurement result, the answer is clear. In the last analysis, it is the customer for whom measurements are ultimately made and it is the customer who determines whether traceability requirements are met. Thus it is in the interest of the customers and suppliers to address and resolve issues concerning traceability in CMM measurements early in the customer-supplier relationship.

3 Specification of Design Intent

Introduction to Dimensioning and Tolerancing

Edward Morse

CONTENTS

For the purposes of this chapter, it will be assumed that a designer has thought carefully about a part and how it interacts with the other parts in an assembly. The designer has decided not only on the optimal material, surface characteristics, and dimensions of the part but also on the allowable variations in these quantities. From a design standpoint, this complete set of information captures the intent of the designer, or the *design intent*. In this chapter, the geometric information (size, shape, form, and orientation of part features) will be inspected using the coordinate measuring machine (CMM).

The specification of a part begins with the nominal geometry. This is often thought of as a part's "design," whether it is shown on a drawing or stored electronically as a solid model in a computer-aided design system. However, the design is not complete until the allowable variation from the nominal geometry is specified. Because it is not possible to make perfect parts, the design must state clearly how close a manufactured part must be to the nominal model to be considered acceptable.

The link between dimensioning and tolerancing has traditionally been very strong. When the dimensions are attached to the nominal model, the model is explicitly "parameterized" by the placement of these values. A full description of the methods used to parameterize models may be found in Srinivasan (2004). In Figure 3.1a, the length of the feature is described by a parameter, and the nominal value for a parameter (10) is given on the drawing. In addition to the assignment of parameter values, dimensioning a part also implicitly includes adherence to a set of drawing rules. For instance, corners that appear to be right angles in the drawing do not require the angle to be written down; 90° is assumed.

The natural way to control the variability in the part is simply to place ranges of allowable values on each of the parameters (e.g., 10 ± 0.1, or $9.9 - 10.1$). This method of variation control is referred to as parametric tolerancing, limit tolerancing, or plus–minus tolerancing. There are two main problems with this method: (1) although there can be many ways to describe the same nominal part (Figure 3.1), the addition of tolerances to the parameters will make different choices of parameters into different designs and (2) the variation of a parameter may look clear on a drawing, but the meaning of this variability on a real, imperfect part may be ambiguous (Figure 3.2).

Figure 3.2 shows a fundamental problem with the parametric method of tolerancing. When the features in question are not perfect parallel lines or planes, the distance between lines or planes is not suitably represented by a single parameter that can then be compared to the tolerance. Should the length in the figure below be measured parallel to the top or bottom face? Should the maximum length be reported? The minimum? Even if one inspector were to measure the part and report the length in the same way each time, this does not guarantee that someone else would not measure the part differently and report a different length, even if the measuring equipment has very low uncertainty.

One solution that overcomes many of the ambiguities present in parametric tolerancing is the use of geometric tolerancing. Geometric tolerancing, as discussed in Section 3.1, contains the methods of variation control described in the American

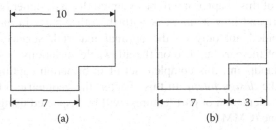

FIGURE 3.1 Two possible methods for parameterizing the spacing of the vertical features on a part.

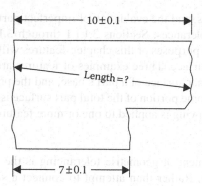

FIGURE 3.2 Ambiguity in the evaluation of dimensions for an imperfect part.

Society of Mechanical Engineers (ASME) standard Y14.5M-1994 (ASME 1994a). There are international (International Organization for Standardization [ISO]) standards that describe similar methods (ISO 2004a), and the majority of the notation and meaning are the same. The differences between the U.S. and ISO standards can be quite subtle and will not be discussed here.

3.1   GEOMETRIC TOLERANCING

The power of geometric tolerancing is in its ability to unambiguously determine conformance or nonconformance to a tolerance, even if the part has imperfect form. The geometric variability of mechanical parts is usually divided into surface finish (very short spatial wavelength variation), waviness (longer wavelength), and form (shape). Although the Y14.5M standard implies that "all points on the surface of the part" are considered in the evaluation of tolerances, the constraints of inspection and function dictate that the surface finish of the part does not usually have any influence on the geometric tolerances, and the form and waviness of the part surface are the types of variability controlled with the geometric tolerances.

Geometric tolerancing uses special symbols, or syntax, to communicate the information about the allowable variability of the parts. This chapter provides an introduction to these symbols and their meaning, or semantics. It is important to note that the same tolerancing symbols are used by the ISO tolerancing standards, but these symbols may have different meaning in an ISO-toleranced drawing (Concheri et al. 2001). More advanced discussion of the ASME tolerancing semantics is found in the ASME Y14.5M standard and its mathematical counterpart, Y14.5.1M (ASME 1994b).

3.1.1   PRINCIPLE ELEMENTS OF GEOMETRIC TOLERANCING

As stated in Section 3.1, the ability to associate tolerances with imperfect parts in an unambiguous manner is the strength of geometric tolerancing. To allow this association to work, there are three principle elements of geometric tolerancing, which will be discussed subsequently. These elements allow an ideal, orthogonal coordinate system or systems to be related to imperfect features, ideal distances to be defined in these coordinate systems, and the evaluation of imperfect part surfaces with respect to theoretically exact tolerances. Sections 3.1.1.1 through 3.1.1.3 will refer repeatedly to *features*. For the purposes of this chapter, features will simply be considered as a subset of a part's surface. Three examples of features are the outer cylindrical surface of a pin, the planar surface of a part's base, and the profile around a bearing race. In each example, some portion of the total part surface is being examined: each tolerance in a design drawing is applied to one or more features.

3.1.1.1 Zones

The first important element of geometric tolerancing is the use of zones in space to contain part features. Rather than attempt to connect a single number (such as length) with an imperfect feature, a theoretical zone in space is constructed; if the feature is contained in that zone, it satisfies the tolerance specification for which the zone is defined. This means that regardless of the part surface characteristics, a definitive result can be associated with the question of tolerance conformance: containment in the zone corresponds to satisfying the tolerance.

A practical example of a tolerance zone is shown in the use of a flatness tolerance in Figure 3.3. Here, the drawing indication tells how "thick" the flatness tolerance zone is. If the zone can be oriented so that it contains the entire indicated surface, then the tolerance is satisfied. The zone has perfect geometry (the space between two theoretical planes, separated by the tolerance value) and thus can be described by a single tolerance value, while variations in the surface do not introduce ambiguity in the conformance assessment.

The flatness tolerance zone is not constrained in its location or orientation because the flatness tolerance is independent of other features,* controlling the feature regardless of its relation to other features on the part. The next two principle elements are used to locate tolerance zones in space.

3.1.1.2 Datums

Datum features are features that are used to set up a coordinate system (or parts of a coordinate system) related to an actual part. What is most confusing about datums is that there are three different things that are often referred to as datums in casual conversation. First is the *datum feature*, which is the actual part surface that is identified by a datum symbol on the part drawing. The datum feature itself will not be perfect; after all, it is a real part. Second is a *datum simulator*, which is a high-precision (higher precision than the part) artifact, such as a surface plate, that is used to mimic a perfect-form feature. Finally, a *datum* is a theoretically exact geometric element— usually a plane—that corresponds to the datum feature on the part.

An example of different aspects of datums is shown in Figure 3.4. The bottom surface of a part is indicated as Datum feature A and, although well made, is not perfect. The part is placed on a surface plate, which *simulates* the perfect Datum plane A.

---

* This tolerance specification may be thought of as "existential" because the part is in tolerance *if a zone exists* that contains the surface. No prior information is given that guarantees the existence of such a zone. This is different from other tolerances where the zone is well defined, and a part is either contained or not contained in the zone.

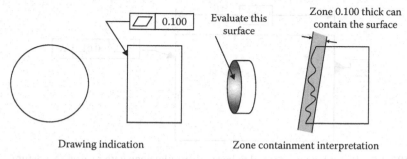

FIGURE 3.3 Flatness tolerance, as described in Y14.5M-1994.

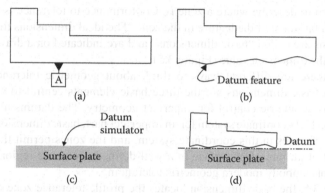

FIGURE 3.4 (a) Datum indication, (b) datum feature, (c) datum simulator, and (d) datum.

The theoretical element (Datum A) is represented physically by the surface plate and allows one to infer a "perfect" plane that corresponds to the Datum feature A.

An important aspect of datum use is the ability to unambiguously and repeatably relate a perfect theoretical datum surface to an imperfect actual part. Each time the part is placed on the datum simulator, the part will have the same rest position, and the measurements on the part made with respect to the datum will have the same result. Of course, a clever reader will likely see that there are parts that may not have a stable rest position on the surface plate (i.e., the part "rocks" on the simulator). Section 4.5.1 of the Y14.5M-1994 standard directs the user to *adjust* the part: "If irregularities on the surface of a primary or secondary datum feature are such that the part is unstable (that is, it wobbles) when brought into contact with the corresponding surface of a fixture, the part may be adjusted to an optimum position, if necessary, to simulate the datum." The fixture in the preceding statement is the datum simulator. This is—at present—a source of ambiguity in the standard: different users might adjust the part differently.

3.1.1.3 Basic Dimensions

Once a coordinate system has been established using the datum concepts, the next step is to position zones in this coordinate system to evaluate different tolerances.

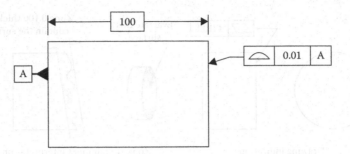

FIGURE 3.5 Basic dimensions used to locate a (profile) zone with respect to a datum.

Rather than have a varying dimension describe where a feature *might* be, ideal dimensions are used to describe where a zone *is*. Conformance to tolerance is then determined by containment of the feature in the zone. The ideal dimensions that describe zone locations are called "basic dimensions" and are indicated on a drawing with a box around the dimension value (Figure 3.5).

Though there are different ways to think about geometric tolerances, zones, datums, and basic dimensions are the three basic elements central to establishing unambiguous geometric control for imperfect geometry. The datums allow one to establish a perfect coordinate system for an imperfect part, basic dimensions describe the location of zones in this coordinate system, and the zones permit the notion of containment of an imperfect surface in a well-defined geometric region. Together, these concepts embody modern geometric tolerancing.

In Figure 3.5, the basic dimension locates the profile tolerance zone exactly 100 units from the datum determined by Datum feature A. Both the datum and the tolerance zone are theoretical, exact geometric entities and can therefore be located exactly with respect to one another. To reiterate, the strength of geometric tolerancing is the ability to relate the perfect entities to the real part, which is imperfect.

3.1.2 TYPES OF TOLERANCES

Tolerances may be divided into six different types depending on the nature of the control they exert on the part: form, orientation, profile, runout, size, and location. Each type of tolerance will be discussed briefly with an example and a discussion of how this geometric control fits into the larger picture of geometric tolerancing.

3.1.2.1 Form

Form tolerances are applied to individual features and control the allowable deviation of that feature from a perfect geometric element. The deviations are with respect to a perfect plane (flatness), a perfect line (straightness), a perfect circle (roundness), or a perfect cylinder (cylindricity). In each case, the zone is bounded by perfect geometric elements. For flatness, the zone between two parallel planes must contain the feature for it to be in tolerance, as discussed in the example in Section 3.1.1.1.

These form controls are an important part of geometric tolerancing, particularly for datum features. Datum features are usually required to have good form;

this reduces the uncertainty in the coordinate system that is determined from these features. The influence of this uncertainty is discussed further in Chapter 14, Measurement Uncertainty for Coordinate Measuring Systems. Form control is also important in articulated assemblies, where component surfaces are required to rotate or slide with respect to one another.

3.1.2.2 Orientation

Orientation tolerances control the geometric relationship of a feature to another feature or features (specifically datums), without controlling the location of the toleranced feature. The specific controls are angularity, perpendicularity, and parallelism. Each of these tolerances describes a zone whose orientation is fixed and that must contain the toleranced feature if it is to be in tolerance. In every case, the zone is oriented with respect to one or more datums. If a planar feature has a perpendicularity tolerance, the zone within which that feature must lie is constrained to be perpendicular to one or more datums (Figure 3.6).

Orientation tolerances exert more control than form tolerances as they control the orientation or the feature in addition to the form. For example, a planar feature that satisfies a perpendicularity tolerance of 0.01 must also satisfy a flatness tolerance of 0.01 because the tolerance zone is only 0.01 thick.

Although Figure 3.6 is shown in two dimensions, orientation tolerances are applied in the three-dimensional space of the part. It is possible to specify perpendicularity to two datums simultaneously. The angularity tolerance is used to control the orientation of features that are not at 0° or 90° to the datum feature(s) used as reference. It can be (and has been) argued that parallelism and perpendicularity are special cases of angularity, and that it would be sensible for angularity to be used in every case where orientation is to be controlled. In the example in Figure 3.6, there is an implied basic angle of 90° from Datum A, but the addition of this explicit angle would clutter the drawing. Whether using angularity for all orientation requirements is a good idea remains open for debate. An excellent summary of the overlap between these orientation controls is found in Appendix A of ASME Y14.5.1M-1994.

3.1.2.3 Profile

The profile tolerances (profile of a surface and profile of a line) are used to control individual features and collections of features. These tolerances can be either

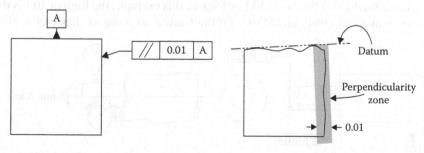

FIGURE 3.6 Specification of perpendicularity and the evaluation of this tolerance.

constrained or not constrained with respect to datums. The use of a profile tolerance without datum references is essentially a form control for a surface where the
nominal shape need not be a regular (line, circle, etc.) feature. A zone of uniform
thickness is created based on the nominal geometry, and if the zone can be placed
so that it contains the actual surface, the feature is in tolerance. If the location of the
zone is controlled with a datum reference frame, then the zone is fixed in space and
the surface must lie in the zone. Figure 3.5 is an example of a profile specification
controlled by a single datum reference.

3.1.2.4 Runout

Runout tolerances relate the toleranced feature to a datum axis about which the tolerance zone is established. A runout tolerance controls form, location, and orientation
in a manner similar to a profile tolerance, but the zone used for assessment is a surface
(or volume) of rotation about a datum, rather than a zone relative to a fixed datum
system. Consider the example shown in Figure 3.7. Here two surfaces have runout
tolerances specified with respect to the cylindrical Datum A. The circular runout tolerance (with the single arrow in the tolerance frame) requires that each "slice" of the
cylindrical surface must exhibit no more than 0.01 variation through a full rotation of
the part about the Datum A axis. The zone in this case has an annular shape, where
the diameter of the annulus is not specified but its width cannot exceed 0.01. For each
different slice of the part, a different zone is created. This means that if the smaller
cylindrical surface is tapered (i.e., it has a conical shape), it can still satisfy the runout
tolerance if each cross section is circular and centered on the datum axis.

The total runout tolerance (with the double arrow in the tolerance frame) requires
that the entire surface indicated (the "end" of the larger cylinder) must vary by no
more than the tolerance as the part is rotated about the Datum A axis. In each of the
profile cases, the deviations of the surface are measured perpendicular to the nominal surface. In the first (runout) case, deviation is perpendicular to the Datum A axis;
in the second (total runout) case, deviation is parallel to the Datum A axis.

3.1.2.5 Size

Size tolerances may only be applied to a small group of feature types called *features
of size* (FOS). These features are spherical surfaces, cylindrical surfaces, and pairs
of parallel opposing planar surfaces. These FOS are specified with limits within
which the feature's size must be contained. In the two-dimensional example shown
in Figure 3.8a, the size limits are 10.1 and 9.9. In this example, the limit of 10.1 is the
maximum material condition (MMC) for the feature, as when its diameter is 10.1,

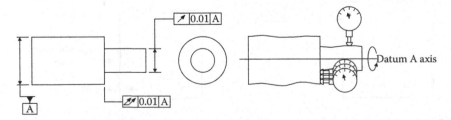

FIGURE 3.7   Specification and illustrative inspection method for runout tolerances.

the part has the most material. The least material condition (LMC) for this feature is at the 9.9 limit. The test for conformance to the MMC size limit is that the feature be contained in a perfect boundary with the MMC diameter. This is shown by the circle of diameter 10.1 in Figure 3.8b. Often in practice, the LMC diameter is checked by a series of two-point measurements such as those made by calipers or micrometers, and as shown in Figure 3.8c. However, the correct interpretation of the minimum size is described by the inscribed diameter shown in Figure 3.8d. This interpretation is described in detail in ASME Y14.5.1M-1994, Section 2.3.

In the case of an "internal" FOS such as a hole, the definitions of conformance are identical with respect to the MMC and LMC limits of size. Figure 3.9 shows the analogous specification for a part with a toleranced hole feature. The MMC in this case is 10.1 diameter, whereas the LMC is 10.3 diameter. The MMC is checked by seeing if a perfect boundary of diameter 10.1 (as shown in Figure 3.9b) can fit inside the hole, whereas the LMC case considers a circumscribed diameter to see that it does not exceed size 10.3.

These very simple, two-dimensional examples reveal how closely the standard is tied to the notion of parts fitting together in assemblies. The MMC test for the external feature is that the feature fits within a perfect circle of diameter 10.1. The MMC test for the internal feature is that a perfect circle of diameter 10.1 fits within the feature. If both these conditions are met, the parts can be guaranteed to go together. This default rule (referred to as Rule 1 in the standard) that "perfect form at MMC is required" for FOS is a strength of the Y14.5M standard when it comes to tolerancing for assembly.

The ideas shown in the examples above apply to all of the "real" three-dimensional FOS, namely internal and external cylinders (holes and pins), spheres (sockets and balls), and parallel plane features (slots and tabs or keys). The key difference in the extension to three-dimensional features is in the application of Rule 1, requiring that

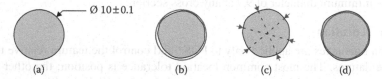

FIGURE 3.8   Size specification and interpretation.

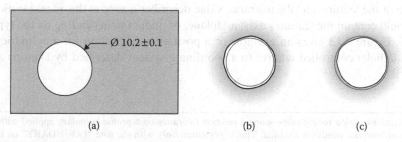

FIGURE 3.9   Specification and interpretation for "internal" size tolerance.

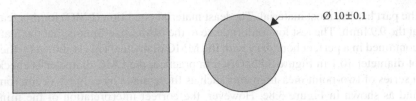

FIGURE 3.10   A size specification on a pin.

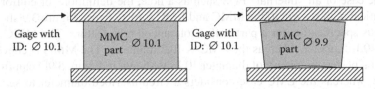

FIGURE 3.11   Gaging interpretation of Rule 1.

an FOS be of perfect form at MMC. For a cylindrical pin as specified in Figure 3.10, a part of diameter 10.1 must be perfectly straight and round, whereas the LMC boundary requires that the local size (at any cross section) still be greater than 9.9, as shown in Figure 3.8d.

In addition, even when the part is smaller than the MMC size of 10.1, the bend of the cylinder axis is not allowed to be so great that the part violates the perfect form boundary at MMC. This is easy to visualize when thinking of a gage to inspect the cylinder size tolerance. An ideal gage, shown in Figure 3.11, would be a block with a "perfect" hole of diameter 10.1 that is at least the length of the part. A part of diameter 10.1 would have to be perfectly straight in order to fit through the gage. If the part was smaller than diameter 10.1, it could be bent a bit but not too much; otherwise, it would not fit through the gage. Of course, the part could be no smaller than the minimum diameter of 9.9 at any cross section.

3.1.2.6   Location

Location tolerances are applied only to FOS* and control the feature relative to one or more datums. The most common location tolerance is position; the other tolerances that may control location are concentricity and coaxiality, or symmetry. The *correct* use of these other tolerances is rare, and it will not be discussed here. A position tolerance specification will contain the basic dimensions that show the ideal location of the feature, and the tolerance value describes a zone at this ideal location that should contain the feature's axis, midplane, or midpoint, depending on the type of FOS. Figure 3.12 gives an example of a position specification for a cylindrical feature (a hole) controlled relative to a coordinate system described by Datums A, C, and B.

---

* There is an exception to this rule—using a position tolerance on a profile boundary applied with a maximum material condition modifier. This is permitted only with the note "BOUNDARY" on the drawing and is beyond the scope of this introductory chapter.

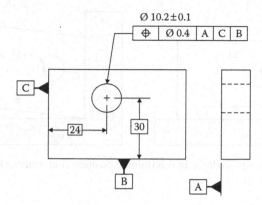

FIGURE 3.12   Specification of position tolerance.

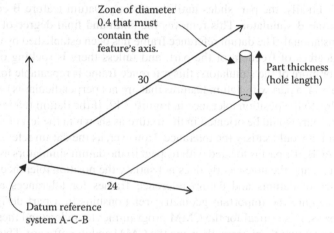

FIGURE 3.13   Tolerance zone for conformance to specification.

The interpretation of the position tolerance is as follows: the axis of the hole must be contained in a cylindrical tolerance zone of diameter 0.4, located at the basic dimensions of 24 and 30 in a datum reference frame consisting of the mutually perpendicular Datums A, B, and C. The zone is shown in Figure 3.13.

One might wonder why the datum order ACB in the position feature control frame is any different from ABC, as the datums (the theoretical planes) must all be mutually perpendicular. The ordering in the feature control frame is important because it controls the order in which the datum features are associated with the datums. This ordering is known as "datum precedence."

To make this example easier to understand, assume there is a physical fixture consisting of three planes, all very flat and square to one another, that are datum simulators for A, B, and C. In this example, the primary (first) Datum A surface of the real part is associated with the simulator by resting the bottom surface of the part on the A simulator surface of the fixture. This will remove three degrees of freedom: two rotational and one translational. Next, following the order of the datum reference

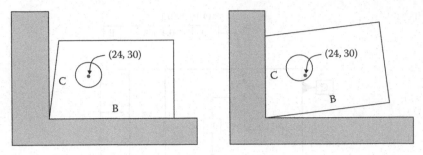

FIGURE 3.14 Datum precedence in determining position conformance (position tolerance zone shown by the dots).

frame, the part slides until the secondary Datum feature C is pressed against the C simulator. This removes two additional degrees of freedom: one rotational and one translational. Finally, the part slides until the tertiary Datum feature B comes into contact with the B simulator. This removes the sixth and final degree of freedom, which is translational. The datum reference frame has been established by constraining all six degrees of freedom of the part, and unless there is rocking of the part against one or more of the simulators, this reference frame is repeatable for the part.

In Figure 3.14, a part with datum features that are not perpendicular is to be evaluated according to the position tolerance in Figure 3.12. If the datum reference frame was ABC, the part would be oriented in the fixture as shown in the left of figure, and the hole position would satisfy the tolerance. However, as the datum reference frame specified is ACB, the part is located with respect to the datum simulators as shown in the right figure, and the hole clearly does not satisfy the position tolerance.

The choice of datums and datum reference frames for tolerances allows the designer to capture the important geometric relationships in a part design. When inspecting parts, it is critical for the CMM programmer to understand these specifications and to capture them correctly using the CMM and its software. The measurement of datums is an important step in accurately determining position conformance.

3.2 COORDINATE MEASURING MACHINE INSPECTION OF GEOMETRIC TOLERANCES

The primary difficulty in using a CMM to inspect geometric tolerances is that the tolerances are defined for the entire surface of a feature, and a CMM in general measures only a discrete subset of the surface (only a few points). If a CMM inspection of a part is to claim conformance to a tolerance specification, two things must be true: (1) the points taken must adequately represent the feature in question and (2) the algorithm that processes these points must do so in a way that is consistent with the tolerance specification. This subject is addressed in detail in Chapter 8, Coordinate Measuring System Algorithms and Filters.

In order to perform an adequate inspection of parts specified with geometric dimensioning and tolerancing (GD&T), the people programming (and operating) the CMM should have adequate GD&T knowledge. Once a clear understanding of the structure of datum reference frames and tolerance zones is attained, the CMM can be used very effectively to measure parts specified in accordance with the Y14.5M standard. Two specific topics are discussed briefly as examples of the thought processes that may be required when inspecting parts.

3.2.1 PHYSICAL INTERPRETATION

It is important to have an understanding of the physical "meaning" of geometric tolerance specifications. One important example is the construction of mutually perpendicular axes for a datum reference frame. CMM software will always create a coordinate system with orthogonal axes, but depending on the relationship of the datum features, the coordinate system may move relative to the actual part. As the datum reference frame is supposed to provide a unique, unambiguous coordinate system, this can be a problem.

In the two-dimensional example, shown in Figure 3.15, the datum reference frame AB is to be associated with the actual, imperfect part. Three possible procedures are described to build the AB datum reference frame. In the first procedure (Figure 3.16), a line is fitted (using least squares) to each of the A and B surfaces, and the origin is placed at the intersection of these lines, with the rotation of the xy axes determined by primary Datum A. In the second procedure (Figure 3.17), Datum A is fitted with a least squares line and Datum B is fitted using least squares, but constrained to be perpendicular to A. In the third procedure, both lines are fitted outside of the part, and the line associated with Datum feature B is constrained to be perpendicular to

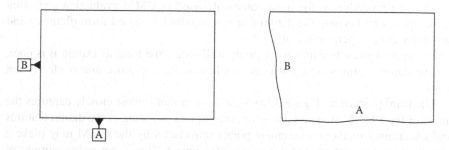

FIGURE 3.15 Datum specification and an imperfect part for which the datum reference frame will be built.

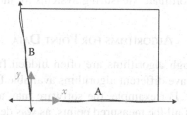

1. Fit least squares line to A.
2. Fit least squares line to B.
3. Find line intersection of A- and B-lines.
4. Build coordinate system with origin at intersection, x axis along A-line.

FIGURE 3.16 Procedure 1: Least squares for A and B; origin at intersection.

1. Fit least squares line to A.
2. Fit least squares line to B, constrained
   to be perpendicular to the A-line.
3. Find line intersection.
4. Build coordinate system with origin at
   intersection, *x* axis along A-line.

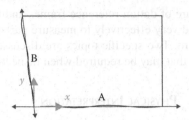

FIGURE 3.17   Procedure 2: Least squares for A and B, B constrained; origin at intersection.

1. Fit A-line outside of part.
2. Fit B-line, outside of part,
   perpendicular to the A-line.
3. Find line intersection.
4. Build coordinate system with origin at
   intersection, *x* axis along A-line.

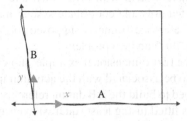

FIGURE 3.18   Procedure 3: A and B fitted outside of part, B constrained; origin at intersection.

A. These different procedures result in very different locations of the origin and thus different results for any position tolerances.

The first procedure is the most commonly used in CMM evaluation and often works quite well because the datum features usually have good form (flatness) and are pretty close to perpendicular.

The second procedure may work pretty well where the form of Datum B is poor, but the datum features are close to perpendicular, but the procedure is clearly not ideal.

The third procedure (Figure 3.18) is the best in that it most closely captures the intent of the Y14.5M standard. However, incomplete sampling of the datum features and uncertainty of the measurement points introduced by the CMM may make a coordinate system constructed this way hard to repeat. That is, several repetitions of the alignment procedure could determine different coordinate system locations relative to the part. This variability in the coordinate system will contribute to the uncertainty of other measurements on the part (See Chapter 14, Measurement Uncertainty for Coordinate Measuring Systems for more on this.).

3.2.2   ALGORITHMS FOR POINT DATA

Although algorithms are often hidden from the user, most CMM software packages have different algorithms available for analyzing the points measured with the CMM. For example, the software may allow the programmer to fit a line on one side of all the measured points, as was described in the third procedure. The default fitting method is usually least squares, as this method gives a unique, stable result that is relatively insensitive to outliers (points far from the "average" surface). The study of different algorithms available for point fitting and their appropriateness and efficacy comprises the field of computational metrology, which is addressed in Chapter 8, Coordinate Measuring System Algorithms and Filters.

3.3 Y14.5-2009

During the writing of this book, an updated version of the ASME Dimensioning and Tolerancing standard (ASME 2009) was released. Although there are many changes and additions in this revised standard, none of them affects the fundamentals of geometric tolerancing, as described in this chapter. The changes from Y14.5M-1994 to Y14.5-2009 are described in Appendix A of Y14.5-2009 standard. The addition of irregular FOS and some new specification methods for creating datum reference frames are probably the most important changes in the standard.

4 Cartesian Coordinate Measuring Machines

Paulo H. Pereira

CONTENTS

As one of the most powerful and versatile metrological instruments, coordinate measuring machines (CMMs) are widely used in most manufacturing plants, large and small. There is hardly a workpiece whose dimensions cannot be measured by CMMs. Flexibility and accuracy, coupled with decreases in time and cost of measurements, account for the wide acceptance of CMMs for industrial metrology. Like other similar systems, CMMs are extremely flexible, which is both an advantage and a disadvantage (refer also to Chapters 15, Application Considerations, and 16, Typical Applications).

A CMM is known as a machine that gives a physical realization of a three-dimensional rectilinear Cartesian coordinate system. There are other types of measuring machines that also produce coordinate measurement results but do not have rectilinear moving axes (see Chapter 17, Non-Cartesian Coordinate Measuring Systems). This chapter covers the basics of CMMs with Cartesian coordinate systems.

In Section 4.1, the concept of coordinate metrology is introduced (see also Chapters 3, Specification of Design Intent: Introduction to Dimensioning and Tolerancing, and 5, Operating a Coordinate Measuring Machine), followed by a general discussion of various design configurations, their characteristics, and typical applications in Section 4.2. In Section 4.3, the basic hardware elements of a typical CMM are presented, including structural elements, supporting bearings, drive systems, feedback elements, and control systems. Software aspects are not covered in this chapter (for software, refer to Chapter 8, Coordinate Measuring System Algorithms and Filters). Section 4.4 covers control systems for CMMs.

Probe heads and probing systems, along with the styli, are used to actually probe the parts to be measured. They are an integral part of the metrology loop of the CMM and, as such, are critical to the overall performance of the system. For detailed considerations regarding probing systems, refer to Chapters 6, Probing Systems for Coordinate Measuring Machines and 7, Multisensor Coordinate Metrology.

4.1 COORDINATE METROLOGY

The primary function of a CMM is to measure the actual shape of a workpiece, compare it against the desired shape, and evaluate the metrological information such as size, form, location, and orientation.

The actual shape of a workpiece is obtained by collecting data on its surface at certain points or areas. Data collection can be carried out through several different sensors, both contact and noncontact and continuously or discreetly. Every measuring point is expressed in terms of its measured coordinates. Some sensors are capable of also collecting the direction vectors of the measured points, which usually allows for better accuracies. However, it is not possible to evaluate the parameters of the workpiece (e.g., diameter, distance, angle, form, and others) directly from the coordinates of the measured points. An analytical model of the workpiece is needed to evaluate the parameters. This model usually consists of ideal geometric elements, the so-called substitute (mathematical) geometric elements (ISO 2004a, 2005b; ASME 2009). Such elements may be determined by applying an appropriate geometric element best-fit algorithm to the measured data set.* Figure 4.1 illustrates

---

* See also Chapter 8, Coordinate Measuring System Algorithms and Filters.

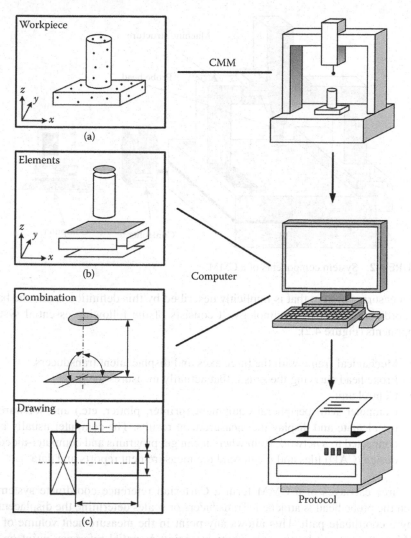

FIGURE 4.1 Nature of coordinate metrology.

the modeling of workpiece geometry for a simple example. The principle of coordinate metrology can be defined as

- Generating the data sets by measuring the actual workpiece with a CMM (Figure 4.1a)
- Calculating the relevant substitute (mathematical) geometric elements in terms of parameters specifying size, form, location, and orientation (Figure 4.1b)
- Evaluating the required workpiece features, for example, by combining the substitute elements and comparing with the drawing dimensions and tolerances (Figure 4.1c)

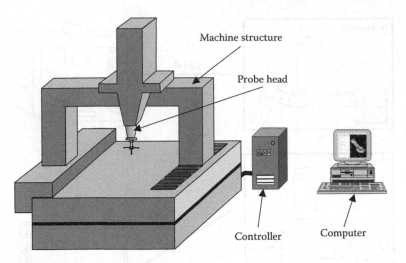

FIGURE 4.2   System components of a CMM.

The measuring system that is implicitly described by this definition forms the basis of coordinate measuring technology. It consists of the following essential system components (Figure 4.2):

- Mechanical frame with the three axes and displacement transducers.
- Probe head carrying the sensor that actually measures the part.
- Control unit.
- Computer with peripheral equipment (printer, plotter, etc.) and software to calculate and display the measurement results. The computer usually is connected to a network from where it can get programs and computer-aided design (CAD) files and it can send the measurement reports and data.

The three carriages of a CMM form a Cartesian reference coordinate system to which the probe head is attached. Transducers or scales determine the displacement along a coordinate path. This allows any point in the measurement volume of the CMM to be covered by the measurements using a spatial reference point on the probe head. This reference point is usually the center of the probe tip for contact sensors. A measurement with a CMM always comprises the following steps:

- Calibration of the stylus or probe tip with respect to the probe head reference point normally using a calibrated sphere (provided an electromechanical three-dimensional probe is used)
- Determination of the workpiece position and orientation (workpiece coordinate system X_w, Y_w, and Z_w) in relation to the machine coordinate system X_m, Y_m, and Z_m (Figure 4.3)
- Measurement of the surface points on the workpiece
- Evaluation of the geometric parameters of the workpiece
- Representation or reporting of the measurement results

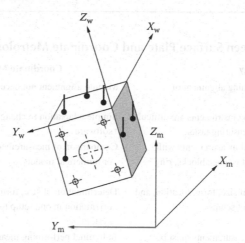

FIGURE 4.3 Workpiece and machine coordinate systems are usually not the same.

A comparison of the measuring principles of coordinate metrology with those of conventional (surface plate) metrology shows some important advantages of CMMs:

- The mechanical alignment to reference coordinates is eliminated because the part is referenced to the machine coordinate system.
- The need for auxiliary tools like adjusting elements, setting plugs, and mounting attachments is all but eliminated.
- The three-dimensional concept of coordinate metrology enables the determination of size, form, position, and orientation usually in one setup on one measuring machine, using one reference system, eliminating individual error sources from multiple measuring devices in conventional metrology.

In conventional metrology, the geometric elements are mostly measured independently from each other on different measuring instruments (i.e., length-measuring comparator; pitch-, angle-, and involute-measuring instruments for gears; form-measuring instruments; and angle-measuring instruments) with different setups having different reference systems. With conventional metrology, the measurements are performed with artifacts such as gage blocks, ring gages, and sine bars, with involute and pitch master gears serving as reference elements. In coordinate metrology, the measured features are compared to the numerical models. Table 4.1 summarizes the main comparison points between conventional and coordinate metrology.

4.2 BASIC COORDINATE MEASURING MACHINE CONFIGURATIONS

A CMM configuration plays an important role in its performance aspects such as accuracy, flexibility, cycle time or throughput, and lifetime costs (initial and

TABLE 4.1
Comparison between Surface Plate and Coordinate Metrology

Surface Plate Metrology	Coordinate Metrology
Manual and time-consuming alignment of workpiece	Manual alignment not necessary
Single-purpose measuring instruments are difficult to adapt to changing measuring tasks	Simple adaptation to changing measuring tasks by software
Comparison of individual measurements with artifact dimensions, that is, gage blocks, ring gages, or others	Comparison of measurements with mathematical or numerical models
Separate determination of size, form, location, and orientation with different setups	Determination of size, form, location, and orientation in one setup using one reference system
Individual performing measurements must be highly skilled	Individual performing measurements need not be skilled if a program is prepared for the task
Manual nature of methods hinders throughput—cannot be made automatic	Throughput less dependent on operator—can be made fully automatic

TABLE 4.2
The Most Common CMM Configurations and Their General Application

Applications/ Configuration	Moving Bridge	Fixed Bridge	Cantilever	Horizontal Arm	Gantry
General-purpose applications	X	X	X	X	
Accuracy	X	X (gage calibration)			
Large parts				X (car bodies, large castings)	X (aerospace structures, large vehicles)

operating*). The commonly used basic CMM configurations[†] are shown in Table 4.2 and described in Sections 4.2.1 through 4.2.6 along with their relative advantages and disadvantages.

4.2.1 MOVING BRIDGE

As the workhorse of the CMM industry, the moving bridge CMM is the most widely used configuration. It has a stationary table to support the workpiece to be measured

---

* Refer also to Chapter 19, Financial Evaluations.
[†] ASME B89.4.1 (ASME 1997) lists the CMM configurations.

and a moving bridge. With this design, the phenomenon of yawing (also called "walking") can occur due to the two columns or legs moving at different speeds, causing the bridge to twist. This affects the accuracy of parts measured at different locations on the CMM table. A design that implements dual drive and position feedback control systems for the two legs, or that drives the moving bridge at its centerline, can reduce this effect.

The advantage of the moving bridge configuration over the cantilever is that the bending effect of the second horizontal axis can be greatly reduced by having two supporting columns. It also has higher natural frequencies than those of cantilever models. The moving bridge design has a small to medium measuring range, with relatively small measuring uncertainties.

Another category of moving bridge CMMs is the ring bridge configuration, which uses a ring to support the two axes of motion. This design enhances the structural rigidity. Compared with other alternative configurations, the ring bridge design has a higher stiffness-to-weight ratio. Other advantages of the ring bridge design in comparison to other types of moving bridge configurations are less Abbe offset and the ability to have a central drive under the table, thus being closer to the center of gravity of the moving elements. Additionally, the measurements with the ring bridge design are not affected by the mass of the workpiece because the metrology frame is decoupled from the structural system that supports the part.

4.2.2 FIXED BRIDGE

In the fixed bridge configuration, the bridge is rigidly attached to the machine bed. The table upon which the workpiece is mounted provides one axis of motion (Figure 4.4). This design eliminates the phenomenon of walking and provides high rigidity. Not surprisingly, almost all of the most accurate CMMs currently offered in the market use this configuration. On the other hand, the operating speed is reduced because it has to move the heavy table with the part on it. This reduces the throughput of the machine, and the lower maximum allowable part weight may become a limitation. It also introduces some bending of the table, which can compromise accuracy if not addressed.*

The main advantage with the fixed bridge design is its very rigid structure. As the table is driven centrally and the feedback element is also located in the center, the Abbe offset error is considerably reduced for the table motion. The need to have extended guideways to support the movement of a long table is a disadvantage of this configuration because it requires a larger footprint.

4.2.3 CANTILEVER

This design has a moving cantilever arm that supports a carrier to move in and out. The carrier supports the probe arm for the vertical movement (Figure 4.5). The part to be measured is placed on a fixed table. Because the table does not include the bearing

---

* See also discussions about quasi-rigid body errors in Chapter 12, Error Compensation of Coordinate Measuring Machines.

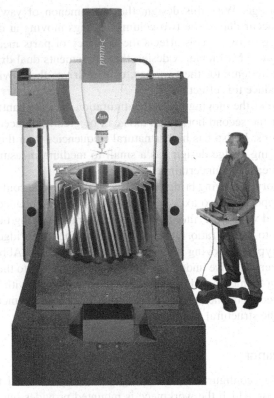

FIGURE 4.4   A fixed bridge CMM. (Courtesy of Hexagon Metrology, Inc.)

guideways, relatively heavy parts can be supported on it without affecting measuring accuracy. This configuration generally has a low moving mass structure for a given measuring volume, making it uniquely agile, thus allowing a higher throughput.

With three sides open, the fixed table cantilever allows good accessibility to the workpiece, but overhead loading is usually not possible. Additionally, high throughput makes this design a favorite for general measurement applications. Its disadvantage is the bending caused by the cantilever design. As the carriage is moved toward its extreme outer position, bending becomes more of a factor. Because of the cantilever effect, it has a lower system natural frequency, which limits the size of the machine. Unique mechanical means to offset this effect have been complemented by most manufacturers with software error correction techniques (see also Chapter 12, Error Compensation of Coordinate Measuring Machines). The cantilever design offers a long table with relatively small measuring ranges in the other two axes, making it particularly suitable for measuring long, thin parts.

4.2.4   Horizontal Arm

Horizontal arm machines are ideal for measuring automobile bodies or other similar sized and toleranced parts. Among many horizontal arm configurations are moving

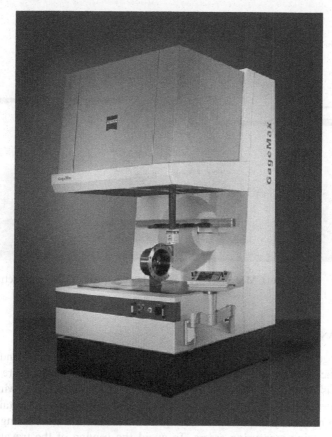

FIGURE 4.5 Cantilever configuration. (Courtesy of Carl Zeiss IMT.)

table, moving ram, and dual-arm designs. This configuration utilizes the table for one of the horizontal axes of motion and the moving column for the other. Like all other CMMs with moving tables, its measuring speed and accuracy depend on the size and weight of the workpiece. In the case of moving ram horizontal arm CMMs (Figure 4.6), the cantilevered design results in low dynamic stiffness and relatively large Abbe offset errors. The dual-arm measuring system design consists of two identical mirror-image sets of horizontal measuring arms and moving axes allowing the simultaneous measurement of both sides of a part, thus enabling higher throughput.

The advantage with all types of horizontal arm CMMs is the excellent accessibility to all sides of the part. High measurement speeds are typical with this configuration. The disadvantage is primarily the limited accuracy.

It is normal for these machines to have large operating ranges, where one axis can be significantly longer than the other two. The moving ram design offers measuring ranges in the long axis, for example, up to 25,000 mm. The relative levels of measuring uncertainties for horizontal arm machines vary considerably.

FIGURE 4.6   A dual-arm CMM checking a car body. (Courtesy of Hexagon Metrology, Inc.)

4.2.5 GANTRY

For very large parts requiring measuring volumes of 10 m³ or more and relatively tight tolerances, the gantry configuration is the most suitable (Figure 4.7). The foundation is designed as a massive structure to limit deflections, which would induce distortions of its structure (see also Chapter 11, Environmental Control). This is also necessary to prevent the weight of the part from distorting the foundation and causing measuring errors. To avoid the yawing of the traveling beam, dual-drive systems (master and slave configurations) are frequently used. Software compensation is effectively used to correct for geometric distortions caused by loading (cross-carriage) and temperature effects.

Besides its large volume (e.g., 25,000 × 6,000 × 4,000 mm), the advantage with the gantry-type CMM is the easy access for the operator to all parts in the machine volume. The moving part of the machine has less weight because only the horizontal beam moves, and hence the accuracy normally achieved is in the medium range even for large measurement volumes. The normal class of machines in this configuration is relatively cost-effective. However, if better uncertainty is required, the costs can increase because of special requirements such as the use of special isolated foundations and special structures (like large granite blocks).

4.2.6 OTHER CONFIGURATIONS

The L-shaped bridge configuration is a modification to the fixed table cantilever design, providing a supporting leg at the opposite end of the cantilever beam (ASME 1997). Because L-shaped bridges are generally adaptations of cantilever machines, they are usually slow and have low resonant frequencies.

FIGURE 4.7 Gantry configuration. (Courtesy of Hexagon Metrology, Inc.)

FIGURE 4.8 Measuring robot. (Courtesy of Hexagon Metrology, Inc.)

There is another type of measuring machine, frequently referred to as a measuring robot (Figure 4.8). It is characterized by very high accelerations and often used alongside production machinery. It usually comes with a rotary table for better accessibility to part features.

4.3 HARDWARE COMPONENTS

While the configuration of a CMM has a definite effect on its performance, there are several other factors that contribute to the overall capability of a CMM, including hardware components. The structural elements, bearing supports, drive systems, displacement and measurement systems, probe head, and control systems constitute the main hardware components. The following subsections deal with technical characteristics of these hardware components and their influence on the performance of a CMM.

4.3.1 STRUCTURAL ELEMENTS

Structural elements serve as the backbone of a CMM. The machine base, table to support the workpiece, machine columns, slideways, and ram are the essential structural elements. Because these support or carry the workpiece, measuring probes, and linear scale reader heads for position feedback systems, their characteristics directly affect the overall performance of the measuring machine. The most important ideal properties of the structural elements are the following:

- Dimensional stability (short and long term)
- Infinite stiffness
- Weightlessness
- High damping capacity
- Low coefficient of thermal expansion
- High thermal conductivity

No material is capable of satisfying all the above listed properties perfectly. Knowing the desirable properties and their influences helps in the selection of materials for the structural elements during the design phase.

Dimensional stability is one of the most important properties of the structural elements. Even with software-corrected machines (see also Chapter 12, Error Compensation of Coordinate Measuring Machines), this property is of utmost importance, as lack of dimensional stability adversely affects CMM performance. Aging and stress-relieving techniques are used for obtaining dimensional stability. Granite, with its natural seasoning of millions of years, is considered by many as the most suitable material for structural elements of CMMs. It is indeed used very often as a basic element, especially in higher accuracy class machines. It was also a natural transition from using granite as the primary material for surface plates to its use for CMMs. Nonetheless, granite is susceptible to liquids if unprotected and changes its shape as a result.

The coefficient of thermal expansion and thermal conductivity of structural materials influence the accuracy of CMMs. It is desirable to have low thermal expansion and high thermal conductivity. Aluminum has a large coefficient of thermal expansion compared with granite or steel, but it is a suitable choice for CMM structural elements because of its extremely high thermal conductivity compared with granite (two orders of magnitude)—120–167 $W \cdot m^{-1} {}^{\circ}C^{-1}$ versus 1–2 $W \cdot m^{-1} {}^{\circ}C^{-1}$ (Slocum 1992a). An aluminum structure can quickly attain thermal stability when disturbed by a temperature change. Granite has low thermal conductivity, which means it responds very slowly to temperature changes. This characteristic attenuates variations due to rapid changes but causes a delay in response to slower changing temperature conditions. Therefore, a CMM with structural elements made of aluminum can follow slower temperature changes in the shop with little distortion, whereas a granite and steel machine can have considerable distortions for longer as a result of their lower thermal conductivity. However, an aluminum structure tries to follow faster temperature changes, making it less stable in these cases.

An alternative to natural granite for structural elements is to use composite materials that provide good properties and are easy to manufacture in different shapes. Certain commercial models offer structures made of composite materials (Hemmelgarn et al. 1997).

Certain CMM models use compensation techniques to minimize the effects of thermally and/or mechanically (due to weight) induced structural deformations. For example, embedded sensors monitoring temperature of a scale or a machine table can feed an algorithm to compensate for deformations due to thermal gradients. Similarly, a cross-carriage may be distorted due to the weight shifting when the machine axes move, and a compensation can be used based on its current position.

To achieve high throughput with sufficient accuracy, the dynamic characteristics of the structural elements are very important. The stiffness-to-weight ratio is a key measure of their dynamic characteristics. Analytical tools, such as finite-element methods and the use of hollow structures, are used for lighter weight and better stiffness of structural elements. The damping characteristics of the overall measuring system are also of concern for good dynamic performance (Weekers and Schellekens 1997), and it can be further enhanced through the servo control system or active vibration cancellation techniques (Lu, Ni, and Wu 1992).

In addition to the above factors, easy manufacturability for higher accuracy and lower overall costs should be considered when evaluating materials for the structural elements.

4.3.2 Bearing Systems

Bearing systems for CMMs are very important because they are part of the structural loop and directly affect measurement accuracy. They can also influence the characteristics of the drive systems. In general, there are two common types of bearing systems in use for CMM applications: (1) noncontact air bearings and (2) mechanical contact bearings. Among many design criteria for bearing systems, the dynamic stiffness, load-carrying capability, damping, and frictional effects are given first consideration.

Aerostatic air bearings, such as the one in Figure 4.9, use a thin film of air under pressure to provide load support. The low viscosity of air requires a close gap of around 1–10 μm.

Contact hard bearings, such as precision roller or ball bearings, and sliding guideways are also used in CMMs. Particularly, they can be found in those CMMs that have developed from precision machine tools (Moore 1970). Hard bearings can normally take higher loads compared with air bearings. Nowadays, hard bearings are primarily used for CMMs designed for harsher factory environments such as foundries.

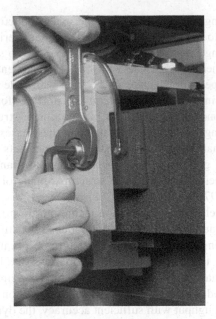

FIGURE 4.9   Air bearing. (Courtesy of Carl Zeiss IMT.)

Both hard bearings and air bearings require maintenance. The hard bearings need to be lubricated and the guideway surfaces on which air bearings operate need to be kept clean. Air bearings also require filtration systems to prevent water and oil in the compressed air lines from getting into the bearings. Air bearings may be more durable in the long term because they have no mechanical contact. Compressed air supply needs to be adequate as pressure variation can cause CMM geometric errors to change and sudden loss of air pressure can lead to catastrophic failure damaging the way surfaces and the bearings. Slocum (1992a) provides more detailed descriptions of different bearing technologies.

4.3.3 Drive Systems

The demand for increased CMM throughput requires fast drive systems and stiff transmission linkages. Drive systems play a significant role in direct computer-controlled (DCC) machines. The purpose of a drive system is only to move the probe and/or the part and not to provide positional information. The displacement transducers or scales provide positional information.

There are many different types of drive system in use, such as rack-and-pinion, belt, friction, leadscrew, and linear motor. The selection of the appropriate drive system is the designer's responsibility based on the intended outcome. CMM users can benefit from an understanding of the operational and performance characteristics of different drive systems.

In general, the natural frequency of a drive systems should be designed to be higher than that of the machine structure, and the dynamic response bandwidth of the servo control system is to be tuned much lower than the machine structural natural frequency to avoid structural resonance. The first resonant frequency of a CMM structure dictates the dynamic response bandwidth of the drive system.

4.3.3.1 Rack-and-Pinion Drive

One relatively simple solution for generating linear motion is the use of rack-and-pinion drives (Figure 4.10). Normally, the pinion gear is the driving element and the rack is driven. This drive system finds application where long length of travel (as in the case of gantry-type CMMs) is needed. Form errors and backlash in gears limit the accuracy of CMMs using this drive system, which is normally used in cost-effective designs.

4.3.3.2 Belt Drive

A belt drive system (Figure 4.11) normally consists of a belt, multistage speed reducer, and a servo motor. The belt can be, for example, a fiber-reinforced notched or a metallic flat tape. The belt also offers a quiet transmission of power to the moving axes. It acts as a low-pass filter preventing high-frequency motor oscillations

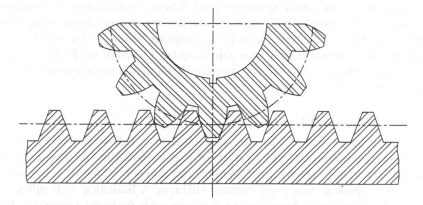

FIGURE 4.10 Rack-and-pinion drive.

FIGURE 4.11 Belt drive. (Courtesy of Hexagon Metrology, Inc.)

from entering the measuring structure. The moving axes can be driven at high accelerations and speeds with belt drives. The disadvantage of the belt drive system is its relatively high compliance or elasticity limiting its use for high-accuracy scanning operations. Manually-driven CMMs usually have disengaging mechanisms, which allow the operator to lock and unlock the moving axes for manual positioning of the probe shaft or machine carriages.

4.3.3.3 Friction Drive

A friction drive (Figure 4.12), sometimes called a "capstan" or "traction drive," is used in some CMMs. It consists of a driving wheel, a flat or round bar, and a supporting backup roller. The driving wheel can be driven directly by a motor or through a speed reducer. A constant preload is applied to the driving wheel so that it maintains a smooth contact with the bar.

The advantages of friction drives include their simplicity in design, low friction force, minimum backlash, deadband due to elastic deformation induced by the preload, and potentially high positioning accuracy when a good position feedback device and an adequate servo system are used. These characteristics are important conditions for designing high-dynamic and stable drive systems. Some of the undesirable properties are their low drive force capability, relatively low stiffness and damping, and minimum transmission gain. Another variant of the friction drive is the design that uses a cable wrapping around a motor-driven capstan to pull the carriage back and forth.

4.3.3.4 Leadscrew Drive

The most common type of leadscrew drive system is ball screws (Figure 4.13). A leadscrew transforms the rotary motion of a servo motor into the linear translational movement of a carriage attached to the nut through rolling balls. Ball screw drive systems provide very good dynamic stiffness. A limitation with ball screw drives is their inability to disengage and reengage easily for manual operation. Ball screw drives can also be noisy and are subject to wear.

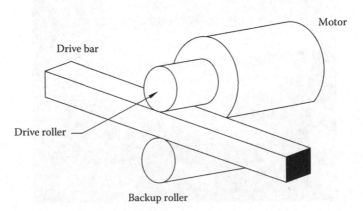

FIGURE 4.12   Friction drive.

FIGURE 4.13 Ball screw drive. (Courtesy of Hexagon Metrology, Inc.)

4.3.3.5 Linear Motor Drive

Linear motors are also used for CMM axis drive applications. These drive systems offer unusually high stiffness. Because of their direct drive, there is no tachometer signal available, which is required for servo velocity control. The speed feedback has to be derived from position feedback signals, which in turn impose the requirement of extremely high resolution on position feedback devices. In addition, no mechanical gear reduction mechanism is available for linear motor drives. Desirable characteristics such as impedance match and torque amplifications do not exist for linear drives.

The oscillation modes due to the motor drive can directly influence the measuring structure. CMMs with linear motor drives often have relatively high-acceleration characteristics. In addition, linear drive systems are well suited for large gantry CMM structures, where a balanced drive through the center of gravity is impossible. To avoid the yaw motion of any axis with a wide span between two supporting members, linear motor drives can be used to provide synchronous drive at each of them. Heat generation from a linear drive system can be another concern because the drive systems are physically embedded in the measuring structure. In some applications, circulating water-cooling systems are designed into the linear drive systems. Yet another concern is the cost of linear drive systems, which is usually higher than other types of drive systems.

4.3.4 Displacement Transducers

When a probe touches the workpiece, its position is actually determined from displacement transducers on each of the three linear axes of the CMM. Certain probe models measure the tip displacement, whereas machine scales measure the probe displacements. In these cases, the actual measurements are combinations of the two readings.

A variety of displacement transducers can be found in CMMs, including optical scales, rotary encoders, inductosyns, magnetic scales, and laser interferometers. The following descriptions are focused on the commonly used optical scales and laser interferometers. More detailed descriptions about other displacement transducers are given by Slocum (1992a).

In principle, optical linear scales consist of a scale element and an electro-optical read head, one of the two is fixed to the moving slide of the CMM. The relative motion between these two components generates the positioning signal. Among the optical scales, there are three general types: (1) phototransmission scale, (2) photoreflection scale, and (3) interferential scale. Laser interferometers are also discussed because of their importance in length measurements, though they are not widely used in CMMs.

4.3.4.1 Transmission Scale

The transmission scale is typically made of glass and is a precision-lined grating with a line pitch of usually 50–100 lines per millimeter. The read head contains a light source, a collimating lens for conditioning the light beam, a scanning reticle with the index gratings, and photocells. Figure 4.14 illustrates the principle of this type of scale. When the read head is moved relative to the scale, the lines of the scale coincide alternately with lines or spaces in the index grating. The periodic fluctuation of light intensity is converted by photocells into electrical signals. These signals result from the averaging of a large number of lines.

The output from these photocells consists of two sinusoidal signals. These signals are phase-shifted by 90°. The signal period is equal to the grating period of the scale graduation. A third signal peak from the reference mark serves as the reference signal (scale zero or home position).

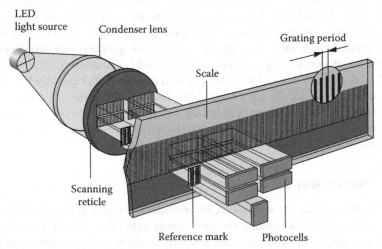

FIGURE 4.14 Photoelectronic measuring principle of glass scale. (Courtesy of HEIDEN-HAIN Corporation.)

4.3.4.2 Reflection Scale

Different from the glass scale in the transmission scale, the reflection scale is made of steel. The interference pattern for the reflection scale is established with alternate reflective lines and diffusely reflective gaps. The read head contains a light source, a collimating lens for light beam conditioning, a scanning reticle with index gratings, and silicon photocells. As shown in Figure 4.15, when the read head is moved, the photocells produce periodic signals similar to those generated from a glass scale.

4.3.4.3 Interferential Scale

This type of displacement transducer uses grating scales. The scale is a precision-lined grating with a line pitch of 100 or more lines per millimeter. The lines are read by a photoelectric head that detects the interference fringes created by light reflected from the scale and a reference grating on the read head as shown in Figure 4.16. The electrical outputs of the read head are sinusoidal waves. The signals are electronically subdivided for finer resolution and then digitized.

4.3.4.4 Laser Interferometer Scale

The laser interferometer is based on the well-known light interference principle. It consists of a laser light source, an interferometer, and two mirrors. Laser interferometer scales provide very high accuracies. The cost of a laser interferometer scale relative to other linear scales is higher.

The interferometer is a device used to split a laser beam into two separate beams, one used as a reference and the other used to create interferences with the reference beam. This concept, illustrated in Figure 4.17, was developed by A. A. Michelson in 1881 (Michelson and Morley 1887). The portion of the beam directed to the fixed mirror is the reference beam, and the other portion is directed to the moving mirror. By displacing the movable mirror, a phase change occurs between the two beams; upon recombination, these beams interfere, producing fringes that can be counted.

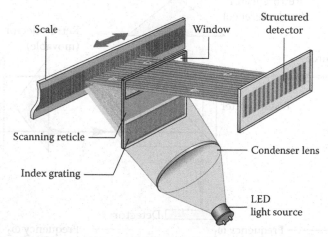

FIGURE 4.15 Photoelectronic measuring principle of steel scale. (Courtesy of HEIDENHAIN Corporation.)

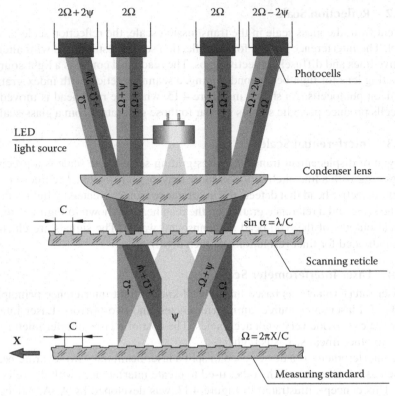

FIGURE 4.16 Principle of an interferential scale. (Courtesy of HEIDENHAIN Corporation.)

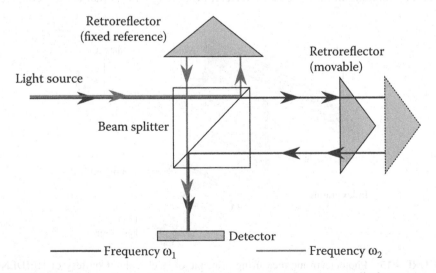

FIGURE 4.17 Schematic of a Michelson interferometer used to measure displacement.

This provides the signal for determining relative displacement. Based on the above fundamental principle, there are many commercially available interferometers.

The major difference between the optical linear scales and laser interferometers lies in the measurement principle (Kunzmann, Pfeifer, and Flügge 1993). Linear scales use optical gratings on material standards as reference, whereas laser interferometers are based on laser wavelength in the medium (e.g., in vacuum or open air). Therefore, the factors that influence the accuracy of both systems are different in nature. Grating accuracy, temperature, and thermal expansion coefficient of the scale material are the main influencing factors in optical scale accuracy. The influencing factor for laser interferometers is mainly the refractive index of air, which is affected primarily by temperature, barometric pressure, and humidity. The resolution obtained with optical scales is typically on the order of 0.1–1 μm. However, with interferential scales and laser interferometers, the resolution can be down to 1 nm or smaller.

4.4 CONTROL SYSTEM FOR COORDINATE MEASURING MACHINES

The control system in a CMM performs the function of a live interaction between various machine components such as machine drives, displacement transducers, probe systems, and peripheral devices. Control systems can be classified according to the following groups of CMMs:

- Manually-driven CMMs
- DCC CMMs
- CMMs linked with CAD, CAM*, FMS†, etc. (also DCC)

The first type (manually driven) is self-explanatory. In the case of DCC machines, the computer control is responsible for the movement of the slides, readout from displacement transducers, and data communication. For movement of the slides, different control strategies are used:

- Point-to-point control
- Continuous-path control
- Vector control

In the case of point-to-point control, the target position is given as point information with its three coordinates. The slide moves toward this position with machine-specific acceleration/speed control until the target values are reached. In this control mode, the slide having the shortest distance to travel to the target position coordinate is reached first.

With the continuous-path control, there are two distinct modes: precalculated scanning control and adaptive scanning control. Other commonly used terms are predefined or open-loop scanning and not predefined or closed-loop scanning. In the

---

* Computer aided manufacturing.
† Flexible manufacturing system.

case of precalculated scanning control, the probe system moves to the target position along a defined path. In this case, the computer is always engaged in monitoring the movement. This kind of control is used when measurement is being performed on known shape features. With the adaptive scanning control, the probe head moves to the target position in one or two axes with freely selected speed. If there is a change in the shape of the feature, the adaptive control ensures that the probe tip maintains contact with the workpiece surface. This control mode is used when scanning undefined features (for which a nominal shape is not available).

Usually, open-loop scanning is faster than closed-loop scanning for the same accuracy because it does not require continuous feedback from the controller. One tactic programmers may use is to create a defined path from an unknown feature by measuring it first in closed-loop scanning. This can save time when the task includes more than one part with the same unknown shape.

With the vector control, the shortest motion (moving on a straight line) between the start and the target positions is achieved. Here the probe head moves to the target position in all three axes under continuously optimized speed. In addition to the vectorial positioning, vectorial probing is possible with this control. These are essential for fast and reproducible measurements when they are carried out on inclined bores and curved surfaces. The vector control is also used for locating free-form workpiece surfaces.

Further tasks that the control system is to carry out include (1) specific control commands (like choosing manual/DCC mode or the control mode and commands for the probing system) and (2) various supervision tasks, like reaching the target position, standstill position of the slides, and collision detection.

The above tasks are fulfilled for newer generation CMMs with modular microprocessor control systems. These modular multiprocessor systems (Figure 4.18), whose single microprocessors are coupled by an internal data bus, do the various tasks such as acquiring information from displacement transducer devices, controlling the probe head, and controlling the CMM driving systems, as well as the data transfer to the computer including peripheral devices. These multiple microprocessor systems disengage the CMM computer more and more from the tasks of process control. The CMM computer is usually left only with high-level tasks of sending the measurement commands (measure feature 1, feature 2, etc.), evaluating measured data, and reporting results, which enhances throughput. For machines with temperature compensation, separate microprocessors usually carry out those calculations for the measured data before sending the calculations to the CMM computer. This is also the case for software error correction methodologies (see also Chapter 12, Error Compensation of Coordinate Measuring Machines).

DCC-driven CMMs mostly use dedicated computers. These systems are cost-effective, compact, and easy to use. More powerful computers, like workstations, may be needed for control activities in multiarm CMM configurations or applications with large data transfers, as in the case of integration with CAD.

4.5 SUMMARY

Coordinate metrology based on the use of CMMs offers tremendous advantages compared with conventional dimensional metrology for many applications. There

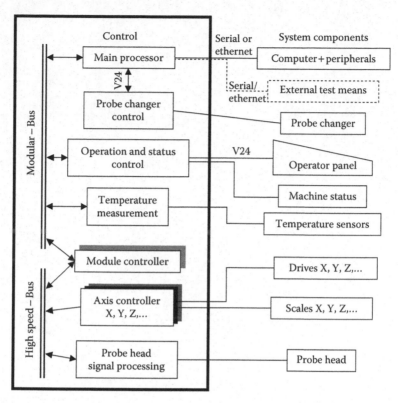

FIGURE 4.18 Modular multiple microprocessor system.

are many different machine configurations and software systems for performing three-dimensional coordinate metrology. These different approaches have their own inherent performance characteristics in terms of achievable measurement accuracy, effective measuring volume, speed of measurement, and cost-effectiveness. Understanding these characteristics and software options makes the informed user better able to define the requirements for specific measurement applications.

ACKNOWLEDGMENTS

Special thanks to Jun Ni and Franz Wäldele who wrote the chapter on which this revision is based (Ni and Wäldele 1995).

5 Operating a Coordinate Measuring Machine

Edward Morse

CONTENTS

This chapter covers the basic operation of a coordinate measuring machine (CMM) for the purposes of part inspection. The actual details will vary depending on the specific equipment and any in-house procedures that may be required. However, the steps needed for successful part measurement are pretty much the same regardless of the details of the machine, the software, or the type of part to be measured. The discussion in this chapter is intended for contact measurement CMMs in which points are collected individually from the part surface and analyzed based on the feature that is measured, as opposed to vision systems that collect a "cloud" of points and then partition this cloud in a postprocessing step. Some steps described in this chapter are only appropriate for direct computer-controlled (DCC) machines, but these are easily identifiable from the context.

5.1 INTRODUCTION

The training threshold for knowing how to use a CMM has been dropping steadily, not because the task of measurement has become easier but because the CMM software has become more user-friendly. This is not to say that one can become proficient in CMM programming in one week but that many of the basic operations have been simplified in more modern software. In order to get the most out of coordinate measuring equipment, a great deal of knowledge and experience must be gained. It is advisable to ask more experienced users for their suggestions and to discuss measurement techniques and strategies with other users, either in person or through electronic bulletin boards (or forums) that are available online. It is easy to stay with a technique that has worked in the past if you never discuss with, or justify that method to, others. The community of CMM programmers is full of knowledgeable people who like to discuss what they do.

This chapter provides an overview of the knowledge that is needed for basic CMM use and also provides some structure for the beginning user as to where certain information is used in the programming process.

5.2 BEFORE STARTING

As with any task, planning and foresight make the job at hand easier and ultimately less expensive. Subsections 5.2.1 through 5.2.6 deal with preparatory tasks that save time once the programming and use of a CMM begins. In environments where CMMs are in almost constant use, these tasks are very important since the time spent on programming the CMM uses productive time when the CMM could be performing useful work (measuring parts).

5.2.1 SAFETY

If one spends a lot of time with machine tools, it is easy to dismiss CMMs as slow and unlikely to cause much damage or injury. This does not mean that safety should not be the first priority when using a CMM. Basic safety precautions for working with a CMM include keeping all manufacturer-supplied covers on the machine, avoiding pinch points, and keeping the pendant (joystick box) with its emergency stop switch within reach. Also, the work table on most moving-bridge CMMs is a bearing surface, and the operator must be aware of objects that may have been placed on this portion of the table. Although the thought of a coffee cup being knocked over by a CMM may seem amusing, the danger and expense involved in a parallel block or holding fixture falling because of contact with the bridge is serious. Figure 5.1 shows the common mistake of leaving a riser block on the machine table after the setup for measurement is complete. This is also hidden from the operator by the leg of the bridge, so it is quite easy to inadvertently drive the machine into the block and knock it onto the floor. Another related point is that safety shoes—or at a minimum, closed toe shoes—should be worn around the CMM. The fixtures used in workholding on a CMM are heavy, as are the riser blocks and parallels that are used to locate parts. It is not uncommon for feet to be injured by dropped fixtures.

FIGURE 5.1 Potential hazard; the unsafe placement of a riser block (even temporary placement).

With many CMMs, access to the measurement table is restricted, and care must be taken in placing parts and fixturing on the table. Use proper lifting techniques to load and unload parts from the table to avoid back injury. If cranes or other types of equipment are used to move parts on and off the table, operators must be trained in their use.

5.2.2 COORDINATE MEASURING MACHINE START-UP

Machine start-up is usually well-documented by laboratories or companies. There is often a checklist that should be followed with weekly and daily maintenance and cleaning tasks. At a minimum, air filters should be checked and the bearing surfaces wiped down before using a CMM. If the air supply to the machine is always on, you may notice that there is a concentration of oil where the air bearings rest on the ways. These areas must be cleaned completely. With bridge-type machines, it is usually not possible to clean the entire z axis ram, as it slides up into the covers of the machine. For this reason, it is important in some machines that the z axis ram is kept in the "up" position when the CMM is not in use. Then, if any oil or other contaminants get on the ram from the bearings, they can be cleaned by moving the axis to the "down" position (Figure 5.2). The ram, rails and ways of the CMM should be cleaned with a lint-free cloth and alcohol. The probe styli and reference sphere

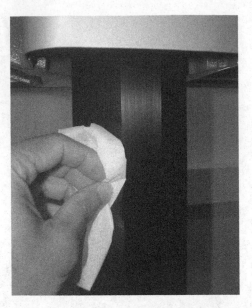

FIGURE 5.2   Cleaning the *z* axis ram of a small measuring machine.

should also be cleaned in this manner. Ensure that all probes and stylus assemblies are tight: Loose styli can cause problems that are difficult to track down after the measurement is complete.

When the CMM is started up from a power-off state, it usually needs to be homed. This is a sequence in which the CMM is driven to a repeatable position in the measurement volume, usually determined by set scale marks or limit switches. Driving to the home position is necessary because many scale systems are incremental (i.e., they measure a change of distance, not absolute position) and need to have an external zero location set to determine where in the CMM volume a probe is located. This information is important to the CMM so that it does not try to drive to a point beyond the limits of the scales, and that it can correctly apply the error map, which is based on the position of the probe in the CMM measurement volume.

5.2.3   REVIEWING THE DRAWING

The knowledge of geometric dimensioning and tolerancing* (GD&T), combined with a programmer's expertise in using the measuring software and his or her experience of measuring many parts, differentiates between the good and the great CMM programmers. The part drawing, or perhaps an electronic computer-aided design (CAD) file, should be annotated with each of the tolerances that need to be inspected. Although there are cases where every single toleranced feature on a part needs to be inspected, often there is a smaller set of "critical features" or "key characteristics"

---

* See also Chapter 3, Specification of Design Intent: Introduction to Dimensioning and Tolerancing.

that must be measured. The programmer must determine several things from the part specification before actually beginning to plan the measurements using the CMM:

- What datum features must be measured in order to construct the necessary datum reference frames for inspection?
- Will it be possible to measure the entire part in a single setup or must the part be reoriented in order to reach all the features?
- Are there groups of features that need to be evaluated as patterns?
- Will it be possible to attain a small enough task-specific measuring uncertainty (see Chapter 14, Measurement Uncertainty for Coordinate Measuring Systems) for each of the tolerances so that a 4:1 tolerance-to-uncertainty ratio (or some other suitable value; refer also to Chapter 15, Application Considerations) can be achieved?

Once these questions are answered, the hardware for probing and fixturing the part can be chosen and a measuring plan can be developed for the part.

5.2.4 CHOOSING PROBES

This section is closely aligned with Section 5.2.5 on fixturing. The programmer must choose the appropriate probes and styli so that each feature to be measured can be reached. It must be possible to contact the feature with the selected probes and styli; in fact, it is preferable to be able to reach the entire feature with a single probe–stylus combination. This is due to the uncertainty that is introduced when changing probes: Using two probes to measure a feature will almost always result in greater uncertainty in the measurand than using a single probe to measure the same feature.

Depending on the available probes and styli, it may make more sense to change the part orientation for better probe access than to use a stylus configuration that is awkward, unsteady, or difficult to calibrate. If you are using an articulating probe head, such as that shown in Figure 5.3, make a list of all the different probe orientations that are necessary for the part measurement. In this way, it is possible to qualify all the probe orientations before measurement, and thus repeated trips back to the reference sphere are not needed. For parts that fill most of the machine volume, it may not be possible to leave the reference sphere on the table, which makes qualifying a forgotten probe tip all the more difficult and time consuming.

Aside from articulating probe head systems, there are also multiple stylus assemblies that can be fitted to a single probe (known as star stylus assemblies). These are usually set at predetermined angles and must also be qualified before use. Finally, other CMM systems utilize a rack with different probes, which can result in still more flexibility for the user. As before, the specific probe and stylus choices should be made ahead of time, so that the full qualification process can be carried out before the part measurement begins.

5.2.5 FIXTURING

The fixturing of parts for CMM measurement is very different from machining fixturing. The parts must be held rigidly, but the forces introduced by the measuring

FIGURE 5.3   Orientations available with articulated probing systems.

process are very low so that the fixtures do not need to be as massive and as restraining as manufacturing fixtures. Although it is often easy to clamp the parts down close to the CMM table, this reduces the accessibility to the part, which would require the part to be refixtured to measure all of the required features. Requiring more than one part setup can increase the total time needed to measure the part, and can also result in a higher uncertainty of measurement, since the features measured in the different setups need to be related to one another.

One common technique for fixturing small parts is to use epoxy or Super Glue® to affix the parts to a parallel sine plate or riser block (it is not recommended to glue parts directly to the table as damage to the table surface can occur). The benefit of using these products is that a strong, rigid bond is created that can be easily broken away at a later time. Two suggestions for the use of epoxy are the following: (1) Place small portions of the epoxy on the region where the part meets the block, as shown in Figure 5.4—if the epoxy is smeared between the part and the block, they may be very difficult to separate, and (2) be aware that the part may lift off the block surface slightly when the epoxy sets, so that the bottom of the part might not be on the same plane as the top of the block.

5.2.6   RECORD KEEPING

Record keeping for a measurement starts with the first steps of planning for how to measure a part. Useful items to include in the record keeping include the marked-up part drawing, photographs of the part fixtured for measurement, descriptions or

FIGURE 5.4 Small parts fixtured with epoxy.

photographs of the probing setup, and a program listing complete with comments that describe what has been done.

5.3 DEVELOPING THE MEASUREMENT PROGRAM

Subsections 5.3.1 through 5.3.7 describe the important characteristics of a measurement program for a CMM. There are some other aspects of measurement planning that are not discussed here but are still important in the business context of stating that manufactured parts comply (or do not comply) with the geometric specifications on the drawing. A good discussion of measurement planning can be found in the American Society of Mechanical Engineers (ASME) standard *B89.7.2—1999: Dimensional Measurement Planning* (ASME 1999). This standard is not specific to CMMs, but it provides a useful discussion of what information may be needed to make a useful measurement plan.

5.3.1 QUALIFICATION

The first part of a measurement program is to qualify the probes that have been identified for the part's measurement. Some software will have this as part of the actual measurement program, whereas other CMM software will keep the probe information in a separate location. In either case, it is important to determine that the necessary probes have been qualified and that the results of the qualification are valid. Of particular importance is the apparent form of the styli and measurement of the effective tip diameter. The form that is reported for the qualification will give a general indication of how well this tip will perform for general point measurements. For most DCC machines, a form of a few micrometers is expected. The form will also depend on the number of points used in measuring the reference sphere. Although many software packages permit the use of as few as 5 points for this measurement, using 25 points or so will give a much better estimate of the probe tip's performance and effective diameter.

For multiple stylus setups, the offsets between the tips are critical to accurate measurement. For this reason, it is good practice to qualify all the tips at once using a single reference sphere position. This will minimize errors in determining the offsets that might be introduced by environmental (e.g., temperature) effects on the reference sphere or the stylus system. After performing this qualification, the form of each tip reported should be examined. The offsets reported will reflect the relative position of the tips for which there is no "correct" reference value. To check the validity of a multiple tip setup, one can use the multiple probe tests that are described in the ASME (ASME 1997) and ISO standards (ISO 2010a; see Chapter 9, Performance Evaluation).

Note: Some software and CMM systems refer to the probe qualification process as *probe calibration*. As this is not a true calibration of the probe or the stylus, qualification is the preferred term. Although not preferred, the use of the term *calibration* is still widely understood.

5.3.2 ALIGNMENT

After the necessary probes are qualified, the reference sphere can be removed from the CMM table if required. The part can now be located on the table using the fixturing scheme developed in Section 5.2.5. Now is the time to take photographs of the fixture, both with the part and without the part. Also, a photograph of the probing setup can be taken at this time. The first measurement step will be to measure the features required for part alignment—the term *alignment* is used to describe the creation of a coordinate system on the part.

This alignment procedure is a two-step process (for DCC machines): (1) Some points are measured manually on the part and an initial "rough" coordinate system is constructed, and (2) the points (either the same as the manually chosen points or different) are measured in automatic (DCC) mode. The points measured in DCC mode are more accurate, as the machine-motion parameters are better controlled and the locations of the points are also programmed. For some applications, a repeated set of measurements in DCC mode is used, either to further refine the coordinate system or to mimic the fitting to a set of locators (such as datum targets) using an iterative technique.

With the completion of this process, a stable and repeatable coordinate system is developed for the part. The next step is to measure each of the datum features in order to establish the datum reference frames (DRFs) needed for part inspection. For simple parts, the alignment step mentioned earlier might use the datum features directly to establish the initial part coordinate system, identical to the DRF needed for inspection. For most parts, however, multiple DRFs are needed, and additional datums must be measured.

One choice that needs to be made when measuring datums stems from the sampling of surfaces used as datum features. The definition (see Chapter 3, Specification of Design Intent: Introduction to Dimensioning and Tolerancing) of a planar datum is usually a stable flat surface that just touches a part's datum feature. Depending on how a part is fixtured, the datum can be determined by measuring points on the planar surface and fitting a plane to those points, or by setting the part on a flat "datum simulator" and measuring the simulator to establish the datum plane. Each of these methods has advantages and disadvantages, depending on part tolerances, machine capability, and available fixturing.

5.3.3 INSPECTION

Once the part alignment is complete and the DRFs are established, the inspection of the part is not difficult. As the measurement program is developed, it is very efficient to measure all features that are near to each other at the same time. This prevents the CMM having to travel around the part more times than necessary. With modern programming software, it is easy to recall different coordinate systems (alignments) and different tips so that all features that are toleranced in a common DRF need not be measured at the same time. As features are added to the measuring program, it is useful to indicate the name of each feature in the software (e.g., CIRCLE1) on the part drawing. Also, if allowed by the software, the programmer should add comments to the program for each feature so that another programmer or operator can easily understand and modify the program, if needed, at a later time.

After all the features are measured, the software can perform the analysis of the part with respect to the tolerances and generate a report. Although these steps are listed separately in this chapter, they may occur at the same time during inspection. For example, after each feature is measured, the substitute geometry could be fitted to the feature and the results reported before moving on to measure the next feature. Whether the analysis and reporting is done following measurement or during measurement depends on the software and the type of reporting that is desired.

5.3.4 ANALYSIS

The analysis of the point data collected with a CMM is a fundamental part of CMM software. In many systems, the analysis is hidden from the user and the "answer" is automatically generated as soon as a feature is measured. This result may include size, form, orientation, and location of the feature, where location and orientation are reported in the current coordinate system. More advanced users learn how to report features in different coordinate systems or how to transform features from one coordinate system to another.

Other advanced techniques may involve choosing the algorithm with which the feature is fitted; it is common for the default fitting of points to a substitute geometry to be done using the least-squares method, but other algorithms may be more appropriate depending on the drawing requirements (see Chapter 3, Specification of Design Intent: Introduction to Dimensioning and Tolerancing). Additionally, the construction of coordinate systems that correspond to DRFs may or may not permit mobility of the reference frame. Although CMM software often provides tools that correspond to the different tolerancing symbols described in geometric tolerancing, it takes a great deal of experience to know what methods of analysis will yield the "most correct" results from a given software package and how to work around cases where the software does not have the desired options for fitting or analysis.

5.3.5 REPORTING

Most CMM software packages have a variety of ways to report the results of measurements. The simplest text-based methods list the feature measured, along with actual values for each attribute of the feature, the tolerances for these attributes, and a bar graph showing where the actual value lies within the tolerance band. More advanced reporting utilizes colors to flag out-of-tolerance features and can also use an image of a CAD model of the part to show where each feature is located. Additional reports may be used to show the deviations of individual points as parts of a form or profile measurement. Although these graphical reports can be very instructive in diagnosing process problems for manufacturing, they can also be time consuming to generate if many parts are being measured.

It is a good idea to include sufficient information in the report so that the conditions of measurement are unambiguous. In addition to reporting the part type and serial number, the date and time of measurement should be included, as well as any external factors that may influence the measurement such as temperature, fixture number (if multiple fixtures exist), CMM identification (if the program runs on multiple CMMs), and CMM operator details.

In manufacturing environments where many parts are measured, a huge volume of reports can be generated. For this reason, multiple report formats can be generated for each measurement program. A full report of all features measured could be created for archival use, whereas an exception report (only for out-of-tolerance features) can be printed for any immediate action. Different electronic report formats are now possible to allow easy archival and access of data. Among these formats are HyperText Markup Language (HTML), Portable Document Format (PDF), Extensible Markup Language (XML), and Excel Spreadsheet (XLS), all of which allow multiple external applications to open and search the reports.

5.3.6 TEST RUN

Once the program is complete, one can run the program from start to finish and ensure that there are no collisions between the part and the probe and also that reporting is in the desired format. It is a good idea to perform this test run slowly so that if there is a collision, the CMM and the part remain undamaged. This is important—even if the operator has a hand poised over the emergency stop (e-stop) switch, it is unlikely that the e-stop can be pressed in time if the program behaves unexpectedly. It is far better to run the program at a slow speed, where damage is less likely. When creating a part program, the programmer will frequently jump around in the program, testing out different sections or remeasuring features with different point patterns or densities. This is normal and encouraged, but there is a risk that the sequence of the program is interrupted and the incorrect probe tip or alignment is in place when the program is executed from top to bottom.

The test run of the program also allows the examination of the reporting method: Are all the features reported? Are the correct tolerances entered into the software for reporting? Is the report format compatible with the standard practice for the company or laboratory? Often, the first complete report will reveal items that need to be corrected.

5.3.7 ENHANCING THE PROGRAM

Once a program is complete it can be run again and again. However, it is hard for a programmer to resist tinkering with the program to give better-looking reports, to allow the operator to enter specific information into the program, or to incorporate myriad other "improvements." Each improvement should be evaluated by the following criteria before it is implemented: Will it save time? Will it prevent an error? Does it provide additional, useful information? Often, the most useful changes to a program are simply improvements in the comments and documentation. When writing a program each step will seem obvious to the programmer, but after some time even the original programmer may have difficulty recalling why certain steps were taken or the reasoning for choosing a certain number of points on a particular feature. The value of complete and accurate documentation for measurement programs cannot be overemphasized.

5.4 OTHER RESOURCES

The National Physical Laboratory (NPL), United Kingdom, has several good practice guides that can be downloaded free of charge (see http://www.npl.co.uk/):

- *GPG(041): CMM Measurement Strategies* (Flack 2001a)
- *GPG(042): CMM Verification* (Flack 2001b)
- *GPG(043): CMM Probing* (Flack 2001c)

5.5 CONCLUSION

Although this chapter gives a basic outline of the information and techniques needed for CMM program development, there is no substitute for the experience gained from results-driven programming. When parts need to be inspected and decisions are to be based on the outcome of the inspection, the steps described in this chapter will be brought into much clearer focus. If readers have the opportunity to receive formal training on their CMMs, it is highly recommended that they bring a part and a drawing along to the class. Although many training classes are centered on a "practice part," there will almost certainly be opportunities to ask the instructor how what has been taught relates to a specific part and drawing.

6 Probing Systems for Coordinate Measuring Machines

Albert Weckenmann and Jörg Hoffmann

CONTENTS

Probing systems for coordinate measuring machines (CMMs) are an integral part of the entire measuring system and form the link between machine and the workpiece to be measured (Weckenmann et al. 2004). Proper selection, configuration, qualification, and use are vital to tap the full potential of a coordinate measuring system (CMS).

Users of CMMs require a clear understanding of the characteristics and differences of a huge variety of probing systems and their further development, so that adequate systems can be selected and used properly for given measurement applications in a cost-effective way. This is supported by knowledge about classification of probing systems according to their principle of operation and their performance characteristics. After describing the purpose and basics of probing, the main functional principles and important application aspects are explained. Information on useful accessories for supporting the probing process and improving probing performance rounds off this chapter.

6.1 PURPOSES AND BASICS OF PROBING

Practically all tasks in geometrical metrology are associated with detecting the position of points of a specimen surface and assessing the relative positions between them. Especially in coordinate metrology, all measured geometric quantities are based on distances and positions of a set of detected points on the physical surface of a workpiece. Probing systems are used to assess the positions of these points with respect to the coordinate system of a CMM. In early CMMs, the so-called hard probes, basically rigid styli, were used to contact the workpiece surface manually and the readout of the coordinate axes was also triggered manually. The consequences were that measuring results were sensitive to the undefined contact force, probing was cumbersome especially when large CMMs with a high moving mass were operated and, maybe the most important consequence, automation of the measurement process was impossible because of the need for manual surface detection to stop the machine and read out the coordinates upon workpiece contact.

A probing system able to "sense" the workpiece surface and allow for the overtravel needed to stop machine movement after surface detection was necessary, and outstanding repeatability was a key requirement for improving the accuracy of coordinate metrology. From this perspective, probing systems form the link between

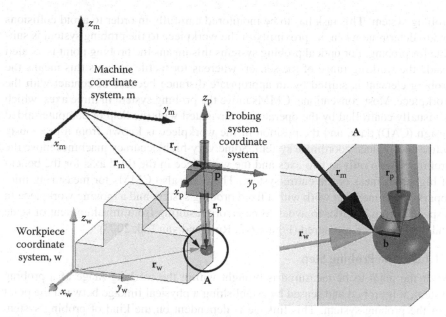

FIGURE 6.1 Vector diagram for measuring a surface point on a workpiece, where x_w, y_w, z_w: Coordinates in workpiece coordinate system; x_m, y_m, z_m: Coordinates in CMM coordinate system; x_p, y_p, z_p: Coordinates in probing system coordinate system; r_m, r_w: Position vector of probing point; r_r: Position vector of probing system reference point; r_p: Position vector of tip ball center point; b: Tip ball correction vector.

workpiece surface and the CMM (Figure 6.1). In principle, many kinds of physical interactions between the workpiece and the probing system can be used to sense the surface of the workpiece, for example, tactile (force), electrical (resistance, capacitance, inductance, eddy current), optical (light reflection), magnetic, and many more, all with different advantages and disadvantages.

In modern CMMs, only tactile and optical probing systems are practically used; most of them are sensitive in three dimensions.

6.1.1 PROBING PROCESS

The determination of position of a surface point of a workpiece can be subdivided into four sequential steps:

1. Positioning
2. Probing
3. Measuring
4. Evaluating

6.1.1.1 The Positioning Step

Positioning involves the task of stationing the surface point to be measured—by moving either the workpiece or the probing system—inside the working volume of the probing system. This task has to be monitored carefully in order to avoid collisions and to determine when the proximity of the workpiece to the probing system is suitable for probing. For optical probing systems this means the probing point is located inside the working range of the sensor, whereas for tactile systems this means the probing element is shifted by an appropriate distance because of contact with the workpiece. Most conventional CMMs move the probing system in three axes, which is visually controlled by the operator or navigated with the help of computer-aided design (CAD) data, and the position of the workpiece is known from a previously defined workpiece coordinate system. Some very-high-accuracy machines move the probing system only in two axes and the workpiece in the third axis for the benefit of higher stiffness of the entire system. There are also CMMs for measuring microparts, machines that work with a fixed probing system and a moving workpiece in a special setup in order to avoid Abbe errors resulting from misalignment of scale and distance to be measured (Figure 6.2; Ruijl and van Eijk 2003).

6.1.1.2 The Probing Step

After the point to be measured is brought within the working range of a probing system, it is probed and sensed by establishing a physical linkage between the point and the probing system. This linkage is dependent on the kind of probing system used and is typically of mechanical or optical nature in the field of coordinate metrology.

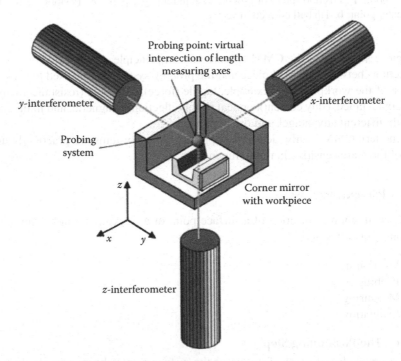

FIGURE 6.2 Abbe error–free setup of three length measuring axes and a probing system.

When mechanically probing a workpiece, it is touched with a solid probing element and a defined static contact force. It is essential to limit the probing force in order to avoid elastic and plastic deformations of the workpiece and to control it carefully to ensure a high degree of repeatability. When applied correctly, tactile probing systems detect points at a place where the workpiece is able to build up a reaction force equal to the probing force.

Optical probing systems, on the other hand, emit light, bring the light to an interaction with the surface to be probed, and analyze the light scattered back or reflected by the workpiece. Hence, they detect points where the workpiece is interacting with electromagnetic waves. These points may differ from points measured through contact.

6.1.1.3 The Measuring Step

Dimensional measuring is the comparison between a standard (e.g., of length) and a measurand (e.g., distance). For the probing process, this means determining the distance between the probed surface point and a reference point of the probing system as a multiple of unit length. This distance might be fixed in the case of touch-trigger probing systems or variable in the case of measuring probing systems. Typically, resolution and repeatability decline with larger distances because various influencing factors and filtering effects increase with the distance to be measured and/or with the application of additional transmitting devices, such as long stylus extensions. On the other hand, large distances between the probing system reference point and the probed point may facilitate accessibility to the workpiece considerably and thus enlarge the area of application and the range of measurable workpiece features. A basic rule is to keep the distance between the reference point and the probed point as small as is practically possible.

6.1.1.4 The Evaluating Step

For evaluating the position of workpiece surface points in the coordinate system of the CMM, the (vectored) distance between the probing system reference point and the workpiece surface point has to be added to the position vector of the probing system reference point in the machine coordinate system (Figure 6.1). Therefore, the CMM measures this position vector with its length measuring axes.

The entire measurement process with a CMM can be regarded as a transformation of the real position $\mathbf{p}$ of a point on the workpiece surface into its measured position $\mathbf{r}_m$ in the coordinate system of a CMM. This can be subdivided into probing the point and measuring the position of the probing system (Figure 6.3). The subprocess of probing is a transformation of the point's position vector $\mathbf{p}$ into the position of the tip ball center in the coordinate system of the probing system $\mathbf{r}_p$ and having the corresponding probing vector $\mathbf{b}$ point from the tip ball center to the probed point. By adding the position vector of the probing system, which is measured by the CMM, the measured point's coordinates in the CMM coordinate system are revealed.

These transformations can be described by a probing matrix $\mathbf{P}$ and a position measuring matrix $\mathbf{M}$. Through $\mathbf{P}$, the surface point's real position vector is transformed

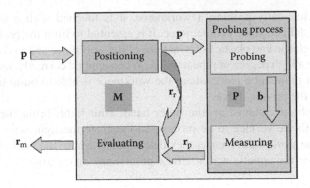

FIGURE 6.3 Probing as a set of transformations.

into the probing system coordinate system; $\mathbf{M}$ describes the measurement of the position of the probing system in the measuring range of the CMM,
   where

$\mathbf{p}$: Position of the probing point in an arbitrary coordinate system
$\mathbf{r}_r$: Position of the probing system reference point in the CMM coordinate system
$\mathbf{b}$: Probing vector
$\mathbf{r}_p$: Position of probing point in the probing system coordinate system
$\mathbf{r}_m$: Probing point's measured coordinates in the CMM coordinate system

$$\mathbf{r}_m = \mathbf{r}_r + \mathbf{r}_p + \mathbf{b} \qquad (6.1)$$

$$\mathbf{r}_p + \mathbf{b} = \mathbf{p} \cdot \mathbf{P} \qquad (6.2)$$

$$\mathbf{r}_r = \mathbf{p}_r(\mathbf{p})\mathbf{M} \qquad (6.3)$$

$$\mathbf{p}_r(\mathbf{p}) = \mathbf{p} - \mathbf{b} - \mathbf{r}_p \qquad (6.4)$$

$$\mathbf{r}_m = \mathbf{p}_r(\mathbf{p})\mathbf{M} + \mathbf{p} \cdot \mathbf{P} = (\mathbf{p} - \mathbf{p} \cdot \mathbf{P})\mathbf{M} = \mathbf{p} \cdot \mathbf{P} \qquad (6.5)$$

$$\mathbf{p} = \mathbf{r}_m \cdot (\mathbf{M} - \mathbf{PM} + \mathbf{P})^{-1} \qquad (6.6)$$

The probing point's real coordinates $\mathbf{p}$ can be derived exactly if the probing matrix $\mathbf{P}$ and the position measuring matrix $\mathbf{M}$ are precisely known and are invertible. Due to various environmental and other not entirely known, controllable, or predictable influences, the matrices cannot be determined exactly, and they vary from probing point to probing point.

6.1.2 History of Probing Systems

The first probing systems for CMMs were just stiff styli, which had to be brought into contact with the workpiece manually by the CMM operator. After establishing the contact, the operator had to trigger a readout of the machine axes, usually done with the help of a foot switch. As the contact detection was carried out by the operator of the CMM, the process was slow, subjective, and prone to error. Static probing force could not be controlled or limited and dynamic probing forces were too large to measure delicate parts accurately and nondestructively. In addition to this, hard probes limited the size and dynamic properties of CMMs, as it was not possible to drive them electrically.

In 1972, Sir David McMurtry of Rolls-Royce Group faced this problem when he had the task of measuring fine fuel pipes; he solved it with his invention of the kinematic resistive probing system (Figure 6.4a; McMurtry 2003). This was the first probing system that allowed some overtravel after automatic contact detection by prestressed kinematic guidance of the probe stylus. When the stylus touched the workpiece, the kinematic mechanism became unseated due to an additional support

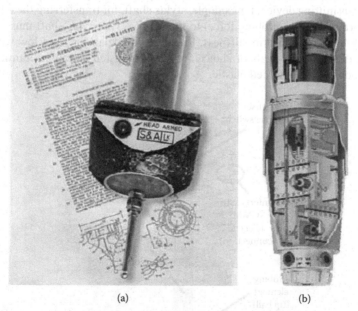

(a)   (b)

FIGURE 6.4 (a) The original three-dimensional (3D) touch-trigger probe invented by Sir David McMurtry in 1972 (Courtesy of Renishaw plc.). (b) The first measuring probing system Zeiss MT, which was developed in 1973. (Courtesy of Carl Zeiss IMT.).

at the probed surface point of the workpiece. The three kinematic supports of the probe were implemented as series-connected electrical switches, allowing an easy but reliable means of automated contact detection by measuring the resistance. The soft spring for prestressing allowed several millimeters of overtravel while limiting the contact force effectively. This invention was a breakthrough in coordinate metrology, and it stimulated the development of more sophisticated probing systems as well as larger, faster, automated and much more accurate CMMs.

Shortly after the first touch-trigger probing system was introduced, the company Carl Zeiss introduced in 1973 their first CMM, the UMM500, with a dedicated three-dimensional (3D) measuring probing system consisting of three stacked axes, each equipped with an actuator for active probing force generation and an inductive sensor for displacement measurement (Figure 6.4b). This was the first 3D measuring probing system for CMMs and also the first one with active probing force generation. Successors of these two systems are still among the most important types of CMM probing systems.

6.1.3 BASIC CONFIGURATION OF A TACTILE PROBING SYSTEM

To fulfill its task as the linkage between CMM and workpiece, a tactile probing system must have at least the following components (Figure 6.5):

* A probing element to establish a mechanical interaction with the workpiece surface, for example, tip ball. Tip balls must exhibit very low form deviation, high stiffness, and low wear, and they are usually made from ruby.
* A transmitting device, for example, stylus shaft, for transferring the contact information (e.g., probing force) to the sensor. The stylus shaft must show very high and uniform stiffness, low thermal expansion, and low weight. For example, materials of choice are steel, tungsten carbide, ceramics, or carbon fiber–reinforced plastic.

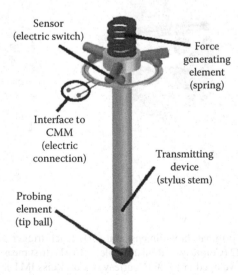

FIGURE 6.5 Basic tactile probing system in kinematic resistive design.

- A force generating and controlling element for producing a defined probing force, for example, a spring. It is important to achieve an isotropic probing force; the amount of force has to be chosen according to workpiece, probing element, and environmental influences such as surface contamination and vibrations. Some probing systems use active probing force, for example, that is generated by voice coil actuators.
- A sensor for evaluating contact information (e.g., switch, force, or displacement sensor). Besides those that merely detect contact, probing systems that measure direction and magnitude of stylus displacement are also common.
- An interface to the CMM for transmitting the measured contact information to the control unit for triggering position measurement of the CMM axes, further processing, and evaluation of the probed point's position in the machine coordinate system.

6.1.4 CLASSIFICATION OF PROBING SYSTEMS FOR COORDINATE MEASURING MACHINES

The range of commercially available probing systems for coordinate metrology is extensive. A classification of the huge variety of systems according to diverse application-oriented aspects is provided in Figure 6.6 for systematic overview and decision guidance. Suppliers of probing systems are usually the major CMM manufacturers who equip their machines with dedicated probing systems; but there are also specialized probing system developers and manufacturers.

6.1.4.1 Contact Detection

Probing systems for CMMs are often categorized according to the kind of interaction with the workpiece to be measured. Even today, the most common are tactile probing systems that use a force interaction between a probing element and a workpiece. Contact with the workpiece is detected when the probing force exceeds an electronically or mechanically controlled value. An example is the kinematic resistive probing system that detects contact with the workpiece when probing force is high enough to open at least one of the kinematic contacts, which are closed by

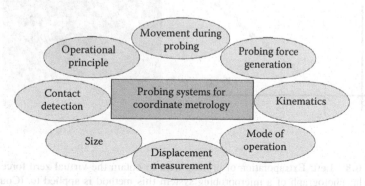

FIGURE 6.6 Classification criteria for probing systems.

the prestressing force of the spring (Figure 6.5). Due to the arrangement with three contact pairs, the levers between probing force and prestressing force vary so that probing force is dependent on probing direction also. This leads to a triangular lobing pattern (Figure 6.7).

More sophisticated systems measure either probing force or displacement of an elastically mounted stylus. This gives the possibility of defining a threshold value for triggering or even measuring one probing point with different probing forces in order to extrapolate its position equivalent to a virtual probing force of zero (Figure 6.8).

Nearly all nontactile probing systems for CMMs are optical ones, that is, they detect the workpiece surface or proximity to it with the help of electromagnetic

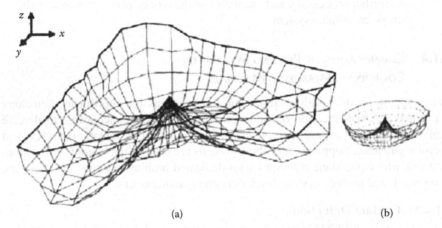

(a)   (b)

FIGURE 6.7   Pretravel variation in the same scale: (a) Renishaw TP6; (b) Renishaw TP200. (Adapted from Wozniak, A., and M. Dobosz, *Measurement*, 34/4:273, 2003 and Dobosz, M., and A. Wozniak, *Measurement*, 34/4:287, 2003.)

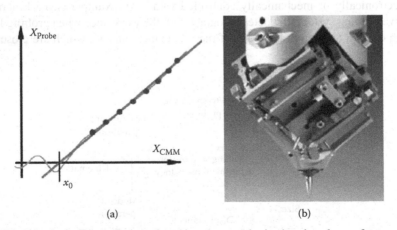

(a)   (b)

FIGURE 6.8   Left: Extrapolation of probing data to obtain the virtual zero force contact point; right: photograph of a microprobing system this method is applied to. (Courtesy of METAS, EPFL.)

waves. Common nontactile systems are point and line sensors, whose working is based on triangulation, and also 3D measuring fringe projection systems. The advantages of optical probing systems compared with tactile ones are as follows:

- The absence of force interaction with the workpiece enables the measurement of delicate parts.
- High measuring speeds lead to a much higher number of probed points.
- Large working distances minimize the risk of collisions between probing system and workpiece.

On the other hand, there are also drawbacks when using optical probing systems:

- Generally higher measurement uncertainty of each probed point.
- Measurement results might be affected by optical properties of the workpiece and/or ambient light.
- Limited tolerance to surface slope, accessibility problems at undercuts and interior features.

There is also a small group of probing systems that are neither tactile nor optical; usually they are utilized for special tasks like the measurement of delicate parts or very small features. Interaction with the workpiece might be electrical (e.g., capacitive, resistive, or even tunneling), by attenuation of a harmonic oscillation by intermittent contact, or through the viscosity of air between a probing element and the workpiece. Any other kind of reversible physical interaction might also be applied.

It is very important to be aware that different principles of contact detection lead to different measured surfaces, as the location where mechanical interaction takes place is generally not identical with the location of electromagnetic interaction. This is referred to by the terms *real mechanical surface* and *real electromagnetic surface* (ISO 2010c).

6.1.4.2 Operational Principle

There are probing systems that only detect contact and hence have a Boolean output (contact yes or no)—*touch-trigger probing systems*. There are also systems that measure a probing vector (norm and direction)—*measuring probing systems*.

Touch-trigger probing systems are suitable for the measurement of standard geometric features, because a relatively small number of discrete probing points is often sufficient and compensation of the influence of tip ball diameter and probe pretravel can easily be done by adding or subtracting the effective tip ball diameter to the size or position being measured with it (tip ball radius compensation at the measured feature). For the measurement of sculptured (free-form) surfaces, measuring probing systems have advantages due to their ability for scanning and the possibility of their directly compensating for the effective tip ball diameter with the help of the measured probing vector (tip ball radius compensation at each probed point).

Generally, advantages of touch-trigger probing systems are low price, weight, and size; ability to use articulating heads; and improvements in dynamic characteristics of the CMM system. In contrast to these systems, measuring probing systems are

usually more accurate, enable the extrapolation of probing points to a virtual probing force of zero, and have the possibility for scanning both known and unknown
objects.

Touch-trigger probing systems always measure dynamically, that is, they detect
a surface point while the CMM is moving. Measuring probing systems can measure
dynamically or statically, that is, the coordinates of the probing system and CMM
are read out during machine standstill.

6.1.4.3 Mode of Operation

Closely connected with operational principle is the mode of operation of probing
systems. Whereas touch-trigger probing systems only collect a few discrete probing
points, measuring systems can keep contact with the workpiece surface and collect
thousands of points on their path along the surface, which is called *scanning*.

In general, measurement uncertainty for the position of a single point is higher
in scanning mode due to dynamic influences; but due to the much higher point rate,
more information about the workpiece is gathered, enabling the testing of more complex properties of the workpiece. Discrete point probing is appropriate for the measurement of standard geometric features, when form deviation is not to be assessed.
In scanning mode, it is also feasible to evaluate form deviation in addition to the
dimensions of geometric features. This is essential when the expected form deviation
of a part to be checked is not negligible compared to the specified tolerance. There is
also the possibility of pseudoscanning with touch-trigger systems where the distance
between the discrete probing points is kept very small.

6.1.4.4 Probing Force Generation

Generating and controlling a defined probing force are crucial for ensuring repeatability and correctness of measurement results due to probing force–dependent
effects such as stylus bending, plastic and elastic deformation of tip ball and workpiece, and also linearity of the contact sensor of the probing system.

Almost all touch-trigger probing systems and also a number of measuring probing systems use passive probing force generation, that is, the probing force is generated by an elastic element (e.g., coil or leaf spring) proportionally to the deformation
of that element. Active probing force generation utilizes an actuator (e.g., voice
coil) that exerts a controlled force onto the stylus, which is independent of tip ball
displacement.

Passive probing force generation systems have the advantages of being cheaper,
more compact, and having less weight due to the absence of actuators and their con-
trollers. A drawback is the influence of varying contact force on measured results
due to varying displacement. Active probing force generation can minimize force
variation and hence measurement uncertainty; but even with a constant probing
force, there are still varying elastic effects due to varying stiffness of the probing
system, stylus, and workpiece according to the direction of load.

6.1.4.5 Kinematics

For setting up a 3-DOF (DOF stands for degree of freedom) kinematic system,
as needed for a variety of probing systems, there are basically two principles of

realization: (1) serial kinematics and (2) parallel kinematics. Serial kinematics consists of several stacked independent axes (translative or rotatory), one for each DOF. To derive the position of the moving part, the displacements in each axis simply have to be added (Figure 6.9a). In parallel kinematics, each translative and rotatory axis is not dedicated to one DOF, so the movement in one DOF may be the combined movement in several kinematic axes. Usually it is not possible to move the system in only one kinematic axis (Figure 6.9b).

As touch-trigger probing systems only detect surface contact without providing any information about the probing vector **b**, there is no necessity for splitting the total tip ball displacement into several coordinate axes; therefore, these systems generally work with parallel kinematics. For measuring probing systems both possibilities are practically used, and, typically, displacement is measured separately in each kinematic axis.

Serial kinematics generally offers the advantages of larger travel range, easier manufacture and calibration, and reduced uncertainty when moved in only one axis. Drawbacks are larger size and mass, lower stiffness, and cross talk between axes.

Due to its advantages regarding size, stiffness, and weight, parallel kinematics is used in most microprobing systems.

6.1.4.6 Size of Probing Element and Feature to Be Measured

For the measurement of most parts in the area of mechanical engineering, such as engine components, gear wheels, shafts, and formed sheet metal parts, conventionally sized probing systems with tip ball diameters between 1 and 8 mm and static probing forces between 50 and 200 mN are most suitable. Repeatability of a single probed point is typically in the range of 0.1 to 10 µm depending on the utilized CMM system, the probing system, workpiece surface quality, and many other factors. Measuring probing systems in this field usually have measuring ranges of a few millimeters in each direction.

The growing field of microsystems technology (MST), however, demands systems with much smaller probing elements, better resolution, and smaller static and

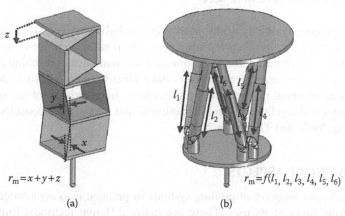

$$r_m = x + y + z$$

(a)

$$r_m = f(l_1, l_2, l_3, l_4, l_5, l_6)$$

(b)

FIGURE 6.9   (a) Serial and (b) parallel kinematics.

dynamic probing forces due to the small dimensions and features to be measured. Such delicate structures, often with high aspect ratios, further increase demands on micro probing systems (Weckenmann, Peggs, and Hoffmann 2006). The type of specimen varies from microlenses, microgears, and fuel-injector nozzles to features and structures on microelectromechanical systems (MEMS). Today, there are several 3D probing systems commercially available for micro-CMMs, most of which show close similarity with conventionally sized systems. Probing forces have to be limited to less than 1 mN in order to avoid plastic deformations of the workpiece due to Hertzian contact stresses, which can reach significant values even at low contact forces because of the low curvature radius of microprobing elements:

$$\sigma = \frac{1}{\pi} \sqrt{\frac{1.5 \cdot F \cdot E^2}{r^2 \cdot (1 - \upsilon^2)^2}} \tag{6.7}$$

$$w_0 = \sqrt[3]{\frac{2.25 \cdot (1 - \upsilon^2)^2 \cdot F^2}{E^2 \cdot r}} \tag{6.8}$$

$$E = 2 \frac{E_1 \cdot E_2}{E_1 + E_2} \tag{6.9}$$

where

σ: Hertzian contact stress
F: Contact force
w_0: Elastic deformation
r: Tip ball radius
E: Effective Young's modulus
υ: Poisson's ratio
E_1: Young's modulus of tip ball
E_2: Young's modulus of workpiece

Microprobing systems are typically used with tip balls between 0.5 and 0.1 mm diameter and stem lengths of a few millimeters. The measuring range is on the order of a few tens of micrometers at a resolution of a few nanometers. Probing reproducibility down to a few tens of nanometers have already been reported, depending on stylus, measurement task, material and surface characteristics of the specimen, and environmental conditions such as cleanliness and temperature constancy (Flack 2001c; Küng, Meli, and Thalmann 2007).

6.2 PRACTICAL ASPECTS

The most relevant property of probing systems in practice is to economically solve the measuring tasks of its user. There are many different technical implementations of probing systems, each with typical strengths and weaknesses for the various kinds of measuring tasks and CMMs. The following section gives information useful when thinking about which probing system is the most appropriate for a certain application.

6.2.1 PRINCIPLES OF DISPLACEMENT MEASUREMENT

Measuring tactile probing systems need a device for accurately measuring the displacement of the probing element in order to determine the vector r_p. This can be done directly or indirectly by measuring the strain or forces at the stylus suspension with different measuring principles. Most common are inductive, capacitive, resistive, optical, and scale-based systems. Optical probing systems directly measure $r_p + b$ using optical means, for example, triangulation.

For the application using measuring probing systems, physical effects with approximately linear responses are preferred due to their constant sensitivity in the whole measuring range. For digital evaluation of a probing signal, principles delivering an electronic output signal are the most convenient ones.

In general, measurement of tip ball displacement must be carried out as directly as possible to shorten the metrological loop. This helps to minimize effects that compromise accuracy, such as elastic deformation or thermal expansion of the transmitting device and other intermediate components.

6.2.1.1 Inductive Systems

The inductive principle has been widely used for displacement measurements for a very long time. A magnetically soft iron core is moved inside a coil, resulting in a change of inductance L according to position of the core inside the coil. The inductance is then a measure of the position s of the core (Figure 6.10).

Because of the nonlinearity of the hyperbolic characteristic $L = f(s)$, usually a differential setup with two coils is used. In order to get a higher output signal another coil is added, forming a differential transformer. The electric circuit for signal evaluation consists of a bridge circuit with a phase-discriminating rectifier and amplifier. The response characteristic of the system is approximately linear in the vicinity of the inflection point. A device like this is called a *linear variable differential transformer* (LVDT); three of them combined in a serial setup form an excellent 3D position measuring system.

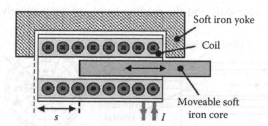

FIGURE 6.10 Inductive displacement measurement (s indicates displacement and I indicates current).

6.2.1.2 Capacitive Systems

Capacitive transducers based on the partial overlap of two conductive surfaces (Figure 6.11) are very sensitive to environmental effects, the required precision of guideways, and disturbing electric fields. Nevertheless, capacitive proximity sensors are widely used as noncontact 1D probing systems. For small measuring ranges of microprobing systems, plane plate capacitors with variable distance between the plates are frequently used. Advantages are simple setup that facilitates miniaturization and the system's high sensitivity to small plate distances. The system's disadvantage is its nonlinear characteristic and hence varying sensitivity.

6.2.1.3 Resistive Systems

Displacement of the probing element can be determined also by measuring the deformation of the stylus suspension or the stylus itself during probing. For deformation measurement, usually strain gages or piezoresistive elements are used. Both alter their resistance according to the elongation of an elastic structure on which they are fixed by an adhesive. With a combination of three elongation sensors, the direction of the strains and, consequently, the probing direction can be derived. The intensity of the effect can be used to create a measuring system of very small measuring range (only elastic deformation is permissible). If piezoresistive sensors are applied on a suitable structure, a very small probing trigger force can be achieved and the system can be made very small by applying manufacturing technology known from microelectronics.

6.2.1.4 Optical Systems

Another principle of tactile probing consists of measuring the position of the probing element optically when it touches the workpiece surface. One possibility is to apply a position-sensitive device (PSD), which detects the beam reflected by a mirror that is rigidly fixed to the tip ball or the stylus. Another possibility consists of measuring the position of a glass tip ball being illuminated through an optical fiber, which serves at the same time as stylus and suspension (see Sections 6.2.5 and 6.2.7.2 and Chapter 8, Coordinate Measuring System Algorithms and Filters).

Optical displacement measurement is fast and offers the possibility of very direct measurement of tip ball position without the influence of stylus bending.

6.2.1.5 Scale-Based Systems

Scale-based transducers consist of a typically incremental scale and devices to read the scale and count the number of increments between the starting point and the end of movement. The scale and the reading head are, respectively, attached to the

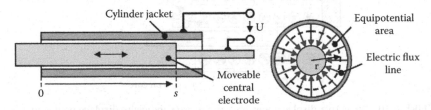

FIGURE 6.11 Capacitive displacement measurement.

stylus mounting and the probing system housing, or vice versa. The scale can be a glass or steel ruler with an etched grating, a ruler with a regular magnetic pattern, or a coherent reference light beam (applied, e.g., in laser interferometers). During the movement of either the scale or the reading head, the number of passing increments detected by the reading head is counted and multiplied by the pitch to get the displacement. In order to get a finer resolution, interpolation to the analog output signal of the reading head is usually applied. To detect the direction of movement and to get unambiguous displacement measurement results, additional arrangements are necessary (e.g., two reading heads with a suitable separation).

6.2.2 Probing Element

The probing element of a tactile probing system is brought into mechanical contact with the workpiece and hence is of enormous importance. The material, form, size, and manufacturing accuracy of the probing element have a significant effect on the reliability and accuracy of measurements performed with it.

The standard probing elements in 3D coordinate metrology are spheres (often referred to as tip balls) due to their symmetrical shape that facilitates highly accurate manufacture leading to isotropic probing characteristics. Most tip balls are made of ruby (monocrystalline aluminum oxide) due to the material's exceptional hardness that limits wear and presents the possibility of synthetic mass production, which leads to a moderate pricing of the product. For comparing the measurement results achieved with different-sized tip balls, the mechanical filtering effect, which is derived from size-dependent accessibility to roughness valleys, is very important (Figure 6.12). Transformation of the true workpiece surface to the measured profile can mathematically be described by the morphological operation of dilation (Minkowski addition); hence the best estimation of the true surface can be calculated by morphological erosion (Minkowski subtraction). However, dilation followed by erosion (i.e., morphological closing) has a smoothing effect suppressing convex surface details of high spatial frequency.

The size of the tip ball has to be chosen carefully taking into consideration the following:

- Size of the smallest interior feature to be probed (accessibility)
- Desired filtering characteristics

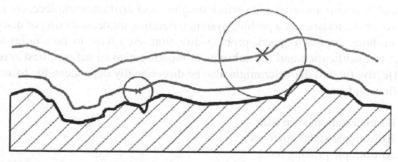

FIGURE 6.12 Mechanical filtering effect of the tip ball.

- Material of workpiece and probing force to limit elastic deformations due to Hertzian stress
- Compatibility with the used probing system
- Availability

For the correct evaluation of the length of vector **b** (see Figure 6.1), the actual size of the used tip ball has to be known. To take into consideration the bending of the stylus stem by probing force, the effective tip ball diameter is determined by probing an artifact of known dimensions (usually a ceramic calibration sphere) and subtracting the diameter of the artifact from the sphere defined by the measured tip ball centers when probing the artifact. This procedure of calibrating the tip ball diameter is called *probing system qualification* (Section 6.2.3.1).

Other shapes of probing elements may have advantages for special tasks, for example, improving accessibility or achieving a well-defined contact situation. Examples for nonspherical probing elements are probing cylinders, probing cones, probing discs, or needlelike probing elements.

6.2.3 Influences on Probing Performance

The practical performance of each probing system is influenced by the kind of application, the ambient conditions and the applied styli and probing elements. The following sections give an overview of the most important factors determining the practically achievable probing performance.

6.2.3.1 Probing System Qualification

Probing system qualification has enormous influence on the performance of a CMM, because its results (tip ball idle position vector r_p and effective length of tip correction vector **b**; Figure 6.1) are used for the calculation of each measurement result with the tip ball and stylus. Measurement errors due to inadequate qualification are very hard to recognize as they are repeatable and consistent. The effective tip ball diameter considers not only the actual physical dimensions of the probing element but also the elastic deformations of stylus stem, probing element, and workpiece caused by probing force (Figure 6.13), as well as the pretravel characteristics of the probing system.

As a direct conclusion, all parameters influencing elastic deformations (probing force and direction, material and surface roughness of artifact/workpiece, etc.) and pretravel characteristics of a probing system (operating mode, scanning or discrete point probing, approach speed, probing direction, etc.) have to be similar with respect to qualification and the subsequent measurement to achieve best results. The effective tip ball diameter might also be directionally dependent due to any of the following factors:

- Form deviation of tip ball
- Anisotropic stylus stiffness
- Anisotropic probing force

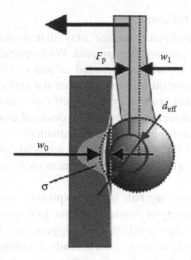

FIGURE 6.13 Effective tip ball diameter, where F_p: Probing force; w_0: Elastic deformation of tip ball and workpiece; w_1: Elastic deformation of stylus stem; σ: Hertzian contact stress; d_{eff}: Effective tip ball diameter.

However, it is not practically possible to provide an artifact similar to all workpieces to be measured; therefore, for 3D tip ball qualification usually a spherical artifact of precisely known dimensions and very low form deviation is probed and measured. A spherical shape is the preferred artifact shape because it can be manufactured and measured very accurately, has normal vectors in each spatial direction, and can be positioned without regard to orientation. The diameter of the sphere defined by the tip ball centers when probing the artifact is the sum of the actual artifact diameter and the effective tip ball diameter d_{eff}. Artifacts for calibrating conventional probing systems are typically made from polycrystalline aluminum oxide, tungsten carbide, stainless steel, and glass or Zerodur. For qualification of microprobing systems with much smaller probing elements and probing forces but much higher demands on accuracy, ruby and silicon nitride spheres are used. In both cases, materials with low thermal expansion and high hardness are preferred due to their better dimensional stability. To perform traceable measurements, the probing system has to be qualified with an artifact calibrated by a national measurement institute.

Qualification strategy includes the choice of the qualified artifact, its position and orientation in the measurement range of the CMM, probing speed, and probing force, as well as a definition of the number, position, and sequence of probing points. Usually, a qualification strategy is suggested by the probing system manufacturer or implemented in the CMM software, considering specific characteristics of the CMM and the probing system. Depending on the kind of probing system and demands on accuracy, the effective tip ball diameter may be directionally dependent.

Monitoring the statistical spread of the qualification process can reveal not only insufficient qualification but also inadequate styli combinations (weak joints, tip ball wear, or pick-up of debris) and deteriorated performance of the probing system and/or CMM.

6.2.3.2 Environmental Influences

Contaminations on the workpiece surface may distort the measurement result severely, depending on the probing system used. With optical probing systems, any kind of visible contamination will be regarded as part of the workpiece; in some cases, these may include even oil or water films on the surface. On the other hand, contaminations of the optical probing system itself can often be avoided easily by increasing the working distance, and they can be detected directly in most cases. In contrast, tactile probing can displace fluid contaminations to some extent and probe the real physical surface of the workpiece; but it tends to collect solid particles on the probing element that might falsify the subsequent measurement of clean surface areas.

Environmental vibrations may blur optical exposures and cause a deteriorated signal-to-noise ratio in a variety of probing systems. In sensitive tactile probing systems, vibrations can cause the so-called false triggers, that is, the probing system falsely detects surface contact when it is not actually in contact with a surface.

Temperature gradients and also deviations from the reference temperature lead to thermal expansion of parts of the probing system that influence the measured result (e.g., tip ball, built-in scales, and charge-coupled devices) and thereby cause measurement deviations. Due to the small size and also the small measuring range of probing systems compared to the CMM, these errors are negligible in most macroscale applications.

6.2.3.3 Wear and Deformation

Although wear and deformation are most important in tactile systems, optical systems with moving parts (e.g., autofocus sensors) also degrade slowly. For tactile probing systems, the probing element can alter its shape and dimensions by the wear caused by mechanical interaction with the workpiece. Under normal circumstances, the wear occurring to the usually very hard and smooth probing elements is very low and does not compromise accuracy significantly; but when scanning hard and rough surfaces, the shape of the tip ball should be monitored by frequent requalification. When aluminum parts are scanned, chemical reactions can cause degradation of ruby tip balls.

The wear and deformation of a workpiece caused by probing force is more practically relevant. At small contact zones between probing element and workpiece (small probing element and/or only pointwise contact on the roughness peaks of the workpiece), Hertzian stress can reach significant values even at relatively low static probing forces and lead to plastic deformation and wear of the workpiece (Figure 6.14).

As a consequence, probing force and size of tip ball have to be carefully chosen according to workpiece material, roughness, Young's modulus, the desired resolution, and uncertainty. It is also important to be aware of dynamic probing forces that often exceed the static force by far. The dynamic probing force is caused by deceleration of the probing element when hitting the workpiece and is thus dependent on the moving mass of the probing system and the approach speed. Especially in microprobing systems, this is a critical issue due to the very small contact zone between probing element and workpiece.

Elastic stylus bending caused by probing force is unavoidable when using tactile probing systems. Stylus bending is usually anisotropic because of the variation in stylus stiffness in the axial and normal directions and also, as in most cases, because of anisotropic probing force caused by asymmetries of the suspension and/or directionally varying sensitivity of displacement measuring systems. To minimize stylus bending, it is essential to use the shortest and thickest applicable stylus, as stylus bending is proportional to the cube of free stylus length and inversely proportional to the fourth power of stylus diameter when probing in the direction normal to the stylus axis with a cylindrical stylus (Figure 6.15).

Stylus bending in normal direction is given by

$$w_s = \frac{64}{3 \cdot \pi} \frac{F \cdot l^3}{E \cdot d^4} \tag{6.10}$$

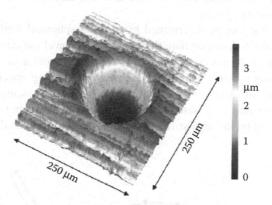

FIGURE 6.14 Plastic deformation of an aluminum surface after probing with a ruby tip ball ($d = 1.5$ mm) and a probing force of 200 mN.

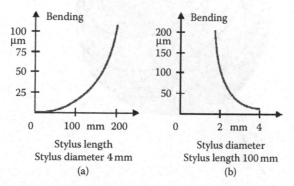

FIGURE 6.15 Bending of a steel stylus upon the action of a probing force of 100 mN.

where

> w_s: Amount of stylus bending
> F: Probing force
> l: Stylus length
> E: Young's modulus
> d: Stylus diameter

6.2.3.4 Pretravel and Overtravel

Pretravel is the stylus displacement upon workpiece contact before contact detection. The isotropic fraction of pretravel is considered by the effective tip ball diameter that is always determined during qualification and always compensated for. More relevant are probing direction-dependent pretravel variations caused by asymmetries of stylus suspension; geometric moment of inertia of the stylus, sensor, and axes arrangement; sensor sensitivity; the direction-dependent behavior of probing force; form deviation of the probing element; and also the direction-dependent dynamic behavior of the probing system.

Pretravel variation can be determined by probing spherical artifacts with a significantly smaller form deviation than the expected pretravel variation. Reproducible pretravel variation can be compensated for by software, if probing direction is known. This is generally not possible with kinematic resistive probing systems, which show much higher pretravel variation than, for example, probing systems with piezoelectric or strain gage-based force sensors (Figure 6.7). The peaks in the middle of the diagrams are caused by the high stiffness of the stylus in the axial direction.

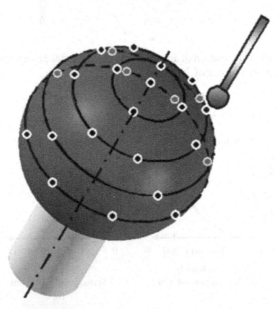

FIGURE 6.16 Evaluation of probing error on a spherical artifact.

Overtravel is the ability of a probing system to deflect further after workpiece contact and is needed to decelerate the CMM axes safely to a standstill without damaging the probing system or workpiece. During overtravel, the probing force has to be limited to small values. Generally, overtravel allowance has to be larger with higher approach speeds. In conventional probing systems overtravel reaches a few millimeters, whereas in microprobing systems overtravel is often only in the range of a few micrometers, posing challenges to the dynamic properties of the CMM and the limiting approach speed.

6.2.4 Probing Error

Like all measurement tasks, probing also has errors that contribute to the measurement uncertainty of each task performed using a CMM.* Usually probing error dominates the fixed part of the maximum permissible error (MPE; ISO 2000a) specified for a particular CMM, whereas length measuring uncertainty is the most relevant contributor to the length-dependent part of the MPE of a CMM system.

The 3D probing error for measurement of single points can be determined according to ISO 10360-5 (ISO 2010a) by probing the northern hemisphere of a spherical standard with 25 evenly distributed probing points and determining the associated least squares sphere. The radial distance between the most exterior and the most interior measured point is taken as a measure for probing error (Figure 6.16).

Probing errors in scanning measurements take into account dynamic effects on the measurement result, which typically grow with higher measurement speeds. The standard ISO 10360-4 (ISO 2000c) specifies two different parameters describing scanning performance*: (1) the MPE_{Tij} (MPE for scanning measurements) and (2) MPE_{τ} (maximum permissible scanning time). The procedure for evaluating these parameters is as follows: A spherical artifact is measured with four scanning paths (around equator, parallel to equator, through the poles, and parallel to the path through the poles). The time needed has to be smaller than MPE_{τ} and the maximum radial distance between the measured points has to be smaller than MPE_{Tij}. Additionally, all measured points have to lie within a tolerance band of MPE_{Tij} within the calibrated diameter of the artifact.

The magnitude and directional characteristic of the probing error are certainly strongly dependent on type, size, and quality of the considered probing system, in addition to the applied qualification procedure. If during qualification only the effective diameter of the tip ball is evaluated, pretravel variation and form deviation of the tip ball contribute to probing error, whereas more sophisticated—and usually more time-consuming—procedures for qualification can also determine and compensate for these systematic direction-dependent errors. The 3D probing errors of 3 μm (for standard touch-trigger probing systems) and those slightly less than 1 μm (for high-quality measuring probing systems) are generated by special state-of-the-art microprobing systems (see Section 6.2.7), which may reach values considerably lower than 1 μm. Probing error is a property of the whole probing system, including stylus and tip ball; therefore, changing the stylus may affect probing error.

---

* Refer also to Chapter 9, Performance Evaluation.

Repeatability is the theoretically reachable limit for 3D probing error, as all repeatable systematic influences may be compensated for by qualification procedures. Unidirectional repeatability may be much smaller than 3D probing error, and it is often specified by probing system manufacturers, although it is not practically relevant for 3D coordinate metrology.

Another important factor for evaluating probing performance is hysteresis. Hysteresis can be determined using a very simple test. By wringing three gage blocks together, an interior feature and an exterior feature of the same size can be created. The measured difference between those features is the hysteresis of the probing system (Figure 6.17). Hysteresis can be caused by backlash or a lack of repeatability of the idle position of the stylus; it also can include machine errors depending on the stylus position inside the volume:

$$H = \left| L_e - L_i \right| \tag{6.11}$$

where

H: Hysteresis
L_e: Measured exterior length
L_i: Measured interior length

Especially when measuring small radii of curvature, factors such as cutting edges of tools, pretravel variation, and form deviation of the probing system may lead to large errors if isotropic behavior of the probing system is presumed and anisotropies are not compensated for by direction-dependent qualification (Figure 6.18).

6.2.5 MULTISENSOR COORDINATE MEASURING MACHINES

Each type of probing system has its own advantages and disadvantages, and there is no one universal probing system that is able to perform all tasks of dimensional metrology satisfactorily. Especially for complex products or parts with a large range of different features, the need for maximum testing quality and efficiency demands a sophisticated combination of different probing techniques (Figure 6.19).

Due to their ability for single point probing and the possibility of applying different styli, conventional tactile probing systems are very flexible and appropriate

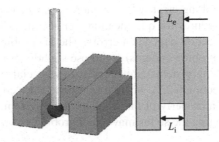

FIGURE 6.17 Hysteresis test for probing systems.

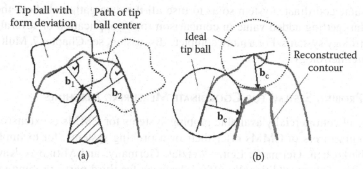

FIGURE 6.18 Effect of tip ball with form deviation a) path of a tip ball with form deviation when scanning a contour; b) reconstruction of the scanned contour from the path presuming perfect sphericity of the tip ball ($\mathbf{b}_1$ and $\mathbf{b}_2$ are real vectors from tip ball center point to probed point, and $\mathbf{b}_c$ is the vector used for tip ball diameter compensation).

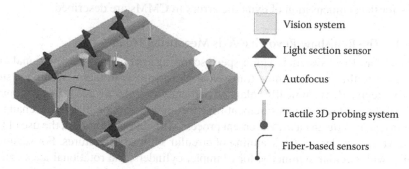

FIGURE 6.19 Application of multisensor probing at a complex workpiece.

for a variety of measuring tasks, like measurement of form, position, dimension, and angles of standard geometric features, if the probing points are accessible by the tip ball and are rigid enough to be touched. However, they are inefficient for the measurement of sculptured surfaces, cannot enter into small interior features, apply a probing force that might destroy or deform delicate structures, and are not capable of special tasks like measuring the thickness of transparent layers.

Optical systems can usually collect a very high number of points quickly and thus are especially suitable for measurement of sculptured surfaces or holistic measurement of clearly visible features. Vision systems with their two-dimensional (2D) characteristics can be beneficially used for the measurement of plane features perpendicular to the optical axis, like edges, or printed features. Laser line and point sensors (triangulation or autofocus) are appropriate for measuring objects or features with distinct height variations like curved surfaces. However, with optical systems, it is not possible to probe surfaces that are nearly parallel to the optical axis, for example, jacket areas of holes. Small holes, narrow grooves, and so on, can be measured by miniaturized tactile probing systems.

When using several probing systems for the measurement of a single component, it is essential that data measured by all the employed systems can be transformed into the same coordinate system so as to fuse all the data gathered about the component, thus getting added value in comparison to the application of several standalone metrology systems. For a more complete discussion, see Chapter 7, Multisensor Coordinate Metrology.

6.2.6 Probing Systems for Coordinate Measuring Machines

The range of commercially available probing systems for CMMs is extensive. Most large manufacturers of CMMs offer their own probing systems, for example, Carl Zeiss, Oberkochen, Germany: Leitz, Wetzlar, Germany: and Mitutoyo, Kawasaki, Japan. Some of them additionally offer interfaces for third-party probing systems. Smaller CMM manufacturers and manufacturers of multisensor CMMs are often not able to develop their own probing systems but use systems developed by dedicated probing system manufacturers.

In Sections 6.2.6.1 and 6.2.6.2, two modern scanning probing systems and techniques for the compensation of dynamic errors in CMMs are described.

6.2.6.1 The Renishaw Revo Five-Axis Measuring Head

The Renishaw Revo system is a high-speed scanning system with two rotational axes in addition to the standard three transversal axes; so it is in fact a probing system with an integrated, dynamically balanced two-axes articulating head. Besides ensuring flexibility and saving measurement time through the option of quickly changing stylus orientation during a measurement process without recalibration, the use of this system vastly accelerates the scanning of circular symmetric features. For scanning features with circular symmetry, for example, cylinders, the rotational axes can be used to perform a circular movement of the stylus tip while the transversal axes are standing still (circular scanning) or are moving at a constant velocity (helix scanning; Figure 6.20). Accelerated movements that inevitably lead to dynamic reactions can be avoided, and scanning speed can be increased without compromising accuracy

FIGURE 6.20   Renishaw Revo five-axis scanning system. (Courtesy of Renishaw.)

(as per manufacturer's specifications*) because of deformations of the CMM caused by dynamic forces.

Another interesting feature of the Revo system is the direct measurement of tip ball displacement when it comes into contact with the workpiece: A reflector is mounted at the back of the tip ball inside the hollow stem. To detect displacements of the tip ball by optical means through the stylus stem, a laser beam is directed onto the reflector and the reflected light is evaluated. Direct measurement of the tip ball position further improves dynamic properties of the probing system as stylus bending by dynamic forces is no longer relevant for the measurement result. With the Revo system, scanning speeds up to 500 mm/s are possible.

6.2.6.2 Zeiss VAST Navigator

The Zeiss VAST Navigator system utilizes a high-speed scanning probe technology with compensation of dynamic errors. High accelerations during scanning introduce heavy dynamic loads on machine axes, probing system, and stylus and cause dynamic bending of these components (van Vliet and Schellekens 1996). Additionally, these dynamic forces may influence the probing force significantly. These two sources of errors compromise scanning accuracy at high scanning speeds and are compensated for by the Navigator system independently of the feature to be measured.

The VAST probing system has active and controlled probing force generation with electromagnetic actuators instead of commonly applied springs (Ruck 1998) and compensates for the changes in probing force caused by centrifugal forces (Pereira and Hocken 2007). Compensation of dynamic bending is achieved by *dynamic bending computer-aided accuracy* (D-CAA): The location- and acceleration-dependent bending behaviors of machine and probe geometry are mapped and embedded into the machine control for correction. With these methods, dynamic errors can be limited to less than 1 μm and scanning speeds of 100 mm/s or more can be achieved according to manufacturer specifications.†

Another feature of the VAST Navigator system for improving scanning efficiency is a software assistant that determines the optimum measuring speed according to the geometric feature to be measured, its tolerance, the stylus, and the used CMM type. The VAST probing system is set up out of three independent stacked axes, each equipped with an inductive displacement measuring system, an active voice coil actuator, and a special damping system to avoid false triggering (Figure 6.21).

6.2.7 PROBING SYSTEMS FOR MEASURING MICROPARTS

The need for reliable quality control in the quickly growing field of MST has much higher demands on resolution and accuracy than the testing of conventionally sized parts; additionally, quality control in MST has to cope with very delicate and small objects that are easily damaged, for example, microlenses, microfluidic channels, microgears, fuel-injector nozzles, and MEMS. Typical nano-scale metrology tools, like scanning probe microscope [SPM] record surface topography maps that can

---

* www.renishaw.com, accessed March 31, 2011.
† http://www.zeiss.de/us/imt/home.nsf, accessed March 31, 2011.

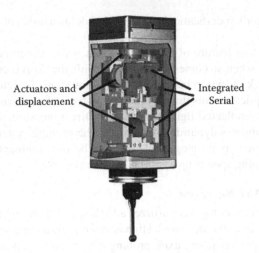

Actuators and displacement

Integrated Serial

FIGURE 6.21 Zeiss VAST probing system with active probing force generation. (Courtesy of Carl Zeiss.)

mathematically be described as z = f(x,y), so there can only be one z coordinate for each point in the x,y plane. With this 2.5D characteristic these tools are not suitable for measuring true 3D features, like cylinders or (entire) spheres. This can only be done using special micro-CMMs and dedicated probing systems (for specific examples, see Chapter 16, Typical Applications).

In many cases, the lateral resolution of optical probing systems is not sufficient for testing microparts. However, optical systems are often used in micro-CMMs to enhance navigation of the main probing system and to measure outer dimensions and 2D micro features. Also, optical 1D probing systems are often used for the measurement of surface contours and layer thickness, so called optical profilers.

6.2.7.1 Scanning Probe Microscope–Based Microprobing Systems

The lateral resolution of scanning optical 1D systems can be vastly enhanced by minimizing the area of interaction between the sensor and the component to be measured. As the spot diameter cannot be reduced to a few nanometers, a mechanical detour can be made via a probe interacting with the surface at a very small zone compared to the focus diameter of optical 1D systems and the movement or position of this probe can be measured by optical means. Such devices are SPMs. Probes used for this purpose are made from silicon by etching, and they reach tip curvature radii below 10 nm, cone angles of about 20°, and tip heights of about 10 to 15 µm. Today, such scanning probe systems are the probing systems having the best resolution; but they suffer from restrictions regarding surface inclination and high aspect ratios caused by limited accessibility of the cantilever tip to the surface, fragility of the probe, and the 2.5D characteristic. At PTB, Germany, an SPM based 3D probing system is experimentally set up that used a micro stylus with tip ball perpendicularly attached to an SPM cantilever. Upon surface contact of the tip ball the cantilever is bent which can be measured with conventional SPM heads. This system potentionally allows for true 3D measurements, though in the experimental setup only two dimensional measurement of the cantilever bending is done.

6.2.7.2 Tactile Microprobing Systems

Conventional tactile probing systems are not appropriate for measuring microparts because of the size of their probing elements (typically 2 to 8 mm), static probing force (50 to 200 mN) and insufficient resolution (0.1 μm). Additionally, the dynamic probing force is very high due to the high moving mass (50 g or more). The need for limiting and controlling the effective probing force—static probing force and dynamic reactions when decelerating the probing element upon workpiece contact— is especially challenging, and it stimulated the development of several commercial and experimental microprobing systems. Today styli with tip balls as small as 50 μm can be manufactured, but the low stiffness and high fragility poses practical difficulties when used with probing systems that measure stylus displacement at the suspension, as by far most measuring CMM probing systems do. However, there are now a couple of different designs offered by commercial companies like Zeiss (Oberkochen, Germany) or XPRESS Precision Engineering (Eindhoven, Netherlands) that use silicon membranes with integrated strain gages as stylus suspension. These designs can reach probing repeatability down to 5 to 50 nanometers at probing forces down to a few ten micro Newtons.

One major drawback of microprobing systems is the lack of availability of styli with tip ball diameters lower than 0.2 mm, their lack of stability, and their high fragility. Another way is to design microprobing systems that are not dependent on force transport via the stylus stem, which can be done, for example, by optically detecting the tip ball itself. This idea has been realized with the Werth FiberProbe, a patented technology developed at PTB* (Christoph, Trapet, and Schwenke 1997), Germany, and then at NIST† (Muralikrishnan, Stone, and Stoup 2006, 2008), United States of America.

The FiberProbe is a microprobing system that was developed to amend optical micro-CMMs, which are based on an optical vision system. A glass tip ball is created at the end of an optical fiber by melting and is brought to the focus of the vision system (Figure 6.22). When it is displaced due to workpiece contact, the amount and direction of the displacement can be measured with the help of the vision system to evaluate the probing vector. To improve the optical measurement of the tip ball position, it can be illuminated via the optical fiber. If the effective diameter of the tip ball is known from a previous qualification, the probed point on the workpiece surface can be calculated from the probing vector and the effective tip ball diameter. As no force transmission via the stem is needed, it is possible to manufacture FiberProbes with very small tip ball diameters (down to 10 μm) and to probe with very small probing forces (a few micronewtons), however it is not capable of 3D measurements, as probing in the stem direction causes practical problems.

---

* Physikalisch-Technische Bundesanstalt
† National Institute of Standards and Technology

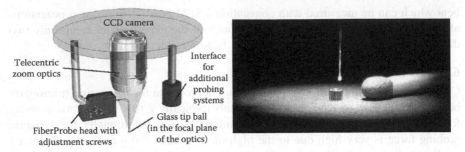

FIGURE 6.22 Werth FiberProbe, left: Schematic setup; right: photograph (measurement of a microgear).

6.3 PROBING SYSTEM ACCESSORIES

The functionality and versatility of probing systems can be enhanced by using various accessory elements. Most of them are used to improve accessibility of the workpiece surface, to facilitate navigation of the probing element to the feature to be probed, and for automation of testing procedures.

6.3.1 ACCESSORY ELEMENTS FOR IMPROVING ACCESSIBILITY

To facilitate accessibility of the probing element to the probing point, stylus extensions are often required, for example, when the feature to be probed lies at the bottom of a deep hole in the workpiece. All elements in the metrology loop contribute to measurement uncertainty. To limit the additional uncertainty derived from stylus extensions, their thermal and mechanical stabilities have to be as high as possible; this leads to the demand for a low coefficient of thermal expansion, very high stiffness, and low weight for styli. For relatively short extensions ceramic materials, titanium, or stainless steel can be used, whereas for long extensions carbon fiber–reinforced plastic is the material of choice due to the combination of high stiffness, low density, and low coefficient of thermal expansion. Stylus extensions are available in a huge variety of different lengths and diameters of the connecting threads. As a rule of thumb, the shortest and thickest extension that can ensure accessibility should always be chosen.

If a probing point is accessible only from a certain direction, articulated joints with fixed or variable angles can be used (Figure 6.23). There are also joints for the fixture of several styli at one probing system that can be beneficially used if several different stylus angles are needed to access all features of the workpiece to be tested. For their use, it is important to qualify each stylus separately and to keep the so-called stylus tree as simple as possible. A complex stylus tree with articulated joints and carbon fiber–reinforced plastic extensions is shown in Figure 6.23.

If the number of different stylus angles needed is too high for practically using articulated joints, it is also possible to utilize articulating heads that alter the direction of the whole probing system together with the stylus. With such a system it is conveniently possible to manually or automatically rotate the probing head so that the stylus points are always in the desired direction, but requalification is needed with many systems after changing the stylus orientation.

FIGURE 6.23 Left: Stylus trees with fixed angles; right: variable angular joint.

6.3.2 ACCESSORY ELEMENTS FOR IMPROVING NAVIGATION

To improve the navigation of the main probing system especially in the field of multisensor and microcoordinate metrology, visual systems are often employed. In the operation of microprobing systems, particularly very small features have to be probed, which are difficult to localize with the naked eye. Additionally, due to the fragility and cost of microprobing systems enhanced navigation is vital for the economic use of such systems.

In the simplest case, a camera can be used for supervising the probing process and for manual navigation of the probing system. More sophisticated systems are based on stereomicroscopy or metrological vision systems that allow for automatic navigation. Especially in the case of multisensor CMMs, it is efficient to use an optical vision system for 2D position measurement and to use this measured information for navigation of other systems, such as tactile probing systems.

6.3.3 ACCESSORY ELEMENTS FOR AUTOMATION

Automation is an issue of great importance especially in coordinate metrology, as the time needed for measuring a part is often significantly longer than that needed for production of the same part. Also in many cases, relatively large lots of parts have to be 100% inspected, so the repetition of metrological tasks is usual. Most modern CMMs feature direct computer control (DCC), but there are some accessories for probing systems that facilitate automation.

To address the challenge of hindered access to probing points, stylus-changing devices are often used. They have advantages compared to stylus trees because there is only one stylus at a time and, therefore, the danger of collisions of idle styli with the workpiece is vastly reduced. The same can be achieved—sometimes even more conveniently and quickly—using automated articulating heads.

If there are measurement tasks to be performed that demand the use of different probing systems, for example, an optical probing system for the measurement of sculptured surfaces with high point densities and a tactile probing system for the measurement of 3D standard geometric features with low uncertainty, probing system changing devices are frequently used.

To protect probing systems from damage, some CMM manufacturers equip their machines with mechanisms for detecting collisions with workpieces as fast as possible to stop the movement and thus limit the harm.

6.4 SUMMARY

Probing systems are an essential part of a CMM as they form the link between the length measuring axes of the CMM and the workpiece to be measured. Due to their task of sensing proximity to a workpiece surface, they have outstanding importance in ensuring high system performance. Depending on measuring task, the material to be probed, the CMM used, specifications regarding required resolution and desired measuring speed, and many more factors, the optimum combination of CMM and probing system has to be chosen. This can also mean the use of several different probing systems on one CMM for the measurement of complex components. Due to a rising demand for measuring speed and accuracy, there is a strong trend toward multisensor systems. Other future trends for probing systems are miniaturization for the measurement of complex microparts, improvements in economy and speed to reduce the cost and time per measured point, and improvement in the reliability of measurement results by collecting holistic information about the workpiece and by improving the system's immunity to environmental influences.

7 Multisensor Coordinate Metrology*

Ralf Christoph and Hans Joachim Neumann

CONTENTS

7.1 FROM PROFILE PROJECTOR TO OPTICAL–TACTILE METROLOGY

Until a few years ago, optical (or noncontact) dimensional metrology was dominated by measuring microscopes and optical projectors. The prerequisites for state-of-the-art optical coordinate metrology include the use of modern image processing techniques and laser sensors, which were developed during the last decade. Multisensor coordinate measuring machines (CMMs) today feature both contact and noncontact sensors, thus combining the advantages of tactile and optical measurements in a single

---

* By kind permission of sv corporate media, Munich, Germany.

system. This sensor combination has made it possible to accomplish most of the measuring tasks encountered in present-day manufacturing. Optoelectronic sensors have gained significance especially because of the growing complexity of part shapes and sizes and the advanced requirements of component miniaturization. The high measuring speed of multisensor CMMs permits economical, near-production measurement.

Although this versatile sensor technology offers the user a wide variety of application possibilities, it also demands a deeper understanding of its inherent productive capabilities and limitations. The purpose of this chapter is to make this new technology both easy to understand and easy to use. It narrows the gap in the previously available technical literature, which primarily dealt with mechanical probing (Weckenmann and Gawande 1999; Neumann 1993, 2000, 2005). Although this chapter is concerned mostly with optoelectronic sensors, it also touches on the important aspects of tactile sensors and combined optical–tactile measurement approaches.

7.2 VISUAL SENSORS FOR COORDINATE MEASURING MACHINES

The sensors of a CMM* are used to pick up the primary signal from a workpiece. They are designed using mechanical and, in some cases, optoelectronic and software components of varying complexity. The sensors must be selected on the basis of the conditions on and near the workpiece, touch sensitivity of the object, size of the features to be measured, requirements of the measurement plan and the number of measured points. The CMMs can be equipped with tactile or optical trigger and measuring (usually called dynamic or scanning) sensors (Figure 7.1).

Trigger sensors produce a trigger signal after detecting a measuring point. This causes the scales of each machine axis to be read out, determining the coordinates of the point in space. Measuring sensors have an internal measuring range of up to several millimeters. An object point is determined by superimposing the measured values of the sensor over the coordinates read out by the measuring machine. It is thus possible to determine a point even when the CMM is standing still (static measurement principle) as long as the magnitude of the object point is located within the measuring range of the sensor.

Another important criterion for differentiating between sensors is the physical principle of transmission of the primary signal. In accordance with this principle, the sensors commonly used today can be divided into two groups: (1) optical and (2) tactile. The location information of a measured point is transmitted to an optical sensor by light in such a way that it can be used to determine the corresponding coordinates. In the case of a tactile sensor, this information is generated by touching the workpiece with a probing element, which in most cases is a stylus tip.

Another important application-specific feature is the number of dimensions of a sensor. This factor determines whether the sensor can pick up information in one, two, or three coordinate axes. For sensors with fewer than three degrees of probing freedom, the remaining coordinates are determined from the previously measured position of the sensor probing point within the machine coordinate system. However, this approach restricts the system's applicability in connection with complex, three-dimensional (3-D)

---

* See also Chapter 6, Probing Systems for Coordinate Measuring Machines.

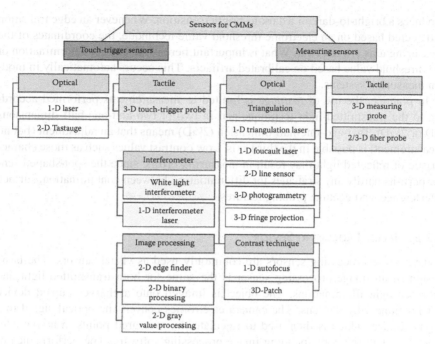

FIGURE 7.1 Sensors for CMMs classified by principle of operation.

objects (e.g., a one-dimensional [1-D] laser cannot measure the cylindrical form of a bore, or an x–y touch probe cannot measure the flatness of a plane in the z direction).

The term *visual sensor* denotes all sensors that, similar to the human eye, can pick up at least a two-dimensional (2-D) image of the object being measured. The intensity distribution of this optical image is detected and evaluated by a sensor.

For many decades, the human eye was the only visual "sensor" available for optical coordinate measuring instruments such as measuring microscopes and measuring projectors. Subjective error sources influencing measurements of this type include parallax (oblique sighting) and faulty measurement of bright-to-dark transitions on edges due to the logarithmic light sensitivity of the human eye. The results of such measurements, therefore, basically depend on the operator and are comparable only to a limited degree. The maximum measuring speed is also limited.

Because of all its drawbacks, visual probing represents the last possible alternative for modern image processing systems. It is used in cases where the object structures to be measured show poor visibility and the geometric features can only be probed intuitively. Assuming that the human eye can resolve several tenths of a millimeter when sighting with a reticle, a final resolution of several micrometers can be attained by using this technique, for example, in conjunction with a 100× optical magnification.

7.2.1 OPTICAL EDGE SENSOR

The *tastauge* or "probing eye" is a trigger sensor for optical measuring projectors. A thin glass fiber picks up a light signal in the beam path of the projector and guides it to a photomultiplier. When an object is moved through the beam path, each edge produces a bright-to-dark or a dark-to-bright transition. Whenever an edge transition is detected based on an electronic threshold value technique, the coordinates of the measuring axis are read out. What is important here is the correct determination of the threshold value based on calibrated artifacts. This occurs automatically in modern measuring systems.

In practice, the use of *tastauge* is limited to measurements performed according to the transmitted-light technique and in two or two-and-one-half dimensions (2D or 2½D). Two-and-one-half dimension (2½D) means that an adjustment (but no measurement) is possible in the third axis. Low contrast values such as those characteristic of reflected light may result in measuring errors, since the spot-shaped sensor permits hardly any strategies for differentiating between contamination, surface interference, and genuine probing features.

7.2.2 Image Processing Sensor

Today, image processing sensors are commonly used as visual sensors. The basic design of an image processing sensor is the same for both transmitted-light and reflected-light illuminations. The object is imaged onto a charge-coupled device (CCD) camera by the lens. The camera electronics convert the optical signal to a digital image, which is then used to calculate the measured points in a computer equipped with the corresponding image processing software. The performance of such sensors is heavily influenced by several individual factors including illumination, the lens system, the sensor chip, the electronics and computing algorithms (Christoph 1989).

A low measurement uncertainty can be achieved using telecentric lens systems. The advantage of telecentrics is that lateral magnification remains constant when the working distance is altered within the telecentric range, thus preventing errors of dimension. This is especially important when working with lower magnifications. The best quality can be attained using telecentric lenses with a fixed magnification.

From an applications standpoint, it makes sense to combine high and low magnifications. This is especially true in cases where features with less stringent tolerances should be measured in one image quickly and, in addition, high-accuracy measurement of closely toleranced features in small image fields should be possible or the positions of the elements to be measured must still be better located after coarse positioning (of the workpiece) on the measuring machine. Using telecentric lenses with a fixed lateral magnification, this can be achieved in two different ways: The first way is simply to change the lens. This can be done automatically (e.g., with a revolving lens changer). The main problem here is the high repeatability required for a lens change. This approach thus has a negative effect on uncertainty. Since only two different magnifications are required in most cases, the simplest approach is to switch back and forth between two different image processing sensors of varying magnification. To increase flexibility, a zoom lens system can be used.

Due to the positioning movement of one design, the optical components in the lens can be moved with slight losses of accuracy (Christoph and Neumann 2007). Linear guides are normally used to ensure high positioning repeatability. The movements of the lens packages required for the zooming processes are motorized. This design enables a 1× to 10× magnification and working distances ranging from 30 mm to a maximum of 250 mm. Optimization between the measuring range of the sensor and the measuring uncertainty can thus be attained (via magnification). Regardless of this, the working distance can be adapted to the specific requirements of the workpiece (to prevent collision problems).

Today, digitization of images is usually achieved using CCD cameras in conjunction with personal computer (PC) components suitable for image acquisition (frame grabber boards, GigE interfaces, etc.). The main advantage offered by CCD cameras over the competing Complementary metal–oxide–semiconductor (CMOS) chips lies in their high metrological quality. For example, the very linear relationship between the light-intensity input signal and the digital output signal is essential for accurate measurement.

The software used to detect the measured points from digitized image data also greatly influences the quality of the measured results. Many different algorithms exist. There is a basic distinction between two different software concepts: (1) the "edge finder" and (2) the processing of image contours (Christoph and Neumann 2007).

With the edge finder, the intersecting points of predefined lines in an image are determined via the visible contours of the object. This is repeated successively at different locations in a predetermined evaluation range or window. The result is a number of measured points, which are then combined by the window to form a group. A separate 1-D evaluation is performed for the determination of each discrete point. The comprehensive 2-D information contained in the image is therefore not taken into account. This causes problems especially for measurements in reflected light. Interference contours caused by surface structures, pits, and contamination can be detected and therefore compensated only under certain conditions.

During the processing of image contours (Figures 7.2 and 7.3), the image is viewed as a whole surface inside an evaluation window. Contours are extracted from this image using suitable mathematical algorithms. In Figure 7.2, the image processing sensor "sees" the object as a grayscale image (Figure 7.2a). The pixels of the grayscale image are converted to digital amplitudes (Figure 7.2b). A pixel contour is calculated from the digital image with a threshold operator (Figure 7.2c). A subpixel point is interpolated from the adjacent values for each point of the pixel contour (Figure 7.2d). An associated element is calculated from the subpixel contour by a number of methods including the Gaussian best-fit method (Figure 7.2e) and displayed in the gray image for visual inspection (Figure 7.2f).

One measured point is thus allocated to each pixel (or picture element) of a contour. The measured points are then lined up in a manner resembling a string of pearls. This makes it possible to detect and filter out interfering influences during measurement without changing the shape of the contours. Several different contours can be distinguished within a single capture range. This is important for practical applications.

The resolution is limited directly by the pixel distance. In the second step, high-quality systems interpolate within the pixel grid (subpixeling), thus enabling improved resolution (Woschni, Christoph, and Reinsch 1984).

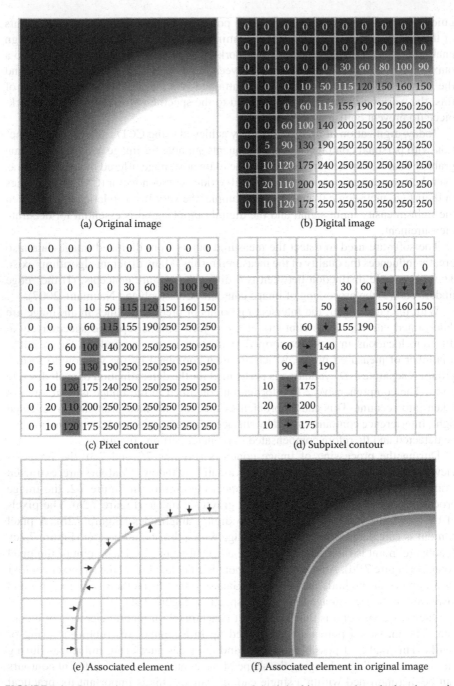

FIGURE 7.2 Processing of image contours from the original image to the calculated associated element.

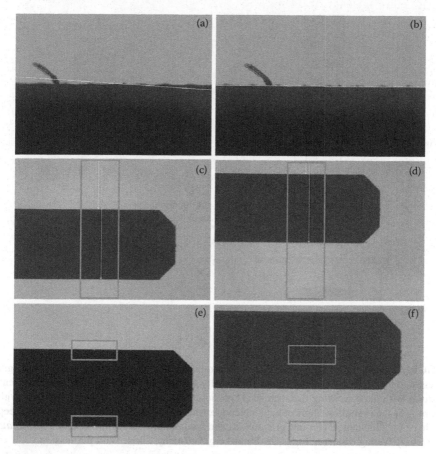

FIGURE 7.3 Contour image processing. (a) Measuring errors due to contamination (b) Correct measurement including form deviation with image processing filter. (c), (d) Large window—sure location of edges through contour selection. (e) Edge finder. (f) Measuring error.

7.2.3 ILLUMINATION FOR VISUAL SENSORS

The basis for every optical measurement is to display the features being measured with a highly accentuated contrast. This can be best achieved on the outer edges of objects. In this case, measurement can be performed in transmitted light (Figure 7.4a).

Ideal conditions are offered by flat objects. On the other hand, where wide edges are involved, the interrelationship among illumination, the test object, and the imaging beam path must be taken into consideration. The manufacturer of a CMM must ensure that the aperture angles of the individual optical systems are mutually aligned and that the image processing software enables adequate calibration of the edge position algorithms. Other transmitted-light units featuring adjustable apertures are used for special applications such as the measurement of rotationally symmetrical parts (e.g., cylinders in a horizontal position).

The use of visual sensors usually requires reflected-light illumination as well as transmitted-light illumination. A distinction must be made here between the

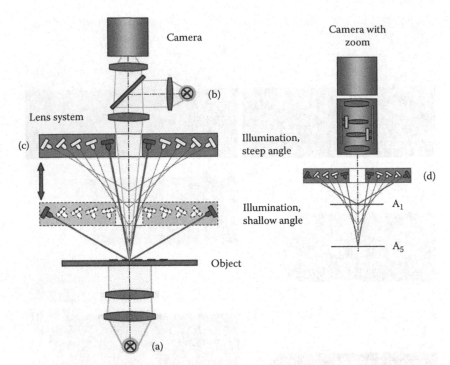

FIGURE 7.4 Types of illumination. (a) Transmitted light. (b) Bright field reflected light integrated in the lens system. (c) Dark field, usually a ring light with adjustable height for lens systems with fixed working distance. (d) Dark field reflected ring light in combination with zoom—A_1 shows shallow angle of light incidence, small working distance, and A_5 shows steep angle of light incidence. (From Christoph, R. and Neumann, *H. J., Multi-Sensor Coordinate Metrology*, Verlag Moderne Industrie, Landsberg, Germany, 2004. With permission.)

two types: The bright field reflected light (Figure 7.4b) is projected onto the object parallel to the axis of the beam path. Ideally, this is achieved directly by the lens system of the imaging system. This type of illumination results, for example, in the direct reflection of light from metal surfaces. The object appears bright. Inclined surfaces reflect the light past the lens system and therefore appear dark (Figure 7.5).

The dark field reflected light strikes the test object at an oblique angle to the imaging beam path. Ring-shaped configurations (or annular ring fiber-optic illuminators) are normally used for this purpose. The advantage of these components is the low amount of heat they add to the measuring volume. Via segmentation, illumination effects can be created from different spatial directions. Light-emitting diodes (LEDs) make it possible to optimally adapt dark field reflected light to the measuring job at hand. Different angles of illumination can be achieved by switching on various diode arrays (Figure 7.4c). By using the illuminator combined with an adjustable working distance zoom lens, it is also possible to vary the angle from the optical axis within a wide range. In addition, this also makes it possible to measure at a sufficiently large working distance away from the object (Figure 7.4d).

In modern multisensor CMMs, all light sources can be controlled automatically via the measuring software. The brightness can be controlled using the light

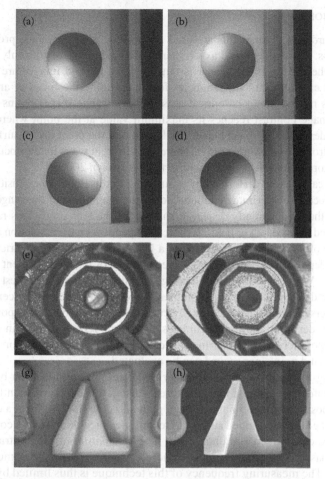

FIGURE 7.5 Test objects under different types of illumination. (a–d) Dark field reflected light from different directions. (e), (f) Bright field and dark field reflected light on the same object. (g), (h) Enhanced contrast through shallow angle of illumination with a ring light.

reflected by the object. This enables practice-oriented application when measuring the surfaces of different materials. Mathematical correction of the lamp characteristics may enable the continued use of the same programs even after changing lamps.

7.2.4 DISTANCE SENSORS

Measurements can be performed only in 2-D planes using the visual sensors described in Sections 7.2.1 through 7.2.3. It is thus possible to measure only 2-D or stepped (2½-D) objects. In order to perform a 3-D measurement of workpieces using optical sensors, an additional process is required to measure the third coordinate. Since the sensors used for this purpose measure the distance to the surface of the workpiece, some companies refer to them as "distance sensors."

7.2.5 AUTOFOCUS

The same hardware components are used for autofocus as for image processing in some systems. Along the optical axis a sharply defined image results only at a single position of the sensor. If the sensor is out of focus, blurred images are produced. The contrast can be used as a parameter for determining the focus of an image. If the sensor is moved along its optical axis within a range that contains the object plane, the image contrast reaches its maximum value at the point where the focal plane coincides with the object plane. The location of the point on the surface can be determined from this sensor position. There are different ways to autofocus, some of which are proprietary (Hocken, Chakraborty, and Brown 2005).

Focusing can then be performed by positioning to this point. The sensitivity of the described procedure is primarily influenced by the magnitude of the range along the optical axis that is apparently displayed sharply by the lens used. This range, commonly referred to as "depth of field," is directly dependent on the resolution and numerical aperture of the lens used. Selection of a lens having a higher numerical aperture reduces the depth of field and increases the accuracy of the measurement taken with the autofocus. With conventional lenses, high magnifications yield the best results.

The main disadvantage of the aforementioned technique is that a certain range must be traversed along the optical axis. Several seconds per measured point are thus required in order to attain high accuracies. This time requirement is in direct contrast to the extremely high measuring speeds of image processing sensors, for which several thousand points per second can be evaluated.

According to one manufacturer, faster distance measurement can be achieved using an autofocus technique in which three sensor chips are permanently mounted on various positions in the imaging beam path and combined to form a camera. As opposed to the aforementioned autofocus technique, three points of the contrast-path curve can be determined simultaneously. The calculation of the contrast curve is performed based on these three points and the known principal parameters of the curve itself. The measuring frequency of this technique is thus limited by the frame rate of the image processing unit (to several tens of points per second).

7.2.6 LASER POINT SENSORS

The measuring principle of a laser point sensor is based on the projection of a beam of light produced by a laser (light amplification by stimulated emission of radiation; usually a laser diode) onto the object to be measured. The reflected beam spot is imaged on an optoelectronic sensor. The position of the point to be measured is then determined via a suitable technique. The best known techniques generally belong to one of two categories: (1) triangulation and (2) interferometric techniques (see Figure 7.1).

The triangulation sensors often used in automation technology function according to the following principle: The laser beam and the axis of the sensor's imaging optics encompass an angle measuring several tens of degrees. A triangle is thus formed between the laser transmitter, the measured point, and the sensor, which can then be used to determine the distance via trigonometric relationships (or triangulation). The measured result depends heavily on the structure and angle of inclination of the surface. This leads to relatively large measuring uncertainties, making the use of this technique suitable only for less demanding applications.

Better results can be achieved with laser sensors that function according to the Foucault principle (Figure 7.6). This principle uses the aperture angle of the sensor imaging optics as the triangulation angle. Instead of a laser spot, a Foucault knife edge located in the beam path is imaged onto the object. Signal evaluation is performed via a differential photodiode. Deviations from the zero position of the laser sensor determined in this way are used for readjustment in the corresponding axis of a CMM. The measured result is obtained by superimposing the values measured by the laser sensor over those measured by the CMM. As was the case before, the material and inclination of the surface also influence the measured result considerably with this type of sensor. A correction of these influencing variables is therefore required. However, if suitable software is used the uncertainty can be reduced to a point where it satisfies the requirements of high-accuracy CMMs. In practical use, a Foucault laser sensor of this type is generally integrated in the beam path of an image processing sensor. This arrangement makes it possible to switch back and forth between the two sensors without any mechanical movement. Moreover, the laser probing process can be visually observed.

The biggest advantage of the laser point sensor in comparison with the focusing technique described in Section 7.2.5 is its considerably higher measuring speed.

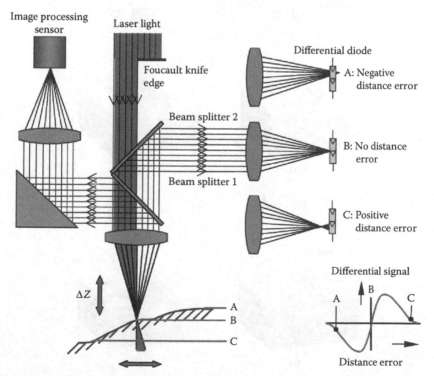

FIGURE 7.6 Laser sensor utilizing the Foucault principle with integrated image processing sensor (illumination system of sensor is not shown).

Several hundreds to a thousand points per second can be measured in this way. These sensors are thus suitable for scanning surface profiles.

7.2.7 MULTIDIMENSIONAL DISTANCE SENSORS

Line sensors (2-D sensors) and area sensors (3-D sensors) function similarly to the spot-shaped distance sensors (1-D sensors) mentioned above (Section 7.2.6). In the laser light sectioning technique (Figure 7.7a), the conventional laser triangulation technique is extended to 2-D measurement by displacing the laser beam with a moving (e.g., a rotating polygonal) mirror.

The evaluation is then performed using a matrix camera to obtain a measured result for multiple points by means of triangulation. A (light) section on the surface of the object is thus measured. A 3-D surface can also be measured by simply moving the CMM perpendicular to the section plane.

Fringe projection sensors (Figure 7.7b) also function according to the triangulation principle (see also Chapter 17, Non-Cartesian Coordinate Measuring Systems). A striped pattern is projected onto the material surface and evaluated in a manner analogous to the light sectioning technique. If the entire 3-D surface is located inside

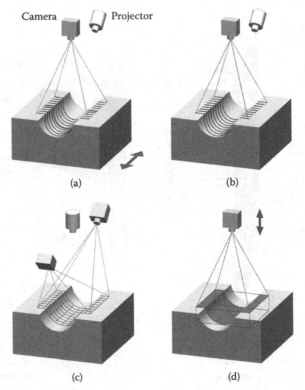

FIGURE 7.7 Examples of multidimensional distance sensors. (a) Laser light sectioning. (b) Fringe projection. (c) Photogrammetry. (d) Werth 3D patch.

the measuring range, no movement along the coordinate axes is required. In order to achieve a higher resolution with a unique allocation of points to their spatial coordinates, different patterns are usually projected and evaluated in succession. A sort of subpixeling can be realized via the "phase shift technique." In principle, the patterns are shifted incrementally and evaluated in each case.

Photogrammetric techniques (Figure 7.7c) are based on acquisition of the object surface from two different directions using one image sensor for each direction (see also Chapter 17, Non-Cartesian Coordinate Measuring Systems). According to the triangulation principle, the space coordinates of each object feature recognized are calculated via the angular relationships. Since the object is not usually sufficiently structured, a 2-D grid is projected over its surface. The resulting pattern is captured by the two cameras and then evaluated. In contrast to a stripe sensor, the accuracy of the projection does not influence the measured result here. Many companies supply software for photogrammetry systems.

A white light interferometer moved along the optical axis also enables 3-D measurements. The number of object points located at a predefined distance away from the sensor is determined for each position of the sensor via a special interference technique. While the interferometer is being moved, point clouds are determined for various section planes and then combined.

With the aforementioned sensor principles, a multidimensional measurement of point clouds occurs. This is comparable to the "measurement in the image" technique performed with an image processing sensor. The uncertainty attainable with the prescribed measuring range is, however, limited. This is due especially to the finite resolution of sensors. A distinction must be made between structural resolution and spatial resolution.

It should be noted here that a much better spatial resolution is required to attain the targeted uncertainty. This means, for example, that a sensor spatial resolution of much less than 1 μm is necessary to attain an uncertainty of several micrometers. Moreover, only sensor measuring ranges of less than 10 mm are attainable. Therefore, the measurement of complex parts with large measuring ranges makes it necessary to position the sensors with the CMM. This corresponds to the "measuring on the image" technique described earlier in this section. In practice, 3-D sensors of this type are provided with larger measuring ranges (several tens of millimeters in length) and are used to measure free-form surfaces and other features with large tolerances.

7.2.8 MULTISENSOR TECHNOLOGY

Multisensor CMMs use a combination of several of the sensors available (described above in 7.2.1–7.2.7 and in Chapter 6, Probing Systems for Coordinate Measuring Machines). The properties of these sensors usually depend on their various primary applications (Figure 7.8).

Regarding applications, their distinguishing characteristics include the size and type of object features they can probe (edge, surface) and their suitability for rapidly acquiring large numbers of measured points (scanning). In order to perform complex measuring jobs, it is usually necessary to use several different sensors for a single measuring run.

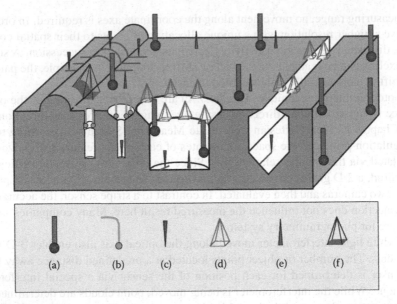

FIGURE 7.8   Multisensor technology: Typical applications of different sensors. (a) Mechanical stylus. (b) Fiber probe. (c) Laser. (d) Image processing. (e) Autofocus. (f) 3D patch.

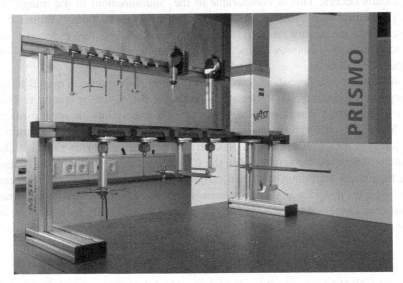

FIGURE 7.9   CMMs with multiple tactile sensors and an optical distance sensor. (Courtesy of Zeiss IMT.)

Starting with a traditional tactile CMM, additional optical sensors such as laser scanners can be simply added (Figure 7.9). These additional sensors can be changed by the usual probe changers. However, this concept limits the weight and therefore their flexibility. Specially designed multisensor CMMs are prepared for the seamless integration of high-performing optical sensors with the

FIGURE 7.10 Multisensor CMM with two separate z rams, vision sensor with integrated laser (TTL), touch probe, and Werth fiber probe—additional rotary/tilt axis for flexible feature access. (Courtesy of Werth Messtechnik.)

necessary light sources in combination with classical touch probes and other sensors (Figure 7.10).

7.3 COMPUTER TOMOGRAPHY

Computer tomography (CT) allows the complete assessment of components, regardless of their complexity. Exterior as well as interior geometries are captured. The industrial use of CT was previously limited to material inspection, due to insufficient accuracy. Combination with other sensors enabled CT to be used in CMMs (see also Chapter 16, Typical Applications).

7.3.1 PRINCIPLE OF X-RAY TOMOGRAPHY

X-ray tomography uses the ability of X-rays to penetrate objects. On its way through an object, a part of the impinging radiation is absorbed (Kak and Slaney 2001). The longer the radiographic length of an object the lower the radiation that escapes from the opposite side. The absorption also depends on the material. An X-ray detector (sensor) captures the escaping X-ray radiation as a 2-D radiographic image. At detector sizes of approximately 50–400 mm, a large portion of the measured object can be captured in a single image.

In order to use tomography on an object, several hundred 2-D radiographic images are made, with the object being measured in various rotated positions (Figure 7.11a). Similar to a pinhole camera, the radiation emitted by a point X-ray

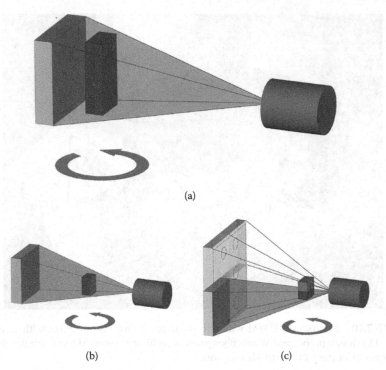

FIGURE 7.11 X-ray tomography. (a) Basic principle. (b) Magnification adjustment. (c) Stitching of images.

source travels through the object to a flat sensor; images are taken for various rotated positions.

The object is located on a rotating table for this purpose, which is incrementally rotated in succession. The 3-D information about the measured object contained in this series of images is extracted using a suitable mathematical process and made available as a voxel image. Each voxel (volume pixel) embodies the X-ray absorption by the measured object for a defined location in the measured volume. Similar to 2-D image processing, the actual measured points are calculated from the voxel data using a suitable threshold process. In the first four steps in Figure 7.12, this process is explained.

The sensors currently used on some systems capture up to 4 million image points. Typically, several hundred to a few million measurement points are derived in the measured volume. These points are distributed evenly across the surface of the part being measured.

Structures in the interior of the measured object, such as hollow cavities or undercuts, are also captured. The measurement points can be evaluated using the familiar methods of coordinate measuring technology.

Similar to measurement using image processing, it is possible to change the magnification using tomography (Figure 7.11b) in order to capture small parts with higher magnification or larger parts completely with lower magnification. To do this, either

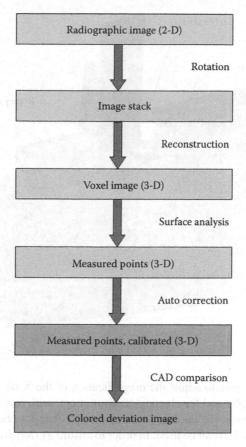

FIGURE 7.12 Processing steps for X-ray tomography measurement.

the measured object is positioned in the radiation path or the X-ray components (X-ray source and detector) are moved in an axial direction relative to the measured object.

In some cases, the size of the sensor or the number of pixels available is still not enough to meet the requirements of the measuring task. In such cases, several images are stitched together (Figure 7.11c) by moving the rotary table with the measured object relative to the X-ray components. Reconstruction of the voxel volume image is then accomplished on the basis of the stitched 2-D images.

7.3.2 MULTISENSOR COORDINATE MEASURING MACHINES WITH X-RAY TOMOGRAPHY

The base of the machine frame of multisensor CMMs with X-ray tomography is normally designed from solid granite. Other system components, such as scales, linear and rotary axes, drives, and guide components, are similar to those of other CMMs (see also Chapter 4, Cartesian Coordinate Measuring Machines). In this construction, calibration data such as magnification, rotary axis position, and geometric corrections are stable over long periods. The basic machine construction is shown in Figure 7.13.

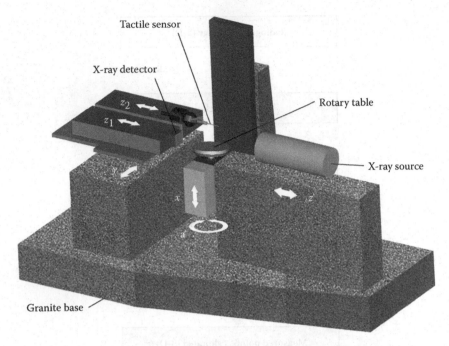

Tactile sensor

X-ray detector

z_2

z_1

Rotary table

X-ray source

y

x

z

a

Granite base

FIGURE 7.13   Design of a multisensor CMM with components for X-ray tomography.

The linear axes serve to adjust the magnification of the X-ray sensors for rastering during tomography and for classical measurement, with tactile and optical sensors. For the latter task, the various sensors and associated probe changers already described (Section 7.2.2) can be used. In order to ensure a crash-free operation, X-ray sensors and the other sensors are mounted on separate z axes.

By selecting the appropriate X-ray components (selecting voltage range of X-ray tubes or detector type), the device can be configured optimally for a variety of different materials. Low voltages, for example, are required for measuring plastic parts that have low densities. High voltages are needed for measuring high-density metal parts.

The Werth TomoCheck shown in Figure 7.14, for example, can be configured to suit any application. In order to obtain lower measurement uncertainty, air bearings and other components are used.

For X-ray sensors, tubes with particularly small focal points (nanofocus) are utilized. This increases resolution and allows measurement of the smallest features. The measuring device is operated by the measurement software, which can also be used for optical–tactile CMMs.

7.3.3 ACCURATE MEASUREMENTS WITH COMPUTER TOMOGRAPHY*

In order to measure the shape and position of a measured object with sufficient accuracy, it is necessary to correct the systematic errors in tomography. Several process-

---

* See also Chapter 16, Typical Applications.

FIGURE 7.14 Detailed view of an air-bearing multisensor CMM (the Werth TomoCheck) with X-ray tomography, image processing, and probe.

related effects lead to such systematic deviations. Common to all of them is the dependence on various parameters, such as cathode voltage of the X-ray tubes, the radiation spectrum that depends on the cathode voltage, as well as material and geometry of the measured object itself.

An example of a process-related effect is beam hardening. This effect can be traced to the fact that the radiation spectrum of an X-ray tube is made up of various frequencies. The high-frequency high-energy radiation components are absorbed by the irradiated material at a lower proportion than the low-frequency ones. As shown in Figure 7.15, this has the effect that for a large material thickness, low-energy sections of the X-ray spectrum are completely absorbed and thus only high-energy radiation can make it through to the detector.

Since the mathematical algorithm for 3-D reconstruction is based on the dependence on thickness of absorption of the entire X-ray spectrum, material areas with large radiographic length will systematically be measured as too large. This effect is known as a *beam hardening artifact* (Srivastava and Fessler 2005). Other geometric artifacts arise from scattered radiation, orientation of the rotary axis in the image, and other effects (see Figure 7.16).

Analytical capture and correction of these complex interrelationships is sufficient for many applications. However, in the case of dense materials or applications requiring the highest accuracy, these methods have limitations. The associated parameters are either partially known or completely unknown. This also applies

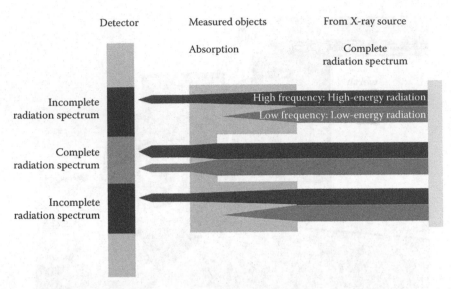

FIGURE 7.15 Filtering out soft radiation: Portions of the radiation spectrum are absorbed differently, due to material and geometric influences.

FIGURE 7.16 Geometric artifact: In the area of the sidewall of the measured object (see arrows), a geometric deviation is visible that does not exist in reality.

to part geometry, material density, and the exact X-ray spectrum. Through the integration of additional tactile or optical sensors, these systematic measurement deviations can be measured with sufficient accuracy and corrected. This measurement must be carried out only once for each type of measured object and only at the relevant or demanding features. The correction data are stored in the measurement program and are used again for new measurements of the same object type.

Figure 7.17 illustrates the principle of the described correction method. The geometric artifact arises because when rotating the rectangular measured object, shorter radiographic lengths occur at the corners than in the middle. This leads to an apparent spherical shape during measurement. Using, for example, tactile measurement of calibration points, these deviations are correctly captured. Building on

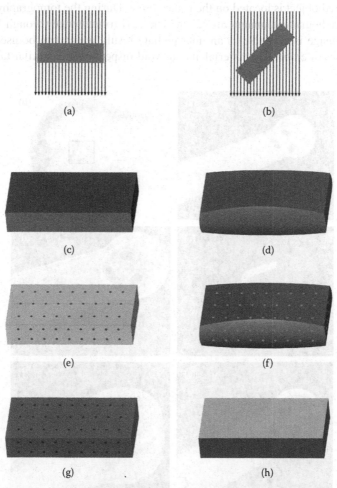

FIGURE 7.17 Autocorrection method. (a) The measured object in X-ray path. (b) The object rotated in X-ray path—artifact generation. (c) The workpiece. (d) Tomography result with apparent bulge. (e) Calibration points on the measured object. (f) Calibration points with the tomography result. (g) Corrected tomography result. (h) Final measurement results.

these reference points, the tomographic point cloud is corrected geometrically as a single entity. Due to the constant nature of the error curve, relatively few calibration points are sufficient. In Figure 7.12, these process steps are represented. Using the autocorrection method, traceable measurement results with specified measurement uncertainties are obtained when using CT. Independent of the tomography process, it is possible to use these devices to measure important dimensions directly by using optical or tactile sensors. This can also be used for measurement of very small features with tight tolerances.

7.3.4 MEASURING A PLASTIC PART

The process for measuring a workpiece using a TomoScope is shown in Figure 7.18. The measured object is located on the rotary table. During the tomography process, many 2-D radiographic images are taken. The part rotates once through 360°. The 3-D voxel image is available as an intermediate result, which can be used for 2-D slice analyses or embedded material and air void inspection and similar tasks.

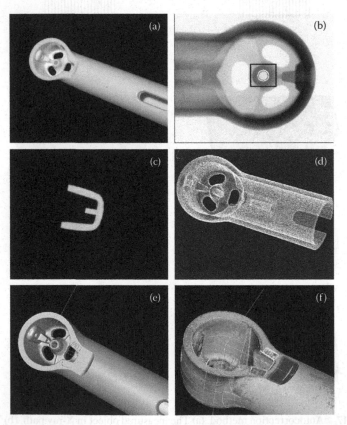

FIGURE 7.18 Measurement example using a TomoScope. (a) Measured object photograph. (b) 2-D radiographic image. (c) Section through voxel volumes. (d) Point clouds generated by tomography. (e) Computer-aided design (CAD) model. (f) Colored deviation image.

The measurement points obtained from the voxel data are geometrically corrected using the autocorrect feature and compared with the part's computer-aided design (CAD) model. In the model, deviations are shown graphically in various colors. In addition, features such as lengths, angles, diameters, and position tolerances are evaluated.

The aforementioned measurement process is particularly cost-effective, for example, for first piece inspection of plastic molded parts. Inspections that previously required several days to verify multicavity molds have been reduced to a few hours. The previously required time of several days to inspect the geometry for the release of the associated multipurpose tools has been reduced to a few hours. The measurement results can be imported directly to generate corrected CAD data for mold rework. Similar advantages result for sample measurement inspections. The same method can also be used in a similar form for other materials and production processes.

7.4 MEASURING ACCURACY

Colloquially speaking, measuring accuracy refers to everything that characterizes the accuracy of measured results. However, a closer look shows that a distinction must be made between various categories based on the following:

- Definition of the characteristics of a CMM in order to specify its measuring performance and to define processes for checking the measuring performance
- Processes for determining the uncertainty of measured results when measuring part features taking all influencing factors into account
- Defining and checking a suitable value for the relationship between measuring uncertainty and feature tolerance

The traceability of measured results to international standards is an indispensable aspect of all these observations. It ensures that measurements can be performed in the same manner throughout the world. The aforementioned aspects will be covered in more detail in the following sections (7.4.1–7.4.3).

7.4.1 SPECIFICATION AND ACCEPTANCE TESTING

The most important attribute of a CMM is its contribution to the uncertainty attainable in a measuring process. The user must be able to compare different machines, define their conditions of purchase, and check their performance.* The International Organization for Standardization (ISO) 10360 series of standards (ISO 2000a, 2000b, 2000c, 2009a, 2010a) have defined specifications and methods for checking these factors. However, these standards currently apply only to tactile sensors. The integration of optical sensors is based on the same fundamental principles and is currently in preparation. The verification of CMMs basically concentrates on two parameters: (1) probing error and (2) length measuring error.

The inspection of probing error has the purpose of characterizing the behavior of the sensors used and the repeatability of a measurement. This is done by measuring

---

* See also Chapter 9, Performance Evaluation.

a calibrated sphere with a specified number of probing points and then determining a limiting value (probing error) from the range of the individual points around the sphere as an associated element. Special considerations regarding the testing of optical sensors* are dealt with in the VDI/VDE[†] Directives 2617, Series 6 (VDI/VDE 2003).

The top factors limiting the performance of mechanical styli are the stylus sphere or the tip itself, bending of the stylus shaft, and nonlinearities or hysteresis on reversal in the probe. For optical sensors, these factors include resolution of the sensors, optical magnification of the lens systems, and depth of field when measuring with the autofocus and, for laser sensors, the radiant reflectance of the material surface as well. Influences caused by the machine itself primarily result from the resolution of its scales and its vibration behavior. Whereas the sphere can be probed bidirectionally from all sides with tactile sensors, only unidirectional probing is possible with many optical sensors. A two-axis indexing probe head, which also requires checking, must then be used to enable bidirectional probing.

The length measuring error takes several factors into account. These include the probing behavior of the sensor, the length-dependent measuring error resulting from mechanical guideway errors, software geometry correction, and the length-dependent measuring error resulting from thermal behavior. It is checked according to ISO 10360-2 (ISO 2009a) by measuring lengths on gage blocks or step gages.

Measurement of gage blocks is possible with all tactile sensors and with fiber probes as well. When checking the length measuring error with other optical sensors, a similar procedure is followed; however, different standards are used. For measurement with an image processing system, the gage blocks are replaced by glass scales with vacuum-deposited chrome graduation lines. Measurement is performed similar to the way in which this is done with a step gage. The required bidirectional probing of gage blocks is not possible with 1-D and multidimensional distance sensors. In this case, ball plates or ball bars must be used. However, in order to ensure the comparability to tactile measurements performed on gage blocks, a mathematical correction must be performed when using this method of measurement. This takes into account the fact that a sensor calibration error (caused by, e.g., an incorrect probe sphere diameter, incorrect origin of laser coordinates, or incorrect magnification) influences the measured results when measuring the gage blocks but not when measuring the sphere. At the same time, the averaging effect achieved by probing the sphere with a large number of measured points is corrected in the results. The machine specification largely depends on the sensor system used.

7.4.2 MEASUREMENT UNCERTAINTY

Every measurement of dimensions such as size, angle, radius, form, and position of workpieces is subject to uncertainty.[‡] The entire measuring process, including machine technology, attributes of the part measured, geometry of the features

---

* Another standard dealing with these matters is in the works of ISO TC 213 (ISO 2011).
† Verein Deutscher Ingenieure/Verband der Elektrotechnik.
‡ Refer also to Chapter 14, Measurement Uncertainty for Coordinate Measuring Systems.

measured, the environment, and operators all influence the magnitude of this uncertainty. The geometry of the features has an especially strong influence on the results of real measurements. Thus, using identical machine technology, the radius of a sector, for example, can be measured much less accurately than that of a full circle.

Other part attributes such as form, roughness, and contamination exert additional influence. For multisensor CMMs, parameters of the sensors are especially important for determining the attainable uncertainty and they must be added to the other machine attributes. Table 7.1 summarizes the parameters that influence the measurement uncertainty of CMMs and of the entire process, classified according to five important sensor types.

Various methods can be used to determine the measurement uncertainty (Wäldele 2002; Wilhelm, Hocken, and Schwenke 2001). If measures of length only are used, the maximum permissible error, $E_{L,MPE}$, can be used for assessment purposes. However, this value does not actually constitute the uncertainty and is only used to evaluate a particular case. Improvements of results (e.g., such as those achieved by measuring a large number of points or through mathematical best fit) and the negative influence of attributes of the workpiece are not taken into consideration.

According to the *Guide to the Expression of Uncertainty in Measurement* (also known as GUM; JCGM 2008a), the measuring uncertainty should be determined by mathematically superimposing the individually evaluated error components (error budget). The following procedure is based on this principle.

Uncertainty can be assessed for a subdomain of tactile coordinate metrology by means of mathematical simulation. This process is described in the technical specification ISO/TS 15530-4 (ISO 2008b). This procedure is not yet available for optical or multisensor CMMs, since reliable error simulation has not yet been mastered for such systems.

TABLE 7.1
Optical Sensor Parameters Influencing Attainable Measurement Uncertainty

Image Processing	Autofocus	Stylus	Laser Probe	Fiber Probe
Type of illumination; optical magnification imaging optics, telecentricity; intensity of illumination; camera resolution; edge detection technique	Depth of field of lens systems; contrast on material surface; intensity of illumination; type of illumination	Diameter of stylus tip (sphere); length of stylus; diameter of stylus shaft; weight of stylus system; probing speed; probing direction	Material of reference sphere; surface tilt (degree of latitude on the reference sphere); optical magnification of Foucault laser sensor; scanning speed; laser intensity	Transmitted light or self-illuminating mode; optical magnification of image processing sensor; contrast factors due to insertion depth; qualification process

The technical specification ISO/TS 15530-3 (ISO 2004b) contains a procedure for determining the measuring uncertainty by measuring calibrated workpieces. This technique can also be used to determine correction values (substitution method), which can be used to substantially reduce the systematic portion of the measuring uncertainty. It is commonly used in the measurement of gages and shafts, for example.

This method does not take into account the influence of changing workpiece surface attributes such as the position of ghost lines, color, and radiant reflectivity. Testing of real parts is still the most reliable method. This method has often been used to assess the overall measurement uncertainty. It is described in numerous company standards and has been introduced under the term *measurement system capability*. Through representative measurements, both the repeatability and the traceability of the measurement to individual externally calibrated components are checked. The repeatability of the measurement is checked by continuously measuring different parts of the same type (representatives of a typical manufacturing process) and jointly evaluating them. Ambient influences, influences of the workpiece itself (surface, color), and influences caused by the operator (clamping and unclamping) can all be examined in conjunction with random errors caused by the measuring machine. However, in order to obtain the total measurement uncertainty, influencing parameters not taken into account during the test phase (i.e., long-term temperature fluctuations) must also be assessed.

With multisensor CMMs, it is also possible to alternatively perform measurements with other sensors or calibrate parts on the same CMM. Systematic errors of dimension in optical measurements can thus be checked.

7.4.3 CORRELATION AMONG SENSORS

One of the most complicated aspects of understanding the errors and estimating the uncertainty for multisensor CMMs is the relationship of measurements among the sensors. Since each sensor may employ a different physical principle, it is difficult to correlate the separate measurements carried out with them. The issue is that only some artifacts (VDI/VDE 2007) can properly test several of the sensors, making it more difficult to correlate them (Charlton 2003). When a single part needs to have features measured by different sensors, it complicates matters even further.

It is also difficult to verify the performance of such CMMs. Work is being performed at standards developing bodies to address these challenges. Specifically, the American Society of Mechanical Engineers (ASME) B89 Dimensional Metrology Committee and the ISO Technical Committee (TC) 213 on dimensional and geometrical product specifications and verification are both working on the matter.

7.5 OUTLOOK

In the course of the last couple of decades, the first prototypes of multisensor CMMs were further developed to systems suitable for serial production. Today, a wide variety of sensors is available for almost all measuring tasks. The corresponding software integrates all the required functions and also takes ergonomic requirements into consideration.

One of the major tasks of research in the field during the coming years will be to minimize the relatively high costs of operator training, developing measurement plans, and programming. An ever-increasing number of activities will be transferred from the operator to the computer based on the existing CAD data interface. Future software packages featuring artificial intelligence will define measurement plans based on CAD models and prompt the user to enter answers to specific questions (e.g., lot number or similar details). Geometric dimensioning and tolerancing (GD&T) can also be embedded within part 3-D models.

An effort is now under way to further reduce the sensitivity of optical sensors to negative workpiece influences (such as surface topography and color). Intelligent software modules will likely be integrated for this purpose. The focus is also shifting to new sensor principles based on parameters such as the takt time.

One important priority in the current development of measuring machines is to increase their accuracy. This is necessary due to the general trend toward closer manufacturing tolerances, especially in electronics and nanotechnology. Machines featuring higher-scale resolutions and corresponding mechanical and sensor concepts are being developed for such applications. The large variety of possibilities offered by multisensor coordinate metrology will ensure its more widespread use.

8 Coordinate Measuring System Algorithms and Filters*

Craig M. Shakarji

CONTENTS

"Coordinate measuring machines (CMMs) can measure anything." Of course, that is an exaggeration, but compared with more elementary dimensional measuring instruments (e.g., a caliper), CMMs can measure much more variety including sizes,

---

* Certain commercial equipment, instruments, or materials are identified in this chapter. Such identification neither imply recommendation or endorsement by the National Institute of Standards and Technology (NIST), nor does it imply that the products identified are necessarily the best available for the purpose.

forms, and locations for an extremely wide array of features simply provided the part and its features meet certain requirements of the CMM (e.g., size), and provided the CMM has the necessary access to the features.

What is it about a CMM that gives it such immense flexibility? From appearance, the CMM seems to only detect a collection of individual points. But it is, in fact, the software that processes these points that turns the CMM from a mere point collector into an immensely flexible, powerful, measuring instrument.

Any book giving a comprehensive treatment of CMMs would be lacking without some consideration given to the software that processes data behind the scenes. The user who has no intention of ever programming algorithms still benefits from an understanding of the basic actions of CMM software. This chapter discusses some aspects of CMM software, but it is written for those who are not software experts. Some of the material will be tutorial in nature. The goal of the chapter is to gently acquaint the unfamiliar reader with some of the key areas in this rich field. Note that no single CMM software package is expected to include all the types of software functionality discussed in this chapter.

8.1 CURVE AND SURFACE FITTING

A key component at the heart of CMM software is curve and surface fitting. Such fitting of CMM data points is necessary in order to assess feature size, location, or form deviation, or to establish a local coordinate system from datum features.

Consider the problem of being given a set of points and optimally fitting, say, a plane to them. The optimal fit will be determined by the sizes of the residuals. The *residuals* are the (orthogonal) distances from points to the plane that has been fit. Although distances cannot be negative, residuals can be thought of as positive (above the plane) or negative (below the plane).

Consider three ways to define an optimal fit, that is, three *fit objectives*. The optimally fitting plane could be defined as the plane that

1. Minimizes the sum of the squares of the residuals
2. Minimizes the size of the worst-case (largest) absolute value residual
3. Minimizes the sum of the absolute values of the residuals

Fit objective (1) is called a *least squares* fit and is also known as a *Gaussian* fit or an L^2 (or l_2) fit. Fit objective (2) is called a *minimum-zone* fit and is also known as a *Chebyshev* fit (with various spellings), a *minimax* fit, a *min–max* fit, or an L^∞ (or l_∞) fit. Fit objective (3) is called the *minimum-total-distance* fit and is also known as an L^1 (or l_1) fit. To avoid confusion, the term "best fit" should be avoided or should only be used when the fit objective has been made clear, because "best fit" is not used exclusively for one particular fit objective across all mathematical literature.

A quick, simplified example will help in understanding these different fit objectives and some of their characteristics. Instead of fitting a plane, imagine fitting a horizontal line to data points that appear in the xy plane. Because the line is constrained to be horizontal, the only thing that can change is its height. So the goal is to simply find the correct height of the line that fits the data.

Imagine the data consist of two points, one at height 0 and one at height 1 (Figure 8.1). The least squares line and the minimum-zone line give the same answer. That is, the optimally fitting line is the horizontal line at a height of ½. For the least squares case, the residuals are +½ and −½, and the sum of squares of the residuals is ¼ + ¼, which equals ½. No other horizontal line placement will result in a sum-of-squares of residuals being that small. For the case of the minimum zone, the residuals are also +½ and −½, which makes ½ the value of the largest absolute value residual. No other horizontal line placement will result in a smaller value for the largest absolute value residual. More unusual is the minimum-total-distance fit, which can be any height between 0 and 1, because any height yields the same sum of distances, namely 1.

It gets more interesting when a third point is added—this point also having a height of zero (see Figure 8.2). In this case, the least squares fit line has height 1/3. It is interesting that this is the same as the average value of the heights of the points. The minimum-zone line has not changed at all. The addition of another point at zero did not change the extreme values, so no change occurred in the fit. In the case of the minimum total distance line (the L^1 fit objective), the optimally fitting line went all the way to zero. This is because any distance from zero is counted twice (reflecting two points) yet any distance from 1 is counted only once (because there is only one point there). The L^1 fit assumed the median value of the heights of the points.

The least squares fit "feels" the new points and adjusts accordingly. In fact, the least squares line will always have a height equal to the average height of the data points. In contrast, the minimum-zone line did not change at all. No matter where the third point appeared (between 0 and 1), the minimum-zone fit would not have

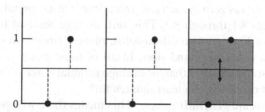

FIGURE 8.1 Fitting a horizontal line to two data points. Left: Least squares fit. Center: Minimum-zone fit. Right: Minimum-total-distance fit. The minimum values of the fit objectives are achieved as ½, ½, and 1, respectively.

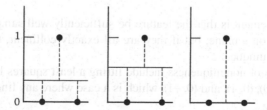

FIGURE 8.2 Fitting a horizontal line to three data points. Left: Least squares fit. Center: Minimum-zone fit. Right: Minimum-total-distance fit. The minimum value of the fit objectives is achieved as ⅔, ½, and 1, respectively.

changed. The minimum-zone line is only affected by what happens at the extreme points and not by the number of points there or other points in between. This is true, in general. *The least squares fit "feels" the effect of all the points, while the minimum-zone fit is only affected by a few extreme points.*

Most CMM software packages perform least squares and minimum-zone fits to standard shapes like lines, planes, circles, spheres, cylinders, cones, and tori. For the cases of lines and circles, the fitting problems can be two- or three-dimensional. For a two-dimensional (2D) fit, all the data points lie in a plane (say the *xy* plane) and the fit of the line or circle takes place within that plane only. A 3D fit allows for the data to lie outside of any one plane and allows the line or circle to be at any angle in the 3D space.

In the cases of circles, cylinders, and spheres, it makes sense to talk about two other fit objectives: minimum-circumscribed and maximum-inscribed. They are almost self-defining. The *minimum-circumscribed* circle for a set of points in a plane is the circle of smallest diameter that has no points lying outside it. A *maximum-inscribed* circle is a circle of largest diameter—surrounded by the points—that contains no data points within it. Without seeking to be more rigorous, the maximum-inscribed definition only makes sense when the phrase, "surrounded by the points," is clear.

Given a basic geometric shape and a set of points, the calculation of the residuals might not be obvious. The Annex of this chapter gives formulae for such calculations for seven basic geometric shapes.

8.1.1 Least Squares Fits

For the least squares fits in the earlier examples, the sum of the residuals (and, thus, the average residual) was zero in each case (recalling that a residual can be negative), as shown in Figures 8.1 through 8.3. This zero average residual is always true for least squares fits of lines (in two dimensions), planes, circles (in two dimensions), spheres, cylinders, cones, tori, and slots. In all of these cases, an attribute of the unconstrained least squares fit is that the average residual is zero. This is a necessary but not a sufficient condition of a least squares fit.*

Least squares fits are *generally* unique. This means there will generally be a single optimal fit to any reasonable data set. There are exceptions to this, but they do not arise very often. For instance, fitting to an insufficient number of points is one exception. There are an infinite number of circles that pass through two data points. So, one requirement of having a "reasonable data set" is having a sufficient number of points.

Another requirement is that the feature be sufficiently well sampled. One could probe 100 points on a plane, but if they are all exactly collinear, the fitting plane would still not be unique.

More examples of nonuniqueness include fitting a least squares line in 2D to the points (1, 0), (−1, 0), (0, 1), and (0, −1), which is a case where any line in the *xy* plane

---

* A much stronger necessary (yet still not sufficient) condition that a fit is the least squares fit is that the gradient of the fit objective is zero. This is a calculation that can be performed without too much difficulty when given a fit and its data set; see Shakarji 1998.

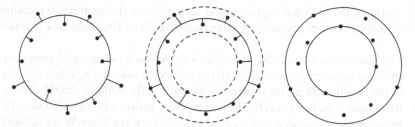

FIGURE 8.3 The same data with four different fit objectives. Left: Least squares fit. Center: Minimum-zone fit. Right: Maximum-inscribed and minimum-circumscribed fits. Note that the centers of the fits are generally different from each other. All fits can be seen overlapped in Chapter 9, Performance Evaluation. The L^1 fit objective is not shown here but is discussed in Section 8.1.3.

passing through the origin qualifies as a least squares fit. But note that in this example of nonuniqueness, the data points are extremely far from lying along any line, hence the earlier qualification of a "reasonable data set." Similarly, fitting a circle in two dimensions to the points $(1, 0)$, $(-1, 0)$, $(0, 1)$, $(0, -1)$, and $(0, 0)$ yields a nonunique least squares circle, but again, the data set was not close to being a circle at all.

A case that needs to be mentioned is fitting a cylinder to a data set consisting of eight points, when those points are located on two rings in a way that they coincide with the eight corners of a box. In this case, there are three least squares (and minimum zone and L^1) fits—orthogonal to each other—each fitting the data perfectly. The axes of the three perfectly fitting cylinders coincide with the axes of rotation of the box. If the box were in fact a cube, then the minimum-circumscribed and maximum-inscribed fits would not be unique either. In contrast to the previous examples of unrealistic data, these points do, in fact, lie perfectly on a cylinder. A user should be aware of this case, as this problem can be easily solved with a better sampling strategy, for example, additional ring(s) and/or additional point(s) per ring.

To sum up then, for reasonable data sets that are well sampled, one should expect a unique least squares solution.

When fitting a circle in 3D, the reason the points do not lie perfectly within a plane is usually due to measurement error (for example, measuring a short cylinder by simply probing a circle of points at one height of the cylinder). Accordingly, the fit is usually performed as follows: Project the data points into the least squares fitting plane, and then fit a 2D circle within that plane. (Even in the case of a minimum-zone fit, the plane of projection is found in a least squares sense as one is seeking to "average out" the measurement errors.) Of course, it is possible to fit the circle to the 3D points directly, meaning that the residuals are the 3D distances from the data points to the fit circle. This rarely encountered approach gives a slightly different result than the projection method just described.

8.1.2 MINIMUM-ZONE FITS

As seen in Figure 8.3, the minimum-zone fit can also be thought of as the circle half-way between the two concentric circles of minimal separation that contain all the points between them. When applied to the actual curve, the minimal separation is the roundness. For other features, the minimum-zone fit can be thought of as follows:

- Minimum-zone plane: The plane halfway between two parallel planes of minimal separation that contain all the points between them. When applied to the actual surface, this minimal separation is the flatness (ASME 2009).
- Minimum-zone line (in 2D): The line halfway between two parallel lines of minimal separation that contain all the points between them. When applied to the actual curve, this minimal separation is the straightness.
- Minimum-zone line (in 3D): The line that is the axis of the minimum-circumscribed cylinder of the points.
- Minimum-zone sphere: The sphere halfway between two concentric spheres of minimal separation that contain all the points between them. When applied to the actual surface, this minimal separation is the sphericity.
- Minimum-zone cylinder: The cylinder halfway between two coaxial cylinders of minimal separation that contain all the points between them. When applied to the actual surface, this minimal separation is the cylindricity.
- Minimum-zone cone: The cone halfway between two coaxial cones, having the same apex angle, of minimal separation that contain all the points between them. When applied to the actual surface, this minimal separation is the conicity.
- Minimum-zone torus: The torus halfway between two concentric, coaxial tori, having equal major radii, of minimal separation that contain all the points between them. (In other words, the two tori only differ in their minor radii.)

For the minimum-zone fits to these shapes, the largest positive and largest negative residuals will always have equal magnitude (except for the 3D line, where residuals are never negative). This can be seen in the earlier examples in Figures 8.1 through 8.3. But in Figure 8.3, it can be seen that there are actually four points whose residuals share the largest magnitude (i.e., four points touch the outer or inner enveloping circles). This is true, in general. That is, a minimum-zone fit circle (2D) will have at least four points that attain the largest magnitude residual. For other geometries, there are other minimum numbers of points that attain the largest-magnitude residual. This is a necessary but not sufficient condition of a minimum-zone fit.

8.1.3 Minimum Total Distance (L^1) Fits

Recall that, in Figure 8.2, the least squares fit had an average residual of zero. In contrast, the minimum total distance (L^1) fit had a median residual of zero. This zero-median residual is always true for L^1 fits of lines (in two dimensions), planes, circles (in two dimensions), spheres, cylinders, cones, tori, and slots. In all of these cases, an attribute of the unconstrained L^1 fit—when using an odd number of data points—is that the median residual is zero. (In the case of an even number of points—as in Figure 8.1—the common convention can be adopted that the fit lie halfway between the two middle-most values.)

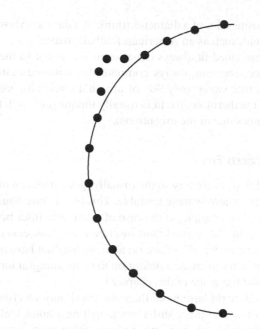

FIGURE 8.4   The L^1 circle fit to the data shown completely ignores the four points not on the circle. The median residual is zero.

This characteristic gives the L^1 fit an ability to ignore outliers—a feature that is helpful in many situations where spurious points should be ignored, whether those spurious points are from the measuring instrument or from an anomaly like dirt on the surface being measured. Figure 8.4 shows an example of this feature.*

It will be seen in Section 8.4.2 that the L^1 fit might also prove useful in cases of establishing a datum plane to a given set of points. In this case, the fit would be constrained to one side of the points, emulating the contact between the planar feature of a workpiece and a surface plate (datum simulator). Datums will be discussed later in the chapter.

8.1.4   MINIMUM-CIRCUMSCRIBED FITS

For the minimum-circumscribed fit, Figure 8.3 shows the circle touching three points. This is true, in general, but the minimum-circumscribing circle might touch only two points (e.g., think of a data set whose points are taken on an ellipse). In this case, the two points would occur at opposite ends of a diameter. A minimum-circumscribed sphere touches four points, or three that lie on a great circle of the sphere (e.g., think of a data set whose points are taken on an oblate spheroid) or two

---

* The figure is correct and has been verified numerically, showing the advantage of the L^1 fit in ignoring outliers. But there are more extreme cases—dependant on location, magnitude, and number—where a minority of outliers can affect the L^1 fit result. Viewing the fit together with the points could show if outliers have created an unexpected effect on the fit.

points that occur on opposite sides of a diameter (think of a data set whose points are taken on a prolate spheroid, such as an American football shape).

The minimum-circumscribed fit always exists even if it is not so meaningful in some cases. For instance, one can always compute the minimum-circumscribed circle for a set of points that sweep only 90° of arc on a circle although the result may have limited value. Furthermore, the fit is usually unique (although the cylinder example given earlier shows one of the exceptions).

8.1.5 MAXIMUM-INSCRIBED FITS

The maximum-inscribed fit is, in contrast to the (usually) good behavior of the earlier-mentioned least squares fit, relatively more unstable. This fit can have issues with both existence and uniqueness. For example, in the case of maximum-inscribed circles (in 2D), not every set of data points has a maximum-inscribed circle associated with it. For instance, a set of points that sweep 90° of arc on a circle does not have a maximum-inscribed circle. (One might try to create a definition to be meaningful for this type of situation, but that is beyond the scope of this chapter.)

Furthermore, a data set might have more than one maximum-inscribed circle, as Figure 8.5 shows (e.g., if the workpiece shifts when drilling a hole). Unlike the special cases identified earlier with other fits, which were either unrealistic or resolved with better sampling, this case is realistic and is not resolved with better sampling. For the left-hand picture in Figure 8.5, data points selected around the shape could yield two maximum-inscribed fits having the same radii but different centers. This means the same data points submitted to two different algorithms could yield different results—yet both are equally valid.

The right-hand picture in Figure 8.5 shows a more common case. That is, there really is only one maximum-inscribed circle, but another candidate might be almost as large, and an algorithm could easily make a mistake in reporting the wrong inscribed circle. This means that in the search for the globally optimal fit, it is hidden among other nearby fits that are locally optimal. This property (which is shared to some degree by minimum-zone and minimum-circumscribed fits) makes it possible for an algorithm to cease searching for an optimal fit before the truly optimal fit is found.

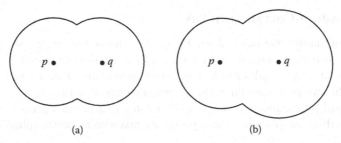

(a)   (b)

FIGURE 8.5 Points selected around the left-hand shape might have more than one maximum-inscribed circle—one centered at p and one at q. Points selected around the shape on the right have one unique maximum-inscribed circle, but without additional searching an algorithm could mistakenly find a circle centered at p and think it had found the maximum-inscribed circle.

This property of the absolute optimal fit being "hidden" among locally optimal fits is one important reason why these fits are computationally more intensive. The minimum-zone, minimum-circumscribed, and maximum-inscribed fits can easily take 100 times more computational time than a least squares fit. (Another reason for the difference in computational times is the smoothness of the fit objective as fit parameters vary.) See Shakarji and Clement 2004.

8.1.6 OTHER FIT OBJECTIVES

Another fit objective is to maximize the minimum residual. Given a rigid shape, this fit seeks to find the location and orientation of that shape within the points such that it contains none of them and keeps the shape as far as possible from any data point. This fit is useful for finding a shape within a casting. See Figure 8.6.

8.1.7 GENERAL SHAPES

Fits are applied to shapes besides the standard ones already discussed. General shapes can be represented as mathematical functions (e.g., in the case of an ellipsoid) or in computer-aided design (CAD) systems as unions of patches, each patch being a—usually simple—function (e.g., shapes may consist of planar patches, splines, etc.). In these cases, the shapes are fit as rigid shapes (unlike a sphere fit, for example, where the sphere can have variable size in an unconstrained fit). The optimal fit to the data points is sought by allowing rotations and translations of the surface without allowing changes to its size or shape. This is sometimes called a *six-degrees-of-freedom* fit or a *rigid body* fit. Given a point and a complex shape, the residual, as before, is defined to be the length from the point to the closest point on the surface. The shortest line

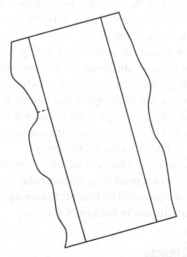

FIGURE 8.6 An algorithm can find a rigid cylindrical shape within a rough casting such that the cylinder stays as far from the rough surface as possible. The dashed line shows the closest the cylinder comes to the surface. Only the location and orientation of the cylinder can vary in seeking the optimal fit. The diameter of the cylinder is considered fixed in this fit.

segment from a data point to a surface will either be perpendicular to the surface, or it will be at an edge or corner of the surface. This holds true for the earlier-discussed standard shapes, as well.

8.1.8 WEIGHTED FITTING

More advanced is the case of weighted fitting. This is a case in which not all residuals are treated equally. Suppose one needs to find a shape within a casting. And suppose when the points are acquired on the surface, it is known that for some reason certain points have a higher uncertainty with regard to their location than other points. One might weigh the residuals differently when defining the fit objective. That is, the objective function might be defined to cause the optimal fit to be closer to points with lower uncertainties because it is "safer" to do so.

Another case where weighted fitting might be appropriate is when the data points are far from evenly spaced. In such a case, data points in sparse regions may be weighted more heavily than those in dense regions. See Section 8.5.

8.1.9 CONSTRAINED FITTING

It is shown later in this chapter that there are times when constrained fitting is needed (finding datums is such a case). For example, one might want to fit a plane to a set of points with the constraint that the plane is perpendicular to another plane. This limits the field of choices for the optimally fitting plane. In fact, the first example in this chapter involving fitting horizontal lines to data was an example of constrained fitting (Figures 8.1 and 8.2). The line was restricted to be perpendicular to the y-axis; it was not allowed to tilt to achieve a better fit. Another case of constrained fitting is fitting a cylinder to points when the cylinder is constrained to be perpendicular to a given plane. (This constrained cylinder fit is actually equivalent to a 2D circle fit with the points projected into the plane.)

Yet another example of constrained fitting is fitting a sphere of a fixed diameter to a set of points. Imagine establishing a local coordinate system using three high-quality (low form error) spheres whose diameters have been calibrated and well known. It could be advantageous to compute the center locations by fitting fixed-diameter spheres to the data. This strategy is more helpful in a situation where only a small patch of the sphere can be probed because the unconstrained diameter fit could vary greatly (see Chapter 9, Performance Evaluation).

Constrained fitting is also used when establishing datum planes. One constraint on a datum plane might be that it must be perpendicular to a previously established datum plane. Another constraint could be that all the data points lie to one side of the plane. See datum planes discussion in Section 8.4.2.

8.1.10 FIT OBJECTIVE CHOICES

Why are there so many fit objectives? In many cases, the physical problem determines the fit needed. If a user was interested in knowing if a pin would fit in a certain size hole, a minimum-circumscribed fit to the pin makes sense. In the opposite

TABLE 8.1

Differences between the Least Squares and Minimum-Zone Fit Objectives

Least Squares	Minimum Zone
Only an approximation to much of the language used in ASME 2009	Matches much of the language used in ASME 2009
Less sensitive to outliers	Very sensitive to outliers
Usually less affected by CMM measurement noise	Usually more affected by CMM measurement noise
Algorithm testing exists for basic shapes	Algorithm testing is only in the primitive stages
Computationally fast	Computationally intensive

case, if a user needed to know if a bore was large enough for a certain sized pin, a maximum-inscribed fit to the bore is appropriate. If it is most important to ignore a small minority of mildly anomalous points, an L^1 fit might be the most appropriate.

Deciding when to use least squares fitting and minimum-zone fitting can be more complicated. The least squares fit has an averaging effect. This effect tends to average out measurement noise (instrument-induced measurement errors from random effects), whereas measurement noise can significantly affect the minimum-zone fit if it occurs at critical, extreme points. But the minimum-zone fit yields a lower (or equal) peak-to-valley distance of the residuals and can be a better indicator of form, particularly if the data points are "good," meaning fairly dense, with no outliers, and not much affected by measurement noise.

Some differences to keep in mind are listed in Table 8.1.

8.2 STYLUS TIP COMPENSATION

Another task of CMM software is stylus tip compensation for contacting systems. For this section, a single (perfectly) spherical stylus tip on a contact CMM is assumed. When contact with the surface is made, the center of the stylus tip is located away from the surface by a distance equal to the radius of the stylus tip (assuming no pre-travel or probe bending).

In the cases of circles, spheres, and cylinders, the effect is to make the diameter seem larger (or smaller) by an amount equal to the diameter of the stylus tip; see Figure 8.7. In the case of a cone, it makes the vertex seem shifted or, equivalently, causes a change in the radius of the cone at a given point on the axis (the orientation and apex angle are unaffected). For a plane, the effect is a translation (without affecting the orientation). For a torus, the effect is seen as a change in the size of the minor radius.

CMM software has to account for this effect. For basic shapes mentioned earlier, the easiest and most accurate solution for stylus tip compensation is to account for the stylus tip radius after both the data collection and fit have been completed. So a sphere, for instance, would be fit with the uncorrected data. The fit would then be altered afterward by a simple subtraction to the diameter. (The user can also do this manually.)

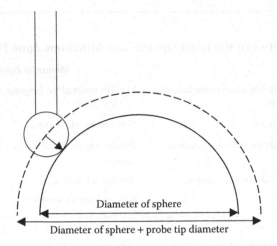

Diameter of sphere

Diameter of sphere + probe tip diameter

FIGURE 8.7 The effect of the sphere tip size on the realized sphere diameter before correction.

This works well, but there are problems with more complicated surfaces. For example, perhaps contrary to intuition, this does not work for an ellipse. It is not correct to simply fit an ellipse to uncorrected points and then to subtract the stylus tip radius from the sizes of the semimajor and semiminor axes. For this case, as well as for complex surfaces, the stylus tip compensation has to be included in the fit itself. That is, when computing the residuals during the fit process, the residuals themselves have to be compensated for the stylus tip radius.

If a star stylus system with five stylus tips (of different sizes) is used and multiple stylus tips are used in acquiring points, the stylus tip compensation needs to be applied during the fit process—now even for the cases of simpler shapes discussed earlier.

The above-mentioned process could make one ask if there can be point-by-point stylus tip compensation. In other words, can the stylus tip radius be compensated for during the acquisition of each point itself? The answer is yes, and it is sometimes done, although the approach is not without drawbacks: The stylus tip correction needs to be made in the direction orthogonal to the surface, and, before fitting, the surface might not be very well located. Also, if the nominal surface shape is not known ahead of time, there might not be sufficient points nearby to characterize the surface locally to get a good approximation of the normal direction.

8.3 DATA FILTERING

Filtering—a process widely used in signal and image processing—seeks to diminish the effect of noise in data and, in other cases, seeks to sharpen edges.* Many filtering techniques can cross over disciplines to be applied in coordinate metrology. The first examples will deal with data that are equally spaced, for simplicity. The principles

---

* A more general way to describe filters is that they seek to differentiate effects that occur on different scales.

will then be applied to the more realistic cases in coordinate metrology having non-uniform sampling.

8.3.1 CONVOLUTION FILTERS

Suppose there is a list of coordinates, (x_i, y_i), where—for now—the x-values are assumed to be equally spaced. An application of a simple filter (not necessarily optimal) would be to replace each y-value with the average of itself and its left- and right-neighbors (with the exception of the first and last y-value). Applying this filter to the y-values [0, 1, 0, 1, 0, 1, 0, 1] would yield [1/3, 2/3, 1/3, 2/3, 1/3, 2/3]. (The new list has only six values; the first and last were dropped.) The application of the filter smoothed out the rough jumps, as shown in Figure 8.8. This averaging technique is an example of a *mean filter*. A mean filter is an averaging (with equal weights) of neighboring points, and it does not need to be restricted to just one point to the right and left.

In coordinate metrology, densely sampled data with measurement error (noise) are often filled with such "jumps," which are a result of the measuring equipment rather than the surface itself. By diminishing these jumps, filtering can help separate out the actual surface from among the measurement noise included in the data points.

The mean filter described earlier can be thought of as a kernel $\boxed{\frac{1}{3}\,|\,\frac{1}{3}\,|\,\frac{1}{3}}$ that is *convolved* with the vector of y-values, $\boxed{0\,|\,1\,|\,0\,|\,1\,|\,0\,|\,1\,|\,0\,|\,1}$. The convolution is performed by sliding the kernel over the y-values, starting at the left and moving the kernel through all the positions to the end of the data (always keeping the entirety of the kernel covering y-values). Each kernel position corresponds to a single new y-value, which is calculated by multiplying together the kernel value with its underlying y-value for each of the cells in the kernel, and then adding all these numbers together.

Although a mean filter will diminish noise, it will also decrease the height of peaks and raise the height of valleys (perhaps only slightly). The effect of a mean filter on a step function will produce a ramp.

One may filter the filtered data. Adding a second application of the kernel is actually the same as a single application of the weighted average given by the longer kernel, $1/9\ \boxed{1\,|\,2\,|\,3\,|\,2\,|\,1}$. Three applications would be equivalent to a single application

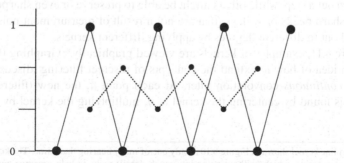

FIGURE 8.8 The simple three-point averaging filter has a flattening effect on the more severe jumps between the original points. The dashed lines represent the filtered graph.

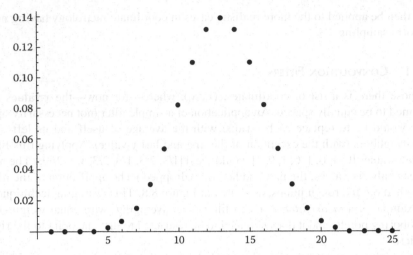

FIGURE 8.9 Repeated application of a mean filter yields a weighted filter that approaches a Gaussian distribution. A 25-cell kernel is pictured, the equivalent of repeated application of the three-celled mean filter kernel. The *x*-axis represents the cell number, and the *y*-axis indicates the value (weight) assigned to a cell.

of the kernel 1/27 $\boxed{1\,|\,3\,|\,6\,|\,7\,|\,6\,|\,3\,|\,1}$. It turns out that repeating this process develops a kernel that approaches a normal (or Gaussian) distribution, as shown in Figure 8.9, where the kernel weights are plotted on a graph. A *Gaussian filter* is a filter whose kernel is obtained by a Gaussian curve. Figure 8.10 shows an example using mean and Gaussian filtering.

The mean filter and the Gaussian filter are just two of a large class of filters called *convolution filters*. Convolution filters work by using a kernel to produce a moving, weighted average. The specific weighting function is different from one filter to another. Although Gaussian and mean filters are the most commonly used convolution filters, one might also see others (i.e., using different kernels) such as cubic-spline or wavelet-based filters (see Figure 8.11). Different kernels yield different effects. For instance, one filter that is effective at smoothing out noise might also smooth out a step, while others might be able to preserve or even sharpen a step. Similarly, sharp peaks or valleys (that are not a result of measurement noise) might be rounded out to different degrees by applying different kernels.

In Figure 8.11, examples of kernels are viewed graphically.* Graphing the kernel conveys an idea of how to extend the concepts of discreet filtering (discussed until now) to a *continuous* convolution filter. At each point *a*, the new (filtered) function value is found by centering the kernel at *a*, multiplying the kernel by *y*(*x*), and

---

* The cubic-spline kernel shown in Figure 8.10 is only one of many bearing that name. The kernel shown is sometimes negative, which (without going into much detail) seeks to help preserve genuine steps as well as the true heights of peaks and valleys while at the same time the kernel seeks to smooth out noise.

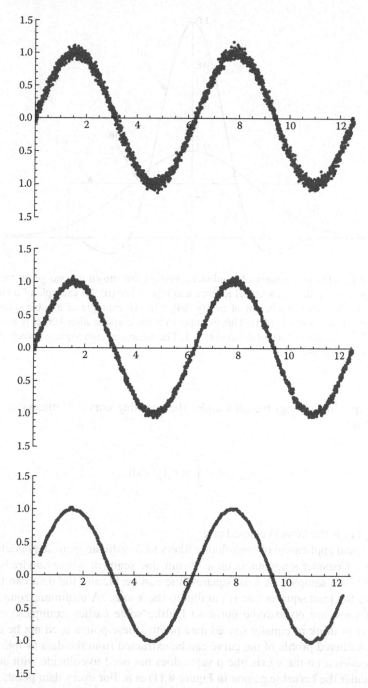

FIGURE 8.10 Random noise was added to a sine curve to get the data pictured in the first graph. The second graph shows the data after a simple mean filter ⅓ ⅓ ⅓ has been applied. The third graph shows the data after an (approximate) Gaussian filter has been applied. The vertical axes can be thought of in micrometers, the horizontal in millimeters.

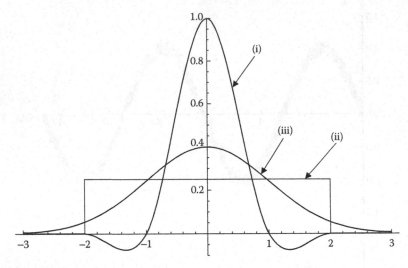

FIGURE 8.11 Three examples of continuous kernels are shown on one graph for (i) An example of a mean filter (this kernel reaches a height of ¼), (ii) An example of a Gaussian filter (this kernel reaches a height of nearly 0.4), (iii) An example of a cubic-spline filter (this kernel reaches a height of 1). This example of a cubic-spline filter kernel is sometimes negative. The area under each of the curves is 1. The mean and cubic-spline kernels are zero outside the interval [−2, 2].

"adding up" (integrating) the area under the resulting curve. Mathematically, the calculation is

$$y_{new}(a) = \int_{-\infty}^{\infty} y(x)k_a(x)dx,$$

where $k_a(x)$ is the kernel centered at a.

A practical application of convolution filters to coordinate metrology could work as follows: Consider straightness on a 12-mm line segment where hundreds of 2D data points are sampled. A least squares line can be fit, and the data can then be rotated so the least squares line is parallel to the x-axis. (A minimum-zone fit also works if there are no extreme outliers.) Unlike some earlier examples, where it was easier to think of equally spaced data points, these points need not be equally spaced. A filtered profile of the curve can be extracted from the data as follows: At every location a on the x-axis (the a value does not need to coincide with any data point), center the kernel (e.g., one in Figure 8.11) at a. For every data point, associate a temporary weight to be the height of the kernel at the x-coordinate of the data point. Normalize the weights by dividing them all by the total sum of all the weights (so the new weights all sum up to 1). Now the new y-value at a is calculated to be the sum of the y-values of the data points times their associated weights. Next, shift the

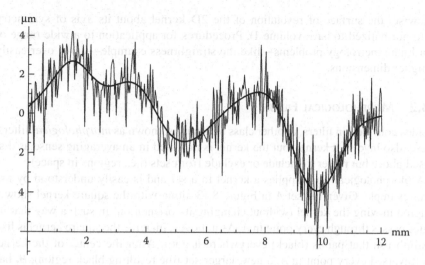

FIGURE 8.12 Output from a cubic-spline filter is shown superposed on an unfiltered input profile. The kernel of the filter used here is shown in Figure 8.11.

kernel slightly down the line segment and repeat the process. Figure 8.12 shows the outcome of this procedure in an example application of a cubic-spline filter.

The filtered profile in Figure 8.12 appears to have captured a better representation of the actual curve than the individual collection of the data points. The peak to valley distance of this filtered profile appears to better represent the straightness than that obtained using the raw data. If the unfiltered data contained extreme outliers (unlike the data pictured), then the difference between the filtered and unfiltered straightness results would, also, have been extreme. Although the filter's averaging effect would have mitigated the effect of outliers, the convolution filter does not perform actual outlier removal. A different action could have been taken to actually remove any extreme outliers from the data set.

Convolution filters exist for higher dimensions. In image processing, a convolution filter could use a kernel that moves over a 2D grid of pixels. An example of a 2D kernel for a mean filter is a simple blurring mean filter kernel:

$\frac{1}{9}$	$\frac{1}{9}$	$\frac{1}{9}$
$\frac{1}{9}$	$\frac{1}{9}$	$\frac{1}{9}$
$\frac{1}{9}$	$\frac{1}{9}$	$\frac{1}{9}$

While in image processing, the data points (i.e., pixels) are usually equally spaced, they are generally not so in coordinate metrology. But this is not a problem for 3D data just as it was not for the 2D straightness example above. The kernels also carry over easily from the 2D case to the 3D case. The kernel for a mean filter for 3D data points is a disc (cylinder), and the kernel for the Gaussian filter is (likewise) the surface of revolution of the 2D kernel about its axis of symmetry (both normalized to have volume 1). Procedures for application to a wide range of coordinate metrology problems—like the straightness example—carry over easily to higher dimensions.

8.3.2 MORPHOLOGICAL FILTERS

Besides convolution filters, another class of filters is known as *morphological* filters. These also involve a kernel, but the kernel is not used in an averaging sense as discussed above but rather to include or exclude from sets (i.e., regions in space).

A morphological filter applies a kernel to a set and is easily understood by use of an example. Given the set A in Figure 8.13 along with the square kernel shown, imagine moving the kernel (without changing its orientation) in such a way that its center passes through every point in A. As it moves, imagine the square kernel is like a paintbrush that paints (black) everywhere it goes. After the center of the kernel has traversed every point in A, a new, larger set (the resulting black region), B, has been created. This process is *dilation*. The next step is to have the kernel act not as a paintbrush but as an eraser. Now the center of the square kernel is imagined to move through every point *outside* of B, with the kernel now erasing everything it touches. This creates a new set, C.

Dilation followed by erosion is *closing*, and erosion followed by dilation is *opening*. Figure 8.13 shows closing, and it is notable that the process has closed the hole in that was in the original shape. Figure 8.14 shows opening, which removed the protrusion from the original shape. This is characteristic of morphological filters and an application of both opening and closing in the example would have resulted in a solid, square shape having neither hole nor protrusion.

Unlike the convolution filters described earlier, these morphological filters change information in irreversible ways. Once the hole is removed (in Figure 8.13),

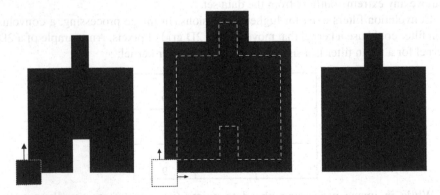

FIGURE 8.13 Left: Set A. The square kernel "paints" everywhere it goes as its center moves within the set. Center: Set B. The resulting set from the dilation (painting) process. The kernel then "erases" everywhere it goes as its center moves outside the (dilated) set. The dashed lines represent the previous set. Right: Set C. The resulting set from the erosion (erasing) process.

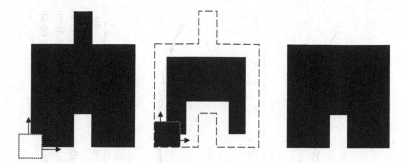

FIGURE 8.14 Left: Set *A*. The square kernel "erases" everywhere it goes as its center moves outside the set. Center: Set *B*. The resulting set from the erosion (erasing) process. The kernel then "paints" everywhere it goes as its center moves inside the (eroded) set. The dashed lines represent the previous set. Right: Set *C*. The result of the dilating (painting) process.

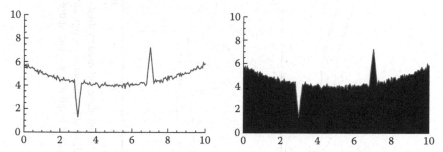

FIGURE 8.15 A graph can be thought of as a set (region) by considering the area under the graph. Then the same procedure can be used for morphological filtering, as described earlier.

the information of its earlier existence is gone for good. It is indistinguishable from the case that the hole never existed in the first place.

After the iteration of dilation followed by erosion (or vice-versa), it is found that *C* is a filtered version of *A*. The strength of a filter like the one shown above depends on the size of the square. The hole and protrusion would not have been removed with a somewhat smaller kernel.

In general, the kernel does not need to be square, and the center does not have to be chosen to be the reference point (i.e., that point that traversed every point in *A* during dilation). The square kernel was chosen for easy visualization, but a circular (or spherical) kernel is much more common for applications in coordinate metrology.

How do such morphological filters apply to coordinate metrology? Suppose one has the graph in Figure 8.15, showing probed points (here in a 2D case). One can define the set *A* to be the area under the graph. Then an erosion and dilation (or vice-versa) could be applied (Figure 8.16) using a kernel that is a disc of a certain radius (the greater the radius, the greater the effect of the filtering).

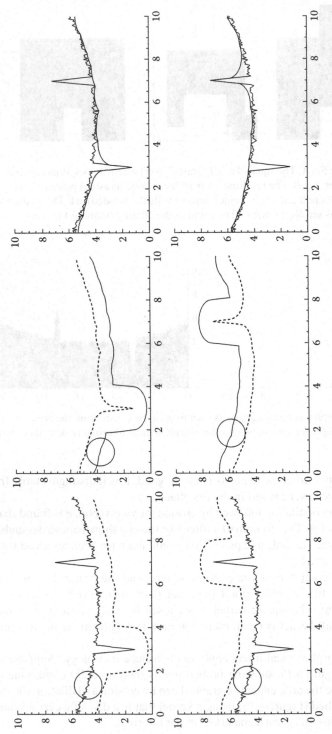

FIGURE 8.16 The first row of graphs shows the process of erosion taking place followed by dilation (opening), along with the original and final graphs together. The second row of graphs shows the process of dilation followed by erosion (closing), along with the original and final graphs together. Both the vertical and horizontal axes are in millimeters, so the noise pictured is quite large. The figure also shows the effects of opening and closing on an extremely large peak or valley.

Opening has the effect of removing sharp peaks, while closing has the effect of filling in sharp pits when applied to a graph of measured values. One or both can be applied.

All these concepts can be extended in a straightforward manner to three dimensions. The kernel of a disc would correspond to a spherical kernel with all other procedures being the same.

One morphological filter is very commonly used in coordinate metrology—even if it occurs without some users intending it or even aware of it. It is a result of the use of a stylus probe tip. A spherical probe tip, for example, applied to a surface will not enter crevices that are small relative to it. Even though the stylus tip radius would be accounted for as described earlier, its filtering effect remains. The larger the spherical tip, the greater the amount of filtering. This filtering procedure differs slightly from the example above in that it is the sphere center that traces out the new shape while the tip boundary touches the surface. It is the opposite of the example above, but the effect is the same.

Although not used as widely as convolution filters, the morphological filters can be used in cases where data points gathered have sharp spikes that do not represent the intended surface. If these spikes occur on one side of the surface, a convolution filter would include their effect (although smoothed out) while a morphological filter might be able to completely eliminate them.

See also Srinivasan 2005 and several filtering techniques described in the many parts of the ISO 16610 Series (2006, 2009, 2010).

8.4 GD&T, DATUMS, AND LOCAL COORDINATE SYSTEMS

CMM software is used to apply the tools mentioned thus far to real problems—often specified by means of geometric dimensioning and tolerancing. A few examples will illustrate various issues of some theoretical definitions versus the practical applications of CMM software to these problems.

8.4.1 FLATNESS EXAMPLE

The first example is a simple flatness determination.

The flatness of a plane is defined as the minimum separation possible between two parallel planes that contain the surface between them (ASME 2009). Consider a flatness callout of 0.1 mm indicating that the flatness of a planar feature on a part must be no greater than 0.1 mm, or put another way, that the entire nominally planar surface must lie between two parallel planes that are 0.1 mm apart.

Given a set of points, the minimum-zone fitting plane is the plane that is midplane between the two planes of minimal separation that contain all the points between them. In other words, this description of the minimum-zone plane exactly matches the definition of flatness. So, given a set of points representing the planar surface, one can compute the minimum-zone plane and then compute the maximum residual minus the minimum residual (known as the *peak-to-valley* distance) to get the flatness. The flatness can then be compared with the tolerance.

Algorithms to find the minimum-zone plane usually go through iterations of candidate planes in search of the optimal one. It should be noted that if the tolerance is known, and if the only reason for the optimal plane search is the tolerance check, some algorithms might (justifiably) stop searching before finding the optimal plane, if any candidate plane is found to yield a peak-to-valley distance less than the tolerance. The peak-to-valley distance using the true minimum-zone plane would be assured to be less than or equal to that value.

In an idealized world, the steps to use a CMM to check the flatness callout would be as follows:

1. Measure the planar surface of the part with an infinite, dense number of points using a perfect (error-free) CMM.
2. Find the minimum-zone plane to the data points.
3. Calculate the flatness of the plane as the peak-to-valley distance, that is, the greatest residual minus the least residual.
4. Compare the flatness of the plane to the flatness tolerance to see if the planar surface is within specification.

In the real world, life is different: CMMs are not perfect, an infinitely dense sampling is impossible, and the minimum-zone algorithm is often not the one chosen (Table 8.2).

8.4.2 DATUM REFERENCE FRAMES

Figure 8.17 shows a GD&T example from Chapter 3, Specification of Design Intent: Introduction to Dimensioning and Tolerancing, which will be used here to understand some additional aspects of CMM software. The specification and its intent can be reviewed from Chapter 3, if needed.

The tolerance zone of the bore is located and oriented in relation to the datum reference frame (DRF) specified in the feature control frame. Realizing the DRF in practice raises several issues.

The first task in creating the DRF is to associate a (perfect) plane with the (imperfect) planar surface A. Physically, this could be realized by the contact of the surface with a surface plate (a datum simulator). CMM software can mimic this contacting behavior. Given a set of points representing the (imperfect) surface A, a "supporting plane" can be found, which corresponds to a mating surface plate.

It turns out this computational process is not so easy in light of the fact that cases exist where the supporting plane is not unique. That is, there could be multiple supporting planes to a given set of points on the planar datum feature.*

In practice, the datum plane corresponding to the supporting plane definition described earlier is replaced by an approximating substitute. This is often the least squares plane or a shifted least squares plane (a least squares plane that is translated—away from

---

* The case of multiple supporting planes (or "rocker" condition) introduces a complicated aspect of DRFs. The set of all supporting planes creates what is called a *Gauss map*. Theoretically, tolerance specifications are satisfied if they are met using any datum plane within the Gauss map. While this is true, the most common current procedures used with CMM software is to simply find a single plane in one of the ways described in this section.

TABLE 8.2

A Contrast between Idealized and Practical Measurement Procedures

The Idealized Measurement Procedure Would Be:	But in Practice, the Following Might Take Place Instead:	And the Reason Practice does not Match the Idealized Case is:
Probe the planar surface with an infinite number of points.	Probe the surface with a limited number of points.	Infinite sampling is impossible. Beyond that, fewer points are sampled than what is possible for practical, economic reasons.
Using a perfect CMM, environment, and so on.	Using a CMM with various sources of error.	A perfect CMM, environment, and so on does not exist. Furthermore, for economic reasons, the CMM, environmental conditions, and so on used are generally inferior to the best possible.
Keep all data points.	Data filtering that might include outlier removal, data reduction, or smoothing.	Filtered data might better represent the actual surface, because CMM error could be removed. But filtering also has some risk that real workpiece surface detail could be removed.
Find the minimum-zone plane and calculate the flatness (peak-to-valley distance).	Find the minimum-zone or the less-than-ideal least squares plane and calculate the flatness (peak-to-valley distance).	Least squares algorithms are ubiquitous in the CMM field; sometimes, minimum-zone algorithms are not available; the least squares algorithms have been better tested (generally) and might therefore be more reliable. They are computationally faster. However, they produce a peak-to-valley distance greater or equal to that of a true minimum-zone fit.
The measurement uncertainty under the idealized conditions is zero.[a]	Evaluate the measurement uncertainty.	All the above deviations from the idealized case introduce uncertainty sources that should be accounted for.
Compare the flatness with the tolerance specified.	Compare flatness with specification taking into account the decision rule for acceptance.	Because of the nonzero uncertainty in the real world, a decision rule is needed—even if that rule is simple acceptance/rejection. Decision rules are discussed in ASME 2001b or in ISO 1998.

[a]   Strictly speaking, there is slight, nonzero uncertainty because of the inexact specification of the measurand, but the point of the reasoning remains nonetheless.

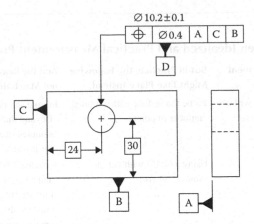

FIGURE 8.17 A specification example (similar to the one in Chapter 3, Specification of Design Intent: Introduction to Dimensioning and Tolerancing) serves to illustrate the application of CMM software to a particular problem.

the material—just far enough so all the data points lie to one side). The effect of this substitute should be accounted for in the measurement uncertainty.*

Another way to derive a datum plane from the imperfect datum feature is to compute the L^1 or L^2 (least squares) plane constrained to lie on one side of the points (the side of the points that is away from the material). As for such an L^1 fit, it closely mimics the physical process of finding a supporting plane by juxtaposing a surface plate (although some differences remain). The least squares fit is not computationally burdensome, but the L^1 fit can also be achieved without being too computationally burdensome, because, in this case, the fit is equivalent to the simplified problem of finding the plane—constrained to the one side of the data—that comes closest to the centroid of the data points.[†]

Understanding the various approaches to calculate a datum plane, consider creating the DRF ACB as called for in the feature control frame in Figure 8.17. In the following paragraphs, "fit" is used understanding the various fits just described.

Sample points (ideally densely and evenly) on surfaces A, C, and B using an (ideally perfect) CMM. For the points corresponding to A, fit the plane (according to a method described in Section 8.4.2). Next, for the points corresponding to C, fit the plane with the constraint that the plane be perpendicular to the previously computed plane. Finally, for the points corresponding to B, fit the plane with the constraint that the plane be perpendicular to both of the previously computed planes. This DRF

---

* The use of the shifted least squares plane instead of the supporting plane that corresponds to the definition is a departure that contributes to the measurement uncertainty. However, the fact that the least squares algorithm is less sensitive to outliers could result in a lower overall measurement uncertainty in some cases—especially, in cases, when the datum feature is very flat, but the points have significant measurement error.

† Because this one-sided L^1 fit closely mimics the supporting plane process, and because it can be computed without extreme computational burden, it appears likely (although not yet official) to be a preferred method for calculating a datum plane in the emerging update to the ASME Y14.5.1 standard (ASME 1994b) to correspond to ASME Y14.5-2009 (ASME 2009).

establishes a local coordinate system to locate and orient the tolerance zone called out in the feature control frame.

As in the previous flatness example, departures from the idealized practice need to be considered in the measurement uncertainty (see Chapter 14, Measurement for Coordinate Measuring Systems; ASME 1999). And again, a decision rule is needed when ascertaining if a part is within tolerance, taking into account the uncertainty introduced by these various steps (see ASME 2001b on decision rules).

8.4.3 More Complex Datum Reference Frames

In Figure 8.17, the bore with the letter "D" is identified. How would one compute the DRF ADC given points sampled on the surfaces? For the points corresponding to A, fit the plane (according to a method described in Section 8.4.2). Next, for the points corresponding to the cylinder D, find the maximum-inscribed cylinder with the constraint that the cylinder axis be perpendicular to the previously computed datum plane.* Finally, fit a plane to the points corresponding to C constrained such that it is perpendicular to the previously computed datum plane. In the example, A was used to remove two rotational and one translational degree of freedom. B then removed two translational degrees of freedom, and C removed the remaining rotational degree of freedom.

DRF DAC would be generated by first computing an unconstrained maximum-inscribed cylinder fit to the points corresponding to D, which would eliminate two translational degrees of freedom and two rotational degrees of freedom. A plane would then be fit to the points corresponding to A with the constraint that the plane be perpendicular to the axis of the previously computed cylinder. This would eliminate the remaining translational degree of freedom. Finally, the remaining rotational degree of freedom would be removed by fitting a plane to the points corresponding to C with the constraint that the plane be perpendicular to the earlier computed plane.

Each of these DRFs establishes a local coordinate system, which can be used to locate and orient the tolerance zone called out in the feature control frame. However, things become more complicated with the use of maximum material modifiers.

DRF ADC, say, would be more complicated if a maximum material modifier is applied to D. The DRF could be computed as in Section 8.4.2, but if datum feature D departed sufficiently from its maximum material condition, then a mobile DRF could result, which would result in mobility in the tolerance zone. CMM software would have to be sophisticated to include such abilities.

Datum features can be more complex still, for instance, a pattern of holes, and to handle this case, the software has to be more sophisticated.

It turns out (as the reader might suspect by now) that most CMM software departs from the exact definitions at some point or another in DRF establishment. It could be that the software does not consider all supporting planes, or it might substitute a least squares cylinder for a maximum-inscribed cylinder, or perhaps mobile DRFs are not exploited to their full potential benefit.

---

* This constrained cylinder fit is tantamount to a maximum-inscribed circle fit to data points projected into the previously computed datum plane.

A user can benefit in the first place from knowing the limitations of any CMM software being used and—especially in critical situations—should have some grasp on the uncertainty introduced by these aspects of software.

8.5 DATA REDUCTION

Certain coordinate measuring systems (CMSs) can acquire millions of data points in very little time. In many cases, there is a need to reduce the number of data points before analysis.

There are various reasons for such reductions. One reason is simply to have a more manageably-sized data set. A second reason might be to even out the points. That is, to have approximately the same data density over various parts of the surface. A third reason is to reduce the set into fewer—yet more accurate—points.

Software accompanying such systems might have tools available to reduce the data. For the first two reasons, a simple strategy of data reduction is to divide the working volume into cubes. If there is more than one data point in any cube, one is randomly selected to keep and the rest are deleted. The user can determine the cube size.

Figure 8.18 shows why such evened out data sets can be advantageous. A greater density of points over some regions tends to overweight the effect of those regions in a least squares fit. Evening out the data can mitigate this effect. Alternatively, a careful application of weighted least squares fitting can also accomplish this. Using the weighted approach, each point would get a weight in the sum-of-squares calculation, the weight being inversely related to the local point density.

One downside to the data reduction described earlier is that the data are reduced most where it is the densest. Often the data points are clustered close to the surface and the ones far from the surface are more isolated. Hence, the data reduction can end up removing more of the accurate points, leaving all the outliers alone.

The data reduction strategy described reduces the number of points but does not improve the accuracy of any data point; it is merely a process to thin them out. An alternative data reduction strategy is to use the filtering techniques described to apply, say, a convolution filter to create a smaller, regularly spaced set of derived points that are more accurate than the points in the original data set.

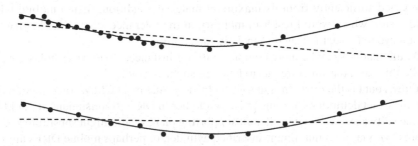

FIGURE 8.18 Top: The solid curve represents the workpiece. Points on the left side are sampled with a different point density than the right side. The least squares fitting line to the data points is skewed reflecting the data. Bottom: The data has been reduced by removing points. The least squares fitting line to the reduced data better matches the workpiece.

Another software tool that can be thought of as a kind of data reduction is edge detection software. This software reduces a large grid of data (maybe acquired from a vision system) into a smaller set of points that represent the edge of an object.

Figure 8.19 shows a grid overlaying an object. Suppose each grid location had a value of 0 if it covered none of the object, a value of 1 if it is completely filled by the object, and otherwise contains a fractional value equal to the fraction of the grid square occupied by the object.

Software can then take this grid information and "guess" the location of the edge of the surface, representing this edge by a set of points.

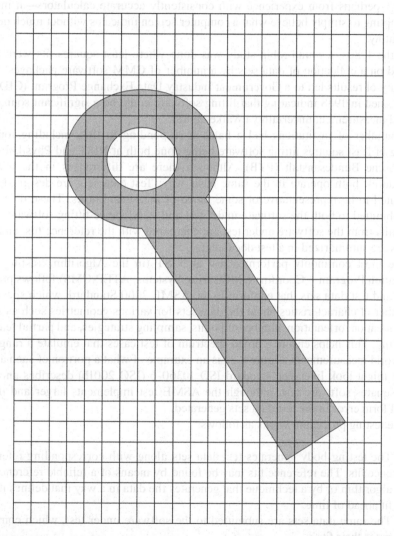

FIGURE 8.19 A vision system gathers a grid of information indicating a measure of the detection of material in each grid square. Software seeks to convert these data into a set of points representing the edge of the object.

Of course, in reality, the grid values obtained by a vision system will include various measurement errors, making the problem more challenging for the software. The software can be helped by reducing these instrument-induced errors with care given to measuring conditions, backlighting, and so on to increase the contrast between the presence and absence of material.

8.6 SOFTWARE TESTING

With such a breadth of measurement responsibility put on the shoulders of software, a question of software testing might naturally come to mind for some. For others—perhaps from experience with consistently accurate calculators—it might be tempting to simply believe what a computer screen indicates without much doubt or scrutiny.

In the 1980s, a round-robin study was done where least squares fits were performed on a collection of data sets by a number of CMM software packages. The diversity of results led to a Government Industry Data Exchange Program (GIDEP) alert issued in 1988 indicating that fitting software could be a significant source of overall measurement uncertainty (Walker 1988).

A number of institutions had a hand in responding to this, including formal testing of least squares fitting software being done both at NIST and Physikalisch-Technische Bundesanstalt (PTB). Although there are differences in the testing procedures, both operate in the same basic way: Test data sets are designed and generated to be representative of some range of measuring tasks. These data sets are submitted to both the software under test and to reference fitting software. The fit results from the software under test are compared with the reference fits, and the results are summarized in a test document.

The most commonly performed test at NIST (in the Algorithm Testing and Evaluation Program—Coordinate Measuring Systems, ATEP-CMS) follows procedures and test data sets that conform to the ASME 2000 Standard, which specifies a number of characteristics about the data sets for various geometries such as size, form, location, orientation, number of points, sampling strategies, and partial feature sampling. This helps ensure a broad spectrum of test cases that emulate a range of real-world measurements. Figure 8.20 is an example of what a portion of such a test report might look like. The standard ISO 10360-6 (ISO 2001b) describes another least squares software test, although the ASME test implements larger and more varied form error in the test data sets generated.

The testing procedure works as follows:

1. The testing body generates test data sets along with corresponding reference fits. The reference fits may be found by means of a reliable reference algorithm or by a technique that generates the data in a way that defines its fit ahead of time.
2. The same data sets are submitted to the software under test, which computes their fits.
3. The fits from the software under test are then compared with the reference fits by the testing body.

ASME B89.4.10-2000 Standard Default Test

Geometry Type	Mean (RMS) Deviation			
	Separation (μm)	Tilt (arc seconds)	Radius/dist (μm)	Apex (arc seconds)
Lines	$<10^{-5}$	2.1×10^{-7}	—	—
Lines 2D	$<0^{-5}$	$<10^{-7}$	—	—
Planes	$<0^{-5}$	$<10^{-7}$	—	—
Circles	8.4×10^{-5}	3.3×10^{-7}	1.5×10^{-5}	—
Circles 2D	9.6×10^{-4}	$<10^{-7}$	9.0×10^{-4}	—
Spheres	7.0×10^{-4}	—	$<10^{-5}$	—
Cylinders	1.3×10^{-3}	7.0×10^{-4}	1.0×10^{-3}	—
Cones	1.7×10^{-2}	2.7×10^{-3}	5.1×10^{-3}	5.5×10^{-3}

FIGURE 8.20 A part of a sample test report following the ASME B89.4.10 Standard. The mean (RMS) deviations shown serve as reasonable valuations for the uncertainty contributed by the least squares fitting software.

One important detail a reader should take away from reading this chapter is the vast number of complex software functions that need to be performed to make seemingly simple GD&T tolerance inspections. In light of this, it is exceedingly important to understand exactly what any software testing does and does not cover. Currently, the national measurement institutes involved in formal CMM software testing services cover only the least squares fits and only for the simple geometries and only for unconstrained fitting. Although such testing is admirable and extremely valuable, as it has led to significant software errors having been corrected, it is nonetheless incorrect to extend those results to other functionality of the software, of which there are many.

Some data files with reference fits are available for download from the NIST Web site (Algorithm Testing; Physical Measurements Laboratory). Also some online resources are available at the Software Support for Metrology, www.npl.co.uk/ssfm.

The often asked question, "Has that CMM software been tested?" is vague and can lead to a very misleading answer. If the answer is, "yes," one might assume that meant that all the functionality of the software has been put under test. Worse yet is to mistakenly assume that software that has undergone such a small degree of testing (albeit valuable) has been "verified," "validated," or "certified" because these terms connote a sense of security and comprehensiveness.

Software that correctly performs least squares fits might perform Chebyshev fits poorly (Shakarji 2002). In fact, least squares testing of simple geometric shapes does not reveal important information about such things as fitting to complex surfaces, or CAD shapes, DRF construction, conformance to specifications (e.g., position, profile tolerances, etc.), outlier rejection, filtering techniques, or any of a number of other vital features that software implements.

8.7 CONCLUSION

The reader of this chapter is likely not a CMM software programmer. However, having a better handle on software behavior can improve various steps in a measurement procedure, including choices in fitting algorithm, sampling strategies, data density, outlier handling, filters, testing, and so on. The knowledge gained can aid in improved buying decisions for future software purchases. Additionally, the knowledge gained from the chapter should make one aware of sources of measurement uncertainty arising from various components of software involved in the measurement, and which components have and have not undergone formal testing. Although the chapter has been intentionally written as an introduction to this vast field with minimal mathematics, the author's desire is for readers to take away a familiarity of what is happening in CMM software, making what may have been thought of as a black box somewhat more transparent.

8.8 ANNEX—RESIDUAL FUNCTIONS FOR BASIC GEOMETRIC SHAPES

For greater detail, see Shakarji 1998.

Given the geometry:	The (signed) orthogonal distance between a point, (x_i, y_i, z_i), and the geometry is given by:
A line defined by a point on the line (x, y, z) and a unit vector in the direction of the line (a, b, c)	$\lvert (a,b,c) \times (x_i - x, y_i - y, z_i - z) \rvert$, which is equal to: $\sqrt{u^2 + v^2 + w^2}$, where $u = c(y_i - y) - b(z_i - z)$, $v = a(z_i - z) - c(x_i - x)$, $w = b(x_i - x) - a(y_i - y)$. This (never negative) distance is f_i for purposes later in this table.
A 2D line defined by a point on the line (x, y) and a unit vector in the direction of the line (a, b)	$b(x_i - x) - a(y_i - y)$
A plane defined by a point on the plane (x, y, z) and a unit vector normal to the plane (a, b, c)	$(a,b,c) \times (x_i - x, y_i - y, z_i - z)$, which is equal to $a(x_i - x) + b(y_i - y) + c(z_i - z)$. This (signed) distance is g_i for purposes later in this table.
A circle (in 2D) defined by its center, (x, y), and its radius, r	$\sqrt{(x_i - x)^2 + (y_i - y)^2} - r$
A sphere defined by its center, (x, y, z), and its radius, r	$\sqrt{(x_i - x)^2 + (y_i - y)^2 + (z_i - z)^2} - r$
A cylinder defined by a point on its axis (x, y, z), a unit vector in the direction of its axis (a, b, c), and its radius, r	$f_i - r$, where f_i is the distance from the point (x_i, y_i, z_i) to the axis, as given above in this table.
A cone defined by a point on its axis (x, y, z), a unit vector in the direction of its axis (pointing toward the apex), (a, b, c), the orthogonal distance from (x, y, z) to the cone surface, called s, and its apex semi-angle, ψ	$f_i \cos \psi + g_i \sin \psi - s$, where f_i and g_i are defined above in this table.
A torus defined by its center, (x, y, z), its major radius, r, and its minor radius, R	$\sqrt{g_i^2 + (f_i - r)^2} - R$, where f_i and g_i are defined above in this table.

9 Performance Evaluation

Steven D. Phillips

CONTENTS

The subject of coordinate measuring machine (CMM) evaluation is a broad and multifaceted one. The central theme of this chapter is the evaluation of CMM measurement uncertainty.* It is described using the terminology and methodology in the international guide recommended by the Joint Committee for Guides in Metrology (JCGM; JCGM 2008a). This procedure is a U.S. national standard (ANSI/NCSL 1997) and is commonly referred to by the acronym GUM derived from its title *Guide to the Expression of Uncertainty in Measurement.* This terminology has also been adopted by the International Organization for Standardization (ISO) and by national laboratories including the National Institute of Standards and Technology (NIST; Taylor and Kuyatt 1994).

This chapter elucidates the sources of uncertainty and discusses methods for quantifying CMM performance. This information leads to a discussion of various national and international standards regarding CMM performance evaluation. The chapter ends with a consideration of issues that users can undertake to increase their confidence in CMM measurements. In keeping with the spirit of this book, it should be an educational resource to the CMM user as opposed to an academic review of the literature; only references that are of direct utility to the typical users are cited.

9.1 MEASUREMENT ERROR AND UNCERTAINTY

Before discussing the factors that affect CMM measurements, a brief review of some fundamental concepts is presented to facilitate the topic. Figure 9.1 illustrates an important difference between measurement error and measurement uncertainty. The term *measurement error* means the difference between the value found by a measurement and the "true value," that is, the measured value minus the true value. For most measurements performed by CMMs, the true value, and hence the measurement error, is unknown. (If the true value were known, then a measurement would not be needed.)

In contrast, the special case of a CMM performance evaluation (sometimes called a CMM calibration) does evaluate measurement errors. In this case, well-calibrated dimensional artifacts, for example, gage blocks, are measured. The uncertainty of the calibrated artifact is usually sufficiently small so that the difference between the CMM measurement value and the calibrated value can be considered to be the measurement error. This is the principle of the "gage maker's rule," in which the uncertainty of the gage is to be ≤10% of typical workpiece tolerances. This allows the measurement error to be estimated by computing the difference between the measured and calibrated values with minimal concern for the uncertainty of the calibrated gage. Similarly, many national and international standards impose a constraint on the maximum allowable uncertainty of artifacts used in CMM evaluations. This often permits the uncertainty of the artifact calibration to be ignored. However, recent trends within the international standards community have explicitly taken the uncertainty of testing artifacts into account by reducing the CMM manufacturer's specification by this amount, as described in Section 9.3.2.3.

---

* Although not explicitly shown here, the concepts in this chapter are, for the most part, also applicable to non-Cartesian CMMs.

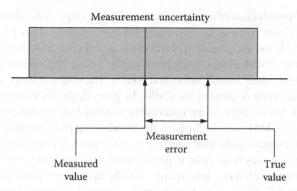

FIGURE 9.1 Illustration of the difference between measurement uncertainty and measurement error. In this case, the error is a negative value.

It is important to point out that even when using well-calibrated artifacts, a series of repeated CMM measurements will yield a series of (in general) different results and therefore different measurement errors. The collection of (all possible) measurement errors describes the measurement uncertainty. The measurement uncertainty is positioned about the measured value and is usually centered symmetrically as shown in Figure 9.1, unless there is a reason to assign greater uncertainty in one direction than in the other. Hence, when measuring a production workpiece, measurement uncertainty indicates one's ignorance of both the measurement error (since the true value is unknown) and highlights the variability of the measurement value (since repeated measurements yield different results).

A few comments regarding the difference between errors and uncertainty may be illustrative. An error is a quantitative statement of the difference between the measured and the true values. In contrast, an uncertainty statement expresses one's ignorance about the true value of a measurement. Errors are a signed value, that is, they may be either positive or negative depending on whether the measured value is larger or smaller than the true value. In contrast, an uncertainty statement is always a positive value. Errors are typically measured only during the special case of a calibration when calibrated artifacts (representing the true value) are measured. The accuracy of a measurement on an uncalibrated object, for example, a workpiece, is described using a measurement uncertainty statement.

The uncertainty of a measurement includes all factors that influence the result of the measurement. The estimation of the magnitude of the uncertainty sources and their effect on the measurement result is required for this task. This process may use many sources of information, including analysis of repeated measurements, data provided in calibration reports, uncertainties assigned to reference data taken from handbooks, manufacturer's specifications, and knowledge of behavior and property of relevant materials and instruments.

Many measurement results, such as the length of a gage block or the diameter of a ring gage, can be expressed as a single number. In these cases, a one-dimensional (1D) uncertainty region shown in Figure 9.1 adequately expresses the measurement uncertainty. Some measurement results, however, require several values to specify their results, for example, x, y, z coordinates of a point in space. This example has a three-dimensional (3D) uncertainty region about the measured point coordinate. This paradigm can be extended to include the uncertainty in the coordinates of each point within the measuring volume of the CMM. Hence, each point in the measuring volume can be viewed as having an "uncertainty cloud" associated with it. The size and shape of this cloud is determined by the error sources of the CMM. In general, these uncertainty clouds will not be spherical. For example, some uncertainty sources may be associated with a particular axis of the CMM, which elongates the cloud along that direction. Similarly, the size of the cloud (which can be characterized by a standard deviation method described in Section 9.1.1) will vary from point to point within the measuring zone.

A sensible model of these uncertainty clouds may view them as ellipsoids, as shown in Figure 9.2. When a probing point is recorded, a particular error vector is "precipitated" out of the cloud. This error vector is the difference between the coordinates the CMM records at a measurement point and the true coordinates in a perfect coordinate system. If the same physical point measurement is repeated, a slightly different (x, y, z) measurement coordinate will usually be recorded. Therefore, a different error vector will be precipitated from the cloud. The actual situation is more complex than the static cloud paradigm. The size and shape of the uncertainty cloud evolve in time as the CMM responds to such changes as the thermal environment, structural distortions due to workpiece loading, and wear or damage to the machine structure. In addition, the uncertainty depends not only on quasi-static errors, for example, those associated with the geometry of the CMM, but also depends on many measurement-specific (and user-selectable) factors such as the probe approach direction, probe approach velocity, and probing force. Hence, the measurement uncertainty is not completely determined until the user actually specifies all details of the measurement.

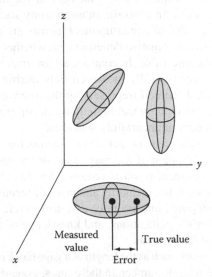

FIGURE 9.2 Schematic depicting the three-dimensional (3D) uncertainty in the coordinates of a few points in the CMM work zone.

9.1.1 COMBINED STANDARD UNCERTAINTY

So far, the term *measurement uncertainty* has been used rather loosely. The GUM terminology, however, has a precise definition that is applicable to all measured quantities, including dimensional measurements. The GUM approach specifically addresses measurement uncertainty and assumes that there is no known measurement bias (i.e., systematic error). This requires that the measurement be corrected for all known error sources, leaving only uncertainty sources that combine to produce the measurement uncertainty. If this is not the case, then the measurement result must include a statement of this bias, in addition to the measurement uncertainty. Note that the result of combining a bias with an uncertainty is no longer a true GUM uncertainty quantity and cannot be stated as such on calibration reports; however, this hybrid quantity can be used in workpiece acceptance/rejection decisions in which guard banding is employed (ASME 2005b). Measurement uncertainty and measurement bias are discussed in the literature (Phillips, Eberhardt, and Parry 1997).

The result of an uncertainty evaluation is the combined standard uncertainty, denoted u_c. This can be thought of as representing one standard deviation of the measurement uncertainty resulting from combining all known sources of uncertainties in a root sum of squares (RSS) manner, described later by Equation 9.2. While the details of an uncertainty evaluation are beyond the scope of this chapter (see Taylor and Kuyatt 1994; ANSI/NCSL 1997; ASME 2007; JCGM 2008a; and Chapter 14, Measurement Uncertainty for Coordinate Measuring Systems for details), a few general concepts are considered to provide a flavor of the evaluation process.

Central to the GUM method is that each uncertainty source is quantified by assigning it a variance, the square root of the variance is the standard deviation of that uncertainty source. This is denoted by $u(x_i)$, where the quantity x_i is the ith source of uncertainty. The sources of uncertainty, as represented by their standard deviations, include the calculation of the standard deviation of a quantity based on a series of repeated measurements using statistical techniques; this is called a type A uncertainty evaluation in the GUM approach. In addition, uncertainties that are not associated with a series of observations are assigned a variance; the square root of this variance is called a type B uncertainty evaluation in the GUM terminology. Type B evaluations are usually based on expert judgments or reference documents. The classification of uncertainty into categories A and B is based on the method of assessing the uncertainty source and not on whether the source is considered, "random" or "systematic."

As an example of these two types of uncertainty, consider the following case. Suppose the coefficient of thermal expansion (CTE), denoted α, is required in order to correct a measurement for thermal expansion. In one situation, the manufacturer of a given type of steel may have made a series of experimental measurements of the CTE and produced a histogram of the results, as shown in Figure 9.3a. The standard deviation of this data is easily computed; this is a type A evaluation of uncertainty. Alternatively, the manufacturer may have specified that $\alpha = 11.5 \pm 1 \times 10^{-6}/°C$, based on its prior knowledge of steel CTE values. A model for this uncertainty source could be a uniform distribution of CTE values of total width $2 \times 10^{-6}/°C$ centered about $11.5 \times 10^{-6}/°C$, as shown in Figure 9.3b. (A uniform distribution is selected because no additional knowledge of where the actual value might lie is available.)

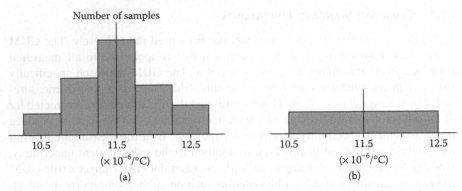

FIGURE 9.3 Example of uncertainty in the CTE of steel. (a) Experimental measurements allow a standard deviation to be calculated (type A evaluation). (b) Modeled uncertainty distribution allows a standard deviation to be calculated (type B evaluation).

The standard deviation of a uniform distribution is a well-known statistical quantity (Taylor and Kuyatt 1994; ANSI/NCSL 1997; ASME 2007; JCGM 2008a). This type of calculation is known as a type B evaluation of uncertainty and in the above example $u(\alpha) = 0.58 \times 10^{-6}/°C$. The choice of method is usually dictated by the type of information available regarding the uncertainty source.

The combined standard uncertainty u_c can be found by the law of propagation of uncertainty. The quantity being measured, for example, the length of a workpiece, is known as the measurand Y and has N different sources of uncertainty X_i ($i = 1-N$), as shown in Equation 9.1. A measurement of the quantity Y yields a specific measurement result y. The combined standard uncertainty $u_c(y)$ of this measurement, using the best estimated values of the uncertainty sources associated with each x_i, is given by the positive square root of its variance u_c^2, as described by Equations 9.2 and 9.3.

$$Y = f(X_1, X_2, ..., X_N) \tag{9.1}$$

$$u_c^2 = \sum_{i=1}^{N}\left(\frac{\partial f}{\partial x_i}\right)^2 u^2(x_i) + 2\sum_{i=1}^{N-1}\sum_{j=i+1}^{N}\frac{\partial f}{\partial x_i}\frac{\partial f}{\partial x_j}u(x_i, x_j) \tag{9.2}$$

$$u_c^2 = \sum_{i=1}^{N}\left(\frac{\partial f}{\partial x_i}\right)^2 u^2(x_i) + 2\sum_{i=1}^{N-1}\sum_{j=i+1}^{N}\frac{\partial f}{\partial x_i}\frac{\partial f}{\partial x_j}u(x_i)u(x_j)r(x_i, x_j) \tag{9.3}$$

The first term in Equation 9.3 (loosely speaking) states that the combined uncertainty (characterized by its variance u_c^2) is the summation of each individual uncertainty source (characterized by its variance $u^2(x_i)$), multiplied by the square of a quantity known as the sensitivity coefficient (denoted $\partial f/\partial x_i$) whose magnitude describes the importance of the uncertainty source relative to the total measurement uncertainty. For example, the importance of a large uncertainty in the temperature of a workpiece is more significant for a workpiece made of aluminum (which has a large CTE) than for a workpiece made of invar (which has a low CTE). This effect would be reflected in different sensitivity coefficients associated with the uncertainty of the workpiece temperature in the two cases.

The quantity $u(x_i, x_j)$ in Equation 9.2 is the covariance of x_i and x_j and is a measure of the correlation between these two uncertainty sources. An equivalent form of the combined standard uncertainty is given in Equation 9.3 in which the covariance is expressed as the product of the standard uncertainties and the correlation coefficient $r(x_i, x_j)$ (which ranges from −1 to 1).

The following example should help clarify the concept of correlated uncertainties. Suppose a gage block stack is constructed from three equal length gage blocks wrung together and each of these three blocks was calibrated using the same master gage block (mgb). Suppose further that the primary source of uncertainty of each of the three blocks is from the uncertainty in the length of the master gage block, whose variance is denoted $u^2(mgb)$. That is, the uncertainty in the transfer of length from the master to the test block is relatively small compared to the uncertainty in the master. An error in the master block will transfer one-to-one to an error in the test block, hence the sensitivity coefficient is unity ($\partial f / \partial mgb = 1$). What is the uncertainty of the final gage? Because each of the three blocks is calibrated against the *same* master, the error in the master block is transferred to each of the three test blocks. Consequently, these errors add directly, and the uncertainty of the three blocks is highly correlated (correlation coefficient = 1), so $u(x_i, x_j) \approx u^2 (mgb)$. From Equation 9.2, the combined standard uncertainty of the gage in this case is $u_c = 3\,u(mgb)$.

Alternatively, if the three gage blocks had been calibrated using three *different* masters, then the uncertainties of each block would be uncorrelated, $u(x_i, x_j) = 0$ and then $u_c = \sqrt{3}\,u(mgb)$. In this example, the effect of correlated errors nearly doubles the combined standard uncertainty compared to the uncorrelated case. In Section 9.3.4 (Equation 9.10), an example of correlated errors reducing the measurement uncertainty is presented. If the sources of uncertainty are all uncorrelated, then all the covariances are zero, and the second term of Equations 9.2 and 9.3 can be neglected.

As a very simple example of determining u_c, consider measuring the distance between two points carried out using a perfect measuring machine on a workpiece whose uncertainty in length is affected only by the workpiece temperature. A simple model of the measurement is given by Equation 9.4, where L_0 is the best estimate of the workpiece length after corrections of the measured length (L_m) for all systematic effects. In this simple model, the only systematic effect is the expansion of the length due to the workpiece temperature (T) at the time of the measurement, which might be different from the reference temperature ($T_0 = 20°C$) for a workpiece with thermal expansion coefficient (α, also called CTE). The quantity $L_m \alpha (T - T_0)$ is the systematic error due to the thermal growth, and it is assumed this is corrected by the CMM's thermal compensation system using the nominal values of T and α.

$$L_0 = \frac{L_m}{(1 + \alpha(T - T_0))} \approx L_m(1 - \alpha(T - T_0)) \tag{9.4}$$

Following the GUM-recommended procedures, that is, Equation 9.2, for expressing the standard uncertainty, u_c, for uncorrelated input variables (α and T) yields Equation 9.5, where $u^2(\alpha)$ is the variance characterizing the uncertainty in the thermal expansion coefficient and $u^2(T)$ is the variance characterizing the uncertainty of the temperature measurement. (Note in this highly simplified example, a perfect measuring machine implies $u^2(L_m) = 0$; also the fact $\partial T = \partial(T - T_0)$ is used.) Equation 9.5, which can be evaluated using Equation 9.4, yields Equation 9.6, which usually can be simplified to Equation 9.7.

$$u_c = \sqrt{\left(\frac{\partial L_o}{\partial \alpha}\right)^2 u^2(\alpha) + \left(\frac{\partial L_o}{\partial T}\right)^2 u^2(T)} \tag{9.5}$$

$$u_c = \frac{L_m}{(1 + \alpha(T - T_0))^2}\sqrt{(T - T_0)^2 \delta\alpha^2 + \alpha^2 \delta T^2} \tag{9.6}$$

$$u_c = L_m\sqrt{(T - T_0)^2 u^2(\alpha) + \alpha^2 u^2(T)} \text{ provided } \alpha(T - T_0) \ll 1 \tag{9.7}$$

Consider this example for the case of a 1-m-long steel workpiece, having a CTE of $11.5 \pm 1 \times 10^{-6}/°C$ (hence using a uniform distribution $u(\alpha) = 0.58 \times 10^{-6}/°C$), measured at a variety of temperatures by a CMM using a temperature-sensing and compensation system having a combined standard uncertainty of 0.1°C. Figure 9.4 shows the combined standard uncertainty over the temperature range of 15°C–30°C.

9.1.2 EXPANDED UNCERTAINTY AND LEVEL OF CONFIDENCE

The concept of measurement uncertainty is incomplete without an associated level of confidence. From a quality assurance standpoint, most manufacturers are interested in the fraction of defective workpieces they are producing, for example, the number of defective workpieces per million. This implies a confidence interval for the measurand (the specific quantity subject to measurement, such as the length of a workpiece). A given value of measurement uncertainty corresponds to some level of confidence in the measurement result.

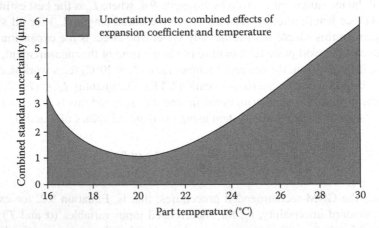

FIGURE 9.4 Combined standard uncertainty decomposed into temperature measurement uncertainty and coefficient of thermal expansion uncertainty for a 1-m-long steel workpiece (using $\alpha = 11.5 \times 10^{-6}/°C$, $u(\alpha) = 0.58 \times 10^{-6}/°C$, and $u(T) = 0.1°C$).

For example, the measurement uncertainty associated with a specific measurement having a 90% confidence level implies statistically that a measurement has a nine-out-of-ten chance of differing from the true value by no more than the measurement uncertainty. This value of measurement uncertainty will be smaller than that assigned if a 99% confidence level is used. A schematic of the relationship between measurement confidence and measurement uncertainty is shown in Figure 9.5.

The term *expanded uncertainty*, as recommended by the GUM, is used to describe an interval in which the measurement can be expected to lie with a specified level of confidence. Equation 9.8 describes the connection between the expanded uncertainty at a level of confidence and the combined standard uncertainty (computed using Equation 9.2). The coverage factor, k, is a positive real number whose value is selected so that Equation 9.8 is true at a desired level of confidence.

$$U = k \times u_c \qquad (9.8)$$

$$y - U \leq Y \leq y + U \qquad (9.9)$$

For example, if u_c is assumed to represent a normal (Gaussian) uncertainty distribution, then a 99.7% confidence level that the true value lies in the uncertainty interval (shown in Figure 9.5) corresponds to using an expanded uncertainty obtained with $k = 3$ in Equation 9.8, that is, three standard deviations. An uncertainty interval (of total width $2U$) can be expressed mathematically, as in Equation 9.9, where y is the result of a measurement and U is the expanded measurement uncertainty for the particular level of confidence (which might be 90%, 95%, 99%, etc.), and Y is the measurand (the quantity of interest). By convention, most expanded uncertainties are reported using a coverage factor of 2, that is, $k = 2$.

As an example of the relative magnitude of the expanded uncertainty and the uncorrected thermal expansion error (i.e., measurement bias), recall the previous thermal expansion problem having a combined standard uncertainty shown in Figure 9.4. An expanded uncertainty (using a coverage factor of 2) is shown in Figure 9.6

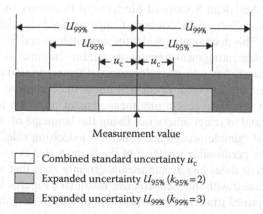

Measurement value

☐ Combined standard uncertainty u_c
▢ Expanded uncertainty $U_{95\%}$ ($k_{95\%}=2$)
▮ Expanded uncertainty $U_{99\%}$ ($k_{99\%}=3$)

FIGURE 9.5 Diagram illustrating the distinction between combined and expanded uncertainty; confidence levels shown assumes the uncertainty is Gaussian distributed.

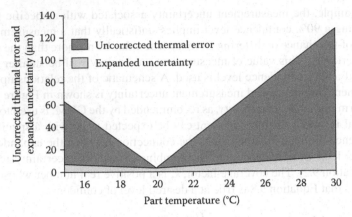

FIGURE 9.6 The expanded uncertainty ($k = 2$) and the absolute value of the measurement bias due to the thermal expansion of a 1-m-long steel workpiece as a function of the workpiece temperature.

together with the absolute value of the measurement bias that would occur if the measured workpiece length was not corrected for its thermal expansion.

9.1.3 IMPACT OF MEASUREMENT UNCERTAINTY

Measurement uncertainty has important economic consequences for calibration and measurement activities. In calibration reports, the magnitude of the uncertainty is often considered an indication of the quality of the laboratory results, and smaller uncertainty values are generally of higher values and of higher costs. In industrial dimensional inspection situations, excessive measurement uncertainty can lead to accepting out-of-specification workpieces or rejecting in-specification workpieces.

To communicate the role that measurement uncertainty plays in accepting or rejecting products, American Society of Mechanical Engineers (ASME) B89.7.3.1, *Guidelines to Decision Rules in Determining Conformance to Specifications* (ASME 2001b) was created. This document provides the terminology and basic decision rules associated with implementing measurement uncertainty in dimensional inspections. For example, if a feature has a 16-μm tolerance and the associated expanded uncertainty is 2 μm, what measurement result will allow the feature to be accepted? A traditional decision rule is to accept any measurement result up to and including the tolerance limit and to reject otherwise. Using the language of ASME B89.7.3.1, this would be called "simple acceptance using a 4:1 decision rule." Here, the ratio 4:1 indicates that the permissible variation of the feature, that is, the tolerance zone of width 16 μm is four times larger than the uncertainty zone that includes possible "true" values associated with the measurement result (see Figures 9.1 and 9.5), that is, 4 μm (derived from ±2 μm about the measurement result). Note that measurement results that happen to be right at the specification limit, and hence will be accepted, will be out of specification 50% of the time. Consequently, if the measured feature was safety-critical, a more conservative approach would be to subtract off some fraction of the measurement uncertainty from the specification limits so that a smaller acceptance zone is formed that has a higher confidence of accepting workpieces that are in the specification limit. Assuming 100% of the expanded uncertainty was subtracted from the specification limits, this would be called "stringent acceptance with a 100% guard band" when using the language of B89.7.3.1.

ASME B89.7.3.2, *Guidelines for the Evaluation of Dimensional Measurement Uncertainty* (ASME 2007), provides a simplified approach (relative to the GUM) of the evaluation of dimensional measurement uncertainty. Since measurement uncertainty has economic implications, B89.7.3.2 is designed to help practitioners understand the GUM and hence effectively evaluate measurement uncertainty for dimensional inspections.

ASME B89.7.3.3, *Guidelines for Assessing the Reliability of Dimensional Measurement Uncertainty Statements in Determining Conformance to Specifications* (ASME 2002b), examines resolving disagreements over the magnitude of the measurement uncertainty statement. Since a customer may have a contract to only purchase workpieces that have been inspected using a stringent acceptance decision rule, they might dispute the measurement uncertainty value determined by the inspectors. This standard is designed to help resolve this type of disagreement.

ASME B89.7.4.1, *Measurement Uncertainty and Conformance Testing: Risk Analysis* (ASME 2005b), provides guidance on the risks involved in any product acceptance/rejection decision. In particular, it describes how to establish the appropriate guard band and hence determine the decision rule to be used, based on the economics of the workpiece under inspection and the production and inspection capability.

Finally, ASME B89.7.5, *Metrological Traceability of Dimensional Measurements to the SI Unit of Length* (ASME 2006b), describes how measurement uncertainty plays a critical role in establishing measurement traceability to the meter.

9.2 OVERVIEW OF THE COORDINATE MEASURING MACHINE MEASUREMENT PROCESS

To understand how measurement uncertainty propagates through the CMM system, it is useful to discuss the measurement process. Ideally, a complete mathematical model of the measurement process, analogous to a very sophisticated version of Equation 9.1, would describe any measurements that can be performed on a CMM. Such a model would include all factors that affect the accuracy of the measurement result. By applying the law of the propagation of uncertainty (an effort analogous to Equation 9.2 but involving a very large number of variables), a combined standard uncertainty of the CMM measurement could be obtained. Such an effort would require a deep understanding of the sources of uncertainty and their impact on CMM measurements. This level of mathematical modeling is at the vanguard of CMM research and is briefly described in Section 9.3.4.

Many factors affect the uncertainty for an actual CMM measurement. To discuss the problem more clearly, the process can be decomposed into several categories of effects that interact, as shown in Figure 9.7. Effects such as those due to the operator

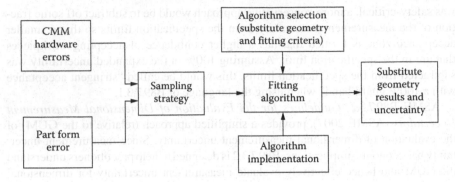

FIGURE 9.7 Schematic of the various factors affecting CMM measurements.

may be separated into these categories. For example, if the operator selects a different probe or stylus configuration, this affects the CMM hardware performance. If the operator uses a different number of probing points, this is a different sampling strategy. These issues are examined below in detail to provide a working knowledge of the various factors and interactions involved in CMM measurements. This information will provide the basis for understanding the various CMM performance evaluation methods discussed in Section 9.3.

A useful starting point in examining the CMM measurement process is to consider exactly what a CMM actually measures. Some CMM users criticize the CMM evaluation procedures specified by national or international standards because the geometric features of artifacts used in the evaluation procedures, for example, spheres on ball bars or small widely separated parallel planes on step gages, do not resemble the features of the user's workpieces and are therefore irrelevant to the particular situation. In actuality, CMMs never directly measure the geometry of workpieces, but only measure individual points in space. This dichotomy of the physical process of taking measurement points and the mathematical process of fitting these points to an ideal geometry is fundamental to understanding CMM measurements. An example of the independence of these two processes is shown in Figure 9.8, where the same set of points is fitted to three entirely different geometries. The geometry produced by the mathematical fitting process is known as the substitute geometry in contrast to the actual physical geometry of the workpiece.

9.2.1 SOFTWARE PERFORMANCE (ALGORITHM IMPLEMENTATION)

Once a series of measurement points has been obtained on a workpiece feature, this information is then analyzed by the CMM software to produce a geometric result, for example, the diameter of a least squares circle. The details of how the software produces a geometric result from the measurement points are usually hidden from the user, but there are some issues that the user must be aware of.

If the coordinates of the measurement points were saved (for example to disk) and then analyzed for the same substitute geometry using another CMM software package, one might expect that the answers would be identical. Unfortunately, this

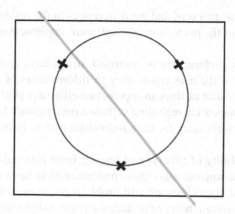

FIGURE 9.8   Three different geometries (circle, line, and plane) fit to the same three measurement points.

is not always the case. It is emphasized here that this discrepancy is not due to the inaccuracies of the CMM hardware (exactly the same point coordinates are used in both cases) but due to the implementation of the software and the computer (see also Chapter 8, Coordinate Measuring System Algorithms and Filters).

The details of the mathematical algorithm used to analyze the points may differ among the software packages because a particular mathematical algorithm is difficult to carry out and requires significant computer time to reach the solution. As an effort to reduce the required computing time, mathematical approximations are sometimes made to make the problem simpler. The degree to which these approximate solutions agree with the desired (mathematically exact) answer depends on both the nature of the approximation and the particular set of measurement points being analyzed. For some measurements, the approximations may work well, but the same approximations may fail for cases in which the measurement points are distributed differently. In addition, the software may contain coding errors, that is, bugs, which lead to spurious results. Furthermore, the implementation of the software on different types of computers may give different results due to factors such as the internal precision the computer uses to represent numbers. All these effects tend to produce a variety of answers between different CMM systems when given the same data to analyze (Wäldele et al. 1993; Hopp 1993; Shakarji 1998).

Fortunately, the metrology community is actively addressing these problems. It is now possible for CMM manufacturers to submit their algorithms for calibration. This process would involve the development (by a national laboratory, for example, NIST, or other calibration facility) of many carefully selected sets of point coordinates, representing different sampling strategies on all the standard substitute geometries (circle, plane, cylinder, etc.). Some of these sets of point coordinates represent very poor sampling strategies that test the software's capability to solve for the correct substitute geometry under these conditions. These data sets are evaluated by the calibration facility using known reference algorithms that have been carefully tested to ensure they produce mathematically correct results. The CMM manufacturers then evaluate the same data sets, and the differences in the results (if any) character-ize the performance of the manufacturers' software implemented on their computer systems.

Once the software performance is evaluated, this information can be distributed by the manufacturer to the user community to inform users of the accuracy of the software (provided the user utilizes an appropriate computer platform). Hence, CMM users may seek documentation regarding software performance from the CMM man-ufacturer, and it is unnecessary for each individual user to have his or her software tested individually.

Due to the availability of this type of testing from national laboratories, errors associated with least squares algorithm implementation have been largely elimi-nated. Other types of algorithms are still problematic because of their complex fit-ting criteria and the unavailability of (nonleast squares) algorithm testing at this time (Shakarji 1998).

9.2.2 FITTING CRITERIA (ALGORITHM SELECTION)

Section 9.2.1 discussed accuracy issues of the CMM software relative to the math-ematical correct answer for that algorithm. In this section, it is assumed that the software converges to the correct answer and address the issue of selecting the fitting algorithm (used in the point fitting process). Many CMM users have access only to fitting algorithms of the least squares type. The least squares algorithm determines the values of the fitting parameters (e.g., the diameter and center location of a circle) of the substitute geometry such that it minimizes the sum of the square of the residu-als to the fit.

The residuals are the distances from each of the measured points to the substitute geometry, as shown in Figure 9.9. In addition to the least squares algorithm, several other types of fitting criteria that yield useful geometric information exist. Figure 9.9

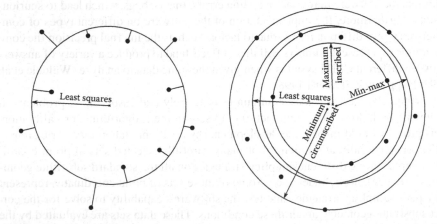

FIGURE 9.9 Left: A least squares circle fit of 10 points taken on a circular feature having form error, showing the residuals to the fit. Right: Four different fitting criteria used to ana-lyze the same measurement points (residuals not shown).

illustrates four different fits for a measurement of a circular feature using the same 10 measurement points. The maximum inscribed circle and minimum circumscribed circle fits are each determined by three extreme points of the measurement data. The mini-max (also called minimum zone or Chebyshev best fit) criterion minimizes the maximum residual of the fit. Consequently, this algorithm always has the largest interior residual of the circle equal in magnitude to the largest exterior residual of the circle.

These different fitting criteria produce significantly different results when measuring a geometrically imperfect workpiece, that is, a workpiece with form error. Furthermore, in the least squares and mini-max circles, a nominally round plug may not physically fit into a hole even though the computed diameters indicate it should. This occurs because these two algorithms always have points (corresponding to the workpiece material, assuming the points coordinates are accurate) outside the diameter of the plug and inside the diameter of the hole. It should be clear that this "problem" is a result of misinterpreting the software result with regard to the workpiece's intended function. Consequently, when inspecting close-fitting mating workpieces using a CMM with a least squares algorithm, it is advisable to examine the residuals of the fit in addition to the computed size of the feature. If the residuals are not available, an upper bound on their magnitude can be obtained by examining the computed form of the feature.

Alternatively, if using an algorithm that computes the minimum circumscribed diameter of a plug determines it to be smaller than the maximum inscribed diameter of a hole, the two workpieces have a better chance of fitting together. Even in this case, unless the measurement point spacing is dense, a "high point" on the workpiece may be missed (not probed) and consequently an interference fit may result. This is a sampling strategy problem, which is discussed in Section 9.2.3.

Similarly, features that have form tolerances, for example, flatness, give different results from the least squares algorithm versus the minimum zone algorithm. Hence, CMM users should be aware that different algorithms, using different fitting criteria, exist and have different geometric interpretations with regard to the function of the workpiece.

Least squares algorithms are ubiquitous in CMM software because they are easy to implement, robust with respect to data "outliers," have short computation times and are generally well-behaved mathematically. Unfortunately, they do not yield a substitute geometry that is fully compliant with Geometric Dimensioning and Tolerancing (GD&T), which is often specified on blueprints. The minimum zone algorithms are more consistent with this geometric interpretation. However, these algorithms are often harder to implement and are much more sensitive to outliers in the measurement points, for example, one bad point can radically change the result.

As a general rule of thumb, least squares algorithms are preferred when the measurement uncertainty from the CMM is significantly larger than the form error of the workpiece. Conversely, minimum zone algorithms are preferred when the workpiece form error is larger compared to the measurement uncertainty from the CMM. Regardless of the algorithm selected, they all must contend with the uncertainty sources shown in Figure 9.7.

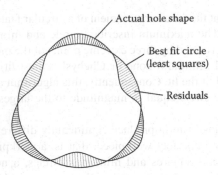

FIGURE 9.10 A circular hole having a three-lobed shape with the best fit least squares circle substitute geometry found using an infinite number of sampling points (analytic result).

A second issue concerns which geometry is used in the fitting process. Figure 9.8 shows three different geometries all fit to the same points using the least squares criterion. Without additional information, it is unclear which of these substitute geometries represents the desired result. This additional information comes from the design specification of the feature, for example, from the workpiece drawing. In Figure 9.8, these three substitute geometries are so radically different that only one of them can represent the intended result, and mistakenly fitting the points to the other geometries can only be considered to be a blunder by the user. In other situations, the issue is not so clearly determined.

Consider the three-lobed circular hole and substitute geometry shown in Figure 9.10. Even though the substitute geometry shown is the correct least squares circle (correct in the sense that the sum of the squares of the residuals is minimized, a result that can be calculated analytically in this case), it clearly represents an approximation to the actual hole shape. If the hole is intended to be circular, then the residuals are considered to be a manufacturing error of the workpiece. If the hole is intended to have a three-lobed shape, then the residuals represent a measurement error as a consequence of not using the proper geometry in the fitting process. If a three-lobed circular geometry had been used as the fitting algorithm, then a much better fit could be obtained; whereas using a pure circle geometry will always represent an approximation to the shape of the three-lobed hole.

For real workpieces, there will always be a difference between the manufactured geometry and the substitute geometry when the workpiece is densely sampled. That is, residuals are both an expected and a natural aspect of the measurement process. The number and magnitude of the residuals are strongly influenced by the sampling strategy, which is discussed in Section 9.2.3.

9.2.3 Sampling Strategy

The effect of selecting a measurement sampling strategy can be a major contributor to the measurement uncertainty (Weckenmann, Heinrichowski, and Mordhorst 1991; Hocken, Raja, and Babu 1993). The sampling strategy is the number and position of the measured points on the workpiece surface. To discuss the impact of the sampling strategy on the substitute geometry, a careful consideration of what constitutes the correct CMM measurement result is useful.

The correct answer to a CMM measurement is obtained if there is zero CMM measurement uncertainty, that is, using a perfect CMM and if the workpiece is measured using an infinite number of points distributed over the entire workpiece feature. In this case, all the information regarding the geometry of the workpiece feature is known (by taking a very large number of points over its entire surface), and this information is uncorrupted (because there is no measurement uncertainty). Consequently, the substitute geometry is the mathematically correct answer and has zero uncertainty, assuming that the substitute geometry and its corresponding algorithm accurately represent the GD&T specification.

Even in the case of a perfect CMM and an infinite number of sampling points, the substitute geometry will still have residuals associated with it, as shown in Figure 9.10. This occurs because a real-world workpiece has form errors resulting from the manufacturing process. In this case, the residuals are a consequence of the imperfect workpiece, not an imperfect measurement. Despite the presence of residuals, the measurement result is the unique and the correct answer. Hence, in the perfect measurement, infinite sampling case, the residuals to the substitute geometry do not represent uncertainty.

When employing a perfect CMM and a sampling strategy using a finite number of measurement points (as is the case in all real measurements), the aforementioned statement is weakened. In this case, the existence of fitting residuals only indicates that the substitute geometry and actual (as manufactured) workpiece geometry are not identical. A perfect CMM with a finite sampling strategy does not ensure that the mathematically correct answer will be obtained. For example, select five randomly located probing points on the actual workpiece surface shown in Figure 9.10 and visualize the corresponding result (circle center and radius). Repeating this process with randomly located points will, in general, yield different results. Each of these results is unlikely to be the correct answer shown in Figure 9.10 (and is found using an infinite number of points). Hence, finite sampling contributes to the measurement uncertainty through two interrelated effects: (1) incomplete knowledge of the workpiece geometry and (2) the sensitivity of the algorithm workpiece form errors. (More generally, this sensitivity also extends to CMM measurement errors when considering an imperfect CMM.)

9.2.3.1 Workpiece Geometry

All manufactured workpieces have some amount of geometric imperfections, such as form error, which deviate from the ideal geometry. In addition, when the workpiece is placed on the CMM, the fixturing process could add further distortions. Since most CMM users employ a relatively small number of points to determine the workpiece geometry, that is, to compute the substitute geometry, it must be recognized that this result is based on incomplete information of the workpiece. For example, consider a perfect CMM, using a perfect algorithm, which measures a workpiece having some form error. If the workpiece is now remeasured using a different sampling strategy (or the same sampling strategy with a change in the relative orientation between the workpiece and the sampling strategy), a different result will be obtained.

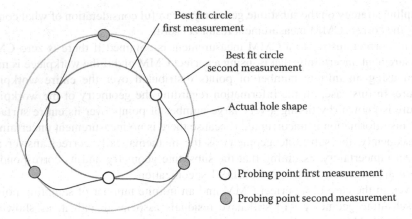

FIGURE 9.11 Two measurements of a three-lobed circle using the same three-point (120° spaced) sampling strategy with different orientations to the workpiece, producing two different substitute geometry results.

Figure 9.11 presents the results of two measurements—using the same sampling strategy—of a circular feature having a three-lobed form error. In this example, each measurement result produces a best fit to the substitute geometry. (In particular, because this sampling strategy uses only three points, the fit is exact, that is, the fitted circle passes through each of the three measured points). The variation in the results is a consequence of different (and incomplete) workpiece information being supplied to the fitting algorithm due to the finite number of measured points.

The form error of a workpiece plays an abstruse role as a source of measurement error. As discussed above and shown in Figure 9.10, a workpiece with form error does have a unique correct answer (with no uncertainty) to a particular CMM measurement (this would require measuring the workpiece with an infinite number of points). However, because the substitute geometry is not identical to the actual geometry, there will exist residuals to this fit, that is, the mathematically correct fit will, in general, have residuals present. When a sampling strategy using a finite number of measurement points is employed, these residuals become a source of measurement uncertainty. For example, a three-point (120° spaced) sampling strategy shown in Figure 9.11 can produce a continuum of measured circle diameters ranging from the maximum inscribed circle diameter to the minimum circumscribed circle diameter. Hence, when using finite sampling strategies (even with a perfect CMM), workpiece form errors are a source of measurement uncertainty.

9.2.3.2 Sensitivity to Measurement and Workpiece Errors

Some sampling strategies can greatly amplify the error introduced at each measurement point, whereas other sampling strategies can reduce this error. This is true whether the error is due to the CMM, that is, the measurement system, or due to the form error of the workpiece (in the latter case, the workpiece does not correspond to the ideal geometry). Since the CMM user selects the sampling strategy, different users of the same CMM, measuring the same workpiece feature, may produce very different measurement results, and these results may have very different

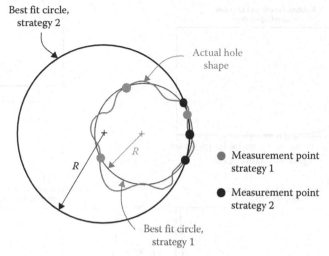

FIGURE 9.12 Two different three-point sampling strategies on a nominally circular hole. Note the substitute geometry perfectly fits the points in both cases.

uncertainties associated with them. Figure 9.12 dramatically illustrates this situation, where sampling strategy 2 has a larger measurement uncertainty than strategy 1. In this example, the workpiece is imperfect, that is, having form error, whereas the CMM is considered perfect, that is, having no uncertainty in the point coordinates.

The sampling strategy can be considered a sensitivity coefficient (described in Section 9.1.2) to the measurement process. A poor sampling strategy corresponds to a large sensitivity coefficient, which can amplify errors (including those associated with incomplete workpiece information) leading to large uncertainties in the computed result. Such is the case of strategy 2 in Figure 9.12.

A more quantitative description of the sensitivity coefficient interpretation of a sampling strategy is given in Figure 9.13. In this example, the workpiece is considered perfect, that is, having no form error, while the CMM is imperfect, that is, having measurement uncertainty in the point coordinates.* This figure displays the variation in the radius and (X, Y) center coordinates of a circle measured using three points, as a function of the angular separation between the points. To each of the three measurement points, a set of uncorrelated random radial errors, characterized by a 1-μm standard deviation, is assigned. (A 1-μm standard deviation is chosen so that the variation, that is, standard deviation, of the computed results can be compared to the input uncertainty and interpreted directly as a sensitivity coefficient).

As shown in Figure 9.13, when each of the three points is separated by 120°, the standard deviation of the radius is small (0.5 μm corresponding to a sensitivity coefficient of ½). As the points become more closely grouped together, the variation in the result increases, for example, at 10°-point spacing, the standard deviation of the

---

* Alternatively, the exact same results will be computed if the CMM is perfect and the workpiece has a random form error of 1-μm radial magnitude. In this case, the workpiece is considered to be measured in numerous different orientations yielding the distribution in results.

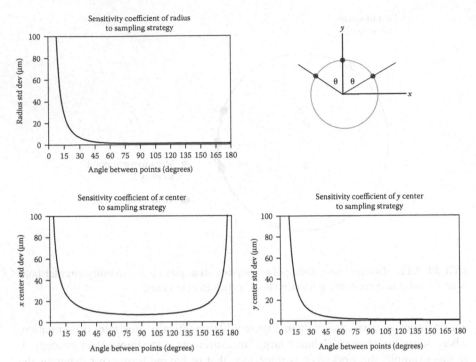

FIGURE 9.13   The variation (as characterized by its standard deviation) in radius and x, y center coordinates of a 10-mm-radius circle, due to measurement errors (having a 1-μm radial standard deviation) at each of the sampling points, for sampling strategies consisting of three points, differing by the angular separation between the points. The expanded uncertainty for each of the three quantities can be found by multiplying the standard deviation by 2 (corresponding to $k = 2$).

radius is 80 μm. (The nature of this singularity varies as $(1 - \cos\theta)^{-2}$ where θ is the angular spacing between the points; see Appendix A for an analytic formulation of these curves.) In this example, the uncertainty in the radius may be orders of magnitude larger than the uncertainty in the individual point coordinates if the sampling strategy consists of points closely grouped together.

The sampling strategy issue is one of the factors that makes the prediction of CMM measurement uncertainty difficult. Even if the uncertainty due to the CMM is known at each measurement point in the CMM work zone, and even if the form error of the workpiece is known exactly, the uncertainty of the measurement result, for example, the diameter of a circle, is unknown until a particular sampling strategy is specified for the measurement.

Figure 9.14 illustrates the effect that the sampling strategy has on determining the best fit center location for a circle using a least squares fitting algorithm. The figure depicts a circular feature having an elliptical form error (which has been greatly amplified for clarity) of 10 μm out-of-roundness, which is measured with three different sampling strategies. If the workpiece is randomly rotated (i.e., the major axis of the ellipse is randomly rotated with respect to the points of the sampling strategy),

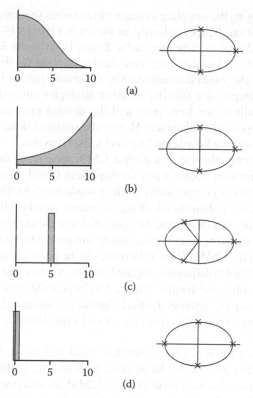

0 5 10

(a)

0 5 10

(b)

0 5 10

(c)

0 5 10

(d)

FIGURE 9.14 Distribution of distances between the best fit center location and the true center position (given by the infinite number of sampling points case) for a circle having a 10-μm elliptical form error. (a) A naive and incorrect guess of a normal distribution for a three-point (90° spaced) sampling strategy. (b) The correct distribution for a three-point (90° spaced) sampling strategy. (c) A three-point (120° spaced) sampling strategy. (Note that the correct center is never found). (d) A four-point (90° spaced) sampling strategy that always locates the correct center.

a different center location may be calculated. Randomly rotating the workpiece a large number of times creates a distribution of center locations. Figure 9.14 displays the distance between the center found using the finite sampling strategy and that of the correct answer (found using an infinite number of points), which is located at the center of symmetry of the ellipse. This procedure creates the distribution of possible errors in the center location associated with relative orientation of a workpiece having elliptical form error and using a finite sampling strategy.

The interaction between the sampling strategy and the form error of the workpiece can be counterintuitive. Figure 9.14a depicts what one might expect at first guess for a three-point sampling strategy. Namely, the correct answer would be most frequently found, and as the magnitude of the error increases, occurrence of this error decreases. This is (half of) the bell-shaped (Gaussian) curve commonly used in statistics. In reality, the distribution of center locations displays the opposite characteristics, as shown in Figure 9.14b, which shows the actual distribution for this case.

By slightly changing the sampling strategy (three points separated by 120° instead of 90°), the distribution changes radically, as shown in Figure 9.14c. In this case, the center location is always in error by exactly 5 μm. Finally, the four-point strategy of Figure 9.14d always produces the correct center location for this particular form error regardless of the workpiece orientation. (It should not be inferred that four-point sampling strategies are superior to other strategies since the results depend heavily on the details of the form error and the desired measurand, for example, center location as opposed to diameter.) Hence, the statistics describing these uncertainty distributions cannot be assumed a priori to be Gaussian.

Figure 9.14 describes the use of a perfect CMM measuring an imperfect workpiece. A similar situation occurs when an imperfect CMM (having measurement uncertainty) measures a geometrically perfect workpiece (having no form error). This will also produce a distribution of measurement results. This occurs because each time the workpiece is measured, the coordinates of the measurement points will usually be slightly different from the preceding measurements due to the measurement uncertainty of the CMM. These differences in the point coordinates of repeated measurements will lead to different computed results. As shown in Figure 9.13, small differences in the point coordinates may lead to large differences in the computed results if a poor sampling strategy is used. Again, the sampling strategy plays the role of the sensitivity coefficient amplifying (or reducing) the effects of measurement uncertainty.

In real workpiece measurements, both the CMM and the workpiece are imperfect. The fitting algorithm cannot know (and does not care) whether the residuals to the fit are due to workpiece form error or CMM measurement error. Both are sources of uncertainty when finite sampling strategies are employed. Consequently, the effect of the sampling strategy must be accounted for when determining the combined standard uncertainty of a CMM measurement.

9.2.4 COORDINATE MEASURING MACHINE HARDWARE PERFORMANCE

The physical structure of a CMM is commonly identified as a source of measurement uncertainty. A CMM is composed of numerous components assembled to provide a typical accuracy specification of 25 μm or less and is designed for a multitude of different geometric measurements in an industrial environment. Although there are many different CMM designs, all CMMs seek to obtain accurate point coordinates taken on the surface of the workpiece. Since most CMMs typically record the (x, y, z) location of the probe, usually at the center of the stylus ball, additional information is needed to determine the coordinates on the workpiece surface. Usually, this involves knowledge of the effective size of the probe, for example, size of the stylus ball and the probe velocity vector, which describes the direction toward which the probe was moving at the time of the measurement. Consequently, CMM hardware performance is determined by the uncertainties in the (x, y, z) coordinate of the stylus ball center, the stylus velocity vector, and the effective size of the stylus at the measurement point. For the purposes of discussion, CMM hardware issues will be separated into those that are associated with the CMM structure and those that are associated with the probing system.

9.2.4.1 Coordinate Measuring Machine Geometry

9.2.4.1.1 Rigid-Body Errors

For most CMMs, the machine frame, composed of three orthogonal linear axes, generates the geometry of a Cartesian coordinate system. This system provides pure one-dimensional motion along each of the axes, and these axes are to be perpendicular to one another. In reality, a carriage moving over a linear guideway will not move purely in one dimension but will undergo extraneous motion. This motion is described by rigid-body kinematics, in which the underlying assumption is that the moving carriage remains rigid during its motion, that is, it does not bend or otherwise distort. Based on this assumption, a complete description of the motion of the carriage can be given by specifying its six degrees of freedom, which include three translational motions (scale and two mutually orthogonal straightness) and three rotational motions (roll, pitch, and yaw), as depicted in Figure 9.15a. Each

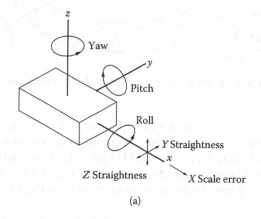

(a)

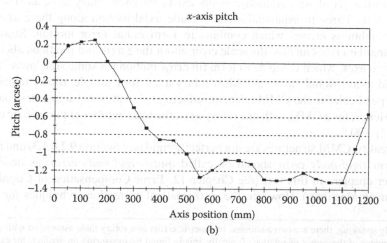

(b)

FIGURE 9.15 The mechanical errors of an axis in the rigid-body model. (a) Schematic of the individual error parameters for the *x* axis. (b) A typical plot of the pitch parameter for a carriage moving along the *x* axis.

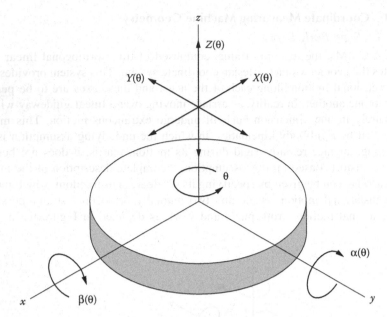

FIGURE 9.16   Schematic of the individual rigid-body error parameters for a rotary axis.

of these parameters is a function of the position of the carriage along the axis, as illustrated in Figure 9.15b. In addition, there are three out-of-squareness values in the axis pairs (X-Y, Z-X, Z-Y). Since a CMM has three independent axes, there are a total of 21 ($3 \times 6 + 3$) parameters that describe the CMM stylus location error in the rigid-body model. These errors are often referred to as the 21 parametric errors of the CMM.

A similar set of six parametric errors exists for each rotary axis, as shown in Figure 9.16. Three translational errors include axial motion along the z axis and two straightness errors, which combine to form radial error motion. Similarly, three angular errors include the scale error about the z axis and rotations about the other two axes, which combine to form tilt error motion (in some standards, this is referred to as "wobble motion"). Most rotary axes, for example, rotary tables, are set perpendicular to the CMM table, introducing two additional squareness parameters. Hence, for a CMM with one rotary axis, the total number of parametric errors is 29 ($21 + 8$).*

In reality, CMM structures are not perfectly rigid (see Section 9.2.4.1.3) and hence the term *rigid-body error* should be called *quasi-rigid-body error*, as described in other chapters in this text (see Chapter 12, Error Compensation of Coordinate Measuring Machines). However, the mathematical correction techniques for these

---

* Strictly speaking, there are two additional parametric errors of a rotary table associated with the x, y coordinates of the center of rotation. Typically, this is found by measuring an artifact, for example, the CMM calibration sphere, in at least three different rotation positions and using a circle fit to locate the center of rotation. This is usually carried out by the CMM users and is somewhat analogous to establishing the datums for a part coordinate system.

mechanical imperfections are very powerful tools, and due to their history, they are still often described as rigid-body error corrections, even though quasi-rigid-body is a more appropriate term.

Many CMMs include an indexable probe head that is capable of finite (usually 7.5°) incremental rotations about two additional axes. In each probe head position used, the probe must be calibrated, and this determines the size and location of the stylus ball. This effectively calibrates out the errors in the probe head rotary axes provided the errors are repeatable. Additional discussion of errors associated with indexable probe heads is provided in the section on multiple styli; a detailed discussion of rotary axis errors can be found in ASME (2005a, 2010).

Historically, accuracy was achieved by mechanically reducing the rigid-body errors by carefully manufacturing and assembling the CMM structure. Since the 1980s, an alternative approach known as error mapping (or accuracy enhancement or computer-aided accuracy), which provides a correction for these errors, has been commercially available. In this method, each of the rigid-body errors is measured and is mathematically combined (with the use of a computer) to determine the positioning error of the CMM stylus at each point in the work zone. This error is then accounted for by either moving the CMM stylus to the correct position (active compensation), which is done in the real time during the measurement process, or mathematically correcting the measurement coordinates (passive compensation), which can be done after all the measurement points are taken. (Chapter 12, Error Compensation of Coordinate Measuring Machines in this text provides the mathematical details of how these parameters are combined to determine the appropriate correction.)

The mechanical structures within a CMM are equally mechanically stable whether or not the CMM is software error corrected. A change in the CMM structure, for example, a change of 2 arc seconds in the X-Y squareness, will cause exactly the same measurement error for the corrected or uncorrected CMM. Hence to ensure accuracy, all CMMs require geometric stability over all the operating conditions encountered during the measurement. Consequently, software error compensation should be viewed as accuracy obtained through measurement (of the rigid-body errors) followed by mathematical compensation, in contrast to accuracy obtained through measurement followed by mechanical adjustments.*

It is important to note that this correction accounts for the effects of rigid-body errors but does not physically remove these motions. Consequently, when carrying out a performance evaluation on a software error-corrected CMM, it is inappropriate to directly measure these rigid-body errors and use their magnitude as a measure of the performance of the CMM. For example, consider a software error-corrected CMM with an axis having a large pitch motion. If the pitch of

---

* Error compensation technology has substantial advantages for recalibration of CMMs in the field. It is usually easier (and more cost effective) to measure a rigid-body error, and then correct for the error using software, than to correct the error by mechanically eliminating it. Mechanically correcting a rigid-body error has the additional difficulty that the physical adjustment to correct for one error often affects other rigid-body errors. This may result in an iterative cycle requiring multiple measurements and corrections.

the CMM is measured (e.g., using an autocollimator or a laser interferometer), its large angular error will still be recorded because the source of the motion is the physical structure of the CMM, which is unchanged due to a mathematical correction. However, the effect of this pitch error on the (X, Y, Z) stylus location is accounted for by the error compensation; therefore, the measured pitch value will have little to do with the actual measuring accuracy of the CMM. This points out an important theme that will be repeated throughout this chapter: CMMs are measuring machines and should be evaluated based on their measuring performance.

Whether or not a particular CMM employs software error compensation, the rigid-body model provides a useful paradigm for discussing CMM geometry. Many common problems associated with the CMM's geometry are angular in nature, that is, out-of-squareness, pitch, roll, or yaw of an axis. These errors typically increase in magnitude in direct proportion to the length of the measurement. For example, an out-of-squareness error of 10 arc seconds can produce an error of 5 μm over a distance of 0.1 m, but it becomes an error of 50 μm over 1.0 m.

This illustrates a useful principle that the magnitude of (and hence the sensitivity to) angular errors increases with the length of the measurement. The orientation and position of the measurement are also important. Certain orientations can maximize the effect of geometry errors and hence allow them to be detected and quantified during an evaluation. As an example, consider the squareness problem shown in Figure 9.17. It is apparent that the measured length of the artifact in the square coordinate system (x_1, y_1) is longer than that in the out-of-square system (x_2, y_2). If the artifact is a known (calibrated) length, then this discrepancy appears as a measurement error. Even if the artifact length is unknown, this property can be exploited by measuring the same artifact in two "crossed" orientations, as shown in Figure 9.18. By this technique, the angular deviation from squareness (shown as α in Figure 9.18) can be determined in the absence of other errors.

$$L_1 = \sqrt{x_1^2 + y_1^2} \quad L_2 = \sqrt{x_2^2 + y_2^2}$$

$$\alpha = \frac{L_1 - L_2}{L_1 \cos\theta \, \sin\theta}$$

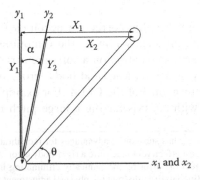

FIGURE 9.17 L_1 is the ball bar length measured in the square coordinate system; L_2 is the ball bar length measured in the out-of-square coordinate system and appears foreshortened.

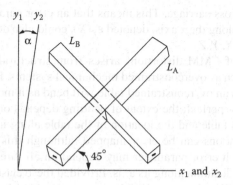

FIGURE 9.18 *X-Y* squareness (in the absence of other errors) can be estimated using the same artifact measured in two crossed positions.

$$\alpha = \frac{2(L_B - L_A)}{L_A + L_B} \text{ radians}$$

9.2.4.1.2 Uncertainty in Rigid-Body Errors

As described earlier, CMM geometry errors need to be measured whether or not a CMM is error corrected. This measurement process also has some uncertainty associated with it, for example, from the uncertainty in the measurements of the rigid-body error parameters. Note the distinction between measuring a mechanical error, for example, the pitch of the *x* axis, and the uncertainty due to the measurement of this value. The consequence of this uncertainty depends on the error mapping coordinate system, whereas the correction itself is independent of the mapping coordinate system. This occurs because the mapping technique uses a set of lines that are defined to have only transitional errors. Consequently, for a typical software-corrected CMM, the measurement uncertainty (in units of length, e.g., micrometers) increases in magnitude with distance from these lines, making the uncertainty dependent on the mapping coordinate system. Such uncertainties are also present in even the most mechanically accurate CMMs since either method of error correction (in software or in hardware, e.g., by scraping the axis ways) requires dimensional measurements, which introduce measurement uncertainty. (A mechanically corrected CMM may also provide corrections for other errors in addition to rigid-body effects.)

9.2.4.1.3 Structural Distortions

Regardless of the method by which the geometric accuracy of the CMM is achieved, this structure must maintain its integrity to retain its accuracy. Strictly speaking, rigid-body behavior means that the moving carriage, not the guideway, remains undistorted throughout its motion. Some deformations are relatively benign. For example, the deflection of the bridge as the ram carriage moves across it appears as an axis straightness (and possibly pitch) error and is easily compensated for by error mapping. Similarly, nonrigid effects may couple the properties of two axes together, for example, the pitch of the bridge as it moves along its axis may depend on the location of the cross carriage. This means that an error parameter, for example, pitch when moving along the x axis, denoted $\varepsilon_Y(X)$, could also depend on its Y and Z locations, that is, $\varepsilon_Y(X, Y, Z)$.

Another source of CMM distortions arises from structural deformations produced by sources such as overconstrained mechanical systems. For example, a moving-table CMM with an overconstrained table will bend as it moves along its axis if its guideways are not perfect; the extent of bending depends on both the measurement location on the table and the location of the table along its axis. In principle, such elastic deformations can be error mapped, although this is a more complex problem because each error parameter may depend on all three coordinate values instead solely on the location along its axis. Provided these elastic deformations are constant over time, they can, in principle, be compensated by the error mapping process.

Other deformations, such as those due to thermal gradients within the CMM structure, are usually transient and are therefore a time-dependent error source. These deformations may create rigid-body error behavior but the error parameters, for example, pitch, are continually changing in time, and mapping them is futile. (An exception to this is for CMMs that have incorporated into their structure sensors, which allow rapid self-mapping in quasi-real time.)

Most CMMs are specified to operate over a narrow temperature range; however, in practice they are often used outside these limits. The simplest part of this problem is the uniform expansion of the workpiece and the CMM scales. This can be treated by a linear correction as described by Equation 9.4 and is discussed more extensively in Chapter 10, Temperature Fundamentals.

A much more difficult issue is quantifying the actual geometric distortions of the CMM structure, which are highly specific to the details of its design, and the exact temperature distribution throughout its structure. Thermal gradients can develop even when the CMM is in a homogenous temperature environment, if the entire environment changes in time as a result of different thermal time constants of the various components within the CMM. It is widely recognized that thermal effects degrade CMM accuracy. Historically, this problem is addressed by the thermal error index method, which combines a thermal drift test and an estimate of the uncertainty in the linear CTE (ASME 1973; ISO 2003d). Unfortunately, this does not provide detailed knowledge of the geometric distortions of the CMM. Hence, the impact of these distortions on CMM measurements may be obtainable only by extensive experimental investigation (Lingard et al. 1991). An approach used to roughly quantify how thermally induced errors affect standardized CMM performance tests and how they are used to derate the specifications described in a recent ASME document (ASME 2008).

A thorough discussion of thermally induced CMM distortions, for example using finite element techniques, is beyond the scope of this chapter. However, these distortions play a major role in the uncertainty of CMM measurements. Although significant progress has been made in designing CMMs to minimize these effects, users must recognize that the same distortions are present in the workpiece under inspection.

Ideally, the workpiece temperature is both constant and uniform during the dimensional measurement. In many applications, this is not possible because production processes may significantly heat up the workpiece and insufficient time is available to allow the workpiece to reach thermal equilibrium. Consequently, the workpiece has a nonuniform, temporally changing temperature distribution. In particular, for low thermal conductivity materials, for example, plastic, the surface temperature may not accurately reflect the internal temperature distribution due to conduction, convection, and radiation coupling to the environment. In addition, stresses and deformations may develop due to a nonuniform temperature distribution throughout the workpiece. Hence, even a CMM designed to be thermally insensitive must contend with the thermally induced uncertainty in the workpiece.

Another significant source of structural deformation can arise from workpiece loading effects if heavy workpieces are inspected. The nature of the deformation depends on the type of CMM structure, the mass loaded, the location of the load, and the details of the workpiece mounting, for example, the number of loading points in contact with the table and the load at each point. Hence loading effects are difficult to predict. Since most CMM performance evaluations are conducted using very light loading, these deformations are often an overlooked and unquantified source of measurement uncertainty. Recent CMM performance evaluation standards (ASME 2008; ISO 2009a) specify the maximum loading allowed under the specified CMM accuracy. If heavy workpieces are inspected, workpiece loading effects should be investigated.

Table 9.1 illustrates the performance degradation resulting from mounting a substantial load (equivalent to that of a heavy steel workpiece occupying about two-thirds of the CMM work zone) on several different CMMs. The test involves measurements, both with and without the load present, of six different length measurement positions, or performing a rotary table test for CMMs. (Additional details of these performance tests are described in Section 9.3.2.) As shown in the table, the degree of performance degradation can vary considerably depending on the particular CMM. Consequently, an investigation of noncompensated structural distortions that introduce errors into the CMM coordinate system is needed to develop an effective uncertainty analysis.

TABLE 9.1
Examples of CMM Loading Effects

CMM Type	Degradation Due to Loading
Four-axis moving ram horizontal arm with rotary table	+54% rotary table test
Four-axis moving ram horizontal arm with rotary table	+66% rotary table test
Four-axis moving bridge with rotary table	+2% rotary table test
Three-axis moving bridge	+58% range of length errors
Three-axis moving bridge	+35% range of length errors

9.2.4.1.4 CMM Dynamic Errors

Machine dynamics create another class of CMM geometry errors. These errors may depend on factors such as probing speed, probing direction, probe approach distance, acceleration settings, and the amount of extension of the CMM axes. Figure 9.19a illustrates a dynamic error (primarily due to machine structure oscillations), which is highly dependent on the probe approach distance. This figure shows the variation in apparent position of a gaging surface as a function of probe approach distance. The CMM used a touch-trigger probe at a constant probe approach velocity of 8 mm/s. The CMM ram behaves like a damped pendulum, and different approach distances cause the probing point coordinates to be recorded at different positions of the ram oscillation. At large approach distances, the mechanical oscillations are damped out prior to the probe triggering.

Errors due to machine dynamics can produce unexpected results, as shown in Figure 9.19b, which depicts the measured length of a gage block as a function of the number of probing points measured at each gaging surface. In this experiment, the block length was evaluated as the perpendicular distance from one line (derived from a least squares fit of the measured points) to the intersection with a similar line on the other end. All the points were obtained using a small probe approach distance of 2.5 mm; this is in the region of significant CMM oscillations, as shown in Figure 9.19a. Such a small approach distance constraint might be encountered in measuring a small hole or slot, which does not provide enough room to achieve the manufacturer's recommended approach distance.

Figure 9.19b shows that the error in the measured gage block length grows steadily as the number of measured points increases beyond four points per gaging surface. Since the width of the gaging surface at each end of the gage block is fixed (about 25 mm), when additional measurement points are added, the distance between these points decreases, for example, with 10 points at each gaging surface, the points are

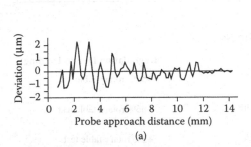

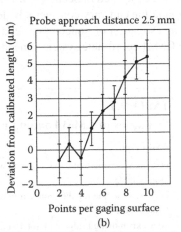

(a)   (b)

FIGURE 9.19 (a) Apparent distance variations due to CMM dynamics as a function of probe approach distance for a CMM using a touch-trigger probe. (b) The change in the measured length (due to CMM dynamic effects) of a gage block as a function of the number of measurement points per gaging surface.

only a few millimeters apart. The short distance between measurement points, combined with the short probe approach distance, does not allow enough time for the CMM oscillations to dampen out. This experiment shows that CMM dynamics in the direction along the gage surface is exciting oscillations in the perpendicular direction along the gage block's length.

Consequently, as additional points are included in the measurement, the effects of the machine dynamics become more pronounced. Even in this case of using a small probe approach distance, these errors can be significantly reduced by adding a short (1 second) pause prior to each probe approach motion. This allows time for oscillations excited by the previous short traverse motion to dampen out, resulting in a more accurate measurement.

Large CMMs in particular may be subject to significant dynamic effects. Their performance should be investigated under several different traverse speeds (the speed used to move between points before changing to probing speed), probing speeds, and (for analog probes) settling times between contacting a measurement point and recording the coordinates.

Dynamic effects are of particular interest in scanning probe CMMs. Since scanning is often along a curved path, the CMM structure must undergo accelerations while the probing points are recorded. The mass of the stylus, probe, and ram may cause the stylus to lag (or lead) the CMM structure, producing errors in the point coordinates. In addition, the CMM structure may have different stiffnesses in different directions, resulting in dynamic errors that are dependent on the direction of scanning. Hence, while checking the ability of a CMM to conduct scanning, all relevant servo motions should be checked.

In some CMMs, these effects are reduced by compensating for dynamic effects in software. In general, the coordinates obtained by discrete point probing are usually more accurate than those obtained by scanning because of these dynamic effects. However, if the form error of the workpiece is larger than the dynamic errors associated with scanning, a large number of scanned points may produce a more accurate result due to more complete information of the workpiece's form, and corresponding improved sampling strategy, as described in Section 9.2.3 and in Figure 9.20.

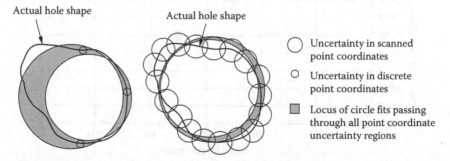

FIGURE 9.20 A schematic illustrating the reduced uncertainty of scanning a nominally circular feature (having significant form error) relative to that obtained using a few high-accuracy measurement points.

Similarly, dynamic effects are one of the principal reasons why manual CMMs using touch-trigger probes are less accurate than their computer-controlled counterparts. The variability of acceleration, velocity, and probe approach distance that are inherent in manual operation (and between different users) often limit the level of accuracy that can be achieved with manual CMMs.

9.2.4.2 Coordinate Measuring Machine Probing System

The probing system on modern CMMs may include probes and styli, an indexable probe head, and an automatic probe or stylus changing system. This additional flexibility extends the reach of the CMM into otherwise inaccessible locations. The following discussion focuses on different sources of probing system-related errors and uncertainty.

9.2.4.2.1 Stylus Ball Size

The size of the stylus ball is often required in order to perform CMM measurements. This value is usually referred to as the effective size of the stylus ball, in contrast to its physical size. The effective size includes any calibration factors (especially stylus bending) that are specific to the probe in use. For example, many common touch-trigger probes require a small displacement of the stylus before the probe triggers and records the point coordinates. This displacement, known as the probe pretravel, results in an effective stylus ball size that is smaller than its physical size. The effective size is determined from the probe calibration process and may be dependent on stylus length (see Figure 9.21).

The CMM geometric structure can accurately measure displacements of the stylus. Hence when calibrating a probe by measuring a reference artifact of a known size, the difference between the size determined by the CMM (displacements of the stylus) and the known size of the artifact determines the effective stylus ball size. Any uncertainty in the size of the reference artifact (usually a sphere of calibrated

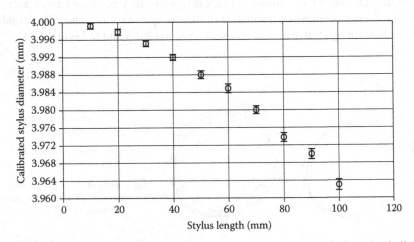

FIGURE 9.21 The effective stylus size of a 4-mm-diameter (physical) stylus ball as a function of stylus length for a common touch-trigger probe as determined by the probe calibration process.

diameter supplied by the CMM manufacturers) transfers directly (sensitivity coefficient of 1) to an uncertainty in the effective size of the stylus ball and hence directly into workpiece size measurements.

For CMMs using analog probes, a more sophisticated stylus ball size calibration procedure is often performed. Since most analog probes are capable of full three-dimensional displacements, the probe calibration factors can be determined on an axis-by-axis basis (Lotze 1994). Typically, this is performed along each CMM axis by advancing the CMM into an immovable object and recording the difference between the CMM translation and the probe's measured displacement over the full range of probe travel. (Often the immovable object is the reference sphere supplied with the CMM.) This allows a correction to be performed for stylus bending and other probe calibration factors for any probe approach direction. The effective stylus ball size can then be either set equal to its physical diameter provided that value is known or determined by measuring the diameter of the reference sphere in the manner described earlier. Which of these two methods is used depends on the relative accuracy in the calibrated values of the reference sphere and stylus ball diameter, residual systematic probe errors not accounted for by the calibration procedure (which would make the stylus ball effective size different from its physical size), the amount of time and effort available, and the recommendations of the CMM manufacturer.

Several CMM measurements are independent of the size of the stylus ball, which Figure 9.22 illustrates by a comparison of two types of length measurements. The uncertainty in the bidirectional length includes the uncertainty of the stylus ball size, while the unidirectional length is independent of this uncertainty source. Similarly, the center location of a circle or sphere is also independent of the stylus ball size. Consequently, the center-to-center distance between two spheres, for example, a ball bar or a ball plate, is independent of the stylus ball size and its uncertainty, as shown in Figure 9.23.

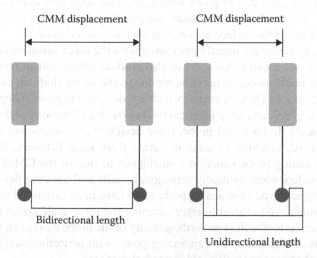

FIGURE 9.22 A comparison of two different types of length measurements. The bidirectional length depends directly on accurate knowledge of the stylus ball size, whereas the unidirectional length measurement is independent of the stylus ball size.

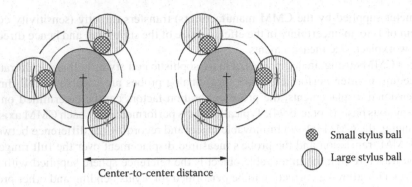

FIGURE 9.23 A schematic illustrating that the center-to-center distance between two circles (or spheres) is independent of the stylus ball size. Note that the correct diameter of the circle (or sphere) depends on accounting for the stylus size, whereas both styli measure the same center location.

9.2.4.2.2 *Probe Lobing*

The effective size of the stylus ball is usually determined through the probe calibration process, which generally involves measuring many points over the reference sphere's surface. This procedure produces a least squares best fit probe diameter, which has an averaging effect over the many different probing directions. In many probes, there still exist direction-specific calibration factors that are not accounted for by this average, that is, by the effective stylus ball size. These direction-specific factors are commonly referred to as *probe lobing* because a measurement of a highly spherical ball results in a computed surface having a spherical structure, that is, a lobed surface structure. (For touch-trigger probes, this effect is also commonly referred to as *pretravel variation*.) Although the form of a measured sphere or circle appears to be poor due to probe lobing, the measured size may be quite accurate because this involves all the measured points, producing an averaging effect, whereas the form of the surface depends on each measured point.

Probe lobing errors are usually associated with the mechanism inside the probe, although an out-of-round stylus ball can also produce effects indistinguishable from lobing, as can nonisotropic stylus bending due to the stylus shaft. In analog probes, there are three stacked flexure stages, which are designed to move independently, be mutually orthogonal, and have identical (and correct) calibrations of their displacement transducers. In an actual probe, these design specifications are only approximately achieved, resulting in some deviation from ideal behavior, that is, probe lobing. (The analog probe situation is analogous to that of the CMM structure in which three independent, mutually orthogonal, calibrated axes are desired but only approximately achieved.) For analog probes that have been calibrated on an axis-by-axis basis (which is equivalent to error mapping the three scale errors in the CMM analogy), other factors such as nonorthogonality of the probe's axes can lead to probe lobing effects. In the example of an analog probe with nonorthogonal axes, a measured ball will appear as an ellipsoid instead of a sphere.

In touch-trigger probes, the mechanical structure supporting the stylus may also often serve as the electrical switch, which is triggered when the stylus is displaced. This commonly results in probe lobing with a three-lobed structure reflecting the triangular mechanical structure within the touch-trigger probe (Estler et al. 1996). One aspect of this probe lobing is easily seen by measuring a ring gage, as shown in Figure 9.24. The ring gage, which has a form error of less than 0.2 μm, typically appears as having three lobes, which is a result of the kinematic stylus mount located inside the probe.

The magnitude of the probe lobing is dependent on the stylus length, as shown in the figure. This effect is also dependent on the orientation of the probe. When the probe is vertical, as when measuring a ring gage in the plane of the table, gravity has little effect on the probe. When the probe and stylus are oriented horizontally, as when measuring a ring gage mounted perpendicular to the table, gravity pulls on the stylus and changes the pretravel. This is most apparent for the longest (100 mm) stylus.

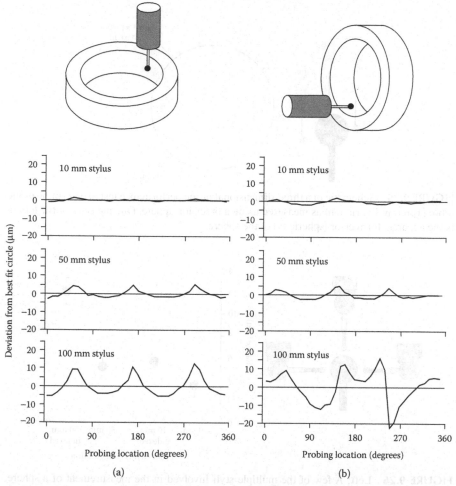

FIGURE 9.24 CMM probe lobing (pretravel variation) as measured using a ring gage for three different stylus lengths on a touch-trigger probe. (a) Probe and stylus are oriented vertically, measuring a ring gage mounted in the plane of the table. (b) Probe and stylus oriented horizontally, measuring a ring gage mounted perpendicular to the plane of the table.

A three-dimensional version of this test includes pretravel variations arising from all possible probe approach directions, for example, measuring a precision sphere, and reporting the apparent form error as outlined in Figure 9.25 and shown in Figure 9.26.

The results of a probe performance test may depend on settings that influence CMM dynamics. For example, for some touch-trigger probes, both the pretravel and the pretravel variation may depend on the probe approach speed. Figure 9.27 displays the result of measuring a short gage block in a manner chosen to illustrate this effect. The block was measured using a touch-trigger probe first by contacting both gage block surfaces with the probe oriented on a low-force (small pretravel) lobe and then remeasured with the probe oriented on a high-force (large pretravel) lobe. (To contact both surfaces of the block using the same probe lobe, it is necessary to rotate

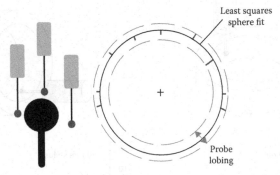

FIGURE 9.25 Schematic of a three-dimensional probe performance test to determine probe lobing (pretravel variation) as measured using a precision sphere. One measure of the lobing is the apparent form error (sphericity) of the sphere.

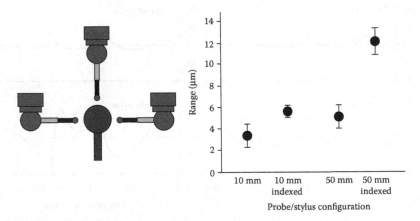

FIGURE 9.26 Left: A few of the multiple styli involved in the measurement of a sphere. Right: The point-to-point probe performance test results from using 10- and 50-mm-long styli. The results show both the single stylus position case and the multiple styli case where the probe head was indexed through eight different positions. The vertical bars represent the range of five repeated tests, and the solid circles represent the mean result.

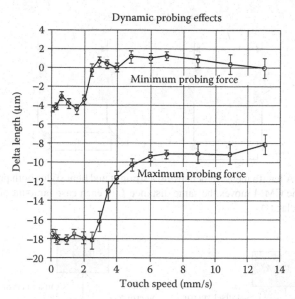

FIGURE 9.27   The differences in the length of a short gage block as a function of probe approach speed and probe pretravel using a touch-trigger probe with a 20-mm stylus.

the probe 180°, which is achieved using an indexable probe head.) The length of the block was arbitrarily set equal to zero on the first (low-force—small pretravel) measurement. This procedure was conducted at several different probe approach speeds, as shown in Figure 9.27.

The difference between the two curves at a given probe approach speed is twice the pretravel variation of the probe. (Twice the pretravel arises because the point at each end of the gage block undergoes a similar pretravel.) This figure illustrates two interesting effects. First, the pretravel variation is a function of both the probe approach speed and the probing force, as shown by the varying distance between the two curves, from a minimum of 8 μm at 13 mm/s to a maximum of 16 μm at 3 mm/s probe approach speed. Second, the effective size of the stylus ball, which is approximately the average of the two curves, decreases as a function of the probe approach speed.

Hence, a gage block will appear shorter at low probing speeds than at high probing speeds assuming the probe is not recalibrated between the two measurements. An obvious corollary when using this type of probe is that the probe should be recalibrated using the new approach speed if a change in the probe approach speed becomes necessary; this will reset the effective stylus size.

9.2.4.2.3   Multiple Styli

The geometric integrity of the CMM frame provides the metrology system, which accurately locates the end of the CMM ram, while the probing system extends this metrology to the stylus ball. For most measurements involving a single stylus, the exact location of the stylus ball relative to the end of the CMM ram is not needed. The stylus should remain at the same fixed location relative to the ram during a

Stylus location issues

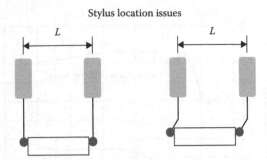

FIGURE 9.28 A length measurement using a straight and bent stylus both produce the same result because the CMM moves the same distance L in each case (amount of stylus bending exaggerated for clarity).

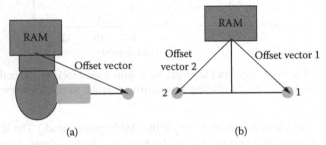

FIGURE 9.29 (a) The probe offset vector for one of many possible probe positions for an indexable probe head. (b) The offset vectors for a two stylus cluster probe.

measurement. Figure 9.28 depicts a workpiece being measured using a straight stylus, followed by the same measurement with a slightly bent stylus. In both cases, the same length is determined since the CMM must undergo the same translation.* Figure 9.28 illustrates that CMM point measurements are relative to other points within the work zone. This principle is often explicitly stated by using a workpiece coordinate system in which the measurements are recorded with reference to measured datums on the workpiece.

When using either multiple styli (as with a cluster probe) or multiple probe orientations (as with an indexable probe head) within a single measurement, it is very important to know the precise position of each stylus ball relative to the other positions. (This also includes measurements that are constructed between different features measured using different styli.) The location of the stylus ball relative to the end of the CMM ram is known as the probe offset vector; Figure 9.29 illustrates this for two common probe configurations. Errors in the offset vectors become measurement errors when multiple styli or multiple probe head positions are used within a measurement. These errors can occur during styli calibration, which determines the location of each stylus ball, due to

---

* Strictly speaking, there could be a small difference between the two cases due to the mechanical imperfections in the CMM structure; however, these errors usually vary slowly such that their magnitude does not change significantly over the distance of a few millimeters.

thermal drift of the relative location of the styli. When multiple styli are used within a single measurement, for example, measuring a precision sphere, then the probe offset errors become point coordinate errors on the sphere's surface.

The plot in Figure 9.26 shows the results of measuring the form of a sphere using both a single stylus and multiple (eight) stylus positions obtained using an indexable probe head. (Each of the eight stylus positions was calibrated prior to the sphere measurement.) The procedure was conducted using a 10-mm-long stylus and was repeated using a 50-mm-long stylus. The range of radial deviations (with respect to the best fit radius) of the 49 points measured on the sphere surface was reported, that is, the apparent form of the sphere surface.

The results of using a single stylus are a measure of the three-dimensional probe lobing effect (convoluted with the repeatability of the CMM and probe), while the indexed results also include errors in the probe offset vectors and the probe head indexing repeatability. In this particular case, the probe head repeatability is about 1 μm, hence the results are dominated by the probe lobing and the probe offset vector errors, which are comparable in magnitude. It is noteworthy to point out that probe lobing itself is a primary source of errors in the probe offset vectors since the lobing can influence the measured location of the stylus ball during the probe calibration process. The situation is often improved by increasing the number of points used in the probe calibration, which then tends to average out much of the lobing leading to a more accurate stylus offset vector measurement.

Probe offset vector errors can also arise if a significant change in the thermal condition of the CMM occurs after the probes are last calibrated. Thermal expansion of the probe or the stylus can change the relative locations of the different probe or stylus positions and hence the probe offset vector. This is clearly a problem for long steel styli, which will typically expand or contract at a rate of $11 \times 10^{-6}/°C$. Frequent probe recalibration is advisable under varying thermal conditions. Similarly, for high-accuracy measurements it may be necessary to allow a thermal stabilization time after handling and loading probes and styli.

A final cautionary note regarding the use of multiple styli: Some CMM software cannot appropriately deal with measurements carried out using different size stylus balls. For example, a cluster probe constructed of 4-, 5-, and 6-mm-diameter stylus balls, each of which is used in the measurement of a circular feature, may produce results (diameter and center location) that are grossly erroneous. Such errors are typically comparable to the differences in stylus ball size, 1 to 2 mm in magnitude for this example, and are indicative that such measurement procedures are inappropriate for that particular CMM. Detecting this problem is simple by measuring a calibrated sphere with multiple styli and examining the computed size and form results.

9.2.4.2.4 Probe/Stylus Changing

Some CMM systems offer the ability to physically remove the probe or stylus and exchange it for another in a probe/stylus rack. The ability to (either automatically or manually) exchange probes/styli increases the flexibility of CMMs. In these systems, it is crucial that the engagement of the probe/stylus be highly repeatable to avoid recalibrating the probe/stylus each time it is changed. (If the probe/stylus is recalibrated each time, it is attached to the CMM, then, provided the coupling is rigid, the repeatability of the exchange is irrelevant.) One method to examine the repeatability of the probe/stylus exchange process is to determine the variation in the location of a small-precision artifact, found by taking repeated measurements, both with and without loading and unloading the same probe/stylus between measurements. If the variation in the artifact is significantly larger when the probe/stylus is changed between measurements, this is attributable to the exchanging process.

9.3   QUANTIFYING COORDINATE MEASURING MACHINE PERFORMANCE

In Section 9.1, the concept of measurement uncertainty was developed. In Section 9.2, the sources of CMM uncertainty were examined. In this section, an overview of specific methodologies used to assess CMM performance will be discussed. The effectiveness of these methods will be considered with respect to the concept of measurement uncertainty due to the various CMM error sources. It is important to remember that CMMs are measuring machines and should be evaluated on their measuring performance.

In Section 9.2.4, it was pointed out that on a software error-corrected CMM, measuring the parametric errors, for example, the pitch of an axis, may produce information that is irrelevant to the measuring performance of the CMM. A similar example would be measuring the positioning capability rather than the measuring capability of a CMM. In a machine tool, the positioning capability directly affects the accuracy of the machined workpiece. Errors in the positioning of the machine tool will cause the cutting tool to remove too much or too little material. In contrast, a CMM must accurately determine the location of the stylus, but for most measurements, the exact physical positioning of the stylus is less important. For example, when measuring a circle the CMM is commanded to probe three points 120° apart. The actual location of the three probed points may not be exactly 120° apart due to positioning errors, but provided the CMM records the correct coordinates of the points that the probe actually touches, the measurement result will be accurate.

This distinction between measuring accuracy and positioning accuracy is the reason that manual CMMs are usable. On a manual CMM, the user grasps the CMM ram, literally pulling the probe to the various measurement points. Typically, the user is doing well to hit the intended location within a few millimeters of the intended target. Nevertheless, manual CMMs commonly measure features with uncertainties of less than 20 μm because the CMM "knows where it is" even though it does not probe the exact intended position. These examples show the value of evaluating a CMM through actual measurements.

The primary CMM performance factors have been discussed in Section 9.2 and are outlined in Figure 9.7. Although all these factors affect CMM performance, each category is often handled independently. For example, the algorithm implementation issue, which determines the uncertainty involving the CMM software, can be assessed by requesting from the CMM manufacturer a certificate of calibration issued by a national laboratory or other qualified organization. Other factors such as the algorithm selection (least squares, mini-max, etc.), the sampling strategy, and the workpiece form error are often highly dependent on the specific workpiece under inspection.

Hence, most standard methods for evaluating performance focus on the CMM hardware, which is discussed in Section 9.3.3. The issue of the uncertainty of a specific measurement, which involves all sources of uncertainty, is discussed in Section 9.3.4.

9.3.1 ARTIFACTS AND THEIR MOUNTING

The artifacts employed in performance evaluations are of usually highly idealized geometries such as spheres and planes. Using these idealized geometries can reduce the error sources shown in Figure 9.7 to those arising from the CMM hardware. Most CMM standards require the form deviations of the test artifacts to be one-fourth or less of the CMM specifications. Fortunately, the mechanical realization of these geometries can be achieved to a remarkable degree, with typical sphericity and flatness values of less than 1 μm. Such idealized geometry can greatly reduce or eliminate the effects of workpiece form error, making the performance test results more repeatable and more focused on the CMM hardware.

Similarly, the use of high-quality artifacts makes the issue of algorithm selection moot since all these algorithms yield the same results (in the absence of measurement errors) for perfect workpiece geometry. Consequently, the least squares algorithm is usually selected since it is robust with respect to the outliers arising due to measurement errors.

Furthermore, the artifacts used in performance evaluations are effective at reducing the influence of the sampling strategy. For example, in the case of a sphere, a large portion of its surface is available to be probed. If four or five probing points are distributed over a hemisphere, the sampling strategy effect (sensitivity coefficient) is reduced to approximately a one-to-one relationship between variations in the probing point coordinates and changes in the sphere center and diameter values. This reduces the amplification effects of measurement uncertainty such as those shown in Figure 9.13. In the case of a step gage or gage blocks, which have parallel planar surfaces, the sampling strategy is often the simple case of a point-to-point distance, which also has approximately a one-to-one relationship between variations in the probing point coordinates and changes in the measured length.

Hence, the artifacts used in performance evaluations have characteristics that mitigate most effects other than the CMM hardware errors. This reduces the error sources of Figure 9.7 to that of Figure 9.30. Note that the measurement results can be interpreted as errors (as opposed to uncertainty), as described in Section 9.1.1, since the artifacts are well calibrated. Performance evaluations, as implemented in various standards, concentrate on producing reliable quantitative tests that evaluate the CMM hardware. Given this focused scope, the use of high-quality artifacts, eliminating form error effects, permitting good sampling strategies, and having simple and minimal amounts of software algorithm involvement is both appropriate and advantageous.

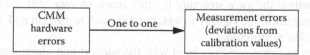

FIGURE 9.30 The goal of CMM hardware performance evaluations is to provide a one-to-one correspondence between CMM errors and measurement errors.

(a)　　(b)

(c)　　(d)

FIGURE 9.31 Artifacts commonly used in CMM performance evaluations. (a) Step gage. (b) Kinematically mounted ball bar. (c) Test sphere. (d) Laser interferometer.

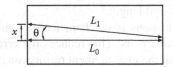

FIGURE 9.32 Misalignment between the calibrated distance L_0 and the measured distance L_1 results in a cosine error, where $L_1 \approx L_0 (1 + \frac{1}{2}\theta^2)$.

Some typical artifacts used in performance evaluations are shown in Figure 9.31. Although significant effort is often expended to obtain well-calibrated artifacts, mounting of these gages is frequently neglected. For high-accuracy measurements normally required in a performance evaluation, poor artifact mounting can amount to a significant source of error. At the micrometer (and submicrometer) level, most artifact mounting mechanisms show nonrigid behavior, for example, bending or hysteresis. To minimize bending, the gage structure is often made of very stiff materials or use designs that maximize the rigidity to weight ratio, such as I-beam and T-beam cross sections and tubular structures.

Two classes of errors are associated with the support and alignment of artifacts. Second-order errors, also called cosine errors, arise when the calibration line and the measurement line are displaced by a small angle, as shown in Figure 9.32. In this case,

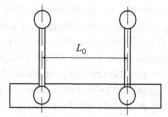 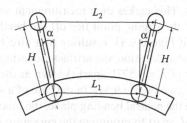

FIGURE 9.33 An artifact having two pairs of balls a distance L_0 apart in the undistorted condition. Upon bending the pair of balls a distance H above the bending plane are separated by a distance $L_2 \approx L_0 + 2H\alpha$, whereas the balls in the bending plane are separated by a distance $L_1 \approx L_0 (1 - \frac{1}{2}\alpha^2)$, a second-order error.*

the measured length is longer than the calibrated value by an amount (using the small angle approximation) of $\frac{1}{2}\theta^2 L_0 = \frac{1}{2}x^2/L_0$, where θ is in radians. For a misalignment of 20 arc seconds (0.0001 rad), the cosine error is less than 0.01×10^{-6} of the measured distance. Consequently, cosine errors are usually negligible provided reasonable alignment is achieved. Artifacts employing spheres, such as ball plates or ball bars, do not require alignment since any reasonable sampling strategy will adequately locate the sphere center, producing a center-to-center, that is, point-to-point, distance calculation.

Direct (or sine) errors contribute in direct proportion to the angular error. This type of error is often associated with angular measurements where the measurement line and the calibrated line are displaced from one another. A common example of this effect occurs when the measurement points are offset relative to the bending axis of the artifact. In Figure 9.33, a gage is shown with two pairs of balls mounted above and on the bending axis of the artifact.

In the undeformed shape, both sets of balls in Figure 9.33 are a distance L_0 apart. Under bending, each ball above the bending axis is displaced a distance of $H \sin\alpha \approx H\alpha$, where α is the angle of bending at the ball location and H is the distance from the bending axis. (This distance is effectively an Abbe offset built into the artifact.) In contrast, the pair of balls on the bending axis is displaced only $\frac{1}{2}\alpha^2 L_0$, a second-order error that is negligible. (This is the same error a ball bar undergoes due to sag of the balls or the bar.) Hence, artifacts constructed with the gaging points near the bending axis (i.e., $H \approx 0$ in Figure 9.33) will reduce measurement errors associated with the distortion of the gage.

Once the material selection and structural shape of the artifact have been chosen, the type of supports should be carefully selected. In general, artifacts should not be overconstrained because this will lead to distorting forces applied to the artifact as the support structure changes, for example, thermally expands. The use of kinematic mounts has the advantage that they avoid overconstraining the artifact, while providing a unique mounting location preventing any rocking or sliding of the artifact in the mount (Slocum 1992b; Slocum and Donmez 1988).

Some artifacts further require the support point to be properly located. For example, gage blocks will bend under their own weight resulting in nonparallel gaging

---

* The exact value of the coefficient on the second-order correction for L_1 depends on the assumptions and approximations made; in any case this second-order term is usually negligible.

surfaces. This makes their measurement very sensitive to the location of the gaging point. If the gaging point lies off the bending axis, then the situation is analogous to that of Figure 9.33, resulting in a direct error. To decrease the sensitivity to the gaging point location, the artifact mounting points can be symmetrically located at a distance of $X = 0.577L$ apart (known as the Airy points) resulting in parallel gaging surfaces (see Figure 9.34). In the case of a step gage composed of a series of parallel faces on the neutral bending plane, the mounting points should be located a distance of $0.586L$ apart to minimize the curvature of the neutral bending plane.

To emphasize the issue of artifact mounting, an interesting case study of the free-standing ball bar is now described. Free-standing ball bars, having each ball cantilevered in space from a central supporting post, have long been used for performance evaluations (see Figure 9.35). As CMMs became more accurate, it became evident that something was faulty with this artifact design when used with certain types of CMMs.

Figure 9.36a plots the measured length of a free-standing ball bar in a single fixed position as a function of probe approach speed for a CMM using a touch-trigger probe. Since the intended use of the ball bar is to position it in many different work zone locations and interpret changes in the measured ball bar length as CMM errors, it is critical that the ball bar itself actually remains a fixed constant length.

For a perfect CMM and a ball bar artifact, the plot in Figure 9.36a would appear as a straight line representing a constant measured length for all probe approach speeds. This is roughly the case for speeds above 7 mm/s. Between 7 and 4 mm/s, there is a marked decrease in the apparent ball bar length due to dynamic probing effects discussed in Section 9.2.4 (Figure 9.27), which results in a smaller effective stylus ball size as the probe approach speed decreases. This effect is due to the CMM under evaluation and is therefore correctly detected by the ball bar measurements.

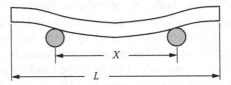

FIGURE 9.34 Side view of a gage block of length L supported at its Airy points ($X = 0.577L$) by parallel cylinders to maintain parallel gaging surfaces.

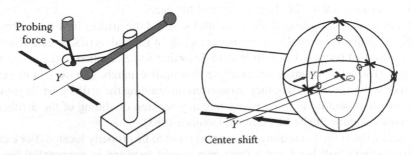

FIGURE 9.35 Left: The deflection Y of the bar due to the probing force. Right: The resulting sphere fit that converts this deflection into a direct error; crosses represent the zero deflection case and circles represent the actual case having deflection.

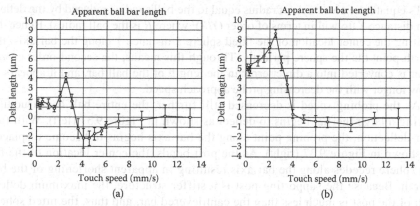

FIGURE 9.36   The measured change in length of a 0.5-m ball bar as a function of the probe approach speed for a CMM using a touch-trigger probe with a triggering force of 0.13 N (13 g). (a) The length as measured without probe recalibration. (b) The measured length with probe recalibration at each new probe approach speed.

Unexpectedly, there is a sudden increase in the measured bar length as the speed slows from 4 to 2.5 mm/s, followed by a decrease and then stabilization of the measured length at 1.5 mm/s and slower probing speeds. The apparent length changes of the ball bar below 4 mm/s are, in fact, due to the mounting of the ball bar and have nothing to do with the CMM's performance. This can be seen even more clearly in Figure 9.36b where the probe has been recalibrated at each new probe approach speed to nullify the probe's effective stylus size dependence on probing speed and hence remove the apparent decrease in the bar length around 4 mm/s. (The sudden increase in the bar length at 4 mm/s is now even more pronounced in Figure 9.36b since the decrease in the effective stylus size in Figure 9.36a is partially canceling out this effect in the 2 to 4 mm/s region.)

The effect in Figure 9.36 is explained by recalling the operating principle of this type of touch-trigger probe. The distance the stylus travels between contacting the surface and triggering the probe (which records the point coordinates) is the pretravel distance. During this time, a force is applied to the ball, the duration of which increases as the probe approach speed decreases. At high probing speeds, this time is very short (about a millisecond), and consequently the inertia of the bar prevents any significant bar deflection. At slower speeds, the force is applied for a longer time and the bar undergoes a deflection Y, in the direction perpendicular to its axis as shown in Figure 9.35.

For the four-point sampling strategy shown in the figure, each of the three points located about the ball's equator will undergo a similar deflection, whereas the point located on the bar axis resists deflection since the bar is in compression and hence much stiffer. It has long been known that the CMM probing force could cause a significant deflection of a cantilevered ball bar in the direction perpendicular to its axis; however, this appears at first glance to be a cosine error, that is, a second-order error, and consequently negligible.

Recall that the sphere fitting algorithm's objective is to fit a sphere through these four points. It is apparent from Figure 9.35 that the three probing points about the ball's equator lie on a circle of radius equal to the ball's radius reduced by the deflection distance Y (to within terms of order $(Y/R)^2$ where R is the ball radius). Hence, by shifting the center location of the fitted sphere a distance Y along the bar axis, the fourth point fits the sphere perfectly. Through this means, the second-order cosine error is converted into a direct error in the length of the ball bar, and it appears to grow longer with the decreasing probe approach speed.

As the probing speed is decreased still further, the entire ball bar structure, having a large inertia, begins to deflect as shown in Figure 9.37 (left). With further deflection, the probing point along the bar axis begins to become displaced as shown in Figure 9.37 (right). As the post bends, the center location of the fitted sphere recedes along the bar axis resulting in apparent shortening of the bar length. Because the supporting post is a stiffer structure, the maximum deflection of the post is much less than the cantilevered bar, and thus, the fitted sphere center never fully returns to the ball's geometric center. This is the reason why the measured length of the ball bar in Figure 9.36 never fully returns to its undeflected value.

These observations can be mathematically quantified and the corresponding results agree with the observed ball bar deflections. The actual deflections of the ball bar can be directly measured using capacitance gages and a digital oscilloscope. Figure 9.38 shows the measurement of the support post deflection (X in Figure 9.37) during the probing of the axial point, for three different probe approach speeds. At time zero, the probe triggers and the point coordinates are recorded. The post deflection is clearly seen to increase as the probe approach speed decreases, exceeding 1 μm at 1 mm/s. This deflection occurs at each end of the bar yielding the observed 2 to 3 μm decrease in the ball bar length at slow probing speeds.

During an actual performance evaluation, the CMM usually maintains a constant probe approach speed that ideally would eliminate the effects shown in Figure 9.36. For touch-trigger probe CMMs used at speeds greater than 6 mm/s, this is usually the case. However, for analog probe CMMs and touch-trigger probe CMMs used at slow speeds, a constant probe approach speed (even if it is zero as is the case for some analog probe CMMs) will not entirely eliminate this effect. The problem occurs

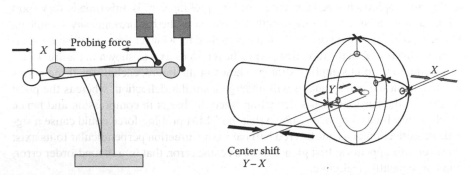

FIGURE 9.37 Left: Deflection of the entire ball bar structure under the probing force. Right: The resulting shift in the probing points and sphere fit.

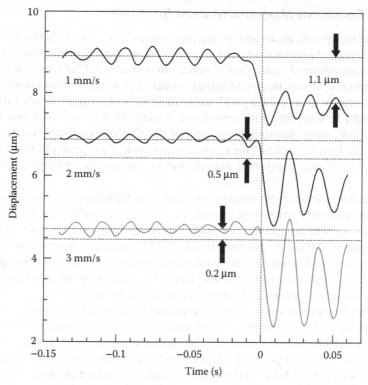

FIGURE 9.38 The deflection of the supporting post shown in Figure 9.37 as a function of time at three different probe approach speeds; time zero corresponds to the triggering of the probe.

because the probing force may be different in different approach directions, and/or the probing points on the balls (the sampling strategy) may vary from repositioning the ball bar within the work zone resulting in different amounts of deflection. During a performance evaluation that requires multiple positions and orientations, this variation in the probing force can result in a collection of apparently different length ball bars due to the mechanism just discussed.

This case study illustrates several important points. Poor artifact mounting can significantly degrade the results of a CMM performance test. The effect shown in Figure 9.36 is larger than the stated accuracy of some CMMs, making this implementation of the ball bar test useless for high-accuracy testing for CMMs with high probing forces or low probe approach speeds. By properly mounting the ball bar, as shown in Figure 9.31, the deflection problem is reduced nearly two orders of magnitude (from 10 to less than 0.2 μm for a 500-mm ball bar).

Similar situations can occur for other high-accuracy artifacts. An example is the mounting of a step gage, whose accuracy is degraded by a mounting mechanism that bends the gage; the amount of distortion depends on how tightly the user fastens the mounting bolts. This distortion can result in a systematic error larger than the calibration uncertainty of the step gage. Hence, artifact mounting may be equally as important as the accuracy of the artifact's calibration.

9.3.2 Coordinate Measuring Machine Standards

Standardized specifications and testing procedures have many benefits. For CMM manufacturers, standardized tests for specifications are less expensive than idiosyncratic user-defined tests. This is particularly true in CMM metrology, where (in the absence of standardized testing) potential customers often present CMM manufacturers with a "golden part" and request their best effort. While this workpiece might be relevant to the customers, it may often be complex and require much measurement planning on the part of the CMM manufacturers. Since this "test" is likely to be the basis for a purchase decision, significant effort is required on the part of each competing manufacturer, thus driving up costs across the industry.

In addition, using a single standard eliminates the high fixed costs associated with redundant standards and testing procedures. These costs include creating separate specifications, service manuals, technician training programs, and custom metrology equipment. All these costs must be absorbed regardless of whether the standard supports one CMM or a thousand. Standardization also reduces costs through scale economies by aggregating the customer base, thus increasing the volume of testing artifacts, manuals, fixtures, and hence decreasing the per-unit cost.

Another benefit to manufacturers is obtained by clearly defining what is expected regarding the CMM's performance and therefore allowing commerce to transpire. It is every manufacturer's nightmare to deliver an expensive CMM to the customer only to hear "I thought you said …" or to find out that the customer's site environment is hostile to the CMM's performance. Widely recognized and well-documented standardized specifications help avoid this situation and protect the manufacturer from "interpretation" of how the CMM should perform. This forms a clear basis for the acceptance of the CMM by the customer and thus allows revenue recognition by the CMM manufacturer.

For CMM users, standardization lowers their costs by efficient capital equipment spending, that is, getting the right tool for the job. Acquiring the right equipment on the first purchase is critical from both cost and time perspectives. Having well-thought-out CMM specifications that objectively evaluate the equipment provides better targeted purchases.

Capital costs are also lowered by promoting fair competition on specifications. Without standards, it will always be an "apples and oranges" world. Most potential CMM customers are almost powerless relative to the global CMM manufacturers. Having standardized specifications is one of the few means small customers have to encourage competition between major CMM suppliers. Additionally, standards provide clear conditions for postpurchase warranty issues by clearly defining when the instrument is out of specification and hence serve as a basis for warranty claims.

The entire CMM industry benefits from standardization because it lowers barriers to adopting new technology by providing expert information and specifications. When a new technology emerges in the marketplace, the manufacturer has all the knowledge and experience regarding its capability and reliability and the potential customer has none. This leads to hesitancy in adopting the technology, sometimes forcing manufacturers into lower (i.e., "introductory") pricing to overcome the barrier. (Economists call this a market failure due to informational problems.) A well-vetted standard with high intellectual content lowers this barrier. The customer believes that the experience, expertise, and integrity of the standards writing committee provide protection against their own inexperience.

Finally, if meaningful metrological tests are specified in a standard, competition will drive these specifications to improve because they are part of the purchase decision. Eventually, the entire CMM industry significantly improves its metrological capability. Industries do not exist in isolation, and as new technology and capability are developed, new market opportunities become available. For example, when CMM standards were first published, the CMM industry as a whole became more accurate and eventually overtook and eliminated most of the high-accuracy (tool room) surface plate-based inspection industry.

9.3.2.1 Interpreting Standardized Specifications

Over the past three decades, various national and international standards have been developed to assess the hardware performance of CMMs. These standards are often limited to CMMs composed of three linear axes and possibly one rotary axis. (Spherical coordinate measurement systems such as articulated arm CMMs and laser trackers are addressed through their own standards [ASME 2004, 2006b].) This section provides an overview of these standardized methods, with emphasis on the metrological concepts. For detailed instructions on performing these tests, the appropriate standard should be consulted.

The goal of all these standards is to ensure that the CMM hardware is in good working order and to operationally provide the definition of good working order. It is noteworthy to point out that none of the standards involve the word "calibration" in their titles since this would imply the capability to accurately predict CMM measurement uncertainty. In recognition of this, these standards are usually considered "performance evaluation methods" designed to allow comparison between different CMMs, particularly for the initial purchase. Consequently, these standards focus on the CMM hardware and use artifacts and procedures that minimize the effect of the operator and testing equipment, as shown in Figure 9.30.

Historically, different standards have adopted distinct testing methodologies, often using different artifacts, to achieve the goal of CMM hardware evaluation. This can make comparing CMMs, each specified by a different standard, difficult and confusing. The situation is further compounded by the different methods of specifying performance test results, as shown in Figure 9.39.

Some standards cite the full range of test values as the performance specification, whereas others cite the maximum permissible error (MPE) from the calibrated value of the artifact—this method is becoming the most commonly cited means of CMM performance specification. Another common specification is the maximum deviation from the calibrated value of 95% of the measurements (where 5% of the data, i.e., the outliers, are eliminated); this is called the U95 criterion. Sometimes, performance results are cited as the standard deviation of the measurements, which describes the spread of the data about the mean value of the measurements, not the calibrated value. The root mean square (RMS) of the errors is another metric to describe the

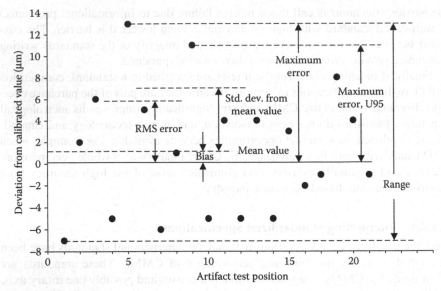

FIGURE 9.39 Various methods of specifying performance test results for the same data.

performance and is similar to a standard deviation of the errors taken about the calibrated values. In the limit of large sample sizes, the square of the RMS errors is equal to the variance of the errors (taken about their mean) plus the square of the mean error (which is also known as the measurement bias or systematic error).*

This distinction in the method of stating performance specifications is crucial to understanding the capability of a CMM since the specified performance results for *the same* CMM and *the same* test can differ significantly depending on which convention is adopted, as shown in Figure 9.39. In the figure, this particular CMM has the following test results: range of errors = 20 μm, maximum error = 13 μm, U95 error = 11 μm, RMS error = 5.78 μm, standard deviation of the errors = 5.80 μm, mean error (bias) = 1.2 μm. Hence, it is advisable to compare different CMMs using specifications from the same standard or to understand how different specification methods are related.

Regardless of the method used to specify performance values, it represents an estimate of the largest value that a CMM in good working order is likely to produce for each test. Since actual results of performing a single test on a particular CMM should be within the manufacturer's specification, the manufacturer's performance specifications do not represent typical test results, but rather an upper limit on the value of the test results.

This fact is often overlooked by users seeking to determine their own performance specifications, who conduct each performance test only once and establish those results as *the* performance specifications for the CMM. Unfortunately, this can

---

* For small sample sizes, the RMS can be smaller than the standard deviation because the mean involves a factor of $\dfrac{1}{N}$ while the standard deviation involves a factor of $\dfrac{1}{N-1}$, where N is the number of measurement results.

lead to needless anxiety when the test is repeated at a future time, and the results differ substantially from the previous case.

9.3.2.2 Recent Trends in Coordinate Measuring Machine Standards

The first CMM standard was developed by a group of CMM manufacturers in 1982 (the CMMA standard, described later). Next came a series of standards written by industrial societies; two prominent ones are the ASME and Germany's VDI/VDE society. These documents became the national standards of individual countries. A variety of these standards are described later in this chapter.

Globalization has led to the formation of many large international CMM manufacturers. Maintaining a multitude of different national standard specifications and testing procedures leads to the inefficiencies previously described. Not surprisingly, the use of a single international standard has recently become the goal for most CMM companies.

The first international CMM standard emerged in 1994 denoted ISO 10360-2 and was adopted by many, but not all countries. Notably, the United States preferred its own CMM standard (ASME B89.4.1), in particular because specific performance tests that quantified certain CMM errors (e.g., ram axis roll) were not well defined in the 1994 ISO standard.

Fortunately, a major long-term revision of the ISO 10360-2 standard has concluded. The ISO 10360-2 (2009; ISO 2009a) standard contains additional specifications and testing procedures and represents a significant metrological advancement in CMM standardization.

In the United States, the ASME B89.4 committee on coordinate metrology has adopted the ISO 10360-2 (2009) standard. It has been recently published as an ASME report, designated ASME B89.4.10360.2, which includes the verbatim text* of ISO 10360-2 (2009) and additional material (largely tutorial in nature), which is included in bounded text boxes to clearly distinguish it from the ISO text. The current ASME plan is to adopt all parts of the ISO 10360 series of standards, some to be published with additional U.S. information. The ISO 10360 series of standards and the ASME counterparts are discussed next. However, since many CMMs have been purchased (and are still re-verified) using historical standards, these documents are also discussed in this chapter in some detail.

9.3.2.3 Decision Rules Used in Coordinate Measuring Machine Standards

The choice of the decision rule used for CMM acceptance during performance testing (associated with the initial purchase) is likely to be one of the few areas in which the U.S. (ASME) CMM standards will differ from ISO standards in the coming years. The explicit use of decision rules and the concept of "test uncertainty" first appeared in the ISO 10360 series in 2000. Prior to that time, the ISO standard,

---

* The text of the ASME document is based on the final draft international standard (FDIS) version of ISO 10360-2 (2009); minor wording changes might differ from the published version of the ISO 10360-2 (2009) standard but they are metrologically equivalent.

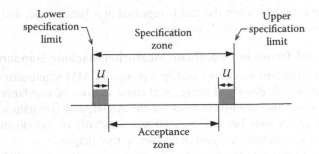

FIGURE 9.40 The use of a stringent acceptance decision rule to reduce the CMM manufacturer's specifications by the test uncertainty, U, yielding a stringent acceptance zone.

like all other CMM standards, used a (in the language of B89.7.3.1 [ASME 2001b]) "simple acceptance" decision rule. The simple acceptance decision rule, which is typically used during performance testing for a CMM purchase, specifies that the CMM is rejected if the test results exceed the manufacturer's specifications and is accepted otherwise. Since 2000, the ISO 10360 standards employ (again using the language of B89.7.3.1) a "stringent acceptance" decision rule. With this rule, the test uncertainty must be subtracted from the manufacturer's specification, yielding an acceptance zone within which the corresponding performance test result must lie for the CMM to be accepted. This change was driven by the ISO 10360 adoption of the ISO 14253-1 (ISO 1998) decision rule standard.[*] In the terminology of ISO 14253-1, the acceptance zone is called the "conformance zone," but this is a misnomer.[†] Indeed, the difference between an accepted product and a conforming product is key to understanding the consumer risk known as "pass errors" (also known as Type II errors [ASME 2005b]), namely the risk of accepting an out-of-specification product. Figure 9.40 illustrates the concept of the stringent acceptance zone (see also Chapter 14, Measurement Uncertainty for Coordinate Measuring Systems for further details).

The "test uncertainty" is the expanded uncertainty of only those factors arising from the CMM testing equipment. This includes the uncertainty in the calibration of the test artifacts (gages) and the uncertainty associated with the test artifacts at the time of use. Factors such as uncertainty in the length of the testing gage due to uncertainty in its thermal expansion coefficient and the uncertainty in the test artifact due to its fixturing (e.g., bending effects) are components of the test uncertainty and must be evaluated under the conditions present at the time of testing.

Test uncertainty, within the context of CMM standards, should not be confused with the accuracy of the CMM. The test uncertainty only refers to the accuracy of the testing equipment at the time of use and has nothing to do with the CMM under test. The

---

[*] Both the ASME B89.7.3.1 and the ISO 14253-1 standards discuss the concept of decision rules; however, the terminology is slightly different and the ASME document emphasizes that decision rules are a business, not technical, decision.

[†] The term "conformance" means that the true value—not the measured value—is within specification (see ISO 9000 [ISO 2005a]). While a test result that lies in the "conformance zone" (as defined by ISO 14253-1) implies a high probability of a CMM conforming to specifications, it is not guaranteed.

CMM accuracy is controlled by the manufacturer's performance specifications—and these are the parameters that are undergoing verification during acceptance testing.

Similarly, the test uncertainty does not refer to issues associated with the comprehensiveness of the testing procedure. A CMM may pass a standardized performance test one day and then may fail it the next. For example, the CMM thermal environment might change day-to-day, while always remaining within the manufacturer's thermal specifications, but these thermal changes could cause a marginal CMM to vary between passing and failing a performance test. Standards writing committees are torn between creating testing procedures that are inexpensive and rapid versus more comprehensive and expensive. Ultimately, the published testing procedure is a compromise between these conflicting goals. Regardless of the consistency of the testing results, the test uncertainty used in the acceptance decision rules does not include shortcomings associated with the testing procedure and is solely confined to the testing equipment. ISO 23165 (ISO 2006) and Annex E of ASME B89.4.10360.2 (ASME 2008) both provide guidance in evaluating the test uncertainty.

The ASME B89 committee, a major standards writing organization for dimensional metrology in the United States, typically limits the test uncertainty interval to be no greater than one-fourth the specification interval under consideration.* Once this upper bound on the test uncertainty is met, the simple acceptance decision rule is employed.

The reason the United States uses the 4:1 simple acceptance decision rule is to prevent undue influence of the test uncertainty in the manufacturer's CMM specifications. Specifically, for every 1-μm reduction in the test uncertainty, a CMM specification that uses stringent acceptance (e.g., ISO 14253-1) will have its specified values reduced by 2 μm. In contrast, a 1-μm improvement in the CMM performance result will yield only a 1-μm reduction in the CMM specification values. Hence, CMM manufacturers will place great value on improving the testing equipment (gages), which are usually the property of the manufacturers and are not left with the CMM after it passes the acceptance test.[†] Thus, the use of simple acceptance promotes improvements in CMM performance, for which the customer is paying, and places less emphasis on the testing artifacts, which are owned by the CMM manufacturer and not left with the customer.

9.3.2.4 International Standard ISO 10360 and ASME B89.4.10360 (2008)[‡]

The ISO standard for CMMs consists of a series of six parts, each addressing some subsystem of the CMM.

---

* The test uncertainty interval has a width of $2U$, where U is the test uncertainty. Since most CMM specifications are defined by a MPE, the specification interval width ($\pm$MPE) is 2MPE; hence, the ratio of the test uncertainty interval to specification interval is U/MPE $\leq$ ¼.

† To understand the factor of 2, consider a CMM manufacturer who tests his or her CMMs with an expensive gage with zero test uncertainty. Suppose the maximum error result for the test is E_{max}. If the CMMs were tested at the customer's site with this gage then the specification limit would equal E_{max}. In practice, the CMMs are tested at the customer's site with a less expensive gage having test uncertainty U; hence, an observed test result could be $E_{max} + U$. However, ISO 14253-1 requires that the specification limit be reduced by the test uncertainty. In order for the CMM manufacturer to sell this CMM, the specification limit must therefore be set equal to $E_{max} + 2U$.

‡ ISO 10360-6 (Part 6) (ISO 2001b) involves CMM software issues and is not addressed in this chapter.

10360-1: Part 1 (ISO 2000a) consists of a collection of definitions of terms that are commonly used throughout the ISO 10360 series of standards. It also includes the format for the CMM specification sheets used to report the performance specifications described in the rest of the series. As described earlier, the B89.4.10360 series mirrors ISO documents, but there is no B89 counterpart to this document because the relevant terminology and speci-fications sheets are included in each part of the B89 standard as needed; this makes the B89 standards self-contained without the need for additional documents. Similarly, the evaluation of the test uncertainty is required for both the ISO and B89 procedures; the relevant evaluation is found in an appendix of each of the B89 documents but is found separately in the ISO 23165 series for each of the corresponding ISO 10360 series of standards.

10360-2: Part 2 (ISO 2001a, 2009a) of the series is focused on the accuracy of the CMM structure. The ISO 10360-2 (2001) standard tests involve, the mea-surement of five calibrated lengths in each of seven different locations and orientations within the CMM working volume. The recommended locations are parallel to the three axes of the CMM and along the four body diagonals of the CMM work volume (see Figure 9.41). Each of the five lengths is mea-sured three times. Hence, there are a total of $(3 \times 5 \times 7)$ 105 length measure-ments. Each of these 105 errors (measured value minus calibrated value), denoted E, is plotted in a figure together with the manufacturer's maximum permissible error specification denoted MPE_E in the 2001 edition or, in the 2009 edition, $E_{0,\text{MPE}}$ (see Figure 9.42).

The testing artifacts used in ISO 10360-2 (2001) are required to have parallel faced gaging surfaces, which typically limits them to gage blocks or step gages. (This restriction has been relaxed in the revised 2009 edition of the

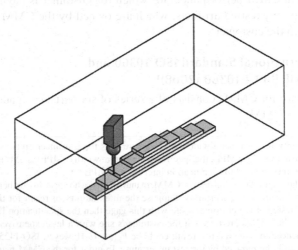

FIGURE 9.41 An example of one measurement line with five calibrated lengths oriented along a CMM axis used to test the CMM structure for accuracy.

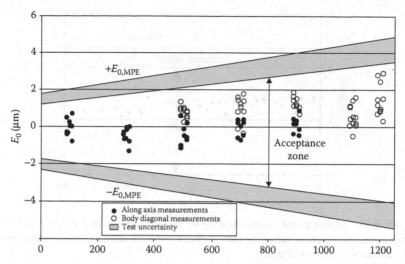

FIGURE 9.42 Measurement results showing E values from the ISO 10360-2 testing procedure; the plot also shows the manufacturer's specification and the acceptance zone, which accounts for test uncertainty.

standard.) These artifacts are also required to be used in a bidirectional manner, that is, the probe must approach the gage from opposite directions, as shown in Figure 9.22. In addition, the longest of the five test lengths is required to be at least 66% as long as the measurement line under test. That is, for a 1-m cubical CMM work volume, the longest length used to test along the axes must be at least 660 mm and the longest length used to test the body diagonals must be at least 1143 mm. A typical performance test result is shown in Figure 9.42. Note that all 105 measurement errors are shown on the plot and that they lie within the acceptance zone; hence, this CMM meets the manufacturer's specifications as described by ISO 10360-2. Finally, the ISO 10360-2 (2001) edition includes a single stylus probing test used to characterize CMM probe performance. In the 2009 edition (and in B89.4.10360.2 [ASME 2008]), this test has been moved to ISO 10360-5 (Part 5 of the series) in order to collect all the discrete point probing tests into one document; this test is described in the 10360 Part 5 discussion found later in this section.

The recent edition of Part 2, ISO 10360-2 (ISO 2009a) and the B89.4.10360.2 document include the CMM structural geometry test of 105 measurements of a calibrated gage, as described earlier. The new notation for this test result is E_0, where the subscript indicates that the ram axis offset* is zero (or the minimum necessary to conduct the measurement), and the CMM manufacturer's specification for the MPE of this test is denoted $E_{0,MPE}$. The 2009 version also allows the use of artifacts that are not gage blocks or step gages,

---

* The ram axis offset is the part of the probe offset vector, shown in Figure 9.28, that is perpendicular to the ram axis.

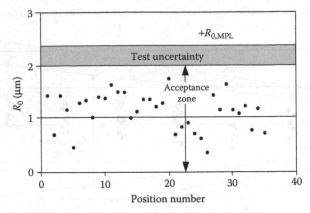

FIGURE 9.43 Measurement results showing R_0 values from the ISO 10360-2 (2009) testing procedure; the plot also shows the manufacturer's specification and the acceptance zone, which accounts for test uncertainty.

(i.e., "unidirectional" artifacts; see Figure 9.22) and is thus more flexible and comprehensive. For example, it is not practical to test large CMMs using physical artifacts. The new edition of the performance test allows the use of laser interferometers (a unidirectional artifact) for this purpose. If unidirectional artifacts are employed, such as ball plates, ball bars, and laser interferometers, each length measurement result is combined with that of a short gage block measurement to yield a result that is equivalent to that of a bidirectional artifact. The equivalent bidirectional results are analyzed in the same manner, as shown in Figure 9.42, which also shows the ISO stringent acceptance zone. In contrast, the B89.4.10360.2 procedure uses a "4:1 simple acceptance" decision rule that limits the test uncertainty to one-fourth the MPE and then sets the acceptance zone equal to the MPE specification zone.

Two new CMM performance test specifications are introduced in the ISO 10360-2 (2009) and the B89.4.10360.2 (2008) documents. First, a repeatability test is included that measures the ability of the CMM to produce consistent results. This test takes advantage of the measurements that are already performed during the 105 calibrated length measurements. Specifically, for each group of three repeated length measurements, the range (longest length minus shortest length) is computed; this is denoted as R_0. Each of the 35 R_0 values is plotted in a diagram together with the maximum permissible limit (MPL)* and acceptance zone, as shown in Figure 9.43.

The second new performance test is designed to measure errors associated with using a probe that is offset with respect to the CMM ram axis (see Figure 9.29). In this test, five calibrated gages are placed in either an *XY* or

---

* Since repeatability is a range of values and is not associated with a calibrated value the term *limit* is used instead of *error*.

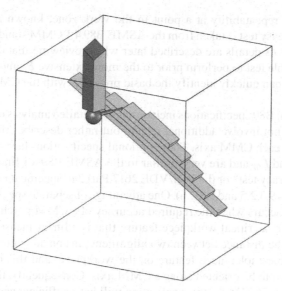

FIGURE 9.44 The measurement procedure for obtaining the E_{150} values using the ISO 10360-2 (2009) testing procedure.

YZ cross-diagonal and are measured with a probe having an offset (default case) of 150 mm. Hence, this test is known as the E_{150} test (see Figure 9.44). This important test evaluates some angular errors in the CMM structural geometry that are not captured by the E_0 test, in particular, ram axis roll for a vertical ram CMM.

There are eight possible measurement combinations (each with five lengths): two crossed diagonals in the XZ plane, two in the YZ plane, and two probe orientations for each of these measurements. The analysis of the E_{150} test results is similar to that shown in Figure 9.42. In the ISO 10360-2 (2009) standard, it states that for time and economy reasons only two of these eight possibilities are tested. This raises the question of how much CMM testing is actually performed, and this is not well addressed in the ISO standard. However, in the B89.4.10360.2 document, this issue is clarified by stating (in the Scope) that "The amount of testing, and which party will bear the cost of testing, is a business decision and must be negotiated between the two parties." The B89 document further states, in the section addressing the E_{150} test, that "it is strongly recommended that the customer negotiate with the CMM manufacturer to test additional positions, ideally all eight positions. This may involve an additional fee to compensate the manufacturer for the additional testing time."

The ASME B89.4.10360.2 document provides considerably more tutorial and informative material than its ISO counterpart. In particular, it also gives additional optional CMM specifications. The first optional specification is the CMM repeatability at a point in the work zone, known as R_{Pt}. This quick and easy test is taken from the ASME B89.4.1 CMM standard (ASME 1997), and its details are described later when reviewing that document. It is a desirable test to perform prior to the more extensive E_0 and E_{150} testing because it can quickly identify the basic problems with the CMM.

Other optional B89 specifications include a more detailed analysis of the E_0 test. This does not involve additional testing but rather describes the E_0 testing results for each CMM axis. These optional specifications have the notation E_{0X}, E_{0Y} and E_{0Z} and are very similar to the ASME B89.4.1 linear displacement accuracy test* or the VDI/VDE 2617 Part 2 u_1 specification (described in Sections 9.3.2.5 and 9.3.2.6). One advantage of separate specifications for each axis occurs when the required accuracy of a CMM purchase decision is driven by a critical workpiece feature that is a linear measurement. For example, the distance between two alignment pins on an engine block may be the tightest toleranced feature on the workpiece, and this distance can be arranged to be oriented along a CMM axis. Consequently, the economically optimal CMM for this application will have sufficient accuracy along a CMM axis but does not require the same level of accuracy in full 3D measurements, which is controlled by the E_0 test.

Another improvement in the ISO 10360-2 (2009), and hence the B89.4.10360.2, documents involves the disclosure of the CTE of the artifacts used in the testing procedure. While the CMM manufacturer is free to choose the CTE of the artifact, its value and the uncertainty in the CTE must be disclosed on the specification form. This eliminates one significant point of ambiguity in all prior CMM standards. Specifically, the situation in which a CMM purchaser is expecting to test the CMM with steel artifacts (because the CMM will be used to measure steel workpieces) but the manufacturer intends to test it with very low thermal expansion artifacts. The difference in these two cases can be a very significant issue for high-accuracy CMMs or those rated to work in a wide range of thermal environments. The new disclosure requirement eliminates this ambiguity.

Of particular significance is the B89.4.10360.2 treatment of thermal specifications and thermal derating. Historically, the B89 CMM standard has led the world on addressing thermally induced CMM errors. In the new B89.4.10360.2 document, thermal issues are given the most comprehensive treatment of any CMM standard. First, a detailed description of the rated thermal conditions under which the CMM performance is assured is presented. For example, the measurement and calculation of the maximum

---

* The measurement procedure is very similar; however, in the analysis the B89.4.1 linear displacement accuracy test reports the range of errors, not the maximum error (see Figure 9.39). Also, the linear displacement accuracy test reports the average of three measurements, whereas a E_{0X} test reports each measurement result independently.

rate of temperature change per hour in the CMM environment are carefully defined and explained. In addition, a derating procedure for CMM specifications is described to address the situation when the customer does not provide a thermal environment for the CMM in compliance with the manufacturer's environmental specifications. This situation is an all too common occurrence, and the B89 derating procedure provides a detailed approach to derating the CMM performance specifications to a level consistent with the actual CMM thermal environment. The B89 derating procedure gives specific guidance for these situations, whereas its ISO 10360-2 (2009) counterpart simply states that if the thermal environment does not satisfy the manufacturer's specification then the CMM performance cannot be verified.

Finally, the B89.4.10360.2 document addresses the issue of metrological traceability. While the ISO counterpart states that artifacts shall be traceable to the SI meter, the B89 version provides specific guidance on achieving and documenting metrological traceability. Specifically, the metrological traceability procedure follows that of ASME B89.7.5 (ASME 2006b) and hence assures a consistent and detailed traceability statement.

10360-3: Part 3 (ISO 2000b) of the series addresses computer-controlled CMMs which include a rotary axis integrated into the metrology system; the rotary table is sometimes called the fourth axis of the CMM. Fortunately, the world has a consistent testing procedure for rotary tables. The ISO 10360-3 standard, ASME B89.4.1, VDI/VDE 2617, and CMMA standards, and the soon-to-be-published ASME B89.4.10360.3 all use the same testing procedure. This test does not use a calibrated artifact but rather checks a complex combination of rotary table geometry, rotary table alignment, probing, CMM accuracy, and coordinate transformations. (Users interested in a more detailed analysis of rotary tables are also referred to ASME B89.3.4 [ASME 2010].)

The procedural details of this test are rather complicated so only an overview will be presented. Two spheres are mounted on the rotary table and located diametrically opposite each other near the edge of the table. Sphere A is near the table surface, while sphere B is mounted above the table. The distance between the spheres relative to the center of the table and the difference in height between the two spheres are standardized and shown in Figure 9.45.

A coordinate system is established such that the origin is at the center of sphere B and the center of sphere A has a y axis coordinate of zero, that is, sphere A lies in the X-Z plane. The table is rotated 13 times to positions specified in the standard. Each time sphere A is measured and its coordinates are recorded. The table is then returned to its initial position, and the coordinates of both spheres are recorded. The table is rotated through another 13 positions (the same as the previous sequence but in the opposite direction), and in each position the coordinates of sphere B are recorded.

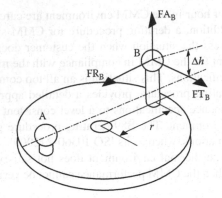

Location of test spheres	
Δh (mm)	r (mm)
200	200
400	200
400	400
800	400
800	800

FIGURE 9.45 The set up for the rotary table test; the fourth axis radial, axial, and tangential errors associated with sphere B are shown.

Finally, the table is returned to its initial position, and the two spheres are measured and again their coordinates are recorded.

The physical basis for interpreting the test is as follows. Since the spheres are rigidly attached to the rotary table and the workpiece coordinate system rotates with the table, the measured locations of both spheres should always be constant. The ranges in the x axis, y axis, and z axis coordinate values for each sphere center are determined. The largest range in the x axis of either sphere is denoted FR for fourth axis radial error. Similarly, the largest range in the y axis is denoted FT for the fourth axis tangential error, and in Z as FA for the fourth axis axial error. In the ISO version of this procedure, the test uncertainty must be accounted for using the stringent acceptance rule of ISO 14253-1, whereas the draft B89.4.10360.3 document (currently under development) uses 4:1 simple acceptance.

10360-4: This part (ISO 2000c) of the series covers the use of contact scanning probes. It consists of a timed test that measures four specified paths on the surface of a precision sphere having approximately a 25-mm-nominal diameter. The specified paths are shown in Figure 9.46. All the paths are inclined 45° with respect to the CMM ram axis. One path is along the equator of the sphere, a second along a plane of latitude that is 8 mm above the equator, a third is 180° of an arc passing through the sphere's pole, and the fourth is a segment of an arc perpendicular to the other three paths and offset by 8 mm from the pole, as shown in the figure.

The test value is the measured form of the sphere, that is, the range of the radial deviations from a least squares sphere fit of all four specified paths and also the total elapsed time to complete the test. There are also four different operating conditions for the test: high-density (0.1 mm between points) and low-density (1 mm between points) data, and with the measurement along a

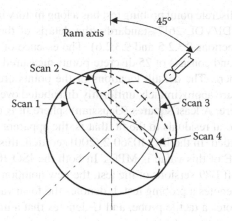

FIGURE 9.46 The specified testing paths on a precision sphere for a scanning CMM.

predefined path or a nonpredefined path. (Manufacturers will provide specifications only for those operating conditions applicable to their CMMs.) For the predefined path measurements, the CMM is programmed with the nominal motion trajectory; for the nonpredefined path, the CMM is given only the start and finish points and an initial direction.

The specifications in the ISO 10360-4 standard are denoted MPE_{Tij} where the T indicates tactile probing and i is either H or L (for high or low point density) and j is either P or N (for predefined or nonpredefined path). Similarly, the duration of the test is given by a maximum permissible time (MPT) specification.*

It should be noted that this test investigates only a subset of the influences that may occur in actual workpiece scanning. The surface of the sphere is hard and smooth and hence does not excite the stylus tip into a vibratory resonance condition as might be the case when scanning over a rough workpiece. Similarly, the test sphere is highly spherical so if the CMM servo control system moves the CMM along the correct path, the stylus deflection may never change and hence the ability of the probe to respond to surface irregularities is poorly tested. Finally, the variation in path curvature (i.e., radius of curvature) in the test is rather limited and in particular does not investigate the CMM's response to surface discontinuities or obstructions.

10360-5: Part 5 (ISO 2010a) of the ISO CMM series addresses contact (tactile) probes used in the discrete point data collection mode. The current version of ISO 10360-5 (2010) and its B89.4.10360.5 counterpart (under development) collect together both the single stylus and multiple styli/probe tests into one document.

---

* The B89.4.10360.4 version currently under development will use the newer notation described in Parts 2 and 5 of the series.

The single stylus discrete point probing test has a long history in both the ASME B89.4.1 and VDI/VDE 2617 standards (the details of these standards are described in Sections 9.3.2.5 and 9.3.2.6). The essence of the test is shown in Figure 9.25 and consists of 25 discrete points measured on the surface of a precision sphere. The default locations of the points are specified in the standard; they are approximately uniformly distributed over one hemisphere of the test sphere. A least squares (Gaussian) sphere fit is obtained, and the range of the radial residuals to the fit, that is, the apparent form error of the sphere, is recorded. In the ISO 10360-2 (2001) edition, this value is denoted P, and the MPE of this value is MPE_P. In both the ISO 10360 Part 5 (ISO 2010a) and draft B89 versions of the test, the new notation is P_{FTU} and $P_{FTU, MPE}$, where P denotes a probing test, F denotes the form value (the test measurand), T denotes a tactile probe, and U denotes that a unique (i.e., single) stylus was used in the test.

Most of the current ISO 10360-5 (2010) is concerned with the specification and testing of multiple styli or multiple probe systems. There are three separate specifications addressing the measuring capability for form, size, and location. Five individual styli (or five different articulation positions) are selected for testing.* The configuration consists of four styli in the plane orthogonal to the ram and one parallel to the ram, all of the same length, as shown in Figure 9.47. A fixed precision test sphere is measured using 25 discrete points with each stylus for a total of 125 points.

The ability to measure the test sphere's size (i.e., the sphere size error) is reported as the difference between a least squares sphere fit using all 125 points and the sphere's calibrated diameter. Similarly, the ability to measure the sphere's form using a multiple stylus probing system is determined by the range of all 125 radial residuals of the least squares sphere fit. To test the ability to measure the sphere's location, each of the five (25 points each) groups is fitted to a least squares sphere fit and the center location recorded. The range of each sphere center coordinate is determined, that is, the range of the five X center values, and similarly for Y and Z. The test result is determined by selecting the largest of the three range values (denoted P_{LTM} as explained later).

The location test is designed to be sensitive to errors in the CMM's ability to accurately determine the probe offset vector. The size test checks the ability of the CMM to correctly account for different size stylus tips used within a single measurement, and the form test investigates the uniformity of the probe's response in various orientations (see Figure 9.24). However, the influence quantities affecting these results are not independently isolated.

---

* Systems with five fixed probes, each with its own stylus, can also be tested—but this configuration is uncommon.

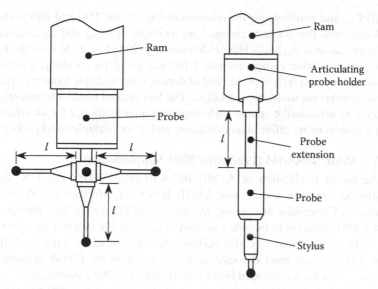

FIGURE 9.47 The specified styli length and probe extension lengths for a fixed styli and articulating probe system; only one of the five articulation positions is shown.

In particular, the size and the form test results will be strongly affected by errors associated with the probe offset vector.

If a probe or a stylus changing system is available on the CMM, then prior to each sphere measurement, the stylus or the probe is exchanged in the rack for another one, for a total of five exchanges. If fewer than five probes or styli are available, then one or more probes or styli will be exchanged more than once—the main issue here is to observe the effect of the repeatability of the exchange.

The MPE specification of the size, form, and location testing is dependent on the stylus length (for the fixed stylus configuration) or on the probe extension length (for the articulation system); see Figure 9.47. For the five fixed styli (each the same length), the suggested styli lengths are 10, 20, 30, 50, 100, 200, and 400 mm. The CMM manufacturers decide which of these lengths are applicable to their CMMs and hence are specified and tested. For articulation systems, a short stylus is used, for example, 20 mm, and the probe extension can be 0, 100, 200, and 300 mm; again the CMM manufacturers decide which of these lengths are applicable to their CMMs.

The notation for the multiple stylus test is rather complicated. In the published (2000) version of Part 5, the notation for the fixed stylus configuration is MF, MS, ML and for the articulation system AF, AS, AL, where M denotes multiple fixed styli, A denotes an articulation system and F, S, L denote form, size, and location. The associated MPE specification is denoted

MPE_{MF} and similarly for the other quantities. In the ISO and B89 revision documents, the notation changes; an example is P_{FTM} and its associated specification is $P_{FTM,MPE}$. Here P denotes a probing test, F denotes the form measurand (other choices include S for size and L for location), T denotes tactile (contact) type of probe, and M denotes that multiple fixed styli (i.e., a star cluster) are under consideration. The last symbol could alternatively be E for an articulating system with empirical qualification, I for an articulating system using inferred qualification, and N for multiple fixed probes.*

9.3.2.5   ASME B89.4.1M (1997 with 2001 Supplements)

Until the recent publication of ASME B89.4.10360.2 (2008), the U.S. standard for CMM performance testing was ASME B89.4.1M, *Methods for Performance Evaluation of Coordinate Measuring Machines*[†] (ASME 1997). This document was the first CMM standard to include a section discussing the thermal environment so that it can be used to arbitrate CMM performance specifications if the user's thermal environment does not meet the specifications stated by the CMM manufacturers. This practice has been continued in the ASME B89.4.10360.2 document.

The principle behind the performance specifications of the B89.4.1 standard is to quantify different categories of CMM uncertainty sources. The 1997 version includes seven performance tests: (1) repeatability, (2) linear displacement accuracy, (3) volumetric performance, (4) volumetric performance using offset probes, (5) volumetric performance using a rotary table, (6) bidirectional length, and (7) point-to-point probing performance using both single and multiple styli.

9.3.2.5.1   *Repeatability*

It should be noted that this useful repeatability test is now an optional specification in the B89.4.10360.2 document and is denoted $R_{Pt,MPE}$. The repeatability test is defined as the range of center coordinates of a sphere determined by 10 successive sphere measurements, each using four probing points. The results can be specified on an axis-by-axis basis; however, a single value that is valid for all three axes is usually cited. Since repeatability involves measuring the same quantity (sphere center coordinates) under the same conditions (fixed sampling strategy, etc.), it is a measure of CMM variability at the particular point where the sphere is located, usually the center of the work zone.[‡] Since all repeatable, that is, systematic, errors are constant during the 10 sphere measurements, this test is designed to check only the nonrepeatable (random) errors at that particular position in the work zone.

---

* In an empirical qualification, each of the five locations of the articulating system is qualified, that is "calibrated" via the CMM's internal algorithm. In an inferred qualification, a set of angular positions of the articulating system (predefined by the CMM manufacturer) are "calibrated," and then other locations (such as those under examination in this test) have their offset vector parameters inferred by the CMM.
† This standard was designated as ASME B89.1.12 prior to 1997; the B89.4.1M 1997 version also had supplements added in 2001.
‡ Note that this repeatability test result is conducted only at the *point* in the center of the CMM work zone (hence, the subscript pt) whereas the R_0 value is evaluated over the majority of the work zone.

Repeatability should not be confused with accuracy since it is possible to repeatedly obtain the same wrong value during successive measurements. Manual CMMs will typically have significantly larger repeatability results than similar computer-controlled machines since a human operator's ability to produce consistent probe approach speeds and probe approach directions is often the most significant source of nonrepeatability.

9.3.2.5.2 Linear Displacement Accuracy

It should be noted that the new B89.4.10360.2 document has the optional specification denoted $E_{0X,MPE}$ (and similar for the Y and Z axes) that controls the same error sources as the linear displacement accuracy test; however, the test results are reported as individual bidirectional errors—not the range of averaged unidirectional length measurements. The linear displacement accuracy test is meant to assess the conformance of the CMM scales to the international standards of length, that is, meter. This test requires the use of a calibrated unidirectional step gage or a laser interferometer with calibrated environmental sensors. The test is performed by measuring many, typically 10, different calibrated lengths oriented along (and spanning the entire length of) each axis of the CMM. The error between the measured value and the calibrated value is determined. The entire process is repeated three times (to reduce repeatability errors), with the average error reported for each position along the axis. The linear displacement accuracy specification is defined as the full range of these (averaged) errors. The intent of the test is to check the scale error for each axis by requiring the orientation of the gage to be parallel to the CMM axis, thus reducing many (but not all) machine geometry effects, and by averaging three runs to reduce the variation due to nonrepeatability.

9.3.2.5.3 Volumetric Performance (General)

Having established the accuracy of the CMM scales using the linear displacement accuracy test, the volumetric performance assesses the geometry of the CMM structure. This procedure includes the volumetric performance test, volumetric performance using an offset probe, and (if applicable) the volumetric performance using a rotary table. Both the volumetric performance test and the volumetric offset probe performance test are usually conducted using an uncalibrated ball bar. This provides an inexpensive and convenient artifact that can be rapidly fixtured at many locations throughout the work zone. Alternatively, a gage block of an appropriate length may be used. For both tests, the length of the artifact should be approximately 100 mm shorter than the shortest axis of the CMM.

9.3.2.5.4 Volumetric Performance Test

Note that the counterpart to this test in the B89.4.10360.2 document is specified by $E_{0,MPE}$. The volumetric test examines deviations from a perfect geometric structure by measuring the length of a ball bar (or other artifacts such as gage blocks) in 20 different locations aligned with the edges, face diagonals, and body diagonals of the work zone, as shown in Figure 9.48. For CMMs with a noncubical work zone, for example, with axis ratios of 2:1:2 and 2:2:1, additional positions are required,

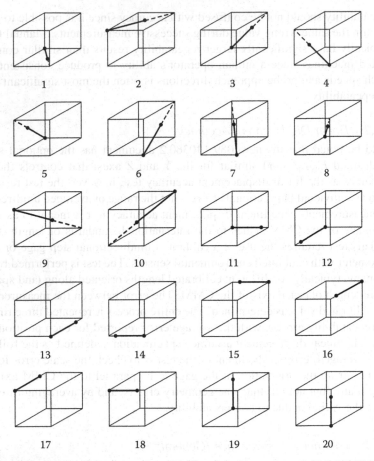

FIGURE 9.48 The 20 ball bar locations for a nearly cubic work zone CMM (ASME 1997).

which yield a total of 30 and 35 positions, respectively. Although the ball bar is an uncalibrated length, the same (fixed but unknown) length should be measured in each of the positions. If the CMM has temperature compensation capability, the ball bar temperature may be measured at each location and an appropriate compensation may be applied to the length measurement. If one of the ball bar lengths appears to be an outlier, that position may be repeated up to six times, with the first of any two successive length measurements that agree (within twice the repeatability test value) taken as the measurement result for that position. The range of all the measured ball bar lengths is designated as the volumetric performance value.

This test is designed to be insensitive to most other nongeometric errors. For example, the test is insensitive to errors in the stylus ball size because sphere center, not diameter, is examined. It is made to be insensitive to multiple stylus effects by using a single stylus/probe position for each ball bar measurement; by using a short stylus, probe lobing effects can be minimized. Similarly, by using 5 to 10 probing points on the surface of each sphere repeatability effects can be minimized. These error sources are tested by additional separate testing procedures.

9.3.2.5.5 *Volumetric Performance Using Offset Probes*

Note that the counterpart to this test in the B89.4.10360.2 document is specified by $E_{150,MPE}$. The angular motion of the ram axis, commonly called z-axis roll, is not tested in the volumetric performance test. This motion is important when probes with large offsets are used. The effect can be assessed by measuring the ball bar inclined at 45° in the X-Z and Y-Z planes. In each position, the ball bar is measured twice using a 150-mm-offset probe (see Figure 9.49). The two measurement results differ because the direction of the offset probe has been reversed 180°. The difference in the two measured ball bar lengths is insensitive to most geometric errors except a convolution of z-axis roll (the error of interest) and two different yaw motions. This pair of measurements is performed at each of the four specified ball bar positions. The largest of the four length differences is divided by 300 mm (twice the probe offset), and this ratio (error/probe offset in micrometers per meter) is the volumetric performance result using an offset probe.

9.3.2.5.6 *Volumetric Performance for Rotary Table Coordinate Measuring Machines*

This test is identical to that of ISO 10360 Part 3 and its B89.4.10360.3 counterpart, which have previously been discussed—see Figure 9.45 for test details. The names of the test results used in the B89.4.1 standard are slightly different. The largest range in the x axis of either sphere is designated as the 3D/alpha radial error. Similarly, the

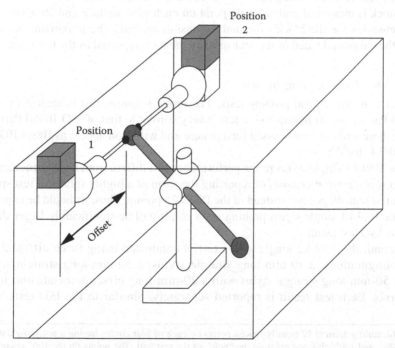

FIGURE 9.49 One method of assessing z-axis roll is to determine the difference in length of a ball bar measured twice using two diametrically opposed offset probes.

largest range in the y axis is the 3D/alpha tangential error, and in Z as the 3D/alpha axial error (see Figure 9.45). The name 3D/alpha simply refers to a CMM with three linear axes (3D) and one rotary axis (alpha).

9.3.2.5.7 Bidirectional Length Test

Note that the new B89.4.10360.2 E_0 test explicitly requires bidirectional length measurements, if a unidirectional length measurement (such as ball bar center-to-center length) is used it must be supplemented with a short bidirectional measurement. Using similar thinking, B89.4.1 requires a bidirectional length test be conducted in addition to the unidirectional linear displacement accuracy test; however, the results are reported separately and are not arithmetically combined with the unidirectional measurement results. This test is explicitly designed to check the bidirectional length measuring capability of the CMM, which, in turn, checks the effective stylus ball size that is derived from the probe calibration artifact (usually a calibrated sphere supplied with the CMM). The test is also sensitive to probe or CMM hysteresis. As shown in Figure 9.23, the effective stylus ball size does not enter into measurements of the (unidirectional) ball center-to-center distance of a ball bar, nor in the length measurement of a unidirectional step gage. Unless a bidirectional step gage is used in the linear displacement accuracy test, the effective stylus ball size has not yet been checked. To meet this need, the bidirectional length test was added to the 1990 version of the B89.4.1 standard. This test consists of measuring a short, typically 25-mm-gage block in four positions, three along the CMM axes and a fourth in a user-selectable position. The block is measured with a single point on each gage surface and an appropriate compensation for the block's thermal expansion applied. The maximum deviation from the calibrated value of the four measurements is reported as the test result.

9.3.2.5.8 Probing Performance

B89.4.1 contains several probing tests. The probe scanning test is identical to ISO 10360 Part 4, and its multiple styli test is very similar to that of ISO 10360 Part 5— hence these will not be discussed further here and will soon appear as B89.4.10360.4 and B89.4.10360.5.

The B89.4.1 single stylus probe performance test differs from its ISO counterpart in two ways. First, it consists of reporting the form of a highly spherical test sphere measured with 49 points* instead of the ISO's 25 points. Thus, it should be expected that the B89.4.1 single stylus probing error result will be significantly larger than a similar ISO test result.

Second, the B89.4.1 single stylus test is conducted using three different stylus configurations: a 10-mm-long straight stylus, a 50-mm-long straight stylus, and a 50-mm-long straight stylus with a 20-mm-long offset perpendicular to the ram axis. Each test result is reported separately. Similar to the ISO tests, each

---

* The (default) pattern is 12 equally spaced points on each of four circles having a polar angle of 30°, 60°, 90°, and 100°, plus one point on the "pole" of the test ball. The points on the 100° polar angle circle are deliberately below the test ball's equator so that the test exercises all three CMM axes in both directions.

probe/stylus configuration is calibrated according to the manufacturer's recommendations prior to the test, and testing using the calibration ball is specifically forbidden.

9.3.2.6 VDI/VDE 2617 Standard

Prior to the international standard on CMMs (ISO 10360 series), most CMMs sold in Europe were specified to the VDI/VDE 2617 series of standards, *Accuracy of Coordinate Measuring Machines, Characteristics and their Checking* (VDI/VDE 1986). This standard, which is not a national standard but rather that of a society of professional engineers, is divided into several separate documents. The standards that have been used to specify the performance of CMMs include *Part 2: Measurement Task-specific Measurement Uncertainty; Length Measurement Uncertainty* (testing methods analogous to the B89.4.1 and ISO 10360 Part 2), 1986; *Part 3: Components of Measurement Deviation of the Machine* (measurement of the rigid-body parametric errors and of probing errors), 1989; and *Part 4: Rotary Tables on Coordinate Measuring Machines*, 1989. Today, these standards are not used to specify new CMMs as the ISO 10360 series currently takes precedence.

The evaluation of the rigid-body parametric errors (outlined in Figure 9.15) is described in Part 3 of the VDI standard and will not be discussed here. The VDI length measuring uncertainty tests (u_1, u_2, u_3) are conducted using calibrated artifacts, either gage blocks, a (unidirectional or bidirectional) step gage, or a laser interferometer (recommended for an axis over 1 m). When using gage blocks, each measurement line is checked with five different length blocks, each measured 10 times for a total of 50 measurements. If a step gage or laser is used, 10 different lengths are measured along the line, each five times, also yielding a total of 50 measurements. The length of the calibrated artifacts and the degree of coverage over the measurement line are not specified; however, the measurements performed are required to conform to the stated uncertainty specifications. Most, but not all, VDI test results are reported by the U95 criterion (see Figure 9.39). For each measurement line, the test is passed if at least 48 of the 50 deviations (from the calibrated values) lie within the specified length measuring uncertainty region.

9.3.2.6.1 *One-Dimensional Length Measuring Uncertainty (u_1)*

This test is analogous to ISO 10360-2 (2009) E_{0X} (or E_{0Y} or E_{0Z}) test and to the linear displacement accuracy test of B89.4.1. Three measurement lines, one parallel to each of the three coordinate axis, are examined using the artifacts and measurement procedures described above, yielding a total of 50 measurements per axis. The 1-D length measuring uncertainty region specified by $u_1 = A_1 + K_1L \leq B_1$, where A_1, K_1, and B_1 are constants specified by the manufacturer, and L is the measured length. The graphical representation of this equation is called the *length measurement uncertainty diagram* and is shown in Figure 9.50.

9.3.2.6.2 *Two-Dimensional Length Measuring Uncertainty (u_2)*

This procedure is similar to the 1D test except the measurement lines are located along the 45° diagonals of the *X-Y, Y-Z, X-Z* planes. The 2-D length measuring

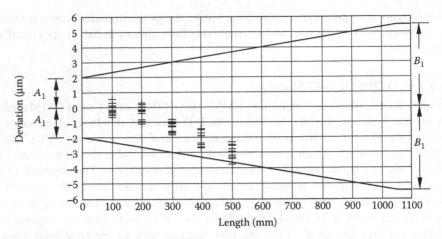

FIGURE 9.50 The u_1 length measurement uncertainty diagram for the x axis as checked with gage blocks, $u_1 = A_1 + K_1L \leq B_1$; this CMM complies with the u_1 test.

uncertainty is given by $u_2 = A_2 + K_2L \leq B_2$. The u_2 test is valid for all six principal plane diagonals, any three of which may be checked with it.

9.3.2.6.3 Three-Dimensional Length Measuring Uncertainty (u_3)

The 3D length measuring uncertainty test u_3 is valid for any measurement line within the CMM work zone. The VDI recommends the measurement lines to be located along at least three of the work zone body diagonals. The test results are specified by $u_3 = A_3 + K_3L \leq B_3$.

9.3.2.6.4 Probe Performance

In Part 3 of the VDI standard, the specifications for probing performance are supplied. Three different probing uncertainty values are reported corresponding to one-, two- and three-dimensional measurements as outlined in Figure 9.51.

V_1 is obtained by taking 50 bidirectional (external or internal) measurements of a parallel-faced gage oriented along one of the three coordinate axes. For each measurement, the deviation from the mean value (of the two probing points) to either of the probing points is recorded. The test is passed if 48 of these 50 deviations are within the stated V_1 value, that is, the U95 criterion.

V_2 is a two-dimensional probe test obtained using a ring gage oriented in one of the principal coordinate planes (X-Y, X-Z, Y-Z). The gage is probed at 50 uniformly spaced points about its circumference. A best fit (least squares) circle is computed, and the set of deviations between each probing point radius (taken from the best fit circle center), and the radius of the best fit circle is calculated. At least 48 of these 50 deviations must be within the V_2 value.

V_3 is a three-dimensional probe test similar to that in ISO 10360 and in B89.4.1 previously discussed. Fifty points are taken over one hemisphere of a test ball roughly uniformly distributed. A best fit (least squares) sphere is computed, and the set of deviations between each probing point radius (taken from the best fit center)

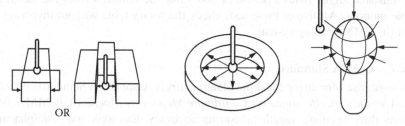

FIGURE 9.51   One-, two-, and three-dimensional probe performance testing artifacts.

and the radius of the best fit sphere is calculated. At least 48 of these 50 deviations must be within the V_3 value.

9.3.2.6.5   Rotary Table Tests

Part 4 of the VDI standard is concerned with rotary table testing. Two approaches are described with one approach being very similar to the ISO 10360-3 test previously discussed with several noteworthy differences. Only a single sphere is involved in the test. A second sphere is used to determine the table alignment, but is not directly involved in the testing procedure. The test sphere (corresponding to sphere B in Figure 9.45) is always located at a fixed radial distance of 206 mm from the table axis and 206 mm above the table surface. The number of rotations and their positions are not specified; however, at least four rotations of 90° each are recommended. The maximum deviations in the x axis, y axis, and z axis coordinate values of the sphere center must be within the specified radial, tangential, and axial performance values, respectively. Note that the actual deviations (in micrometers) are reported, similar to the ISO 10360-3 test but unlike the B89.4.1 version, which reports the ratio of the deviations to R_S.

The other VDI/VDE rotary table approach is a method for determining the rigid-body error parameters (as illustrated in Figure 9.16). The rigid-body error parameter tests consist of checking the angular position deviation (angular scale error), deviations owing to axial movement, deviations owing to radial movement, and deviations owing to wobble (also known as tilt).

The wobble test checks a combination of the two degrees of freedom of axis tilt. The angular deviations are measured using a polygon and an autocollimator or a plane mirror on a precision indexing table and a laser interferometer. In this test, at least 12 measurement positions are chosen with each position measured at least three times from both the positive and negative directions. The difference (in arc seconds) between the measuring device and the angle indicated by the rotary table is plotted from 0 to 360°. At least 95% (the U95 criterion) of the position deviations must lie within the specified annular position deviation to pass the test.

The tests for the axial, radial, and wobble motions can all be performed using electronic indicators, and for all these tests the maximum measurement deviation must lie within the stated specifications. In the case of measuring the axial motion, alignment errors of the indicator cannot be distinguished from axial motion and, therefore, cannot be calculated out of the results using Fourier analysis. In contrast, this sinusoidal signal (with a period of 360°) may be removed from the radial and wobble motions. All five of these tests check the rotary table without involving the use of the CMM probing system.

9.3.2.7 CMMA Standard

The Coordinate Measuring Machine Manufacturers Association Standard (1989 edition), *Accuracy Specifications for Coordinate Measuring Machines*, (CMMA 1989) contains three sections: length measuring accuracy, four-axis and 3D/alpha measuring accuracy, and supplemental testing (which primarily involves determining rigid-body parametric errors and will not be examined here). The CMMA standard is no longer used to specify new CMMs but may occasionally be used to re-verify existing machines.

The length measuring accuracy test is conducted using three gages of length approximately equal to 1/3, 1/2, and 3/4 of the full travel of the longest axis, up to a maximum of 1000 mm. Although not explicitly stated, these gages appear in figures as long gage blocks and hence are bidirectional. The three gages are measured in four different positions, which may be selected from any of the four body diagonals, and the three linear axes passing through the center of the work zone. In each of the four positions, the three gages are measured three times for a total of 36 measurements. The measurement deviation, which is the difference between the measured and calibrated value, is calculated for each measurement. The maximum permitted deviation for a body diagonal measurement M and for an axis measurement G are given below, where L is the length of the gage under consideration and parameters K, H, Q_x, Q_y, Q_z, S_x, S_y, and S_z are specified by the manufacturer. For the 36 measurement deviations, 95% of them must lie within these limits. Hence, no more than one of the 36 deviations may exceed the M and G specifications for the test to be successful.

$$M = K + HL \quad \text{and} \quad G_x = Q_x + S_x L \quad \text{and} \quad G_y = Q_y + S_y L \quad \text{and} \quad G_z = Q_z + S_z L$$

The length measuring repeatability is also determined using the same set of 36 measurements. The repeatability is determined by taking the range of the three repeated measurements at each of the gage positions yielding a set of 12 ranges. In accordance with the U95 criterion, one of the 36 measurements may be excluded from the repeatability calculation, thus at one position the range may be calculated using only two measurements. The 12 ranges should all be less than the manufacturer's stated specification for the length measuring repeatability.

The setup and measurements for rotary table testing are the same as that described previously for the ISO 10360-3 standard with the slight exception that the locations of the test spheres are given by a table in the CMMA standard instead of always being located close to the edge of the rotary table. In addition, the CMMA standard, in accordance with the U95 criterion, allows one sphere coordinate value for each axis and one distance between the two spheres to be excluded in the data analysis. The four rejected values are independent of each other and may come from any four positions and either or both spheres.

9.3.3 MACHINE TOOLS USED AS COORDINATE MEASURING MACHINES

A discussion of the performance of machine tools used in their production mode (metal removal) is outside the scope of this book (see ASME [2005a]). However, with increasing frequency, CNC machine tools are capable of exchanging the cutting tool for a measuring probe, which transforms them into a CMM as described in Chapter 18, Measurement Integration. The inspection of workpieces in situ has the significant advantage of allowing immediate rework of the workpiece based on the inspection results. In addition, the time savings of removing the workpiece from a machine tool and remounting it on a CMM (often in another building) are considerable.

The use of a machine tool as a CMM also has some significant liabilities that need to be carefully considered before employing this measurement technique. In particular, the inspection of workpieces is sensitive only to errors that are not common to both the production and the measurement process. Hence, major sources of error such as those associated with the machine tool's machine geometry are not detected by the inspection. For example, if a machining center's axes are out-of-square, the cutting process will produce an out-of-square workpiece. The inspection process, using the same out-of-square axes, will consequently measure the workpiece to be perfect. Therefore, the machine tool inspection process is oblivious to all error sources common to both the production and the inspection process. This includes all the machine's geometry, thermal distortions, and errors in thermal corrections.

Sources of error that are not common to the two systems may be detected by the inspection process. These include many important sources of workpiece errors such as tool and workpiece deflection under cutting forces, tool offset errors, and tool wear. Additionally (and undesirably), errors associated with the probing system will also be detected and possibly erroneously associated with the production process. For example, errors introduced by the probing system such as a miscalibrated probe, might indicate an incorrect feature of size when, in fact, it is the probe, not the workpiece, that is in error. These are the same concerns as CMM measurements, and they are ever present when measuring on a machine tool.

Particular attention should be applied to machine tools that both produce and inspect the same workpiece. Some form of an interim-testing program (discussed in Section 9.3.5) is needed to ensure the accuracy of the measurement system. This may involve periodically comparing the machine tool's inspection results to those from an independent CMM, or by the inspection of independently calibrated artifacts measured directly on the machine tool. Similarly, the assessment of measurement uncertainty from a machine tool used as a CMM should proceed in the same manner as a true CMM. This assessment may involve the use of CMM standards (Section 9.3.2) and the development of measurement uncertainty statements (Section 9.3.4).

9.3.4 METHODS FOR THE ESTIMATION OF MEASUREMENT UNCERTAINTY

The central theme of this chapter is the description of error and uncertainty sources that affect CMM measurements. Combining these sources together to create a useful estimate of the measurement uncertainty of a specific CMM measurement is seldom an easy task. With the possible exception of the powerful Monte Carlo method, there is no "one size fits all" uncertainty statement that can be applied to all CMM measurement results.

The formulation of an uncertainty statement must include a list of all significant error sources, what assumptions have been made about the measurement, and usually a description of why sources not explicitly included are negligible. This is often begun by determining upper limits on each error source. As a simple example, the repeatability of a large horizontal arm CMM may be largest when all the axes are fully extended (yielding a large cantilevered arm), but could be significantly smaller when all the axes are retracted (producing a stiffer structure). In this example, using the largest repeatability for all positions in the CMM work zone is a conservative approach to an uncertainty statement. However, employing this line of reasoning often leads to an uncertainty statement that is too large to allow the measurement result to be useful. Refining the description of the error sources, for example, determining how the repeatability varies as a function of work zone location, will generally lead to smaller and more refined uncertainty statements at a cost of a more sophisticated uncertainty budget.

In the following sections, different methods for quantifying CMM behavior are described with regard to developing measurement uncertainty estimates. Uncertainty sources associated with the CMM software and computer are not explicitly considered in the following discussion. If this represents a significant source of uncertainty (as determined by an algorithm calibration), then this should also be included in the uncertainty statement.

9.3.4.1 Measurement Uncertainty Using CMM Standards

From a modern metrology viewpoint, the performance evaluation of CMMs should provide enough information to estimate the uncertainty of any valid CMM measurement. This would be valuable for both the inspection of workpieces and for the purchasing selection of a CMM suitable to the intended task. Such a performance evaluation would specify the appropriate combination of uncertainty sources. For example, when the probe is oriented along the ram axis (as is often the case), z-axis roll (a twist in the ram structure) does not affect the measurement results. Whereas when an offset probe is used, the uncertainty associated with z-axis roll becomes important. Similarly, the use of a probe/stylus changer introduces uncertainty primarily when probes/styli are changed during (within) a feature measurement, but may have little effect when changed between unrelated feature measurements. Unfortunately, current published CMM standards are developed primarily to facilitate commerce by allowing users to compare different CMMs to the same set of tests and do not lend themselves directly to the determination of measurement uncertainty.

At best, the information in a performance evaluation can be used to establish the uncertainty in the individual point coordinates. Referring to Figure 9.7, this information must then be combined with that of the workpiece's form error and the sampling strategy used in the measurement to estimate the measurement uncertainty of a specific task. The user may often have a reasonable estimate of the workpiece form error based on the method of production, previous measurements of similar workpieces, or tolerances indicated on the workpiece drawing. The usual procedure is then to combine the CMM coordinate uncertainty variance in a RSS manner with an estimate of the variance of the workpiece form error. (Given only an upper bound on the form error, a uniform distribution to describe the form error could be assumed.) Information on how to determine the sampling strategy's effect on the measurement, for example, as is shown in Figure 9.13, is presently very limited, but can be obtained on a task-specific basis from Monte Carlo simulation.

Acquiring a reasonable estimate of the point coordinate errors from the information obtained from the testing procedure in a CMM standard is quite tenuous. A performance specification of the $A + B \times L$ format describes the uncertainty of a two point length measurement. This uncertainty results from projecting the error vector at each measurement point along the measurement line. To assign point coordinate uncertainties, assumptions must be made. The quantity selected to reflect the largest uncertainty found at a local region of space is usually the probe performance test value associated with the measured form of a test sphere. It is important that the probe performance test value reflect the type of probe and length of stylus used in the measurement. The quantity to represent long-range CMM errors could be taken from the three-dimensional space measuring capability test, that is, $E_{0,MPE}$. Similarly, it is important that this value reflect the nature of the measurement under consideration. For example, if a large offset probe is used in the measurement, then the appropriate performance test, that is, $E_{150,MPE}$ should be used. Hence, the point coordinate uncertainty might include a contribution from the $B \times L$ term of a standardized specification where the length L can be estimated to be a characteristic length (or the largest length) of the feature under measurement, for example, the diameter of a bore.

As an example, consider measuring the radius of a 200-mm-diameter circular feature with three points spaced 30° from one another. Assume the probe used was a straight 50-mm-long stylus oriented along the ram axis on a CMM with a ISO 10360-5 probe test specification (using a 50-mm stylus) of $P_{FTU,MPE} = 4$ μm, and having a length measuring specification from ISO 10360-2 (2009) of $E_{0,MPE} = 1.5$ μm $+ L/200$ μm (where L is in millimeters). The approach is to convert the MPE specifications into point coordinate uncertainties and then use the sampling strategy sensitivity coefficient obtained from Figure 9.13 to compute the combined standard uncertainty of the circle radius.

It is obvious that the E_0 and the P_{FTU} uncertainty sources are correlated, as the sources resulting in the A term in the $A + B \times L$ formulation of the $E_{0,MPE}$ specification are also captured in the $P_{FTU,MPE}$ specification. Additionally, the A term found from the length dependence tests involves only two-point measurements and therefore cannot detect errors composed of odd harmonics (Knapp, Tschudi, and Bucher 1991). Thus, for general 3D measurements, the probe performance test results more accurately reflect the short-range errors of the CMM.

There are several reasonable (but different) assumptions that can be made to account for the correlation of the sources. Taking a simple approach, assume that the $P_{FTU,MPE}$ specification captures all of the uncertainty sources of the CMM at a point in the CMM work zone and that the $B \times L$ term of the $E_{0,MPE}$ specification captures the CMM geometric effects and that the two sources are independent.

Because the probe specification represents the full range of residuals, which is typically four standard deviations, one standard uncertainty for the point coordinates from this sources is 1 μm. Assuming that geometric effects on the $B \times L$ term of the $E_{0,MPE}$ specification are shared equally among two points of the measured length yields a point coordinate standard uncertainty from this source of $0.3 \times (200/200) = 0.3$ μm.* Hence, the uncertainty due to the CMM yields a standard point coordinate uncertainty of $(1^2 + 0.3^2)^{1/2} = 1.05$ μm. Suppose further the workpiece out-of-round-ness is 2 μm, and assuming a uniform distribution of random form deviations gives a standard deviation of 0.6 μm.

The sensitivity coefficient for this sampling strategy is approximately 8 (see Figure 9.13), and this applies to the point coordinate uncertainty arising from both the CMM and the workpiece form error. Therefore, in the absence of all other error sources, the combined standard uncertainty is approximately $(8^2 \times (1.05^2 + 0.6^2))^{1/2} = 9.6$ μm. An expanded uncertainty (using a coverage factor of 2) yields an uncertainty of 19 μm for the radius of the feature.[†]

This example illustrates why performance evaluation results are difficult to convert into measurement uncertainty statements. Furthermore, this estimate lacks any specifics concerning the location of the measurement within the CMM work zone. There may be a region of the work zone in which the CMM is particularly accurate, but the above uncertainty estimation cannot take advantage of this because the performance evaluation factors $P_{FTU,MPE}$ and $E_{0,MPE}$ must apply to all regions of the CMM.

9.3.4.2   Measurement Uncertainty Using Comparison Methods

A more direct route to obtain the measurement uncertainty of a specific feature is through the comparison method (ISO 2004b), which is commonly used in calibration laboratories. This method naturally includes the appropriate combinations of CMM error sources involved in a specific measurement. The comparison method is conducted by inspecting a well-calibrated artifact that is essentially identical in all respects to the workpiece feature under consideration (this includes the method of fixturing the workpiece). For example, the uncertainty in the measurement of a nominal 100-mm-diameter ring gage can be obtained by comparison with a calibrated 100-mm-diameter ring gage of similar grade. Numerous repeated measurements, all using the same measurement procedure, are carried out on the calibrated artifact. These measurements are conducted exactly in the same manner as the actual workpiece measurement. This includes the same location within the work zone, the same probe and stylus configuration, and the same sampling strategy.

It is desirable to vary all the factors that will undergo variation in the actual workpiece inspection during the repeated measurements of the calibrated artifact, for example, removing and remounting the artifact, and possibly having multiple operators

---

* Assuming the point coordinate uncertainty is the same at each end of the measured length and is uniformly distributed, the corresponding point coordinate standard uncertainty is approximately $0.3 E_{0,MPE}$.
† If there were additional sources of uncertainty, for example, associated with the thermal expansion of the workpiece, these would also have to be included in the uncertainty of the radius. Note: These sources would not involve the sampling strategy sensitivity coefficient.

to conduct the measurement. The standard deviation of the measurement results for example, the measured diameters, and the measurement bias (the difference between the mean value of the measurements and the calibrated value) are determined. The bias should be used to correct the actual workpiece measurement. If the bias is not used to correct the workpiece measurement, then it must be accounted for in the decision rule (ASME 2001b).*

The inspected ring gage diameter (once corrected for the measurement bias) has a combined standard uncertainty, which is the RSS sum of the standard deviation found from the repeated measurements, the uncertainty of the bias correction, and the (combined standard) uncertainty in the diameter calibration of the known ring gage. (There may be additional uncertainties, such as the thermal expansion of both the calibrated artifact and the workpiece. If the same sensor is used to measure both the artifact and workpiece temperatures, then these two sources of uncertainty are highly correlated.)

The comparison method is a powerful technique to obtain measurement uncertainty provided an identical but well-calibrated gage or workpiece is available. Often the strict requirements of the comparison method are somewhat relaxed to broaden the scope of this method. For example, a 100-mm-diameter calibrated ring gage could be used to assess the measurement uncertainty of a 110-mm nominal diameter gage. Similarly, the workpiece and artifact temperatures could be slightly different, and a correction for the thermal expansion is calculated, with the uncertainty of this correction added (RSS) to the uncertainty statement. One important factor when using the comparison method is the effect of the workpiece form error on the measurement. Most well-calibrated artifacts have very small form and surface roughness deviations from the ideal geometry. Hence, this factor is not assessed by repeated measurements of the calibrated artifact. If the actual workpiece possesses significant form error relative to the errors in the CMM, then the actual workpiece measurement will be influenced (perhaps dominated) by the use of a finite sampling strategy in the presence of this form error, as discussed in Section 9.2.3.

The effect of form error can be included in either of two ways. A distribution for the form error may be assumed yielding a standard deviation that is then multiplied by the sensitivity coefficient of the sampling strategy used in the measurement. This was the case in the example of determining measurement uncertainty using CMM standards. This quantity is then added in an RSS manner to the standard deviation of the repeated measurements, the uncertainty in the bias correction, and the (combined standard) uncertainty in the calibration of the artifact, yielding a combined standard uncertainty for the workpiece measurements.

Alternatively (and more accurately), the standard deviation of the measurement process can include the effect of form error by performing repeated measurements of the actual workpiece with a reorientation of the workpiece (or the sampling strategy) between each measurement; this forces the form error to be included in the measurement variation. The mean value of the workpiece measurements is then corrected

---

* For measurement results that yield calibration reports, a known bias must always be corrected—this is a requirement of the GUM. However, for accept/reject decisions, the bias may be arithmetically added to the expanded uncertainty to form a guard band (Phillips et al. 1997).

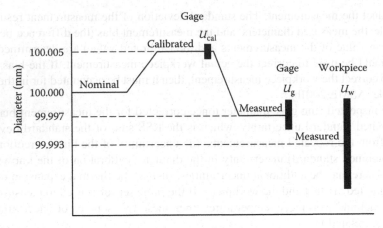

FIGURE 9.52 An outline of the measurements and uncertainties associated with measuring the diameter of a workpiece by comparison to a calibrated ring gage.

for the measurement bias (found from the repeated measurements of the calibrated artifact), and the uncertainty in the result is the RSS sum of the standard deviation of the mean for the repeated measurements, the uncertainty in the bias correction, and the (combined standard) uncertainty of the calibrated artifact.

An example of this process is shown in Figure 9.52, which describes the uncertainty assessment of a circular workpiece feature having a nominal diameter of 100 mm. In this case, a calibrated 100-mm-diameter ring gage is selected as the comparison artifact. The calibrated diameter of the ring gage is 100.005 mm, with a combined standard uncertainty $u_{cal} = 1$ μm obtained from its calibration report. This gage is measured a total of n_g times on a CMM, giving a mean measured diameter of 99.997 mm, with a standard deviation of $u_g = 2$ μm. Therefore, the estimated bias for this CMM measurement is −8 μm, with a standard uncertainty of $2/n_g^{1/2}$ μm. The actual workpiece is measured a total of n_w times (rotating the sampling strategy relative to the workpiece each time to assess the workpiece form error), giving a mean measured diameter of 99.993 mm and a standard deviation of $u_w = 6$ μm; this is larger than u_g because it includes form error effects. Suppose further that both the gage and the workpiece were corrected for thermal expansion effects using the same temperature sensor, which means the uncertainty of the two thermal corrections will be highly correlated. (In this example, the correlation of the temperature measurement errors reduces the total uncertainty, as in the last term of Equation 9.10.)

The final result of this workpiece measurement would be a diameter of 100.001 μm, with a combined standard uncertainty (as described in Section 9.1.1) given by the square root of Equation 9.10. Here, α_g and α_w are the thermal expansion coefficients of the gage and the workpiece, respectively, having $u(\alpha_g)$ and $u(\alpha_w)$ uncertainties; L_g and L_w are the sizes, for example, the diameters, of the gage and the workpiece, respectively; T_g and T_w are the temperatures of the gage and the workpiece, respectively; T_0 is the standard temperature, that is, 20°C; $u(T)$ is the uncertainty of the temperature sensor; and u_{cal} is the (combined standard) uncertainty in the diameter of the ring gage as obtained from its calibration report.

$$u_c^2 = u_{cal}^2 + \frac{u_g^2}{n_g} + \frac{u_w^2}{n_w} + L_g^2(T_g - T_0)^2 u^2(\alpha_g) + L_w^2(T_w - T_0)^2 u^2(\alpha_w)$$

$$+ \left(L_g^2 \alpha_g^2 + L_w^2 \alpha_w^2 - 2L_g L_w \alpha_g \alpha_w \right) u^2(T) \tag{9.10}$$

The obvious disadvantage of the comparison method is the requirement of having a similar and well-calibrated artifact available to carry out the procedure. The less similar the artifact, the less relevant is the comparison. Important nongeometric factors, such as workpiece loading, must be included in the comparison procedure. For example, the results of measuring a lightweight artifact will completely omit the geometric distortions of the CMM under a 1000-kg workpiece load. Another drawback is the requirement to perform numerous measurements on the artifact in addition to those on the workpiece. Nevertheless, the comparison method offers a powerful and direct technique of obtaining the correct combination of error sources whose interaction may not be known or easily modeled. The comparison method also offers a direct means of testing the effectiveness of other methods of uncertainty assessment for the special cases in which well-calibrated artifacts are available.

9.3.4.2.1 Gage Repeatability and Reproducibility Issues

Gage repeatability and reproducibility (GR&R) studies are widely used as an indicator of CMM measurement variability. This procedure is similar to the comparison method; however, these two techniques are not equivalent. Indeed, the GR&R study can be viewed as an important subset of the comparison method. In particular, a well-designed GR&R study can capture many sources of uncertainty by a type A evaluation provided these uncertainty sources are appropriately varied during the GR&R study. Unfortunately, most GR&R studies use uncalibrated workpieces, and hence the connection to the true value of the measurand is lost, that is, the observed variations are not that of measurement errors but rather are just deviations computed from the mean value of an uncalibrated workpiece.

The details of conducting a GR&R study (AIAG 2002) will not be considered here because some of the issues such as identifying poorly trained operators are not of direct interest to producing an uncertainty statement (but obviously will impact it). In a GR&R study, typically 10 randomly selected workpieces (in contrast to a calibrated artifact) are measured many times, varying all the factors that are expected to be present in production measurements. An important aspect of GR&R data analysis is the ability to extract what portion of the observed variation is due to the measurement system and what portion is due to the workpiece. Some aspects of workpiece variation such as size or location are due to different workpieces having different values of the measurand and are not of interest with regard to a measurement uncertainty evaluation; these differences are captured by the result of the measurement. Other aspects such as variation in workpiece form error are of interest since (when combined with a finite sampling strategy) they result in measurement uncertainty.

Since the measurement method used on production workpieces may not be tightly specified, there may be numerous sources of variation. For example, different operators may use different sampling strategies, which produce significantly different results. This is a significant difference between the GR&R procedure and the comparison method.

To be useful for CMM measurement uncertainty evaluation, a GR&R study must clearly state what measurement factors are variable, and the limits of variability. Specifically, it must vary all the factors that will be present during production measurements and vary them over their full limits of the production environment. For example, if the location and orientation of the workpiece within the CMM work zone are not specified for production measurements, then the GR&R study should vary both the location and the orientation of the workpiece over all positions that might be practically realized. In particular, the workpieces should be removed and remounted in a somewhat different orientation between each repeated measurement to assess the influence of workpiece form error and fixturing effects on the results. If widely varying environmental conditions are allowed during production measurements, then these conditions should be varied during the gage R&R test. Since the thermal environment may not be under the user's control, it may be necessary to conduct GR&R measurements for a few hours weekly for several weeks or longer to include all allowable environmental conditions. Even simple factors such as the allowable probe and stylus configurations must be stated since they can have a substantial impact on the measurement result.

Unfortunately, most GR&R studies lack the rigor of explicitly denoting the range of each variable influence quantity and the comprehensiveness of collecting measurements over the full range of these conditions. In this case, a rigorous statement regarding the measurement variation of production workpieces cannot be made for (future) production measurements.

Provided the GR&R study does reflect the variation of production measurements, then the standard deviation of the GR&R results is a type A evaluation of many of the uncertainty sources. Hence, a rigorous and comprehensive GR&R study can be a powerful* input quantity to a measurement uncertainty budget. A complete uncertainty evaluation must include measurement errors, not just measurement variations. In the comparison method, many potential type B uncertainties are accounted for in the bias correction found from the difference of the mean values of the calibrated artifact and workpiece measurements. Additional type B uncertainties, for example, the uncertainty in the artifact's calibration, are explicitly accounted for in the comparison method's uncertainty statement, as in Equation 9.10.

To address this issue, some GR&R procedures specify that the gage under study, that is, the CMM, should be calibrated prior to conducting the study. In the context of measurement uncertainty, this "calibration" means that the measurement bias and its uncertainty for a CMM measurement, which is identical to the workpiece measurement, must be known. As discussed in the comparison method, the link between CMM performance evaluation results and task-specific measurement uncertainty is difficult. Consequently, "gage calibration" should not be interpreted as a performance evaluation specified by a national standard. Rather, it is the difference between the calibrated value and the mean of many CMM measurements on a well-calibrated artifact similar to a workpiece measured in the same manner, this yields the measurement bias.

---

* In general, type A uncertainty evaluations do not rely on the assumptions required for type B evaluations and hence are generally preferred.

The most direct means of addressing the missing link between measurement variation and measurement error is to conduct the GR&R using at least one[*] workpiece that has been independently calibrated (for the measurands of interest and with sufficiently low uncertainty). Doing so establishes the connection to the SI meter and hence allows both the uncertainty evaluation and the measurement traceability to be established (ASME 2006b).

As in the comparison method, the preferred case is calibrated artifacts that are identical to the workpiece, including form error. Such a case could be realized if actual workpieces are calibrated using another CMM (and environment) that is much more accurate than the CMM under evaluation. Additionally, when using calibrated workpieces, only a few are necessary, which reduces the duration of the gage R&R significantly. This is a powerful but often underutilized technique in GR&R studies.

By specifying the permissible measurement conditions, using calibrated workpieces, and accounting for other sources of uncertainty, the GR&R method approaches equivalence with the comparison method. An uncertainty statement analogous to Equation 9.10 can be constructed. The corresponding measurement uncertainty will generally be larger than the task-specific comparison value since the measurement conditions are allowed to vary within the production limits. From this perspective, a rigorous GR&R study is equivalent to a broadened comparison method. Unfortunately, many GR&R studies are not performed at this level of thoroughness and do not capture the full range of variation possible in production measurements or the bias estimation obtained with the use of calibrated workpieces. Hence, most GR&R results cannot be used to determine task-specific measurement uncertainty.

This illustrates a distinction between the comparison and GR&R approaches. The comparison method requires using exactly the same sampling strategy (and all other factors) when measuring the calibrated artifact and the workpiece; thus, these effects are well quantified. Many GR&R procedures do not restrict users on which sampling strategy is used. Consequently, during production measurements, if a user selects a sampling strategy other than those that are used in the gage R&R study, its effect on the measurement results have not been quantified. A comprehensive GR&R study will include measurements taken over all possible sources of variation. If, in addition, the GR&R study is conducted with well-calibrated workpieces, then the measurement errors (in contrast to measurement deviations) can be established and the measurement uncertainty can be evaluated.

9.3.4.3 Measurement Uncertainty Using Monte Carlo Software

This chapter began with a formal description of measurement uncertainty embodied in Equations 9.1 and 9.2. Constructing the mathematical relationship between the measurement result and the sources of error is referred to as the mathematical modeling approach to measurement uncertainty. For this approach to be effective, each uncertainty source must be appropriately modeled and combined through mathematical equations representing the measurement process. That is, for a specific set

---

[*] Preferably all the workpieces under GR&R study will be independently calibrated since this will directly remove *physical* differences in size and location of features between different workpieces leading to a smaller type A evaluation.

of influence quantities* a specific output value is computed, which is a potential measurement result. Once all the uncertainty sources are quantified for the CMM under consideration, the method can predict the uncertainty for any measurement result, which is consistent with the assumptions of the model. Since this approach involves a theoretical model of the measurement system, the uncertainty predictions should be thoroughly checked against known cases. For example, uncertainties found directly from the comparison method may be used to validate the procedure.

Since computers can execute mathematical algorithms extremely rapidly, the mathematical modeling approach is typically encoded into Monte Carlo software. Two examples of this computer simulation technique are the "Virtual CMM" (Schwenke et al. 1994) and "PunditCMM" (Baldwin et al. 2007). The basic idea is that the CMM user inputs (relatively easy via the software interface) both the specifics of the measurement task (measurands, CMM type, measurement strategy, etc.) and the potential variations in the measurement influence quantities (thermal environment, CMM errors, fixturing problems, etc.). Internal to the simulation software are mathematical equations that describe how each of the error sources propagate into point coordinate errors and then into the error in the task-specific measurement result.

The Monte Carlo procedure draws, at random, a possible value for each of the influence quantities and then calculates the corresponding error in the measurement result under investigation. This process is repeated hundreds of times. The final result is a histogram of mathematically calculated errors for the measurement of interest. This histogram represents the possible outcomes of a real physical measurement, and hence the statistics of this distribution, for example, its standard deviation, represent the measurement's combined standard uncertainty. The concept is schematically illustrated in Figure 9.53.

The sophistication of the commercially available CMM Monte Carlo uncertainty simulation systems is impressive. For example, the software can evaluate CMM error sources from either performance evaluation specifications (e.g., MPEs from national or international standards) or the detailed knowledge of the CMM errors sources such as the systematic deviation of the pitch of an axis (see Figure 9.15) as well as its uncertainty in this measurement (which may also be a function of axis position). Additionally, correlations between the individual uncertainty sources can be considered. For example, a large uncertainty in the straightness of an axis at a particular location may require a corresponding large uncertainty in an angular error, for example, pitch, at that location.

The effectiveness of this approach depends on the ability of the model to accurately simulate the real measurement process. This may become quite sophisticated if effects such as changing thermal conditions are included in the model of the CMM performance. Consequently, the scope of the model might be limited to certain types of environmental conditions, probe and stylus configurations, and particular types of feature measurements. For example, the measurement of massive workpieces may deform the CMM structure in a way that is not accounted for in the model, possibly resulting in an unrealistic uncertainty prediction. (The effects of workpiece loading are hard to model since they depend not only on the load but also on its location and how the load is mounted to the table.)

---

* In contrast, the previous methods of uncertainty evaluation relied heavily on combining the standard deviations of uncertainty sources and their sensitivity coefficients.

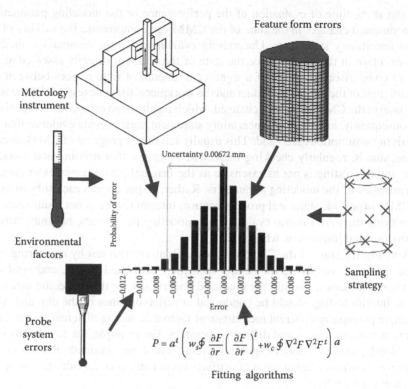

Metrology instrument

Feature form errors

Uncertainty 0.00672 mm

Environmental factors

Probability of error

Error

-0.012 -0.010 -0.008 -0.006 -0.004 -0.002 0.000 0.002 0.004 0.006 0.008 0.010

Sampling strategy

Probe system errors

$$P = a^t \left(w_s \oint \frac{\partial F}{\partial r} \left(\frac{\partial F}{\partial r} \right)^t + w_c \oint \nabla^2 F \nabla^2 F^t \right) a$$

Fitting algorithms

FIGURE 9.53 A depiction of some of the information used in a Monte Carlo uncertainty calculation resulting in a histogram of potential measurement errors.

The mathematical modeling method represents a very powerful means of evaluating measurement uncertainty. The inclusion of these methods into Monte Carlo software creates a flexible and task-specific approach to addressing measurement uncertainty. However, this method includes only those errors that the software has considered. Since these sophisticated mathematical models are usually encapsulated within a software program that is invisible to the user, it is the user's responsibility to ensure that the software is used correctly and includes all the appropriate uncertainty sources.* As the power of computers inevitably increases, the capability of these simulation techniques will grow. Future versions may be able to not only evaluate measurement uncertainty but also help economically optimize the measurement process.

9.3.5 Interim Testing

Uncertainty statements formulated using either CMM performance evaluations or mathematical modeling of the measurement process critically require that the CMM be maintained in good working order. In particular, the CMM and its subsystems must have the same errors and uncertainties during actual workpiece measurements

---

* In particular, if a significant uncertainty source is not evaluated by the software, then it must be separately included (e.g., by paper and pencil) into the uncertainty evaluation.

as it did at the time of evaluation of the performance or the modeling parameters. Undocumented changes in the state of the CMM can compromise the validity of all future uncertainty statements. Uncertainty evaluation via the comparison method is more robust in this regard since the state of the CMM is directly assessed at the time of comparison. However, if a significant amount of time elapses between the measurement of the reference artifact and the workpiece, the state (especially the thermal state) of the CMM may have changed, which results in an erroneous comparison.

Consequently, any rigorous uncertainty statement must provide evidence that the underlying assumptions are valid. This usually requires a program of CMM interim testing, that is, regularly checking the CMM to ensure that it is in good working order. Interim testing is not as extensive as the original performance evaluation or determination of the modeling parameters. Rather, it spot-checks each subsystem of the CMM to provide statistical process control. Interim testing is not a substitute for re-verifying the performance evaluation or modeling parameters, but may provide information and impetus on when to do so.

Typically, the state of the CMM is checked in an interim test by measuring calibrated artifacts, comparing the results with their calibrated values, and verifying that the differences are within the predicted uncertainty for those specific measurements. Interim testing should be conducted at different times of the day and week to sample potentially different measurement factors including different thermal and vibrational environments and different operators. For example, the test results on a cold Monday morning may be quite different from those obtained on a hot Friday afternoon. Several different types of artifacts commonly used for interim testing are shown in Figure 9.54.

Some CMMs are dedicated to measurements of a single type of workpiece or a family of similar workpieces. In this situation, an actual workpiece may be used as the interim testing artifact. This type of artifact will be sensitive to errors that are important to actual workpiece measurements. An additional benefit is that the user is familiar with the required workpiece measurements and, consequently, may have a CMM program available, which can be used for the interim testing. The selected testing workpiece and the measured features on that workpiece should span the largest volume of the CMM work zone that is encountered during actual workpiece measurements to ensure that the relevant volume of the CMM is tested. If an actual workpiece is used, the workpiece's form error and surface finish should not significantly affect the measurement results in order to keep the test focused on the CMM performance.

Since interim testing occurs frequently, a minimum of testing time is highly desirable for economic reasons. An efficient test must concentrate on sources of performance degradation, which commonly occur, testing for as many errors as possible with a minimum number of measurements. Consequently, the location of the artifact is important, with the body diagonals being particularly sensitive to common CMM errors. Similarly, since many CMM angular geometry errors increase in magnitude in direct proportion to the length of the measurement, the artifact should be as long as practical, typically about 60%–75% of the body diagonal of the CMM or alternatively the work zone that encloses the largest workpiece under measurement. On artifacts that can produce several lengths during measurement, for example, ball plates, the longest length present will provide the greatest sensitivity to angular errors. A short

FIGURE 9.54 Some artifacts used for CMM interim testing. Top left: Indexable kinematically mounted ball bar system; top right: ball plate with spheres in the neutral bending plane; bottom left: hole plate; bottom right: pivoting ball ended arm.

artifact positioned in several locations in the CMM work zone is not equivalent and will not have the same sensitivity to angular errors as a long artifact.

An effective interim test checks the total CMM measurement system including subsystem components, which are used in the normal operation of the CMM. This may include components such as probes, probe heads, temperature compensation systems, and rotary tables normally used with the CMM. For example, the CMM probe should be checked for both probe lobing and calibration, that is, effective stylus size. (The probe calibration must be checked using a different artifact than that used to calibrate the probe.) If multiple styli are used, either a stylus cluster, for example, "star probe," or an indexable probe head, then a test that checks the ability to locate one stylus ball relative to another should be included. Such a test would use multiple styli within a single measurement (see Figure 9.26). If a probe-changing rack is available, then this subsystem should be tested by swapping probes in and out of the rack. This not only checks for the repeatability of probe changing but also checks for defective probes in the rack. For CMMs that include a rotary table, an appropriate test such as an abbreviated ISO 10360-3 rotary axis test should be included as part of the interim testing procedure. In summary, an effective interim test artifact should examine the complete CMM measurement system to ensure confidence in the entire measurement process.

The following is a simple example of an interim test for a vertical ram CMM with a nearly cubical work zone having an indexable probe head, a probe-changing rack containing two additional probes, and a temperature compensation system. In this example, a ball bar calibrated for ball roundness, ball size, and center-to-center distance is chosen. A basic test involves measuring the four body diagonals of the CMM. In each position, several points (e.g., eight points) are probed on each ball of the ball bar. The apparent form error of the ball (analogous to a point-to-point-probing test; see Figure 9.25), the difference between the best fit sphere diameter and the calibrated diameter, and the differences between the measured ball bar lengths (after correction for thermal expansion) and the calibrated value are determined (see Figure 9.55).

In the first body diagonal position, the user employs a single probe (oriented along the ram axis). In this position, the measured form error of the balls shows the repeatability of the CMM and of the probe, as well as any probe lobing effects and the results should be less than the $P_{FTU,MPE}$ specification for that stylus length. The ball diameter measurements are bidirectional lengths and check the probe calibration, that is, stylus ball size, and the short range CMM errors. The bar (center-to-center) length measurement checks for long-range (CMM geometry or thermal expansion) errors in that orientation. The sum of the radius errors of the two spheres and the center-to-center length error should be less than the $E_{0,MPE}$ specification. For the second body diagonal, a similar measurement is conducted but with the probe head indexed so that the probe is perpendicular to the ram axis. This measurement will produce similar information to that of the first body diagonal position but will include any z-axis roll error in the CMM geometry. The sum of the radius errors of the two spheres and the center-to-center length error should be less than the $E_{150,MPE}$ specification provided the probe offset vector is less than 150 mm. In the third body diagonal position, each ball of the ball bar is measured with the probe head indexed in several positions; this supplies information on probe head repeatability and the ability to accurately find a stylus ball location relative to others with different probe head orientations. The form error of the spheres should be less than the $P_{FTE,MPE}$ specification.*

The final ball bar position checks the probe-changing process for any defective probes present in the probe rack. The first ball of the ball bar is measured using the second probe obtained from the probe-changing rack, and the second ball of the ball bar is measured with the third probe from the probe rack. Again, the form error of the spheres should be less than the $P_{FTE,MPE}$ value since this specification includes the possibility of probe changing. The form error and the diameter reported for each ball of the ball bar check each of the two probes for probe lobing effects and stylus size calibration, respectively. (If additional probes are available, these could be checked

---

* Maximum permissible multiple stylus form error specification for probing test of an articulating tactile probing system using empirical qualification, as per ISO 10360-5 (ISO 2010a).

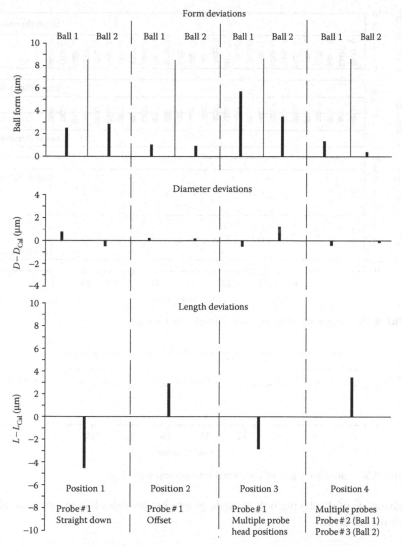

FIGURE 9.55 Results of an interim test using one ball bar in four (body diagonal) positions. The test includes checking the temperature compensation system, the indexable probe head, the probe-changing process, and the different probes available in the rack.

by measuring each ball of the ball bar, in each position, with a different probe.) Figure 9.55 shows the result of this example, and Figure 9.56 shows one possible method of displaying the results of several successive tests. Some users may prefer a single plot representing the test results (instead of the three shown in Figures 9.55 and 9.56). Such a plot can easily be constructed, as shown in Figure 9.57, by combining the largest length deviation, the largest diameter deviation, and one-half the largest form deviation in a RSS manner. (One-half the largest form deviation is used so that each of the three contributions is appropriately weighted.) This method has

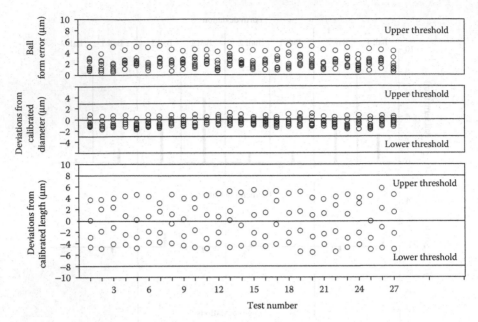

FIGURE 9.56 Summary plots of several interim test results.

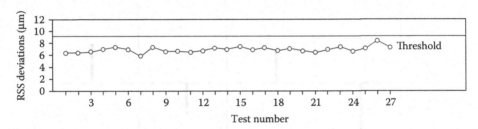

FIGURE 9.57 Summary plot of combined interim testing results.

the advantage of displaying only a single graph, but provides less information about the sources of error.

9.4 SUMMARY

Coordinate Measuring Machines are exactly that—measuring machines—whose primary value is the accuracy of their measurement results. Hence, this chapter reviews CMM performance from a modern perspective of measurement uncertainty, which is a highly task-specific quantity. Primarily, this is due to the large number of uncertainty sources present for any given measurement, each with its own distinct sensitivity coefficient.

Consequently, the uncertainty estimation for any nontrivial measurement must include complications such as numerous CMM error sources, environmental conditions, sampling strategy effects, and workpiece form errors. Each of these need to be propagated into the uncertainty of point coordinates and then into substitute geometry uncertainty. The point coordinate uncertainties themselves depend on many details of the measurement including the location in the work zone, the probe and stylus configurations, and user-selectable quantities such as the CMM probe velocity and probing force and selection of the software algorithm. Only after all these details have been decided can the measurement uncertainty of the point coordinates be precisely determined. Any uncertainty statement that lacks these measurement details will generally have an unreliable uncertainty estimate for the measurement. Therefore, precise uncertainty statements are linked to detailed uncertainty budgets.

Three general approaches to measurement uncertainty are currently available to CMM users: (1) performance test results (typically from national or international standards) combined with sampling strategy rules; (2) comparison to similar well-calibrated artifacts; and (3) the mathematical modeling of the measurement process codified in Monte Carlo software. Most performance tests specified in national or international standards were designed to promote commerce, as opposed to establish point coordinate uncertainties. Consequently, these tests are often piecemeal with regard to addressing measurement uncertainty. Deciding how to combine the various performance test results together and to avoid either omitting error sources or double counting them is frequently perplexing. However the performance results are combined the user must still determine the sensitivity coefficient appropriate to the sampling strategy used in the measurement. Unfortunately, such sampling strategy information is usually incomplete and scattered throughout the literature.

The comparison method can potentially provide the most accurate statement of measurement uncertainty since few assumptions are necessary. The disadvantage is a large number of measurements required and the need for a ready supply of calibrated artifacts similar to all workpiece features under consideration. This effectively reduces the CMM from an absolute measuring machine to a transfer standard, partly defeating the power and flexibility of the CMM.

Mathematical modeling of the measurement process through Mont Carlo simulation software has now become the most powerful method to achieve task-specific uncertainty statements. Although this technique is typically implemented on a desktop computer, a knowledge of the underlying metrology—and the myriad of uncertainty sources—is still needed to produce meaningful results. But, in the hands of a capable metrologist, who knows which uncertainty sources are addressed by the software and which must be separately evaluated, this technique is very powerful and efficient. Additionally, with the continual decrease in computing costs, it offers the most cost-effective means of estimating the accuracy of CMM measurements.

All these methods can benefit from a regular program of CMM interim testing. When using either the performance test or mathematical modeling methods, interim testing is necessary to provide evidence that the underlying assumptions of the uncertainty statement are still valid. In particular, the interim testing program should spot-check all the CMM systems and subsystems that are used in the inspection of production workpieces. The difference between the measured and known interim test results should be within the predicted uncertainty for those specific measurements. If the interim test results lie outside the (expanded) uncertainty interval, it indicates that the assumptions in the method of predicting measurement uncertainty have changed and need to be reassessed.

This chapter has outlined the concept of measurement uncertainty and the various CMM error and uncertainty sources from the hardware, the software, and the user's measurement plan. An overview of national and international standards regarding CMM performance provided a context for addressing task-specific uncertainty. Three methods to quantify task-specific measurement uncertainty were reviewed and methods of employing them were discussed.

Yet, much work remains with regard to implementing measurement uncertainty evaluation in industrial settings. However, in the long run, the efficient use of capital equipment, reducing costs associated with incorrect decisions (either workpiece acceptance/rejection or in adjusting manufacturing processes), and the cost of product quality will all ultimately drive manufacturing toward the rigorous evaluation of measurement uncertainty. For it is the employment and optimization of knowledge that will characterize successful industrial enterprises in the twenty-first century and measurement uncertainty is detailed knowledge about the enterprise's metrological capability.

ACKNOWLEDGMENTS

The author is indebted to the numerous suggestions and recommendations of the editorial review board, which resulted in a more complete and precise manuscript. Bruce Borchardt (NIST), Christopher Blackburn (NIST), and Greg Caskey (formerly NIST, now at UNCC) produced many of the illustrations and also reviewed the chapter. Dr. Keith Eberhardt (formerly NIST, now at Kraft Foods) and Dr. Tyler Estler (NIST) provided many statistical and technical suggestions and comments. Dean Beutel (Caterpillar Inc.) provided the data on workpiece loading effects and a review of the chapter. Similarly, Brian Parry (Boeing) and Jim Henry (Sheffield Measurement, now Hexagon Metrology) reviewed the chapter from a user's perspective. All experimental results (except workpiece loading effects) were produced by the author and his colleagues in the Precision Engineering Division of NIST. This work was supported by NIST's dimensional metrology program and by the Air Force's CCG program.

APPENDIX A

The sensitivity coefficients for a sampling strategy consisting of three points on a circle symmetrically centered about the y axis, as shown in Figure 9.13, can be expressed in closed form as given in Table 9.2 below (Phillips et al. 1998).

Table 9.2

Sensitivity Coefficients

$$\text{Radial sensitivity} = \sqrt{\frac{1 + 2\cos^2\theta}{2(1 - \cos\theta)^2}}$$

$$x \text{ center sensitivity} = \frac{1}{\sqrt{2}\sin\theta}$$

$$y \text{ center sensitivity} = \sqrt{\frac{3}{2(1 - \cos\theta)^2}}$$

10 Temperature Fundamentals

James B. Bryan and Ted Doiron

CONTENTS

Good repeatability has always been important in the performance of coordinate measuring machines (CMMs). It is even more important today because of the development of computer error correction techniques (also refer to Chapter 12, Error Compensation of Coordinate Measuring Machines). These techniques can make the CMM as accurate as the metrology devices used for the mapping of its errors. Repeatability of the machine is the limiting factor.

Thermal effects are the largest single source of apparent nonrepeatability and inaccuracy in most CMMs.* In most industrial settings, where cost is a major concern, one cannot win. However, if one understands the problems and their sources, one can choose how to best minimize the disadvantages for a specific application. This chapter provides some insight into the nature of the thermal problem and suggests some solutions. Much of the information in this chapter is drawn from two works that are very similar in their approach: (1) ASME B89.6.2 (ASME 1973) and (2) ISO TR 16015 (ISO 2003d; ASME stands for American Society of Mechanical Engineers and ISO stands for International Organization for Standardization).

10.1 THERMAL EFFECTS DIAGRAM

The thermal effects diagram shown in Figure 10.1 provides a means for understanding the overall thermal error problem for measuring machines as well as machine tools (Bryan 1968). The diagram divides the problem into two major categories: (1) the effects of uniform temperatures other than 20°C and (2) the effects of non-uniform temperatures. The diagram also illustrates the concept that every measuring and machining operation consists of a three-element system made up of the (1) part, (2) machine frame, and (3) master (or scale).

Six types of thermal influences are shown in the diagram: (1) heating or cooling influence provided by the room environment, (2) heating or cooling influence provided by various cooling systems, (3) the effect of human body heat, (4) heat generated by the machine, (5) heat generated from the cutting process, and (6) thermal memory from any previous environment. Measuring machines are a special case of machine tools in which there is no heat generated by the cutting process and there is a minimum of internally generated heat. Rooms or boxes and their coolant systems are the only influences that can create a uniform temperature environment for a

---

* Other sources of apparent nonrepeatability are inconsistent procedures; hysteresis; vibration; dirt; Coulomb friction; change of position of rolling elements; and variations in the supply of utilities such as electricity, compressed air, and vacuum. The term *apparent nonrepeatability* means that these variables are themselves repeatable if examined closely enough (Bryan 1993).

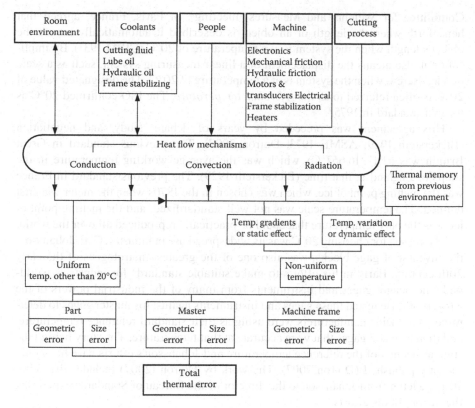

FIGURE 10.1 The thermal effects diagram.

CMM. This is illustrated by the bidirectional arrows in Figure 10.1. The remaining heat sources will cause either steady-state temperature gradients or temperature variations, or both.

All heat sources affect the three-element system through the three possible modes of heat transfer: (1) conduction, (2) convection, and (3) radiation. The errors can be either geometric or dimensional. Geometric errors (squareness, straightness, flatness, and angularity) can be induced and/or exaggerated by nonuniform temperatures or nonuniform coefficients of thermal expansion (CTEs).

10.2   THE 20°C REFERENCE TEMPERATURE

A meter is the distance between two points in space*. A meter does not vary with temperature. This fact is obscured because the lengths of the more common representations of the meter, such as the meter bar, gage blocks, leadscrews, and scales, do vary with temperature. The lengths of most materials also change with temperature. On April 15, 1931, the *Comité International des Poids et Mesures* (International

---

* Refer to Chapter 2, The International Standard of Length for a complete definition of the meter.

Committee for Weights and Measures) meeting* in Paris, France, agreed that henceforth when the length of an object is described it automatically means the object's length when the system is at a temperature of 20°C (CIPM 1931). By implication it also means the displacement of a linear measuring system, such as a scale or a leadscrew, when the system is at a temperature of 20°C. The designated value of 20°C is often referred to as the *reference temperature*. The ISO confirmed 20°C as its own standard in 1975.[†]

This agreement was preceded by years of debate, study, and negotiation (Bickersteth 1929; ASME 1973; Doiron 2007). The previous standard in Great Britain was 62°F (16.667°C), which was the average working temperature in the shops of England at that time (Bickersteth 1929). The previous standard in France was the melting point of ice, which was chosen in the 1870s when the meter was first defined. The temperature scale was not well standardized, and the melting point of ice was the only temperature that could be practically reproduced all over the world.

The reason for choosing 20°C was its widespread use in industry. C. E. Johansson, the inventor of gage blocks, was also one of the greatest metrologists of the early 20th century. Early in his efforts to make suitable standards for industries worldwide, he bought gages and instruments from many of the industrial powers of the early 1900s. He tested these gages and instruments against his master gages to determine what their manufacturers were using as their standard reference temperature, and then he made gages that were accurate at that temperature. The story behind the standardization of the reference temperature and Johansson's efforts have been only recently published (Doiron 2007). The work by Doiron (2007) includes the entire 10-page letter from Johansson to the director of the Bureau of Standards explaining the history of his efforts.

In retrospect, it might have been better for the world if a slightly higher temperature had been chosen. According to the human comfort chart in ASME standard B89.6.2 (p. 13), 20°C is uncomfortable for 70% of the population in the summer (ASME 1973). Only 30% of the population thinks it comfortable at an air velocity of 7.6 m/min and a humidity of 70%. In the winter, 90% of the population thinks the values comfortable. These charts also show, however, that if the air velocity is increased to 122 m/min, the apparent temperature will be 4.4°C cooler. For human comfort reasons, B89.6.2 recommends that air velocities should be less than 6.1 m/min at 20°C.

The possibility of changing the 20°C reference temperature has been considered by ISO committee TC SG-1. On January 26, 1994, the study group recommended that no change should be made to the reference temperature (ISO 2002a). The principal reason was the cost of changing drawings and replacing gages and scales throughout the world (Blaedel 1993a).

---

* Members present at the meeting: P. Janet, France; E. S. Johansen, Denmark; W. Kosters, Germany; J. C. MacLennan, Canada; V. Posejpal, Czechoslovakia; J. E. Sears, United Kingdom; C. Statescu, Romania; S. W. Stratton, National Bureau of Standards, United States; M. P. Zeeman, Netherlands; C. E. Guillaume, BIPM, France; D. Isaachsen, Norway. Members not present in the meeting: M. Chatelain, Russia; C. Casgatchin, Yugoslavia; H. Negasaki, Japan; R. Gautier, Switzerland.

[†] It was reaffirmed in 2002 (ISO 2002a).

If dimensions are correct only when workpieces are at 20°C, how has the world been getting by all these years by measuring at other temperatures? The answer is that if the workpiece is made of steel and the scale is made of steel, then both expand together and the resultant errors tend to cancel each other. Two wrongs traditionally make a right. If, however, the workpiece is made of another material, such as aluminum, the errors are different and they do not cancel each other. This error is referred to as *differential expansion*. Differential expansion problems are also introduced if the workpiece is steel and the scale is made of a low-expansion material, such as Zerodur,* or the scale is a laser interferometer.

Corrections for nominal differential expansion (NDE) must be made if the 1931 agreement is to be achieved. They should be made by the person responsible for making the measurement. Unfortunately, NDE corrections are generally ignored in workshops around the world.

Educating people to make NDE corrections is particularly difficult if a shop normally works with ferrous materials and ferrous gages. Corrections are not usually necessary. The habit of reading measuring instruments directly to establish dimensions is so ingrained that even the most skilled artisan will often forget that corrections are necessary when dealing with nonferrous materials at temperatures other than 20°C. The responsibility involved in changing important dimensions to a different value is another factor leading to the likelihood that corrections are not made. To avoid this responsibility, the inspector or machinist may simply record the temperature at which the measurement was made. This transfers the responsibility of making NDE corrections to someone else who may be even less qualified or less willing to make the corrections.

Most CMMs in the past had steel scales, which perpetuated the NDE correction problem. Steel scales actually worsened the problem. Older instrumentation like height gages and gage blocks were positioned very near to the part being measured, and thus they had similar temperatures. The scales on a CMM are often placed inside covers to protect them from the environment, and in most cases the temperature inside the covers is different from the outside temperature. Although some machines had thermometers to make corrections for this temperature difference, most did not. Most new CMMs have low or high thermal conductivity materials located in critical areas to reduce the sensitivity of the machine to temperature changes, and many have the necessary temperature sensors and computer capability for making NDE corrections automatically. This is a welcome development, one that is long overdue. Laser interferometers have had this capability since they were first marketed some 30 years ago.

One must keep in mind, however, that these corrections are not exact. The coefficients of expansion used to make the NDE calculations are not the same as the actual coefficients of the scale and the part. For this reason, one should avoid the temptation of calling these techniques thermal compensation and refer to them as *automatic NDE* (ANDE) corrections.

---

* Zerodur is a trademark held by Schott (www.us.schott.com) for a ceramic that has a near-zero CTE at room temperature.

10.3 AVERAGE TEMPERATURES OTHER THAN 20°C

Since all materials change dimension when the temperature changes, to specify the length we must also report the temperature at which our measurement was made. The meter bar used as the definition of the meter in 1889 was assigned its standard length at the melting point of ice. This temperature choice seems to be awkward for use, but at the time the temperature scale was still in development and the most reproducible temperature was the ice melting point.

As time went on and industrial precision became better a more convenient temperature was needed, and after a decade of discussion the reference temperature for industrial length measurement was set at 20°C (Doiron 2007).

10.3.1 UNCERTAINTIES OF COEFFICIENTS OF THERMAL EXPANSION

It is often very difficult or inconvenient to determine the precise value of the thermal expansion coefficient for each part. The effect of this lack of precision is called the uncertainty of nominal expansion (UNE) per B89.6.2 (ASME 1973).[*]

The uncertainty of length measurement due to the UNE is zero at 20°C. (Errors associated with the measurement of the 20°C temperature will lead to length measurement uncertainty). The magnitude of uncertainty in coefficients varies for different materials. It is about 10% for gage block steel and up to 25% for other materials. Differences between actual CTEs and nominal CTEs occur because of experimental errors, anisotropy, variations in chemistry, cold working, and heat treatment of the material.

The uncertainty in expansion between the scale and the part is called uncertainty of NDE (UNDE) in B89.6.2. The problem of uncertainty in coefficients was well understood in 1920. A quote from Peters and Boyd of the National Institute of Standards and Technology (NIST, formerly the National Bureau of Standards) in Washington, DC, is as follows (Peters and Boyd 1920; p. 627):

> Another property which must be recognized when considering true accurate length of gages is the thermal expansion of the material. A 1-inch (25.4-mm) steel gage block increases in length about 13 ppm per degree Celsius rise in temperature. The temperature at which the actual length of the gage equals the nominal length must, therefore, be specified and is usually taken as 20°C. At 25°C, the length of a gage, which is 1 inch (25.4 mm) at 20°C is about 1.000065 inches (25.4016 mm). If a gage is measured at a higher temperature, its length at 20°C may be computed if the expansion coefficient is known. If high precision is desired, it is not good policy to use expansion coefficients given in tables because our measurements show that the expansion coefficient of steel may vary from 10.5 to 13.5 ppm depending on the hardness and composition.

This variation was confirmed by Breyer and Pressel (1991); p. 65 of Carl Zeiss: "Two sets of gage blocks measured at Zeiss showed completely different results in the expansion coefficients of the individual blocks."

Set No. 1: Range of alpha (coefficient) values = $10.1–12.3 \times 10^{-6}$/K
Set No. 2: Range of alpha (coefficient) values = $10.3–12.9 \times 10^{-6}$/K

---

[*] Refer also to Chapter 14, Measurement Uncertainty for Coordinate Measuring Systems.

There has also been an increase in the variety of materials used for manufacturing gage blocks. Currently, gage blocks from major manufacturers are made with CTEs (in parts per million per degree Celsius) of 4.6, 8.4, 9.2, 10.8, 11.5, and 12.0. The range of materials used in industry has also become very broad, ranging from carbon composites with very low thermal expansion to aluminum alloys with CTEs higher than 20 ppm/°C and polymer materials with CTEs that are both very high and very variable. From the standpoint of thermal expansion corrections, this is not a good trend.

Johansson was also concerned about the accuracy of NDE corrections. Henry Ford acquired the U.S. subsidiary of C. E. Johansson Ltd. in 1923, and shortly afterwards he moved the business to Detroit, Michigan, where he provided Johansson with a room with constant temperature. A paraphrased quote from Johansson's biographer T. Althin is as follows (Althin1948; p. 143):

> As late as 1926, the International Bureau of Standards in Paris did not have a constant temperature room. Measurements had to be made in a room whose temperature was entirely dependent on the season of the year. All measurements had to be recomputed to apply to 20°C. Since the coefficient of expansion varies due to unavoidable variations in the material, it is easy to see that the final result was not as reliable as Johansson required. Johansson realized from long experience that it was not possible to accurately calculate the consequences of measuring at different temperatures. Measurements should be made in 20°C temperature-controlled rooms.

It should be clear from the aforementioned explanation that some error will exist when measuring or machining operations are carried out at temperatures other than 20°C regardless of the sophistication of attempts at thermal error correction. How significant is this error? The thermal error index (TEI; ASME B89.6.2) is designed to provide an answer to this question. The TEI is discussed in detail in Section 10.6.

10.4 NONUNIFORM TEMPERATURE

The thermal effects diagram (Figure 10.1) shows that the three-element system, consisting the part, scale, and machine frame, is affected by steady-state and dynamic temperature differences originating from the six principal sources of thermal disturbance. Recognition of the three-element system is crucially important in evaluating the effectiveness of different proposed solutions to temperature-related problems. It is helpful to visualize each of the three elements made of Zerodur, one at a time, in pairs, or all together.

If, for example, the part, frame, and scale are all made of high-CTE materials, it would be necessary to measure the temperature of all three and to know their CTEs to reduce the thermal error. Even then it is necessary to provide a uniform temperature environment for the part and to wait for the part to reach a uniform temperature so that the measured temperatures are true characterizations of the thermal states. If the scale were of Zerodur, one would not need a thermometer for the scale, which would not only be convenient but also reduce the overall thermal error. If the frame, scale, and part were all of Zerodur, there would be an optimal solution and temperature effects would be negligible for most applications. All real systems fall somewhere in this spectrum and it is important to consider all the three elements for accurate measurements.

10.4.1 GRADIENTS—THERMAL VARIATIONS IN SPACE

A reasonable understanding of the influence of gradients has been achieved in the granite surface plate industry. The U.S. government federal specification GGG-P-463c (GSA 1973) includes a graph, developed by R. McClure, which shows the magnitude of nonflatness for different sizes of plates and thermal gradients. For example, a 2-m-long and ⅓-m-thick plate will have a flatness error of 10 μm for each degree Celsius of gradient. (It is also of some interest to realize that the radiation from normal shop lighting can create a 1°C temperature rise on the surface of any materials with low thermal conductivity, such as granite and ceramics.)

Encouraging progress has been made in recent years in dealing with the problem of gradients. Finite element analysis (FEA) programs have been developed that can approximate the effect of steady-state gradients.

Trapet and Wäldele (1989) have made a major contribution to the understanding of gradients. They deliberately induced gradients in a measuring machine located in a special room at Physikalisch-Technische Bundesanstalt (PTB) with poor temperature control and then showed excellent agreement between calculations and actual error as measured by the two-dimensional calibrated Zerodur ball plate method developed by Kunzmann et al. (Kunzmann, Trapet, and Wäldele 1990). They used measurement data from 56 temperature sensors with an accuracy of 0.03°C as the basis for their calculations.

Trapet and Wäldele (1989) also call attention to the fact that there is no mechanical stress on a beam made of homogeneous, isotropic material that is subjected to a uniform gradient. This conclusion is not something new, but it has come as a surprise to many people. It means that the term *thermal stress* should not be used on a casual basis. *Thermal deformation* is a better term. Deformation may or may not be associated with mechanical stress.

Another contribution on gradients has been made by Zeiss, Oberkochen, Germany (Breyer and Pressel 1991) and in the past by Sheffield, Dayton, OH (Bosch 2007, pers. comm.). Both companies use aluminum for the guideways of their measuring machines. The high conductivity of aluminum reduces the temperature difference between the top and the bottom of the beam when subjected to a given temperature gradient. This minimizes the thermal distortion of the slideways and the resultant angular motion and Abbe error. Breyer and Pressel's data concerning temperature effects show a factor-of-seven improvement of aluminum compared with granite.

The effect of room temperature gradients on temperature gradients within the machine can also be minimized by the use of insulation (Trapet and Wäldele 1989). The disadvantage of insulation is that it requires more time for the machine to reach equilibrium after a thermal disturbance.

10.4.2 TEMPERATURE VARIATION—THERMAL VARIATIONS IN TIME

The problem of temperature variation is much more complicated than that of gradients. It is not yet possible to predict the response of a machine to temperature variation before it has been built and tested. Finite element programs are not presently rigorous enough to adequately predict the transients. (See the quotations from Spur,

Harary, and Hicks in Sections 10.7.1 and 10.7.1.1.) The main problem is the thermal resistance at joints (Attia and Kops 1979).

McClure was able to demonstrate in 1965 that temperature variation has deterministic* effects on machines by accurately predicting the effect of temperature variation on an existing machine with a given structural loop and workpiece (Bryan et al. 1966; McClure 1969). He was able to make this prediction after measuring the response of the machine and a part to a step change in temperature. Any change in the setup, however, required a new step response measurement. This limitation prevented the technique from being used on a practical basis.

The need to determine a new correction algorithm every time the workpiece or the setup changes is a major obstacle in present-day efforts to compensate for errors caused by temperature variation.

10.4.3 DYNAMIC EFFECTS OF THERMAL VARIATIONS

A quotation from a 1966 ASME paper describes the effect of temperature variation on a three-element system consisting of a comparator, frame, and part (Bryan et al. p. 5, 1966):

> All length-measuring apparatus can be viewed as consisting of a number of individual elements arranged to form a "C." Figure 10.2 shows a schematic of a C-frame comparator measuring the diameter of a short section of hollow tubing. The comparator frame and the part form two elements. If the coefficient of expansion of the comparator is exactly the same as the part, the gage head will read zero after soak out at any uniform temperature that we might select. If we induce a change in temperature, however, the relatively thin section of the tubing will react sooner than the thick section of the comparator frame and the gage head will show a temporary deviation. The amount of the deviation will depend on the rate of change of temperature. If the rate is slow enough to allow both parts to keep up with the temperature changes, there will be a small change in gage-head reading. If the rate is so fast that even the thin tubing cannot respond, there will again be a small change in reading. Somewhere in between these extremes, there will be a frequency of temperature change that results in a maximum change in reading. This frequency is similar to resonance in vibration work.

To confirm the intuition on the nature of these effects, the model shown in Figure 10.2 was further simplified to that shown in Figure 10.3. Sample heat transfer calculations were made for this model and programmed on a computer. The cylinder with the displacement pickup can be considered the comparator. Both cylinders are made of steel and are 100 mm long. Cylinder A is 50 mm in diameter and cylinder B is 12.5 mm in diameter. Figure 10.4 shows the computer-predicted changes in length of the two cylinders as a result of ±1°C sinusoidal change in air temperature at a frequency of 1 cycle per hour (cph). The thick cylinder shows less than one-third of the temperature change of the thin cylinder, and its temperature lags the thin cylinder by about 4 minutes. Figure 10.4 shows the predicted gage-head reading, which is the same as the instantaneous difference in lengths of the two cylinders. This change in length is called *thermally induced drift,* or simply *drift.*

---

* Webster defines "deterministic" as "the doctrine that whatever is or happens is determined by antecedent causes." Another definition is that "the process obeys cause and effect relationships that are within our ability to understand and affordably control" (Bryan 1993).

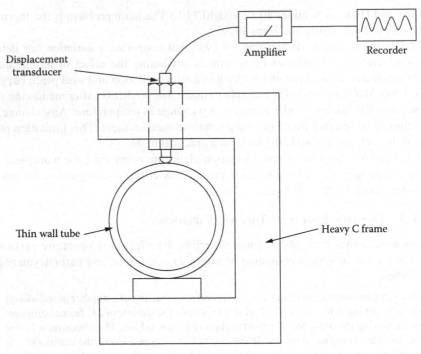

FIGURE 10.2 C-frame comparator and part. (Redrawn from Bryan, J. et al., *ASME 65, Prod. 13*, 1966.)

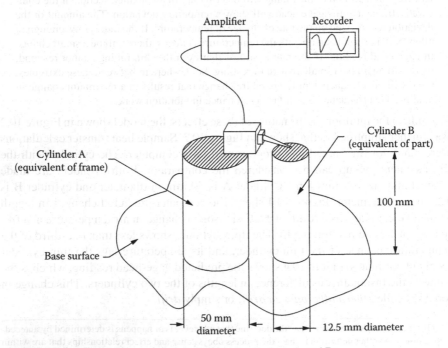

FIGURE 10.3 Comparison of the length of two cylinders, A and B.

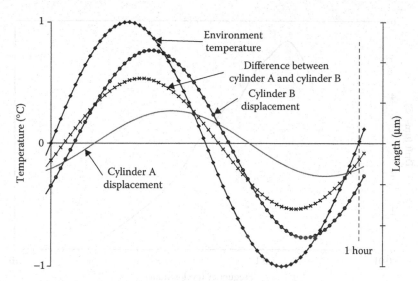

FIGURE 10.4 Plot of changes in length for two cylinders, A and B.

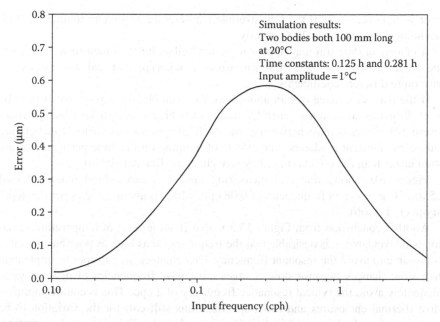

FIGURE 10.5 Frequency response for two cylinders, A and B.

The effect of varying the frequency of temperature variation is plotted in Figure 10.5. Agreeing with intuition, the drift is small for very high or low frequencies and it reaches maximum amplitude at a point in between, which is called *resonance*. Figure 10.5 is called *frequency response*. (Control engineers call it *transfer function*.) In this case, resonance occurs at 0.2 cph and has a value of 0.68 μm. This

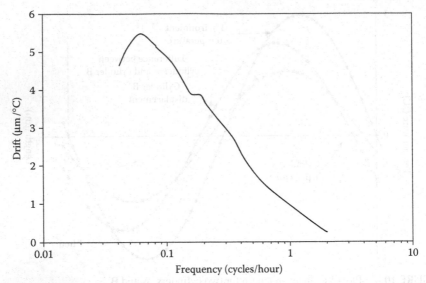

FIGURE 10.6   Frequency response of rotary contour gage, an early measuring machine.

error would occur even if the time average temperature of the environment and all mechanical elements was 20°C exactly.

A change of 0.68 μm may appear to be negligible, but real measuring machines and machine tools are built from a mixture of materials and real workpieces can have quite different coefficients.

If the part were made of thin aluminum, for example, these responses would be much larger. Real systems generally have much bigger overall lengths and more severe differences in mass between elements. Real systems also suffer from bending caused by transient gradients. The effects of bending on machine performance are often larger than the effects caused by size changes (Blaedel 1993b).

Figure 10.6 shows that real measuring machines can exhibit magnitudes of 5.5 μm/°C at resonant frequencies of 0.08 cph, which is about 2 cycles per day (cpd; Bryan et al. 1966).

Another conclusion from Figure 10.5 is that if some level of temperature variation is unavoidable, it is desirable that the frequency be as high as possible or as low as possible to avoid the resonant frequency. Fast changes are cheaper to implement than slow changes because slow changes must have frequencies of many days to adequately avoid the typical resonance frequency of 2 cpd. This is counterintuitive. Most thermal enclosures and room specifications still call for the variation to be as slow and gradual as possible. Specifications should call for the predominant air temperature frequency variation to be no lower than, say, 1 cycle per minute (cpm).

10.4.4   THERMAL MEMORY FROM A PREVIOUS ENVIRONMENT

When a part is moved to a different environment, a certain amount of time must pass before it reaches thermal equilibrium with the new environment. During this "soak-out" time, the part changes in size and may suffer temporary geometric distortion.

Serious errors can result if attempts are made to measure or machine the object during its soak out period. The effect is always dynamic, and its location on the thermal effects diagram (Figure 10.1) reflects this fact. Although the soak out problem is well known, it is frequently forgotten under the pressure of business. It can become a critical bottleneck in the flexible manufacturing system (FMS) and just-in-time (JIT) concepts of modern manufacturing if ignored.

A general equation for heat transfer by conduction or convection is

$$\dot{Q} = -hA(T - T_S) \tag{10.1}$$

Here, T is the temperature of the object, T_s is the temperature of the environment, A is the area of contact, and h is a constant that depends on the details of the heat transfer mechanism.

If heat transfer is given by Equation 10.1, change in temperature of a gage block is given by Equation 10.2. Since the temperature differences being discussed are fairly small, the same equation can be used for all heat transfer mechanisms, varying only in the coefficient h.

$$\frac{T - T_S}{T_0 - T_S} = e^{-\frac{hAt}{\rho CV}} = e^{-\frac{t - t_0}{\tau}} \tag{10.2}$$

Here, T_0 is the initial temperature, T_s is the temperature of the environment, t_0 is the initial time, and τ is the decay time (or time constant) that depends on the thermal transfer (h), area (A), density of the body (ρ), heat capacity per unit mass (C), and volume (V) of the object.

Thus, although there are more than one heat transfer mechanisms at work, it is expected that the approach to equilibrium for a gage block will be very similar to exponential decay, and the soaking time needed is dependent only on the initial temperature and the decay time τ.

The ASME B89.6.2 (p. 2) standard defines soak out as follows:

> One of the characteristics of an object is that it has a thermal "memory." When a change in environment is experienced, such as occurs when an object is transported from one room to another, there will be some period of time before the object completely "forgets" about its previous environment and exhibits a response dependent only on its current environment. The time elapsed following a change in environment until the object is influenced only by the new environment is called soak out time. After soak out, the object is said to be in equilibrium with the new environment. In cases where an environment is time variant, the response of the object is also a variable in time.

The B89.6.2 (pp. 10–11) standard includes a coolant effectiveness chart (Figure 10.7) for air versus water in natural and forced convection for metal thicknesses varying from 2.5 to 2540 mm (ASME 1973). This chart, originally conceived by R. McClure in English units (McClure 1969), shows that the film coefficient for natural convection in still air is about $1.7 \ W \cdot m^{-2} \cdot {}^\circ C^{-1}$ and that the film coefficient can be increased by an order of magnitude using forced convection. It also shows that natural convection in still liquid can achieve another order of magnitude increase, and for forced convection of liquids the film coefficient can increase by three orders of magnitude over natural convection in still air.

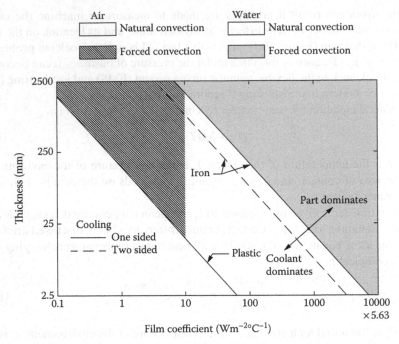

FIGURE 10.7 Coolant effectiveness chart for air versus water in natural and forced convection for material thickness varying from 2.5 to 2500 mm.

Figure 10.7 shows the coolant effectiveness chart for air versus water in natural and forced convections. The solid lines separate regions of part and coolant dominance for iron and a typical plastic. A condition within the area of coolant dominance indicates that improvement in the control of part temperature can be achieved by increasing the flow or changing the coolant. It shows that internal resistance to heat flow in metals becomes equal to film resistance when the thickness approaches 25 mm and the film coefficient is 1360 W·m^{-2}·°C^{-1}. This is the point at which an increase in film coefficient begins to produce diminishing returns. There is, of course, a significant difference between the conductivity, density, and specific heat of different metals, but this chart is intended to give order-of-magnitude values only.

10.4.4.1 Soak Out Experiments

Cooper, Pack, and Cogdell performed some experiments that generally confirmed McClure's chart (Bryan 1990). Their preliminary results showed that a steel roller bearing cup of 200 mm diameter and 25 mm thickness require a soak out time of 6 hours (5τ in Equation 10.2) to reach a temperature of 0.1°C above a 21°C ambient starting from a soaked condition at 40°C in still air. The same bearing soaked out in 25 minutes when subjected to 600 m/min air velocity and in only 2 minutes when submerged in a still, grinding coolant.

Although the most dramatic effects are seen with liquids (high heat capacity) and moderate speed air, there are very significant gains to be made with modest

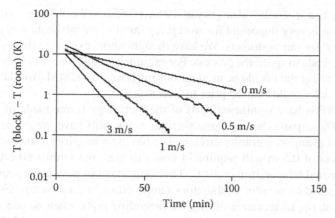

FIGURE 10.8 The change in the time needed to reach room temperature for a 500-mm gage block depends dramatically on air speed.

air speeds that are appropriate to the laboratory environment. DeFelice (1970) and Chakravarthy (Chakravarthy, Cherukuri, and Wilhelm 2002) published analyses of soak out times in air 30 years apart with very similar results. Recent experiments at NIST have shown much the same behavior.

One experiment was to compare the cooling rate of a 500-mm square gage block with various air speeds. Figure 10.8 shows the relaxation time (1/e) for the different cases. The change from still air to an air speed of 1 m/s is substantial, but higher speeds do not yield proportional gains in efficiency. An air speed of 1 m/s at 20°C is uncomfortable for humans, but most high-accuracy CMMs are computer controlled and do not need operator presence.

A few studies over the last 40 years (DeFelice 1970; Chakravarthy, Cherukuri, and Wilhelm 2002; Doiron, Schneider, and McLaughlin 2007) have found very similar effective heat transfer coefficients for the relaxation time for conduction, natural convection, and forced convection, which agree very closely with the available standard references (ASME 1973).* The typical range of the coefficient h for mechanical contact is 100–4000 $W \cdot m^{-2} \cdot K^{-1}$, free convection of air 1–30 $W \cdot m^{-2} \cdot K^{-1}$, and forced convection of air 50–150 $W \cdot m^{-2} \cdot K^{-1}$. The various heat transfer modes are quite different, but secondary effects from surface geometry, orientation, emissivity, and air speed also cause wide differences within each mode.

For parts that are significantly warmer than 20°C, more radical methods may be needed to speed up the soak out process. J. Hicks (Bryan 1990; p. 652) said the following: "The time required for soak out may be reduced by the use of a prewash environment that is set to the new temperature. The prewash may be a liquid (used to remove chips and cutting coolant) or a high velocity air stream."

* Details on laminar flow over flat surfaces can be found at "Forced Laminar Flow over an Isothermal Plate," eFunda: http://www.efunda.com/formulae/heat_transfer/convection_forced/calc_lamflow_isothermalplate.cfm, accessed March 12, 2011.

R. Kegg had a similar idea (Bryan 1990; p. 652): "We believe that adequate soak out time is very important for workpiece quality and advocate this practice at Milacron and for our customers. We have thought about exploring the use of liquid soak out methods to speed the process. For example, on machine tools, temperature-controlled cutting fluid could be used for this purpose. At a CMM installation, water showers or tanks could be used prior to measuring."

Many CMMs have significant parts of the metrology frame made of granite or cast iron. These parts, having large thermal mass, will have very long soak out times. As an example, a granite surface plate having a length of 2 m, width of 1 m, and thickness of 0.5 m will require 74 hours to soak out within ±0.2°C of 20°C from an original temperature of 40°C. This long soak out time is the reason why it is recommended to monitor a laboratory temperature 24 hours a day. For a CMM measurement, the temperature during the preceding night when no one was in the laboratory is important because machine geometry is a function of long thermal history.

10.5 DRIFT TEST

The *drift test* is a simple but powerful tool for evaluating the influence of temperature variation on machines of any size or complexity. The ASME B89.6.2 standard (ASME 1973; p. 2) defines a drift test as follows: "An experiment to determine the drift in a measurement system under normal operating conditions is called a drift test. Since the usual method of monitoring the environment (see Definition 3.13) involves the correlation of one or more temperature recordings with drift, the test will usually consist of simultaneous recordings of drift and environmental temperatures. The recommended procedure for the conduct of a drift test is given in paragraph 20.3.1."

The drift test can provide concrete justification for the need for improvement and can support the effectiveness of procedural solutions, such as working at night. It can provide proof of the enormous influence of the sun's radiation and the dramatic benefit of shielding machines from the sun. It can be used to demonstrate the influence of heat created by people coming near the machine. The effect of room lighting is easy to measure by simply turning the lights on and off and observing the resultant drift. The effect of thermal memory from a previous environment can also be demonstrated by installing a warm part on the machine and measuring its drift.

Memory from a previous environment affects the machine as well as the workpiece. When machine elements are moved from one location to another in the presence of gradients, there will be a soak out effect. The ASME B89.4.1 standard specifies that the temperature variation error (TVE) should include some of the effects of gradients by allowing the machine slides to soak out at one location before starting the drift test at another location (ASME 1997).

The drift test has been incorporated in B89.6.2 as the method to be used for determining the TVE portion of TEI (ASME 1973). Drift tests are now part of both CMM standards (ASME 1997) and machine tools standards (ISO 2002b; ASME 1998, 2005a).

10.5.1 DRIFT OF COORDINATE MEASURING MACHINES

T. Charlton, formerly of Brown & Sharpe, Providence, RI points out that the drift tests described in ASME B89.6.2 and B89.4.1 fail to detect the influence of temperature gradients and temperature variation on squareness, straightness, angular motion, and linear displacement of the axes of CMMs.

The existing drift tests can be described as multiaxis, (x, y, z) single point drift tests. Charlton proposes a multipoint, multiaxis drift test. Such a test can be implemented by repeatedly measuring a ball plate over an extended period of time with the machine operating in its normal mode of operation. Instead of reporting the change in x, y, z coordinates of each ball, Charlton calculates the difference in distances between various pairs of balls. These differences in distances are more representative of the errors introduced in the measurement of real parts on CMMs.

Trapet agrees with Charlton in distinguishing the importance of single point drift in machine tools from single point drift in CMMs (Bryan et al. 1972). The CMM drift is not as important because of the speed with which a measurement procedure can be carried out and the ability of a CMM to quickly return to the reference point. Charlton also suggests that the error caused by temperature measurement uncertainty (ISO 2003d) should be included in TEI as a fourth term.

As an example, the NIST M48 CMM has roller bearing in x and y axes. Although they are extremely reproducible, the rollers do generate small amounts of heat. Early drift tests, which consisted of motions typical of those made during calibration, showed that at speeds higher than 2.5 mm/s the machine did not reach a stable geometry for over 6 hours. To combat this, machine speed was limited and room air was circulated through the large parts of the machine structure, as shown in Figure 10.9. With these two modifications, the machine was made satisfactorily stable in less than 2 hours.

10.5.2 TRANSDUCER DRIFT CHECK

The drift test is made possible by the availability of high-sensitivity drift-free displacement transducers and recorders. The *transducer drift check* provides a simple means of proving the stability of these devices. Transducer drifts of less than 0.05 μm per day for ±1.5°C environments can be expected.

The B89.6.2 standard (p. 3) defines a transducer drift check as follows:

> An experiment conducted to determine the drift in a displacement transducer and its associated amplifier and recorders when it is subjected to a thermal environment similar to that being evaluated by the drift test itself. The transducer drift is the sum of the "pure" amplifier drift and the effect of the environment on the transducer, amplifier, and so on. The transducer drift check is performed by blocking the transducer and observing the output over a period of time at least as long as the duration of the drift test to be performed. Blocking a transducer involves making a transducer effectively indicate on its own frame, base, or cartridge. In the case of a cartridge-type gage head, this is accomplished by mounting a small cap over the end of the cartridge so the plunger registers against the inside of the cap. Finger-type gage heads can be blocked with similar devices. Care must be exercised to see that the blocking is done in such a manner that the influence of temperature on the blocking device is negligible.

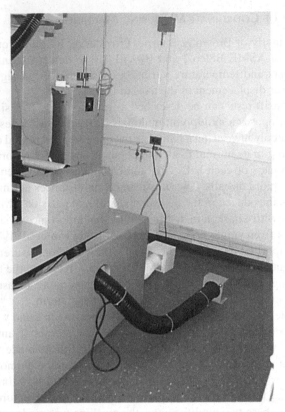

FIGURE 10.9   Room air is pulled through the base and bridge of the CMM to draw away the small amount of heat generated by the roller bearings. (Courtesy of NIST.)

The B89.6.2 standard (p. 24, Figure 7) shows some examples of how to block various types of transducers (ASME 1973).

10.5.3   THERMAL DRIFT OF LASER INTERFEROMETERS

Laser interferometers are also subject to thermal drift due to transient temperature gradients within the glassware of the system. The optical path length is dependent on glass temperature. This relationship is characterized by the parameter ds/dt, where s is the optical path length and t is the temperature. This parameter represents the combined effect of the coefficient of linear thermal expansion and the variation of refractive index of glass with temperature. A simple test for the existence of laser interferometer drift is to bolt the corner cube reflector to the interferometer body and record the interferometer readings as a function of time. Note that because of the very short air path of the interferometer, the drift is not from the changing index of refraction of air but from the actual structure of the interferometer system (appendix I of ASME B5.57; ASME 1998).

10.6   THERMAL ERROR INDEX

The B89.6.2; p. 4 standard offers two options for compliance with the standard: "A calibration, part manufacture, or part acceptance procedure complies with this standard if

1. It is carried out with all pertinent components of the measurement system at 20°C; or,
2. It can be shown that the thermal error index is a reasonable and acceptable percentage of the working tolerance."

The TEI is the estimated overall thermal error expressed as a percentage of the working tolerance. Overall thermal error is the sum of the error caused by average temperatures other than 20°C and the TVE.

The TEI is determined by taking the sum of estimated UNDE of the part and scale; value of NDE, if NDE corrections are not made; and TVE. The sum is expressed as a percentage of the working tolerance of the part. Although not standardized, twice the length uncertainty due to temperature divided by the tolerance multiplied by 100% (ISO 2003d) can be substituted for TEI.

10.6.1   EXAMPLE OF THE THERMAL ERROR INDEX

A 250-mm aluminum part having a tolerance of ±25 μm is being inspected at a temperature of 24°C ± 2°C. A CMM with a steel scale is being used. No corrections are made because the inspector has never been told that they are necessary or that it is the inspector's job to make the corrections. A drift check shows the TVE to be 2.5 μm. The TEI is calculated as follows:

Assuming the uncertainty of the coefficient of aluminum is 20% and that of steel is 10%, TEI comes to 84%. To determine the suitability of this environment for this case, management should ask themselves if they can afford to give up a possible 84% of the working tolerance to this one source of error.

The calculations are as follows:

NDE = 0.25 m (24 ppm − 11.5 ppm)(24 − 20) = 12.5 μm
Uncertainty of expansion of the scale = 0.25 m (11.5)(24 − 20)(10%) = 1.15 μm
Uncertainty of expansion of the part = 0.25 m (24)(24 − 20)(20%) = 4.8 μm
Total for scale and part = 5.95 μm
TVE = thermal drift = 2.5 μm
Total thermal error = 2.5 + 12.5 + 5.95 = 20.95 μm
TEI = 20.95/25 × 100 = 84%

The TEI can be reduced to 33% by teaching inspectors to make NDE corrections and giving them the responsibility to do so. Alternatively, the necessary equipment can be purchased to make ANDE corrections.

The TEI can be reduced to 5% by installing the CMM in an air shower box with a flow rate of 47 m/min and adjusting the temperature control to 20°C ± 0.62°C. The NDE and UNDE portions of the error approach zero. The inspectors will not complain about being too cold because they remain in an environment designed for human comfort. The drift will very likely be reduced to 1.25 μm because the magnitude of temperature variation has been reduced to a third of its original value and the frequency of temperature variation has been increased because of the high flow rate in the box:

$$TEI = 1.25 \ \mu m/25 \ \mu m \times 100 = 5\%$$

It will be necessary to wait for the part to soak out to the box temperature from the shop temperature, but this time can be reduced by a factor of 15 by using a 600 m/min soak out chamber located immediately adjacent to the CMM box.

Suppose the part tolerance is tightened by a factor of 50 to ±0.5 μm instead of ±25 μm. It is still possible to maintain the TEI at less than 5% by adopting a liquid shower instead of an air shower. The temperature of the liquid can be held to 20°C ± 0.006°C (Bryan et al. 1972; DeBra, Victor and Bryan 1986; Chou and DeBra 1990). Based on prior experience, thermal drift will very likely be reduced to 0.05 μm. The soak out time will be reduced to a few minutes because of the high convective film coefficient of the liquid (Bryan 1990).

$$TEI = 0.05 \ \mu m/1.25 \ \mu m \times 100 = 4\%$$

10.6.2 FEATURES OF THERMAL ERROR INDEX

The TEI has the desirable feature of responding to each of the variables that affect the need for temperature control, which are

- The amount of the tolerance
- The size of the workpiece
- The coefficients of the workpiece and the master
- How well the coefficients are known
- Whether or not NDE corrections are made
- The effect of temperature variation

The TEI concept is the heart of ASME B89.6.2. (This standard, like many standards, has unfortunately not yet trickled down to working engineers in industry or trickled up to management's attention.)

On the other hand, ASME B89.4.1 "Methods for Performance Evaluation of Coordinate Measuring Machines" (ASME 1997) has become a working document for the buying and selling of CMMs (Donaldson and Patterson 1983). All provisions of B89.6.2 are automatically incorporated in this standard. In addition, B89.4 calls for mandatory corrections of NDE between the scale of the machine and whatever standards of length are used for testing (laser interferometer, gage blocks, step gages, etc.). The manufacturer is also required to provide a statement giving the effective CTE of the machine scales for the purpose of making the NDE correction.

In the special case of performing acceptance tests, the CMM standard (B89.4.1) recommends the value of UNDE to be two parts in 10^6 (2 ppm) per degree Celsius of thermal offset from 20°C (1 ppm for the scale and 1 ppm for the gage artifact).

Footnote 1 (p. 20) of B89.4.1 says, however, the following: "This approximation does not apply for scales which have a nominal coefficient of expansion of zero, or properly compensated laser interferometer scales. In these cases, the user and supplier shall negotiate a value for the scale uncertainty to be used for calculating the UNDE."

10.6.3 INTERNATIONAL ORGANIZATION FOR STANDARDIZATION USE OF THERMAL ERROR INDEX

The technical committee ISO/TC 213, dimensional and geometrical product specifications and verification, has released Technical Report 16015 (ISO 2003d) that covers much of the material in ASME B89.6.2. The report makes recommendations on how to evaluate and report uncertainties resulting from workpiece and working standard temperatures, CTEs, environmental temperature variation, and other thermal effects. Examples are given in the annexes of the standard.

A subcommittee of ISO TC39 on machine tools, under the chairmanship of Dr. Alkan Donmez of NIST, has published a document (ISO 230-3) that includes the TEI as the means for determining the suitability of an environment when performing acceptance tests on machine tools (ISO 2007).

10.7 REDUCING THE THERMAL ERROR INDEX

Some possible solutions for reducing the TEI consist of compensation of the machine, compensation of the part, and direct temperature control (Sections 10.7.1 through 10.7.3).

10.7.1 COMPENSATION OF THE MACHINE

Balsamo, Marques and Sartori conducted a study on the possibilities of real-time compensation of errors induced by quasi-static gradients on an SIP 560M measuring machine (Balsamo, Marques, and Sartori 1990). This study is the most ambitious and perhaps the most successful of any such effort to date. The thermal model has the capability of handling 1400 equations. The number of thermometers used was more than 100. The authors believe that "... such a high number of thermometers is a minor problem. In fact, once the idea of placing thermometers on a machine (in particular, on its moving parts) is accepted, their number, be it 10 or a hundred, is of small importance when considering the limited cost of sensors and of the acquisition-system cards, and the necessity of solving space and cable movement problems."

A special temperature measuring system was developed, which is capable of scanning platinum resistance thermometers every 3 seconds with an uncertainty of 0.05°C. The temperature readings are fed to the thermal model, which then calculates the deformation by integration along a line from the coordinate origin to the measurement point. It then updates the machine control system every 5–10 minutes (which is the author's definition of quasi-static). The thermal model was developed using a combination of FEA and black box empirical techniques.

A three-dimensional workpiece consisting of a space frame and spheres was used. Some thermometers were placed on the workpiece, and NDE corrections were made. The material and size of the space frame were not mentioned.

Results of the study showed that 80%–85% of the deformations could be explained by the model. Automatic corrections were able to compensate for 65%–75% of the errors. The magnitude of corrected errors was more than 100 times the resolution of the machine (0.1 μm). The gradients, induced by a 2-kW panel about 1 m away, were as high as 10°C/m. Rotations of the part with respect to the machine and rotations within the machine (squareness errors) were the largest source of uncorrected errors. Sartori, however, believes the following (Bryan 1990; p. 650): "Software thermal compensation of CMMs only makes sense where high accuracies are required. It can therefore be assumed that the environmental conditions are reasonably good if (1) the average temperature is 20°C, (2) temperature variation is less than ±2°C, (3) thermal gradients along any direction are less than 2°C/m, and (4) temperature variations with time are such that quasi-stationary conditions are achieved with reference to the thermal constants of the machine."

Sartori is also of the opinion that the design of the machine should be optimized before compensation:

> All the precautions and the methods described in the literature should be adopted in order to reduce thermal effects. This is a very important point. It is in fact deceptive to believe that real-time correction for thermal deformations can eliminate design defects connected with thermal phenomena.
>
> Thermally symmetric structures, circulation of temperature-controlled fluids, appropriate disposition and screening of heat sources, structures capable of uniform and rapid distribution of the heat they produce, these, as well as others, are the essential prerequisites for defining software procedures for thermal deformation correction.

Professor G. Spur of the Institut für Werkzeugmaschinen and Fertigungstechnik in Berlin, Germany, has been working on thermal effects on a continuous basis longer than any other academic researcher in the field and has this realistic assessment of the situation regarding internal as well as environmental influences (Bryan 1990):

> Optimizing the design of the machine tool is the most effective solution, considering energy and additional components. Design measures alone are insufficient, however, to completely avoid thermal displacements. Compensation techniques are necessary for increasing the working accuracy especially of numerically controlled machine tools. Methods using correlation functions between locally measured temperatures and displacements permit only description of the displacements for machine components in steady state. These formal relations between temperatures and displacements do not deliver satisfying results for rotating components such as spindles, because the temperature field causing their displacement cannot be obtained from temperature measurements at steady state. The use of a signal proportional to the average temperature in the direction of displacement gives better results. Such a signal may be obtained by using resistance thermometers as integrated measuring sensors.

10.7.1.1 Limitations of Finite Element Analysis

H. Harary of NIST reports his experience of using a particular FEA code to solve the transient case of the temperature of the center of an imaginary metal sphere when an incremental temperature change is suddenly imposed on the outer skin of the metal (Bryan 1990). An analytical solution is available for this problem, which can be used as a test for the FEA code. The FEA results were in disagreement (with respect to the accuracy needed for measuring machines) and did not tend to converge, as finer meshes and time steps were used.

Spur, in the past, also had poor results with FEA (Bryan 1990): "According to our experience, numerical algorithms such as the finite element (FEA) or differential method are suitable only for a qualitative description of the thermal behavior of machine tools. Empirical equations have to be used for calculating the amount of thermal energy as well as the convective coefficients, as these values cannot be calculated using analytical methods."

Hicks had a similar opinion regarding FEA codes (Bryan 1990): "The finite element method is a very good means of predicting mechanical stress. However, for thermal deformations, we believe it is able to predict only about 20% of the actual deformation."

Modern efforts have concentrated on using FEA to identify the shapes of the dominant "modes" of thermal behavior (Weck et al. 1995) while using temperature sensors to evaluate the amplitude of the modes rather than relying on the FEA predictions. This bypasses the problem of thermal resistance of joints and allows better prediction of thermal effects (Zhu, Ni, and Shih 2008).

10.7.2 Part Compensation versus Temperature Control

Sartori agrees with Trapet and Wäldele that compensating for real parts is the challenge of the future (Balsamo, Marques, and Sartori 1990): "When workpieces are sizable, it is impractical to obtain a model for them, and even if modeling is possible without difficulty (families of similar pieces), it is out of the question to cover the workpiece with thermometers. It is, therefore, essential to develop optical temperature measuring devices capable of coping with the problem of emissivity effects."

Harary of NIST has some experience with optical temperature measurement devices and thinks that they will not be accurate enough (Bryan 1990). They can be useful, however, in determining the optimum locations of contact sensors.

E. Thwaite of the Commonwealth Scientific and Industrial Research Organisation (CSIRO) in Australia is equally pessimistic about part compensation (Bryan 1990): "It is not going to be possible to instrument the workpieces to the degree necessary to determine their temperature condition and, as well, their response to temperature distributions."

Professor W. Lotze of Dresden, Germany, has the following comment: "In the case of very large CMM structures and workpieces, the temperature influence is complicated by the local and time-dependent temperature distributions caused by external and internal heat sources in the CMM, as well as the temperature field in large workpieces with local zone heated by workpiece machining. In this case, the thermal deformation of the CMM and workpiece depend on so many parameters that there is no way for a full numerical error correction."

In spite of these somewhat pessimistic reports, research should continue on achieving a fundamental understanding of the effects of temperature on mechanical structures. Research should also continue on the development of practical temperature measurement systems that are accurate (0.01°C), fast, reliable, easy to install and remove, and inexpensive.

Due to the many difficulties in modeling the thermal behavior of complex geometries, it is still impractical to compensate in real time for thermally induced distortions of parts. Complexities include material differences, clamping forces, fixture design, heat transfer conditions, part geometries, surface finish, and so on. Even today, the best approach is to control the temperature of parts.

10.7.3 DIRECT TEMPERATURE CONTROL

An alternative to compensation is direct control of the temperature of the part, master, and machine frame by the use of temperature-controlled air- or liquid-showered boxes. This alternative is reliable, easy to understand, and has been demonstrated in practice over the past 40 years in diamond turning machines (Bryan 1979; Bryan et al. 1972; Donaldson and Patterson 1983; Roblee 1985; DeBra, Victor and Bryan 1986; Chou and DeBra 1990), ruling machines (Loewen 1978), integrated circuit steppers, computer numerical control (CNC) lathes (Bryan et al. 1973), CNC milling machines (McMurtry 1987), and CMMs (Bryan et al. 1982). These machines, operating in air- or liquid-showered enclosures, have repeatabilities nearly equal to their resolutions.

Some engineers are aware of these developments but regard them too expensive for normal applications. This assumption should be reexamined in the light of recent developments in the design of machine tool enclosures.

10.7.3.1 Modern Machine Tools Boxes

Modern CNC machine tools are on the verge of adopting the temperature-controlled box idea without knowing it. About 90% of recently manufactured machine tools are equipped with enclosures. These enclosures are justified for safety and environmental reasons. With the enclosure justified for other reasons, temperature control by air or liquid shower to a level of ±0.1°C can be provided at relatively small additional costs by using chilled water, which is readily available in the thermally optimized factory.

10.7.3.2 Thermally Optimized Factory

The thermally optimized factory would have temperature control similar to that found in airports, shopping malls, hotels, offices, and restaurants. Temperatures would be about 24°C ± 3°C, depending on the preferences of the employees. This kind of temperature control can be separately justified because of the merits achieved by temperature control, such as better morale, good productivity, improved equipment life, and improved health of machine operators. It can also motivate engineers to spend more time in the shop. Many large manufacturing companies are now air-conditioning their factories.

The problem of human discomfort at 20°C disappears. The machines would no longer be exposed to the influence of heat from the operator's body. In addition to its air temperature control, the thermally optimized factory would have chilled water piped throughout the building for controlling the temperature of the air or liquids used in air- or liquid-showered boxes. The chilled water would also be used to remove heat generated by local sources of heat, such as motors, hydraulic pressure supplies, and control systems.

The total amount of energy used by the factory would be minimized by removing the heat from local sources. The energy saving is due to the reduced fan horsepower required from the building's temperature control system. Fan horsepower is reduced because the air flow required to maintain the upper temperature limit is proportional to heat load. Pump power is expended to circulate the chilled water, but pumps are far more efficient than fans.

If the TEI for the most sensitive part is greater than a reasonable and acceptable percentage of the part tolerance, the machine should be placed in a temperature box. Machine manufacturers would have the responsibility for providing appropriate boxes. Users would no longer have to agonize over the additional cost and custom design of temperature control for each new machine. Users of integrated circuit steppers have had this luxury for many years.

With temperature control applied to the workpiece and the machine, there is minimal concern about mixing different construction materials, which have different coefficients of expansion, and different time constants. The least expensive materials that meet the other machine requirements can be used.

10.7.3.3 Coordinate Measuring Machine Boxes

A. Wirtz, formerly of the University of Buchs, Switzerland, told an interesting story about his experience with the administrative cutoff of close-tolerance (high-flow) air-conditioning for his metrology laboratory, which he used in order to save energy (Bryan 1990). He solved the problem by building a box around his Leitz CMM and installing an air shower, which was able to hold the machine to 20°C ± 0.1°C. He was able to save energy, provide better temperature control than before for the machine that needed it very much, isolate the machine from the operator's body heat, and provide more comfortable temperature control for the students and faculty who use the rest of the laboratory for less thermally critical measurements, such as surface finish.

10.7.3.4 Coordinate Measuring Machine Manufacturers Boxes

Some CMM manufacturers have acknowledged the advantages of boxes. Figure 10.10 shows the photograph of a commercially available, temperature-controlled, air-showered CMM box.*

10.7.3.5 Coordinate Measuring Machines and the Lead of Machine Tools

The best diamond turning machines, operating in their temperature-controlled boxes, are much more accurate than the best CMMs (Bryan 1979; Bryan et al. 1972; Donaldson and Patterson 1983). High-quality CNC production milling machines have accuracies that are nearly equal to those of average-quality CMMs. This is a sad state of affairs, because the traditional rule of "10 to one" calls for inspection machines to be 10 times better than production machines.

---

* Customers also have the option to customize solutions—see Case Study 2 in Chapter 16, Typical Applications.

FIGURE 10.10   Zeiss measuring machine box. (Courtesy of Carl Zeiss IMT.)

When the thermally optimized factory becomes a fact, machine tools, operating in their temperature-controlled boxes, would behave deterministically and demonstrate repeatabilities approaching their resolution. It would then be a relatively simple matter for machine tool builders to adopt the error correction techniques used on CMMs (Zhang et al. 1985).

10.8   DESIGN OF TEMPERATURE-CONTROLLED ROOMS

Temperature-controlled rooms would continue to play a role in the thermally optimized factory for a variety of special circumstances. The design of temperature-controlled rooms is still evolving. At the moment, it is a rather primitive business. There is relatively poor communication between the buyer and the seller about the quality of the product. There are many conflicting theories about design principles. There is a wide range of costs for the same product, and the average cost is higher than it should be. Rooms that are designed and built in-house can be very good or disastrously bad.

This situation is in contrast to the buying, selling, and construction of clean rooms. The clean room industry is now a mature business with good standards, methods of test, and design principles. There is good competition and good value for money.

Some progress in the design of temperature-controlled rooms has been made in recent years. The clean room people are beginning to understand the special requirements for dimensional metrology and precision engineering. The special requirements are, for example, an average temperature of 20°C and an air velocity limitation (for human comfort) of 6 m/min. This conflicts with the standard clean room requirement of 30 m/min air velocity, with no particular emphasis on the absolute temperature.

The use of hanging plastic strips (similar to those found in some supermarkets) for the walls of temperature-controlled enclosures has been pioneered by D. Thompson of Lawrence Livermore National Laboratory (LLNL), California, and others. These walls are now referred to as "soft walls." They are economical and effective. C. Evans, formerly of NIST, reports that these soft walls are amazingly efficient for blocking the influence of infrared radiation.

P. Sideris of LLNL provides some information on the current cost of permanent, temperature-controlled rooms in the western part of the United States (Bryan 1990). His experience (in 1990) is that 80% of the average cost of new office/laboratory buildings comprises the cost of constructing the building and the remaining 20% comprises the cost of providing normal heating and air-conditioning for human comfort (23°C ± 2°C). If high-quality temperature control is required for precision engineering activities, typically 20°C ± 0.2°C, the cost of heating, ventilating, and air-conditioning (HVAC) is between two and three times as much as that for a normal building. This additional cost is primarily for the additional air flow and the uniform distribution that is required (fans, ducts, false ceilings, and extra refrigeration for cooling the extra fan horsepower). The cost of better control is not as great an added expense as most people would expect.

The accuracy gains from a well-designed thermal space can be dramatic. One advantage of a new space is that since most CMMs are programmable and can run without an operator in the laboratory, the comfort of the operator is an unnecessary constraint on air speed. Forced air convection raises the effective thermal coefficient, and provides a very effective way of reducing the temperature variations of a CMM. In the new Advanced Metrology Laboratory (AML) at NIST, the new thermal space for high-accuracy CMMs was designed to have 300 air changes per hour with laminar flow. Coupled with this is the fact that nearly all of the heat sources are outside the room; as a result, the space is remarkably good. But even in this room, the air had small variations in temperature across the machine volume, which affected the stability of machine geometry; the laminar flow preserved the temperature variations down to the machine.

Removing the panels above the machine made the air flow turbulent, which averaged out variations in the air supply and nearly doubled the air flow in the machine space. The temperature at the walls of the room is not as good as the original specifications, but it does not affect the machine geometry or the results of measurements. In the previous space, which had much lower air flow and larger gradients, the performance was significantly enhanced with small fans aimed at critical parts of the machine, as shown in Figure 10.11.

In the new AML space, the reproducibility of the machine is reduced by nearly half, as shown by a number of check standards measured with every calibration. Nearly all of this gain in performance is related to temperature control. The importance of air flow to control the thermal state of instruments cannot be overstated. With large air flow the gradients that remain are usually stable, and as long as the gradients are made small in the volume of the machine that actually moves the measurements would be stable. It is nice to control the whole CMM space, but only the volume of the machine's movement actually needs to be controlled for stable measurements.

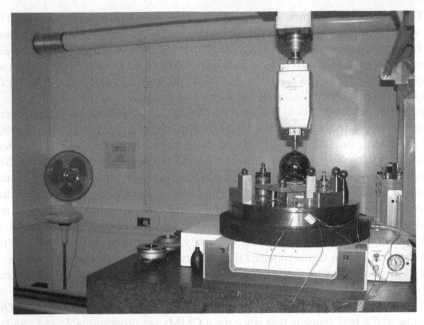

FIGURE 10.11 At the left of the picture, there is a small fan that was used to circulate air through the middle of the machine to reduce small but persistent temperature gradients in measurement volume. (Courtesy of NIST.)

10.9 RELATIONSHIP BETWEEN REFERENCE AND WORKPIECE TEMPERATURES

A question frequently arises concerning the apparent discrepancy between the need for close-tolerance temperature control in manufacturing and the wide temperature fluctuations experienced by workpieces in operation. The answer to this discrepancy is that temperature control in manufacturing is necessary to establish the initial clearances between mating workpieces on an interchangeable basis. Clearances between parts (at one temperature, usually the reference temperature) must be known by the designer so as to manage these clearances over the full range of operating temperatures.

An example of good clearance management is the piston and cylinder of a model airplane engine. These engines generally cost about $20 each. Piston rings are not used in these engines to reduce the cost. The clearance between the 10-mm diameter piston and the cylinder is 1–2 μm at room temperature to achieve sufficient compression to start the engine. When the air-cooled engine is running at 300 revolutions per second (rps), the temperature rises by many hundreds of degrees. The material used for both the cylinder and the piston is free machining steel. The piston, however, is case hardened and heat treated to give the necessary hardness for long life. The case hardening process has an important side effect. It reduces the CTE of the piston. When the engine is hot, the cylinder expands more than the piston, which increases the clearance. The clearance can be much larger when the engine is running, since there is very little time available (3 milliseconds) for leakage.

To maintain interchangeability and control, the clearance must be held between 1.0 and 2.0 µm and the tolerance on the piston and cylinder of these engines must be held to ± 0.5 µm. A temperature change of 4.3°C will change the diameter by 100% of the tolerance. To maintain the TEI at 5%, it will be necessary to control the average temperature at 20°C and keep the thermal drift (TVE) of the manufacturing process to 0.025 µm, which will probably require a liquid shower maintained at 0.01°C. Such a tolerance on the temperature of liquids is not very difficult to achieve.

Machine designers use a wide variety of techniques to manage clearances. In addition to the selection of appropriate materials, they use controlled preload, springs, gaskets, and gaps for thermal expansion as found on railroad rails and bridges. A well-informed machine designer provides an extra margin of safety to take care of the UNDE. A machine that does not have properly managed thermal clearances simply fails in service.

In the days before interchangeable manufacturing became a reality, a workman would adjust the size of one part to fit its mating part, and temperature control in manufacturing was not necessary (see also Chapter 1, Evolution of Measurement).

10.10 SUMMARY

Thermal effects are the largest source of apparent nonrepeatability and inaccuracy in most CMM applications. Some actions can be taken in the short term while other actions must be regarded as long term. Some suggestions are given in Sections 10.10.1 and 10.10.2.

10.10.1 NEAR-TERM ACTIONS TO REDUCE THERMAL ERROR INDEX

The following are some actions that can be followed to reduce TEI:

- Obtain copies of B89.6.2, "Temperature and Humidity for Dimensional Measurement" (ASME 1973); ISO/TR 16015:2003, "Systematic Errors and Contributions to Measurement Uncertainty of Length Measurement Due to Thermal Influences" (ISO 2003d); and ASME B89.4.1, "Methods for Performance Evaluation of Coordinate Measuring Machines" (ASME 1997).
- Begin a training program to teach inspectors, machinists, and manufacturing engineers how to make NDE corrections. Assign the responsibility for performing these corrections.
- Consider buying the necessary CMM upgrades for temperature compensation to permit ANDE corrections.
- Buy the necessary equipment (linear variable differential transformers, recorders, thermographs) and start making single-point, three-axis drift checks to demonstrate the magnitude of the effect of temperature variation. If there are windows in the workshop, the drift check can be useful

in demonstrating the powerful influence of solar radiation, radiation from room lights, and the influence of human bodies.

- Buy a ball plate or similar artifact (see Chapter 9, Performance Evaluation) and measure the plate repetitively over time and calculate the change in distances between pairs of balls to show the influence of changing thermal gradients on straightness and squareness of the axes.
- Determine the TEIs for a variety of manufacturing and necessary procedures. Determine the costs of not controlling the temperature.

10.10.2 LONG-TERM ACTIONS TO REDUCE THE THERMAL ERROR INDEX

Consider adopting the thermally optimized factory concept. It offers the following advantages:

- Eliminates the problem of human discomfort at 20°C
- Allows measuring machines and machine tools to reveal their intrinsic capability for achieving repeatabilities near their resolution
- Allows the achievement of TEIs of less than 5% on most operations at affordable costs
- Allows the use of low-cost CMM construction materials with mixed coefficients and time constants
- Reduces the number of temperature sensors required
- Reduces the consumption of power because of the smaller volumes that need to be temperature controlled at 20°C and the reduced fan horsepower necessary for temperature control in the factory

APPENDIX: HOW MEASUREMENT OF TEMPERATURE AFFECTS THE MEASUREMENT OF LENGTH

Quotations by H. Kunzman 1989; p. 14 provide perspective on the interaction between temperature and length:

Among all the material properties which are important for the accuracy of length measurements, the temperature-proportional expansion in length is the most important. According to the international standard of 1931, the reference is 20°C.

For discussion of accuracy limits, it is important to know that a temperature of 20°C can be realized only with an uncertainty of 0.2 mK.

It is of interest to know that a change to the International Practical Temperature Scale (IPTS-68) was recently agreed upon (1989), which establishes a new standard called the International Temperature Scale-1990 (ITS-90)*, which will (has) result(ed) in the reference temperature of 20°C being changed by −0.005°C (colder) effective January 1, 1990. This change of definition will (has) led to an overnight increase in the length of all objects made of steel (11 ppm/°C) by about 5.5×10^{-8} or 0.55 μm in 10 m.

In addition to the uncertainty of temperature scales discussed above, the uncertainty of temperature measurements and their effect on the accuracy of length measurement must be considered. Under very good conditions, an optimistic assessment

---

* (Preston-Thomas 1990).

allows a measurement uncertainty of 0.001°C to be obtained in the measurement of instrument and workpiece temperatures. From this it follows that for materials having a coefficient on the order of $1 \times 10^{-5}/°C$ (some steels) the present limit of uncertainty of length measurements must be assumed to be on the order of 1×10^{-8}, which corresponds to 0.01 μm/m.

In most cases, the uncertainty of temperature measurement is greater than 0.01°C and the actual values of the coefficients of thermal expansion of the workpiece and/ or the instrument are not known. For materials having a coefficient on the order of $1 \times 10^{-5}/°C$, a realistic uncertainty of length measurement is about $1 \times 10^{-7}/°C$ or 0.1 μm/m.

ITS-90 defines temperatures in the range of 0°C to 30°C as a specially defined, curvilinear interpolation between the triple point of water (0.0100°C) and the melting point of gallium (29.7646°C).

11 Environmental Control

Paulo H. Pereira and Jun Ni

CONTENTS

A coordinate measuring system (CMS) comprises four basic elements: (1) the specimen (workpiece), (2) the operator, (3) the instrument (a coordinate measuring machine [CMM*] or similar†), and (4) the environment in which it operates. Each element is critical to the measurement process.

In order to have a CMS that is accurate despite environmental influences, there are four possible approaches (Moore 1970; Slocum 1992a; Teague 1997; Luttrell 2007). A CMS can be designed to reduce environmental influences (e.g., by employing symmetry), to use materials insensitive to such influences (e.g., materials having low coefficient of thermal expansion), to compensate for the influences, or finally, to control the environment to minimize its variations.

This chapter covers environmental controls, with more emphasis on temperature and vibration aspects. Manufacturers specify these conditions because they affect the performance of the systems‡ (ASME 1997, 2008; ISO 2009a).

11.1 IMPORTANCE OF TEMPERATURE CONTROL

Environmental temperature§ is one of the most obvious yet least understood of CMS elements. Environmental control is consequently often less than acceptable. Compensating for thermally induced errors is another option, but it presents limitations. The linear coefficient of expansion is accurate only to approximately ±10% for steel and ±15% for aluminum and is based on simple linear geometry, homogeneity, and purity of the material with an assumption that temperature is uniform throughout the specimen. The following examples show the potential errors introduced by each of these variables under changing temperature conditions.

A commonly used value for the coefficient of thermal expansion for steel is $11 \ \mu m \cdot m^{-1} \cdot {}^\circ C^{-1}$ with an uncertainty of $\pm 1.1 \ \mu m \cdot m^{-1} \cdot {}^\circ C^{-1}$. If a 1-m linear steel part is to be measured with a temperature variation of ±1°C, the possible error from just the coefficient of expansion uncertainty is ±1.1 μm. For aluminum, a typical coefficient is $22 \ \mu m \cdot m^{-1} \cdot {}^\circ C^{-1}$ and the error under similar conditions is ±3.3 μm. These calculations are for simple linear geometries and do not consider changes in configurations. Since most machine parts have complex three-dimensional geometries and can include alloys of different materials, the errors caused by temperature-induced distortions are considerably larger.

Another critical assumption often used to perform a temperature compensation calculation is that the skin temperature is the same as the temperature at the core of a component. Unless a part has been allowed to soak out, or stabilize, at that temperature, a temperature gradient is present within the part. This gradient is usually of unknown magnitude and distribution, and as a result it cannot be accurately compensated. If a part is brought from an uncontrolled factory environment to a

---

* The term CMM is used generically in this chapter to include all measuring machines capable of measuring coordinates, including traditional Cartesian as well as other types, such as AACMMs and laser trackers.

† See also Chapters 4, Cartesian Coordinate Measuring Machines, and 17, Non-Cartesian Coordinate Measuring Systems.

‡ See also Chapter 9, Performance Evaluation.

§ See also Chapter 10, Temperature Fundamentals.

20°C* metrology laboratory, the part will change continuously with a varying gradient within the part until it reaches equilibrium with the new temperature.

Compensating for the disturbances induced by thermal gradients is extremely complex and beyond the capabilities of most compensation software packages. For example, determining the temperature values of an engine block and their rate of change is very difficult. Even if this could be done with reasonable accuracy, it would still be very complex to estimate, and then compensate for, the thermally induced distortions. When a very low measurement uncertainty is required, even a simple linear geometry (like a step gage) cannot have its temperature gradient properly compensated for and the best approach is to control the temperature (Zurcher 1996).

Temperature effects cause linear and nonlinear strains and deformations of the workpiece as well as the measuring equipment. Temperature effects are also responsible for drift in electronic instruments. Environmental control is an effective way to eliminate many of the thermal error sources. It is necessary not only to control temperature at the measurement point but also to consider all the following factors:

- Temperature (at the control point)
- Temperature as a function of time (temporal variations)
- Temperature—its horizontal and vertical gradients (spatial variations)

Even CMMs designed to be less sensitive to thermal effects cannot operate in all temperature conditions and, in general, they perform better when in a temperature controlled environment. Certain models have different performance specifications for different temperature conditions. A user needs to carefully analyze all the factors before deciding on which model to acquire. A shop floor model may have the same performance specifications as those for a CMM intended for a temperature-controlled environment, but a shop floor machine implies that part temperatures are not controlled, which could negatively influence the measurement results. In that case, the part temperature influences on measurement uncertainty must be carefully considered ahead of time (see also Chapter 14, Measurement Uncertainty for Coordinate Measuring Systems).

11.1.1 TEMPERATURE CONTROL

The three mechanisms of heat transfer do influence temperature stability in a CMM room:

1. Radiation: Walls, windows, ceiling, floor, lamps, control systems, personnel, and other items may have a different temperature from the ambient. In a production environment, neighboring machines are heat sources.
2. Convection: Air conditioning, computers, control systems, drafts.
3. Conduction: Machine internal heat sources, personnel, floor.

In designing a temperature-controlled room, consideration must be given to the factors discussed in Sections 11.1.1.1 through 11.1.1.6.

---

* The standard reference temperature for dimensional measurements is fixed at 20°C (ISO 2002a). Doiron (2007) details how this reference temperature came to be used.

11.1.1.1 Direction of Airflow

Airflow can be directed either vertically or horizontally. For vertical airflow, there is a choice of upward or downward flow, which influences the ducting configuration.

11.1.1.2 Entrance and Egress of Airflow

In both vertical and horizontal airflow designs, it is important not to have large objects obstructing the airflow in a path. If an upward vertical airflow is chosen, the floor needs to be perforated for the air to pass through. For downward airflow, the ceiling has a pattern of holes (pressurized plenum) and the air returns are through the floor or the bottom of peripheral walls.

11.1.1.3 Velocity and Volume of Airflow

The velocity and amount of airflow to be circulated in the room per unit time is another factor considered while designing the room. The system should provide constant airflow. The energy conservation practice of turning the system on and off as required is devastating for maintaining the expected level of temperature control. The volume of air needs to be around 20–30 air changes per hour, but this is dependent on temperature control requirements, traffic, number of people using the room, and heat sources inside the measurement room or laboratory (DoD 1998). A large number of heat sources, increased traffic, or tighter temperature control requires more air changes.

11.1.1.4 Mode of Temperature Regulation

The temperature control should be on constantly. In most temperature-controlled systems, the airflow is forced first through cooling units and then through heating units to produce a constant temperature with low humidity.

11.1.1.5 Location of Temperature Sensors

The selection of locations of temperature sensors does affect the overall temperature field distribution in a controlled environment. There may be a single sensor or a number of sensors at different locations. No localized heat source (like a computer heat exhaust fan) can be placed close to any of the sensors.

11.1.1.6 Air Lock or Soak Out Room

The design of air locks is often used to minimize disturbance of the room's environment as personnel or parts move into or out of the room.

11.1.2 Common Practices to Reduce Thermal Influences

Sections 11.1.2.1 through 11.1.2.4 discuss some general methods for reducing temperature influences.

11.1.2.1 Reducing Radiation Effects

Some commonly applied methods for reducing the effects of heat radiation are as follows:

- Apply shielding to walls to reduce emissivity. Aluminum foil or sheet is well suited for this purpose. Shielding can significantly reduce the need to apply temperature control to walls.

- Apply shielding between the floor and the measuring equipment.
- Keep the lights and electric equipment on at all times: If they are turned on and off, a cyclic thermal error is introduced. Provide filtering of the light from lamps.
- Personnel should not be in the room during measurement of very tight tolerances with direct computer-controlled (DCC) CMMs, or if present they should wear radiation-shielding clothes.
- Apply coating to windows to reflect radiation from the outside and to reduce emissivity to the inside of the room.

11.1.2.2 Reducing Convection Effects

Some commonly applied methods for reducing the effects of heat convection are as follows:

- Provide air locks as the only access to the temperature-controlled area. The temperature within an air lock must be the same as the temperature of the area it protects.
- Use ventilating systems to dispose of heat generated by computers, control units, and other sources of heat. One way of doing this is by enclosing the heat-generating equipment and then passing cooling air through it.
- The air-conditioning system must be designed to produce a stable temperature distribution and to eliminate the influences of operating personnel and other localized heat sources.

11.1.2.3 Reducing Conduction Effects

To reduce the effects of heat conduction, apply thermal isolation between the CMM structure and the foundation.

11.1.2.4 General Methods

Some general methods to ensure temperature control in a CMM room are as follows:

- The CMM should be placed at least 1.5 m from the outer walls. The less temperature-sensitive instruments and accessory equipment should be placed in the outermost zones of the room.
- Temperature-sensitive equipment can be provided with reflecting walls (sandwich constructions from aluminum sheets and Styrofoam™) for additional reduction of radiation influences and to protect against drafts.
- Having the area of interest surrounded by other temperature-controlled areas may be necessary to achieve tighter specifications. Locating the area to be controlled underground is also an option when tighter specifications are required.
- Direct manual contact with workpieces and machine components must be avoided when tight tolerances are to be achieved. When necessary, cotton gloves are to be used and handling time minimized.
- The workpiece to be measured should be in thermal equilibrium with the CMM and the environment before starting the measurement task.

11.1.3 Temperature Recording

Continuous temperature recording of the environment where the CMM is installed is very important to ensure the quality of the results. Not only the current temperature but also the immediate history of variations can affect results. Depending on specifics like size of the room, the CMM model, severity and duration of temperature disturbances, and the required uncertainty, temperature fluctuations can affect measurement results many hours or even days after they occur. The recording device can be tied to an alarm system to warn users and maintenance personnel when temperatures vary beyond specified limits.

Part temperature recording can also be useful. Most CMMs do offer some way to measure part temperatures. On select models, this task can be programmed so that the machine performs the temperature measurement. On other models, operators need to manually place a sensor on the part to be measured. In both cases, part temperature readings can be added to the inspection reports like any of the measured dimensions. Although the use of part temperature readings for compensation can be challenging (as mentioned previously in Section 11.1), having that information on measurement reports can be useful for assessing potential thermal influences on the results.

11.2 HUMIDITY CONTROL

The usual humidity range requirement for metrology laboratories is 40%–60% relative humidity (RH). It is also important to know if any other items of the measuring systems are affected by this condition. For example, certain measuring machines have laser scales that can be significantly affected by humidity variations.

Gage blocks and unfinished steel corrode above 50% RH. The Instrument Society of America (ISA) and the National Council for Standards Laboratories* (NCSL) both recommend a maximum of 45% RH for gage blocks and related items (ISA 1975; NCSL 2000). A stable humidity is also important for the longevity and maintenance of granite. Large humidity variations can be deteriorating to unprotected granite. (It is not uncommon for granite used in metrology applications to be coated with a protective layer making it more resistant to moisture.)

11.3 DUST CONTROL

CMMs are susceptible to accumulation of dust and grime. Their bearing surfaces and scales require periodic cleaning. A controlled environment ensures less dirt and thus reduced cleaning and maintenance requirements. The ISA and NCSL (ISA 1975; NCSL 2000) both recommend a clean room class from 7 to 8[†] for metrology laboratory environments. A class 7 room is 10 times cleaner than a class 8 room. A positive room pressure of at least 12.4–24.8 Pa (Braudaway 1990) is also recommended to

---

* Now National Council for Standards Laboratories International (NCSLi).
[†] Both references use classifications as per the Federal Standard 209E (GSA 1992), which was canceled in 2001 (GSA 2001) in favor of ISO 14644-1 (ISO 1999a) used in this chapter.

provide an outflow through openings, preventing the suction of particulates into the controlled environment.

Measuring equipment that is intended for installation and operation in a shop floor environment without special enclosures can also be affected by dust and other airborne particles. Certain environments can be very harsh when certain welding or forging operations are present. Shop floor CMM models, laser trackers, and articulated arm CMMs (AACMMs), in general, have some sort of limitation with respect to maximum levels of dust and particulates in the air. Users need to consider this and be very specific in explaining to a supplier the actual conditions under which the system is to be installed or used.

11.4   VIBRATION ISOLATION TREATMENTS

The demands for higher measurement accuracy highlight external vibration as a potential problem area for CMMs (Meredith 1996). A reduction in measurement accuracy can occur if a CMM is in a location where it is excited by the floor or the surrounding environment (DeBra 1992; Rivin 1995). As with temperature effects, certain CMM models are less sensitive to vibration influences than others and each environment varies in its vibration characteristics from others.

11.4.1   THE NEED FOR VIBRATION ISOLATION

Ideally, if all components of a CMM, including the part to be measured, were to vibrate in unison at a specific frequency, amplitude, phase, and orientation, no degradation in measurement performance would result due to vibration. To the CMM, this situation represents the same condition as no vibration excitation whatsoever since there would be no relative motions. It is only when components begin to move out of phase with each other, or vibrate, that accuracy problems occur.

For vertical ram CMMs, the influence of vibration on accuracy in the vertical z axis is typically less severe than that in the horizontal x and y axes due to the relatively higher stiffness of the ram. External disturbing vibrations may, therefore, be suspected if a specific vertical ram CMM exhibits acceptable performance in the vertical axis but worse in the horizontal axes.

Some CMM users may attempt to minimize external vibration effects by confining their most accurate measurements to more stable times of the day. For example, a CMM user in a manufacturing facility may elect to conduct critical measurements during a break, shift change, or other similar times during the day when plant activity is at a minimum. Although this fix may provide a temporary solution to the problem of external disturbing vibrations, it obviously compromises the value of the CMM to the user.

11.4.2   SOURCES OF EXTERNAL DISTURBING VIBRATIONS

Vibration transmission components can be categorized into one of three broad categories, as shown in Figure 11.1 (Meredith 1996), as follows: (1) source, (2) path, and (3) receiver.

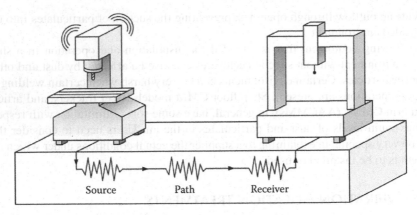

FIGURE 11.1   Schematic diagram of vibration source, path, and receiver.

11.4.2.1 Source

Vibration can be created by a multitude of different sources. In a typical manufacturing facility, some of the more common vibration sources may include

- Production equipment (machine tools, stamping presses, forging hammers, reciprocating air compressors, shaker feeders, conveyor systems, etc.)
- Heating and ventilation equipment (out-of-balance fans, blowers, pumps)
- Lift trucks and product-moving equipment within the plant
- External rail or road traffic in the vicinity of the installation
- Airborne (acoustic) vibrations
- Shipping and receiving areas (pallets being dropped, truck impacts to a dock)

The most effective method of minimizing a vibration problem is to control its source. For example, if the operation of a blanking press causes accuracy problems for a CMM in the immediate area, the best method available to control this vibration is to stop the operation of the press. If this is not possible, the placement of vibration isolation mounts beneath the press reduces the amount of vibrational energy transmitted into the floor structure.

Although source isolation is the most effective method for controlling vibrational energy generated by a piece of production equipment, this solution has limitations when other potential sources are considered. The installation of dock bumpers may help to minimize vibration caused by truck impacts, for example, but source isolation becomes impractical for minimizing vibration due to low-flying aircraft, nearby highway traffic, or oceanic activity (waves, tides).

11.4.2.2 Path

For most installations, the path for vibrational energy is the structure common to both the source and the receiver. The soil beneath and between a forge press (source) and a CMM (receiver) is a common example of the paths that vibration can traverse. Other paths include building structures, piping, and ductwork. Airborne vibration (i.e., noise) may also result in vibration problems in extreme cases. The installation of a CMM within a climate-controlled room or enclosure typically solves most airborne noise problems since the walls of these rooms generally exhibit high airborne-noise-reduction capabilities.

The most common method of minimizing vibration effects through path modification is to simply increase the distance between the source and the receiver. Placing the CMM as far away as possible from potential vibration sources is an effective method of vibration control. Unfortunately, for many installations, exactly the opposite is required. The CMM may need to be located adjacent to a piece of production equipment manufacturing the items it is intended to measure.

Other path modifications may involve the use of inert materials between the source and the receiver. Sand, for example, does not transmit vibratory energy as readily as clay soil. The placement of a CMM on top of a pit filled with sand may provide the required vibration attenuation capabilities (if problems with compaction and settling are properly addressed).

11.4.2.3 Receiver

Controlling the transfer of vibration to the receiver, although less effective than either source or path control, is the most common method for isolating a CMM. The placement of a CMM on an isolation system provides a stable support for it. The cost for doing this can be included in the initial installation budget.* The design of an appropriate CMM foundation must address a number of aspects. Both existing and future potential vibration sources must be considered in the design phase. Long-term stability of the isolation media used (if any) is also critical to the success of the foundation.

It is important to understand that a vibration-free CMM foundation cannot be obtained. Even if an extremely efficient vibration isolation system is used, residual vibration is always present. It is the goal of the engineer to ensure that the residual vibration to the CMM meets the criteria put forth by the machine's manufacturer (see Section 11.4.5).

Several CMM manufacturers have undertaken development projects to harden their machines against vibrations. This concept involves the stiffening and strengthening of internal CMM components in order to render them less susceptible to vibratory motion. The addition of vibration isolation or damping products during CMM assembly can also result in increased capability for the machine to perform properly under more hostile vibration conditions. The CMM manufacturer should always be consulted in the event of a vibration control problem to determine the possibility of a retrofit for the specific unit. This potential modification to a CMM is beyond the scope of most users to install by themselves and should be coordinated through the manufacturer or agent.

11.4.3 Vibration Isolation System Types and Characteristics

Vibration isolation systems can be used in conjunction with a CMM in order to reduce site vibration levels to meet manufacturer requirements. Such systems permit a CMM to be placed in a high-vibration environment and yet perform to its full accuracy and repeatability.

---

* See also Chapter 19, Financial Evaluations.

Active vibration control involves the use of force-generating transducers, which are fed by an amplifier and a vibration-monitoring device (DeBra 1998). The transducers produce a vibratory force equal in frequency and amplitude to the vibration detected by the vibration-monitoring device but 180° out of phase with the latter. In principle, this equal-but-opposite vibratory forces result in a net cancellation of the disturbing vibration at the installation location.

Passive vibration isolators can take several forms, but they share several common characteristics. They are comparatively easy to design, install, and troubleshoot. They have been proved in thousands of installations, and their initial cost is typically modest. It is therefore reasonable to expect that passive isolators will continue to be used for years to come (Rivin 2003). Passive isolators can be grouped into three broad categories: (1) pads, (2) springs, and (3) air springs.

11.4.3.1 Passive Isolators: Pads

Vibration isolation pads are typically manufactured from neoprene, fiberglass, felt, cork, or other similar compressive materials. Their natural frequencies generally fall within a range of 5–30 Hz, and they can be manufactured in a variety of sizes, thicknesses, and load-carrying capacities. Pads (Figure 11.2) are the least expensive type of isolation to purchase, and they exhibit a high damping rate when excited at their natural frequency.

A point of concern with the use of pads is their stability over time. Many types of pads continue to creep and compress when statically loaded, resulting in a change in elevation for the isolated equipment. In addition, some pad materials (primarily neoprene) exhibit age stiffening, so that the material's natural frequency slowly increases over time. This can result in the isolation capability of the material becoming less effective with time. It is obviously important that users carefully review and discuss these issues with pad suppliers and take proper measures.

11.4.3.2 Passive Isolators: Springs

Helical coil springs are available in many different sizes and load-carrying capacities. They are classified in terms of deflection under load. A 25-mm spring, for example, is a spring that compresses by 25 mm when supporting its rated load. The natural frequency range for coil springs is typically 1.5–6 Hz, corresponding to deflections

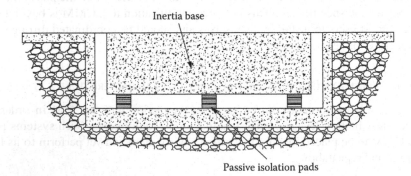

Inertia base

Passive isolation pads

FIGURE 11.2 Passive inertia isolation system.

of 6–100 mm. Springs exhibit fairly low damping rates, so amplification at resonance could be a concern. Although typically more expensive than pads, springs are nevertheless economical. Properly manufactured coil springs do not suffer from creep or settling and are generally manufactured within housings that are equipped with isolator adjustment bolts to facilitate equipment leveling.

11.4.3.3 Passive Isolators: Air Springs

Air springs, the most expensive type of passive isolators, offer the user a number of unique features. With natural frequencies typically falling in the 0.75–4.0 Hz range, they are the softest and the most efficient isolators available. Since the air pressure within the spring is easily varied, a user can quickly adjust the height, load-carrying capacity, and stiffness to meet installation requirements. The air spring system can be equipped with height-sensing valves, which automatically adjust the air pressure to compensate for isolated equipment load changes (due to CMM bridge movement or part of different weights) and to maintain a preset equipment level.

Air spring isolators can also be designed to provide an adjustable damping rate, which permits a high degree of user fine-tuning to match individual installation requirements. Figure 11.3 shows the schematic of a foundation with air springs.

Figure 11.4 shows a set of three such air springs ready to use with a bridge-style CMM. This example does not have an inertia base; rather, it is intended to function on top of a normal factory floor (usually 100 mm thick or so). Additionally, a simple saw cut could separate the concrete slab where the springs are to be mounted from the surrounding areas. This would reduce even further the vibratory energy getting into the system.

11.4.4 Inertia Bases

An important part of some isolation systems is the use of a suitably designed inertia base. An inertia base is the platform supported by vibration isolators, upon which the isolated equipment (CMM) is placed. The inertia base is typically manufactured from concrete or heavy steel and is designed to match the support requirements of the equipment to be isolated (see Figures 11.2 and 11.3). Inertia bases are usually required for larger (gantry-style) CMMs.

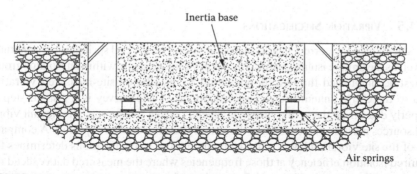

FIGURE 11.3 Inertia base with air spring isolators.

FIGURE 11.4 Air springs for a bridge-style CMM.

Inertia bases provide a number of benefits. The base lowers the overall center of gravity of the isolated equipment, which minimizes the tendency for it to rock on the isolators. When the vertical center of gravity of the isolated equipment and base coincide with the isolator centerline, the isolator rocking modes are decoupled, which results in an extremely stable installation. Usually, the heavier the inertia base the smaller the equipment motion resulting from a given vibratory input. An inertia base also provides a stiff, firm foundation for the isolated equipment to rest upon. Bases can be designed with pockets to accept alternate isolator types in the event that the isolation requirements change in the future. Isolation performance changes can then be readily accomplished without significant rework.

When an inertia base is used with a pit around it (Figures 11.2 and 11.3), one has the choice to apply temperature control on it as well. By circulating air through the pit, for example, and making it a part of the environment, the inertia base becomes a heat sink that helps in attenuating possible disturbances to temperature.

11.4.5 VIBRATION SPECIFICATIONS

The selection of the correct vibration isolator group (pads, coil springs, or air mounts), as well as the unique isolator natural frequency requirements within an isolator group, is best accomplished through the matching of installation-site-measured vibration data to the CMM manufacturer's specifications. A site survey is a critical step to properly determine the actual vibration levels. During the survey, all relevant vibration sources must be present so that the correct baseline can be captured. A comparison of the site vibration levels with the manufacturer's specifications determines the required isolation efficiency at those frequencies where the measured data exceed the criteria, as illustrated in Figure 11.5. A review of a vibration isolator manufacturer's

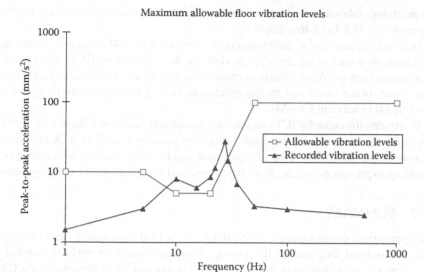

FIGURE 11.5 Measured vibration levels versus manufacturer's criteria. (Courtesy of Hexagon Metrology, Inc.)

technical product literature or discussions with the project's engineers can then pinpoint the specific isolation products that are appropriate for the project.

A safety factor on the order of two or three should be considered in the selection of vibration isolation products for a specific project. Future increases in equipment accuracy requirements or potential changes in site vibration levels should be anticipated. Amplification at isolator resonance can become a concern if the site vibration frequencies should coincide with the isolator's natural frequency. The amount of potential amplification varies with the isolator type, and this characteristic must be reviewed and understood during the design process.

11.5 SOUND LEVEL CONTROL

Sound levels must be controlled for two reasons: (1) The noise is generated by vibration and CMMs are sensitive to it, and (2) human comfort. Most CMMs tend to be susceptible to vibrations of 5–20 Hz, but many more frequencies must be addressed. The American Society of Heating, Refrigerating, and Air-Conditioning Engineers (ASHRAE) provides "noise criteria" (NC) curves. Most professionals recommend a maximum of NC 50 sound level (ASHRAE 1993). Ambient noise and vibration must also be considered. In cases where the ambient noise and vibration levels are extreme, a new location for the CMM should be considered or additional isolation should be provided.

11.6 OTHER CONSIDERATIONS

A metrology room also needs to have proper illumination for people to work, and this must be considered when designing the environment and its temperature control.

For metrology laboratories, the recommended lighting level or illuminance is 1000 lux (lumens · m^{-2}; ISA 1975; Rea 2000).

In some special cases, electromagnetic interference (EMI) may be a concern. Although most measuring machine models are not affected by EMI, some parts to be measured can be. Also, certain operations may produce considerable electromagnetic waves in the vicinity of the equipment. In such special cases, the design of the room needs to account for EMI.

Computed tomography (CT) measuring equipment (see also Chapter 16, Typical Applications) has special requirements due to generation of radiation (X-ray) during measurement. Hence, this kind of equipment needs to be installed with appropriate shielding to prevent personnel from being exposed to radiation during its operation.

11.7 SUMMARY

An appropriate environment is critical for successful measuring machine applications. To avoid long and costly rework of the installation or performance below requirements, environmental factors must be considered when planning for a CMS installation. When estimating the budget for a CMS installation, the cost of environmental control must also be included.* Variations in environmental conditions could result in the recalibration of all measurement equipment and considerable waste due to incorrect measurement results (excessive scrap, rework, internal and external failures). The CMS site needs to be selected carefully with consideration to operational needs as well as limitations arising from vibration or other existing or future environmental conditions. Consider the use of air locks or soak out rooms to minimize temperature fluctuations, especially when no environmental control is applied on the surrounding areas.

---

* See also Chapter 19, Financial Evaluations.

12 Error Compensation of Coordinate Measuring Machines

Guoxiong Zhang

CONTENTS

With the demanding developments of science and technology, higher accuracy requirements are presented to coordinate measuring machines (CMMs). A CMM is a complex piece of equipment. There are many factors causing measurement uncertainty (see also Chapter 14, Measurement Uncertainty for Coordinate Measuring Systems). These include not only 21 geometric errors (on three-axis machines) but also probing errors, errors of various accessories, and those caused by environmental factors, such as temperature, besides the ones caused by operational conditions, such as force deformation and dynamic errors. The accuracy of a CMM can be enhanced by improving its design, increasing its manufacturing accuracy, and by placing it in a controlled environment (see also Chapter 11, Environmental Control). Such an approach is called "error avoidance" and should be the first consideration. However, in many cases, it is difficult to achieve the required high measuring accuracy purely by improving the manufacturing accuracy and the strict control of the environmental and operational conditions. Even if it is feasible technically, it might lead to higher cost of the machine. To economically achieve high accuracy, error compensation has been widely applied to CMMs.

To measure the workpiece accurately, it is required to have high-accuracy measuring instruments. The reference elements of machine tools and measuring instruments must have accuracy higher than the workpieces to be machined or measured. As an example, in a roundness measuring instrument, the rotation of the spindle is often taken as the reference for measurement. When this is done, it is required that the spindle system have higher rotational accuracy than the workpieces to be measured. However, the accuracy of the roundness measuring instrument is often checked by a reference ball. The issue of what the final reference is appears. There are, of course, self calibration methods for the spindle/reference ball system, including the multistep method and Donaldson reversal (see also Chapter 13, "Reversal" Techniques for Coordinate Measuring Machine Calibration; Evans, Hocken, and Estler 1996 and references therein). In fact, researchers are always seeking ways to compensate various types of errors to make machine tools better able to make parts more accurately than the machine itself and to make the instrument be able to measure parts more accurately than the instrument itself. To accomplish this, error compensation is essential and plays an important role in the development of science and technology. This modern era requires the use of error compensation techniques to achieve higher manufacturing and measuring accuracy.

There are several methods to compensate various types of errors. For example, the errors in indexing tables and in optical grating systems are eliminated to a large degree by averaging. In angular measuring systems, the errors caused by mounting eccentricity and error motions of the spindle are also reduced by averaging. In linear variable differential transducers and in differential amplifiers, common mode interferences are reduced by using a differential principle, which is also an example of error compensation.

Error separation methods, such as reversal methods, multistep methods, and multipoint methods, which are widely used in roundness and straightness, as well as rotational and straightness error motion measurements, are extensions of the above-mentioned two methods. In these cases, roundness error of the part and roundness error motion of the spindle, and straightness error of the part and straightness error motion of the guide system are separated using different combinations of these errors. Error separation techniques are also an important means of error compensation.

Closed-loop methods are a fourth type of error compensation, which has found wide application in closed-loop amplifiers and tracking systems. In closed-loop amplifiers, the characteristics of open-loop amplifiers have almost no effect on the final behavior of the amplifier. In closed-loop tracking systems, the accuracy is mainly determined by the zero detection error and error in the feedback loop.

A fifth commonly used error compensation method is calibration. The basic idea consists of calibrating the machine tool or measuring instrument by using a more accurate instrument and introducing corrections in accordance with the calibrated results. Gage blocks are often used in accordance with calibrated values. Lead screws are often calibrated and corrections for pitch errors are introduced. This method is also widely used in CMMs that are calibrated by using more accurate instruments, such as laser interferometers, or reference artifacts, such as ball plates. The calibrated errors are stored in software and recalled during CMM operation. Corrections are introduced to enhance measuring accuracy.

An error forecasting method is often used for dynamic or complicated processes. The machining error of a machine tool can be predicted based on the results of previously machined parts and correction introduced according to the forecast value to improve the accuracy of subsequent parts.

Error compensation can be understood as a procedure for separating all kinds of noise, interferences, and all foreign signals from measurement signals. All signal-processing techniques deal with error compensation in a certain sense. Error compensation techniques will be expected to develop further.

12.1   CLASSIFICATIONS OF ERROR COMPENSATION TECHNIQUES

12.1.1   REAL-TIME AND NON-REAL-TIME ERROR COMPENSATIONS

The basic idea of error compensation of CMMs is as follows. The CMM is calibrated by using more accurate instruments or reference parts, and the corrections are introduced during its operation to enhance measuring accuracy. The error calibration can be conducted in advance. The calibrated data are stored in software and recalled during CMM operation. This type of method is called *non-real-time error compensation*.

It is also possible to calibrate a machine while it is measuring a workpiece. Correction is introduced immediately. This kind of method is called *real-time error compensation*.

In non-real-time error compensation, corrections are introduced in accordance with precalibrated results. Only stable systematic errors can be compensated this way. Time-dependent and pseudo nonsystematic errors cannot be compensated. In real-time error compensation, errors are measured in real time and both systematic and time-varying errors can be compensated.

Mostly non-real-time error compensation techniques are used in CMMs because highly accurate and expensive instruments are needed for calibrating the machine, and the calibration work is not an easy job. In the case of non-real-time error compensation, all the calibration work can be carried out by the CMM manufacturer. The calibration data can just be stored in the computer, and the user can perform error compensation. However, for real-time error compensation, the CMM must be equipped with highly accurate and expensive instruments, which should carry out the calibration work during operation.

In most cases, error compensation aims at achieving high accuracy at low cost. In cases where systematic errors dominate, non-real-time error compensation is a reasonable choice for achieving improved accuracy at low cost.

There are still certain strict accuracy requirements for manufacturing CMMs and environmental conditions for their operation when error compensation techniques are applied. As stated above, due to economic reasons, non-real-time error compensation is mainly applied in CMMs. Because only systematic errors can be compensated, the machine must have high repeatability. Both the construction of the machine and the environmental conditions should assure that the CMM has only simple thermal deformations in the form of expansions, which are relatively easy to compensate. The machine should have small probing and dynamic errors. In this case, the measuring uncertainty is mainly caused by geometric errors and thermal expansion of the machine. The geometric errors are systematic, which can be compensated by non-real-time error compensation. Thermal errors caused by departure from 20°C change with time and should be compensated by real-time error compensation.

On the other hand, there are also cases where the purpose of error compensation is not reducing the cost but achieving a high accuracy, which is difficult to do without using error compensation. For example, two scales, (1) and (5) (Figure 12.1), are often used in high-accuracy CMMs with large y travel Y_m.

The scales can be high-accuracy optical gratings or laser interferometers. By measuring x displacements of gantry (2) along two lines in accordance with the readings of scales (1) and (5), the yaw angle $\varepsilon_z(X)$ of gantry (2) can be determined and a real-time servo control is introduced for compensating the yaw error. Real-time error

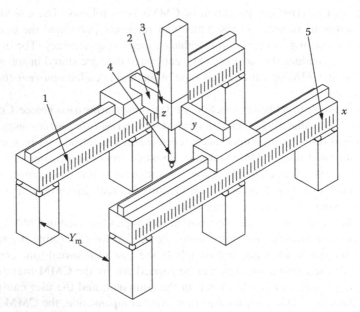

FIGURE 12.1 Diagram of a gantry machine with dual scales. (1) x scale; (2) gantry; (3) carriage; (4) ram; (5) second x scale.

compensation for x displacements of probe is made by interpolating the readings of scales (1) and (5).

Real-time error compensation possesses another distinguishing advantage. Because the calibration and measurement are carried out simultaneously, it complies with the principle of identity, that is, the working conditions of real measurement and calibration are almost the same.

12.1.2 SOFTWARE AND HARDWARE ERROR COMPENSATIONS

After the machine errors have been calibrated, these errors can be corrected by different approaches. For example, if it is discovered that a carriage does not move along a straight line, it could be corrected by rescraping the guideways. This type of correction is called *hardware correction*. However, this could be considered rework for improving manufacturing accuracy and not a case of error compensation. A second case is shown in Figure 12.2 in which certain voltages are applied to piezoelements (PZTs) P_1 through P_4 in accordance with calibrated straightness error motion data. The PZTs change the gaps between the bearings and guideways (2), and the table (1) will move along a straight line. Straightness and yaw error motions are compensated in this fashion. In this case, the error compensation is realized by hardware (piezoelements P_1 through P_4), so it is another way of hardware error compensation.

The second approach does not make any correction for the motion trajectory but only introduces some corrections in the measured data to eliminate the effect of the error motions. This type is called *software error compensation*. It is also called *computer-aided accuracy enhancement*.

Another approach is shown in Figure 12.3. Error compensation software calculates the error motion of the component ΔX in accordance with its command distance X. This value is subtracted from the command value X. The actual command sent to the feeding mechanism equals $X - \Delta X$. This command is compared with the feedback from the scale. The motion component stops when the scale shows $X - \Delta X$. However, there is an error motion ΔX. The actual distance moved by the component is X.

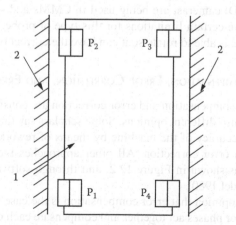

FIGURE 12.2 Error compensation for the trajectory of a working table.

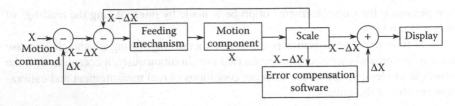

FIGURE 12.3   Error compensation by modifying the motion command.

The display shows the sum of the scale reading and the error motion and equals X. For compensating straightness error motions, the error values should be sent to another motion axis. However, this method cannot be used for compensating for the effects of angular errors.

This approach is the simplest and does not suffer from the problem caused by the dynamic response of the servomechanism. This approach has wide application. Precalibrated errors can also be input directly into the servo equations so that the correct position and the command position are the same. This is preferred by many users (J. Bryan, personal communication 2009).

Software error compensation can be used not only for CMMs but also for machine tools (Hocken 1980; Donmez et al. 1986). Certainly, for machine tools, it is required to change the relative position between the tool and workpiece and to make the machined part have the correct size and form. For this purpose, the compensation signal is sent to the controlling software and added to the command position. This approach can be also used for CMMs as mentioned above.

Software error compensation has certain restrictions in its application. It is mainly applicable for "point" machining and "point" measuring, that is, there is only one point on the cutting tool or measuring probe, which interacts with the workpiece as in the case of turning or measuring with a touch-trigger probe. However, in the case of drilling, it is impossible to change the direction of the drill by introducing a correction to the command position. In this case, it is better to use the servo drive shown in Figure 12.2. More new probes with large sensing area, for example, charge-coupled device (CCD) cameras, are being used in CMMs and software error compensation would have certain limitations for this type of probe. The machine table shown in Figure 12.2 is able to rotate so it can have the correct orientation.

12.1.3   Error Compensation, Error Correction, and Error Separation

In most cases, error compensation and error correction are considered synonymous. However, there are also different opinions. Some scholars say that only the approach for improving the accuracy of the machine by means of rework based on machine calibration is called error correction. All other approaches including those using servomechanisms, as shown in Figure 12.2, and through software are called error compensation (Blaedel 1980).

There is another opinion that error compensation is the case when several errors with different signs or phases act together and compensate each other. Angular measurement with multiple reading heads, index table with all teeth meshing, differential amplifier reducing common mode signals are some examples of applying error compensation. The cases in which some additional measures are taken to eliminate the existing errors, such as the diagram shown in Figure 12.2, and error correction of gage blocks, are examples of error correction.

Because there is no common terminology, such minor differences in these terms will be ignored in this chapter by using the commonly accepted term *error compensation* for all cases.

Error separation is another approach used to minimize errors to increase machine accuracy. Chapter 13, "Reversal" Techniques for Coordinate Measuring Machine Calibration discusses these techniques in more detail.

12.2   KEY TECHNIQUES OF COORDINATE MEASURING MACHINE ERROR COMPENSATION

The key techniques of CMM error compensation include establishment of a mathematical model, machine calibration, and development of compensation software or device.

12.2.1   ESTABLISHMENT OF MATHEMATICAL MODEL

To make a complete error compensation for a CMM, measurement errors within the whole working volume and under different environmental and working conditions should be known. However, it is not a simple task even without considering environmental and working conditions, or the errors caused by probes and accessories. For a three-axis machine with $1 \times 1 \times 1$ m volume, if it is required to know errors every 50 mm in each direction, there are $21^3 = 9261$ points distributed as a three-dimensional (3D) grid that should be measured and the vector error at each point is 3D. So, there is a total of $3 \times 9261 = 27{,}783$ error components. This does not include the direction of approach (reversal errors). It is a very big job to calibrate the machine and to measure all these errors accurately. These errors vary with changes of environmental and working conditions. A realistic approach is to establish a reasonable mathematical model for the CMM. The mathematical model sets up the relationship between the machine functional errors and the individual (or parametric) ones. The functional errors of the CMM at any point and under different environmental and working conditions can be determined and compensated based on its mathematical model by knowing its individual errors, including geometric errors, thermal errors, and so on.

The model should be accurate and simple. A model is an abstraction of the real machine. To express the main behavior of a machine mathematically, some approximations are unavoidable. However, the model ought to reflect the real behavior of the machine, that is, the inaccuracy caused by approximations should be within certain limits. Also, the model should not be too complicated. The model must be easy to use in practice. Some compromises are needed in model selection.

The mathematical model of a CMM depends on the types of error with which it deals. Depending on that, the mathematical model of a CMM can be classified as a geometric model, a thermal model, and so on. The mathematical model of a CMM depends on how functional errors are expressed. In most cases, the functional errors of a CMM are expressed as errors of the measured point in Cartesian coordinates. Conversely, for machines with rotary tables, functional errors are often expressed as errors in a cylindrical coordinate system. The functional errors can also be expressed as task-specific errors.

The mathematical model of a CMM also depends on the machine type. The geometric model of a machine mainly depends on the available motions of the machine components. In this aspect, 3D CMMs can be classified into four basic groups: **FXYZ, XFYZ, YXFZ,** and **ZYXF** (Zhang et al. 1988). In the name of each group, the letters before F show available motions of the workpiece with respect to the machine frame, and the letters after F show the available motions of the probe with respect to the frame. Both the CMM with moving bridge shown in Figure 12.7 and the gantry model shown in Figure 12.1 belong to **FXYZ** type, where a CMM with a moving table and fixed bridge belongs to **XFYZ** type machine. The letters are arranged in accordance with the axes relationship. For example, in the gantry CMM shown in Figure 12.1, which is an **FXYZ** type machine, gantry (2) moves in *x* direction with reference to the machine frame, which is fixed. Carriage (3) moves in *y* direction with reference to gantry (2), which moves in *x* direction. Ram (4) moves in *z* direction with reference to carriage (3), which moves in *y* direction. For a CMM with rotary table, there is an additional rotational motion **R**, and if the workpiece rotates with the rotary table, an extra letter **R** should be put in front of the type name of the machine. Hence, the full names of a machine with a rotary table would be **RFXYZ**, or **RXFYZ**, and so on.

Mathematical models should be established individually for different types of errors such as geometric and thermal errors. Mathematical models should be discussed individually for machines with different types of motions and different functional errors. A discussion of mathematical modeling is detailed in Section 12.3.

In order to create a model, a notation for the individual errors must be established. Figure 12.4 shows a typical translation axis of a machine having six degrees of freedom with six possible errors. For a machine with three Cartesian axes, there would be 18 errors from them plus three squareness errors between the axes, totaling 21. Table 12.1 lists all of them along with their notation.

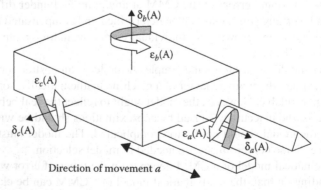

FIGURE 12.4   Six possible errors of a translation axis.

TABLE 12.1

Error Notation for a Three-Axis Machine

Description	Error Notation
Linear displacement error in each of the axes (total 3)	Generic form: $\delta_a(A)$ Errors: $\delta_x(X)$; $\delta_y(Y)$; $\delta_z(Z)$
Straightness error in the "b" direction due to movement in the "A" direction.	Generic form: $\delta_b(A)$ Errors: $\delta_y(X)$; $\delta_z(X)$; $\delta_x(Y)$; $\delta_z(Y)$; $\delta_x(Z)$; $\delta_y(Z)$
Angular error roll (around the axis of movement)	Generic form: $\varepsilon_a(A)$ Errors: $\varepsilon_x(X)$; $\varepsilon_y(Y)$; $\varepsilon_z(Z)$
Angular error pitch	Generic form: $\varepsilon_c(A)$ Errors: $\varepsilon_z(X)$; $\varepsilon_x(Y)$; $\varepsilon_x(Z)$
Angular error yaw	Generic form: $\varepsilon_c(A)$ Errors: $\varepsilon_y(X)$; $\varepsilon_z(Y)$; $\varepsilon_y(Z)$
Squareness error between two axes	Generic form: α_{ab} Errors: α_{xy}; α_{xz}; α_{yz}

12.2.2 MACHINE CALIBRATION

The main requirements for machine calibration are (1) accuracy, (2) identity, (3) completeness, (4) error evaluation, and (5) simplicity.

The purpose of error compensation is to enhance the functional accuracy of a CMM. The machine errors can be compensated only if they have been measured accurately. Generally speaking, it is required to calibrate the machine with an uncertainty less than one-fifth to one-third of its original errors. Otherwise, it would be difficult to achieve any significant effect.

The machine should be calibrated in the same way it functions, which is the main requirement of identity. Only in this case, the calibrated results will give the real errors in CMM function and the error compensation will be fully effective. As mentioned previously, error compensation should be effective for the whole measuring volume and for different environmental and working conditions. The calibration should cover all these aspects and be complete.

The requirement of individual error evaluation comes from when the machine errors can be reduced by rework or servo control, as shown in Figure 12.2, or when a mathematical model can be used for determining functional errors of the CMM and compensating them by software. There are two types of methods in machine calibration—parametric and indirect methods. In parametric methods, all the individual errors are measured directly. When indirect methods are used, all the individual errors are derived from a series of measurements (Kunzmann, Wäldele, and Salje 1983).

In most cases, accuracy enhancement needs to be accomplished at a low cost. Simplicity of calibration work is the basis for achieving low cost. The requirement of simplicity includes two aspects. First, the calibration work should not be time consuming and should not require high qualification of the operator; and second, the devices used for calibration should not be expensive or delicate. The requirement of simplicity is also driven by the following factors. Each CMM has its own error map. The calibration work for error compensation is often carried out at the customer site. In the ideal case, the customers should be able to calibrate the CMM by themselves and recalibrate it periodically during its useful life.

The commonly used methods for machine calibration are discussed in Chapter 9, Performance Evaluation. Here, only some major points, to which special attention should be paid, are explained.

1. To derive the individual errors as required: Quite often more than one error is at work when a machine is being calibrated. For example, temperature errors are involved while linear displacement errors $\delta_x(X)$, $\delta_y(Y)$, and $\delta_z(Z)$ are calibrated. It is important to separate thermal errors from displacement errors by measuring the temperature of the scale or artifact used for machine calibration and introducing necessary compensation for thermal errors in order to obtain correct values of displacement errors for subsequent compensation. For non-real-time error compensation, only repeatable or systematic errors can be compensated and hence the repeatable components of errors should be extracted from calibration. However, measurements of systematic errors are always accompanied by uncertainties that must be computed (JCGM 2008a).

   But it does not mean that all errors must be treated separately. If several errors always act together in the same way in machine calibration, then functionally they can be treated together as a single one. An example is shown in Figure 12.5, in which (1) is the machine table and (2) is the moving bridge. Beam (3) deforms under the weight of carriage (4) and ram (5). This deformation changes when carriage (4) moves along beam (3). Beam (3) also has straightness errors on its guiding surface. Both errors are functions of the Y displacement of the carriage.

   The measured straightness and angular error motions are the sums of errors caused by both these factors. These two always act together in the same way in machine calibration and in CMM function.

   The machine errors are coordinate dependent. A simplifying assumption that is usually made is that each error only depends on the position or movement of a single axis (axis independence assumption). By combining errors that act in conjunction, this can be highly efficient as far as calibration, modeling, and compensating machines (Hocken et al. 1977). It is

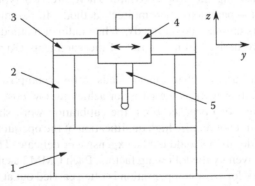

FIGURE 12.5 Effective straightness error motion. (1) machine table; (2) moving bridge; (3) beam; (4) carriage; (5) ram.

important to keep the coordinate system in error compensation the same as that in machine calibration. For carrying out error compensation, the CMM must have a machine coordinate system.

It is well known that the displacement errors measured along different lines have different values due to Abbe offsets and angular error motions. The sensing point during machine calibration should be the same as that in CMM operation. The vertex point of cube corner P_1 is the sensing point during machine calibration when it is calibrated by a laser interferometer, as shown in Figure 12.6. The center of the measuring sphere of probe P_2 is the sensing point during machine function. A tip offset $Z_p = (b - a)$ must be introduced in error compensation. The same should be said for the case when the CMM is calibrated by an artifact. The center of the measuring sphere of probe P_1 should have the same position as P_2 in machine operation. Otherwise, a tip offset equal to the difference between their positions should be introduced. (Note that the angular errors of the ram axis need to be calibrated and applied to the probe offset for a correct calibration.)

2. To reduce the uncertainties of systematic errors: In most cases, the non-real-time error compensation is used for CMMs, which compensates only systematic errors. However, uncertainties are always involved with systematic errors. To get better results, these uncertainties should be reduced. As a common practice, the machine errors should be calibrated repeatedly and their averages should be taken as systematic components and used for error compensation.

3. To select proper interval for error calibration: Theoretically, the machine errors such as positioning, straightness, and angular error motions should be calibrated continuously. However, in reality, all errors can be sampled only at discreet intervals. Increasing the number of sampling points increases the work for machine calibration. When the machine is calibrated by certain artifacts such as step gages and ball plates, the density of sampling points is restricted by the artifacts. Great attention should be paid to short-period errors, such as the subdivision errors of an optical grating scale. A laser interferometer is recommended for checking the short-period errors with very high density (e.g., several micrometers) within a small range. Only

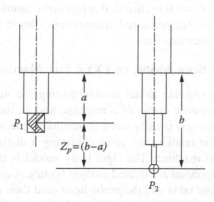

FIGURE 12.6 Tip offset.

after confirming that short-period errors are negligible, the machine can be calibrated with relatively large intervals (e.g., several tens of millimeters) or by artifacts to assure that interpolation between two adjacent sampling points does not cause significant error in compensation.

12.2.3 COMPENSATION SOFTWARE AND DEVICE

Generally speaking, it is not difficult to error compensate a CMM, once the mathematical models have been established and all the individual errors have been measured. Error compensation needs to be carried out in the machine coordinate system. Part coordinate systems are specific to each part and the calculations for compensation need a stable reference. When compensation takes place in software, the coordinate points get transformed to the machine system for that and then back to the part coordinate system before the final output.

In case of problems, such as a collision or power outage for example, a machine may lose its reference (or machine) coordinate system. In that case, it needs to be homed to find the physical references on each of its scales, or "home." To home a machine, one just needs to run a canned routine available in all systems.

Practical approaches for creating and storing the compensation values can take different forms. The geometric errors can be stored in the form of tables or fit equations. In the first case, the errors of points, which are not given in the table, are calculated by interpolation. In the second case, there are always some approximations.

CMMs with software compensation work under certain limitations. This becomes important especially in CMMs corrected off-line and equipped with sensors, such as CCD cameras, which measure the workpiece not by individual points but in an areal mode (see also Chapter 7, Multisensor Coordinate Metrology). For such machines, hardware or real-time software error compensation possesses clear advantages.

12.3 MATHEMATICAL MODEL OF MACHINE ERRORS

A mathematical model describes the relations between the functional volumetric errors of a CMM and its individual (or parametric) errors. Models for geometric and thermal errors should be discussed separately. In most cases, geometric errors, which include 21 component errors, are the major contributors in the total error budget. Once geometric errors are minimized, thermal errors usually become the largest ones (see also Chapter 10, Temperature Fundamentals; Bryan 1990). The geometric model of a machine is discussed first.

12.3.1 QUASI-RIGID BODY MODEL OF FXYZ TYPE MACHINE

The major task of a geometric model is determining the mathematical relationship between final volumetric errors of a machine and its individual components. To simplify the mathematics, usually some assumptions are made. These assumptions should simplify the relationship while obtaining a mathematical relationship close to that of the real machine. The rigid body model is the simplest model in which all machine components are treated as rigid bodies. A quasi-rigid body model assumes that the machine table and the probe head (and their respective coordinate systems) are rigid. For the CMM shown in Figure 12.5 beam (3) may have significant deflections. If the deformation can be expressed as a function of only one argument, the deformation of beam (3) has the same effect as the straightness error of an absolute rigid beam. The CMM behaves like a rigid body one.

It is worth while to start the discussion with a particular type of machine, a moving bridge CMM that is the most commonly used and is of **FXYZ** type.

Figure 12.7 shows a machine containing one body component, table (1), and three moving ones: bridge (3), carriage (5), and ram (7). Four independent coordinate systems are required to describe the system. They are the table system **OXYZ**, bridge system $\mathbf{O_1X_1Y_1Z_1}$, carriage system $\mathbf{O_2X_2Y_2Z_2}$, and ram system $\mathbf{O_3X_3Y_3Z_3}$. All the machine parts, which are fixed together and have no relative motion, can be treated as a single component. For conceptual purposes, bridge system $\mathbf{O_1X_1Y_1Z_1}$ and carriage system $\mathbf{O_2X_2Y_2Z_2}$, shown in Figure 12.7, are being attached to bridge (3) and carriage (5), respectively, through small nonexistent connecting rods to aid in visualization. System $\mathbf{O_3X_3Y_3Z_3}$ is attached to ram (7). It is assumed that at the start position, all four origins coincide and the axes of all four systems are aligned.

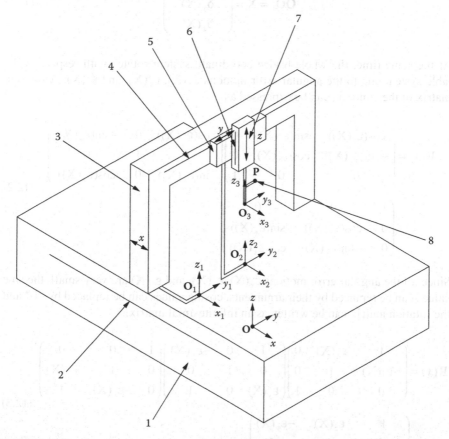

FIGURE 12.7 Coordinate system on CMM with moving bridge. (1) table; (2) x scale; (3) moving bridge; (4) y scale; (5) carriage; (6) z scale. (Redrawn from Zhang et al., *Ann CIRP*, 34(1):445, 1985.)

In CMMs, the relative position of the measuring tip with reference to the ram coordinate system origin should be determined. The measuring tip can have certain offsets X_{p3}, Y_{p3}, Z_{p3} with reference to the origin of the ram coordinate system O_3. Such an offset is often caused by a motorized head and extension rod. Tip offsets X_{p3}, Y_{p3}, Z_{p3} are determined by the configuration of probe (8), the shape of the extension rod used, and the rotation angles of the motorized head (if used), which are independent from machine axis motion. However, CMM error motions may change the tip position in the workpiece system, and this change causes an error in measurement. The task is to determine the functional relationship between tip position change and error motions of the machine.

When the bridge moves a nominal distance X, the displacement of the origin of the coordinate system connected to the bridge with respect to the table system is modeled by a vector.

$$\mathbf{OO_1} = \mathbf{X} = \begin{pmatrix} X + \delta_x(X) \\ \delta_y(X) \\ \delta_z(X) \end{pmatrix} \tag{12.1}$$

At the same time, the whole-bridge coordinate system rotates with respect to the table system due to the angular error motions $\varepsilon_x(X)$, $\varepsilon_y(X)$, and $\varepsilon_z(X)$. A rotation matrix of the x motion can be expressed as

$$\mathbf{R}(x) = \begin{pmatrix} \cos(\varepsilon_z(X)) & \sin(\varepsilon_z(X)) & 0 \\ -\sin(\varepsilon_z(X)) & \cos(\varepsilon_z(X)) & 0 \\ 0 & 0 & 1 \end{pmatrix} \begin{pmatrix} \cos(\varepsilon_y(X)) & 0 & -\sin(\varepsilon_y(X)) \\ 0 & 1 & 0 \\ \sin(\varepsilon_y(X)) & 0 & \cos(\varepsilon_y(X)) \end{pmatrix}$$

$$\begin{pmatrix} 1 & 0 & 0 \\ 0 & \cos(\varepsilon_x(X)) & \sin(\varepsilon_x(X)) \\ 0 & -\sin(\varepsilon_x(X)) & \cos(\varepsilon_x(X)) \end{pmatrix} \tag{12.2}$$

Since all the angular error motions $\varepsilon_x(X)$, $\varepsilon_y(X)$, and $\varepsilon_z(X)$ are very small, the sine values can be replaced by their arguments, cosine values can be replaced by "1," and the rotation matrix can be written as an infinitesimal matrix.

$$\mathbf{R}(x) = \begin{pmatrix} 1 & \varepsilon_z(X) & 0 \\ -\varepsilon_z(X) & 1 & 0 \\ 0 & 0 & 1 \end{pmatrix} \begin{pmatrix} 1 & 0 & -\varepsilon_y(X) \\ 0 & 1 & 0 \\ \varepsilon_y(X) & 0 & 1 \end{pmatrix} \begin{pmatrix} 1 & 0 & 0 \\ 0 & 1 & \varepsilon_x(X) \\ 0 & -\varepsilon_x(X) & 1 \end{pmatrix}$$

$$= \begin{pmatrix} 1 & \varepsilon_z(X) & -\varepsilon_y(X) \\ -\varepsilon_z(X) & 1 & \varepsilon_x(X) \\ \varepsilon_y(X) & -\varepsilon_x(X) & 1 \end{pmatrix} \tag{12.3}$$

For point $\mathbf{P}$ with coordinates (X_p, Y_p, Z_p) in the table system and coordinates (X_{p1}, Y_{p1}, Z_{p1}) in the bridge system,

$$\mathbf{P} = \mathbf{X} + \mathbf{R}^{-1}(x)\mathbf{P}_1 \tag{12.4}$$

where $\mathbf{P} = (X_p, Y_p, Z_p)^T$ and $\mathbf{P}_1 = (X_{p1}, Y_{p1}, Z_{p1})^T$ are vectors of $\mathbf{OP}$ and $\mathbf{O_1P}$, representing coordinates of point $\mathbf{P}$ in the table system and bridge system, respectively. $\mathbf{R}^{-1}$ is the inverse matrix of $\mathbf{R}$ and superscript T indicates the transposition of the vector.

Equation 12.4 can be also written in homogeneous form

$$\begin{pmatrix} X_p \\ Y_p \\ Z_p \\ 1 \end{pmatrix} = \begin{pmatrix} 1 & -\varepsilon_z(X) & \varepsilon_y(X) & X+\delta_x(X) \\ \varepsilon_z(X) & 1 & -\varepsilon_x(X) & \delta_y(X) \\ -\varepsilon_y(X) & \varepsilon_x(X) & 1 & \delta_z(X) \\ 0 & 0 & 0 & 1 \end{pmatrix} \begin{pmatrix} X_{p1} \\ Y_{p1} \\ Z_{p1} \\ 1 \end{pmatrix} \tag{12.5}$$

Similarly, when the carriage moves a nominal distance Y and the ram moves a distance Z, two additional vectors representing translation motions are obtained.

$$\mathbf{O_1O_2} = \mathbf{Y} = \begin{pmatrix} \delta_x(Y) - Y\alpha_{xy} \\ Y + \delta_y(Y) \\ \delta_z(Y) \end{pmatrix} \tag{12.6}$$

$$\mathbf{O_2O_3} = \mathbf{Z} = \begin{pmatrix} \delta_x(Z) - Z\alpha_{xz} \\ \delta_y(Z) - Z\alpha_{yz} \\ Z + \delta_z(Z) \end{pmatrix} \tag{12.7}$$

The infinitesimal rotation matrices for these motions $\mathbf{R}(y)$ and $\mathbf{R}(z)$ can be readily generated from Equation 12.3 by simply replacing the argument X with Y or Z.

$$\mathbf{R}(y) = \begin{pmatrix} 1 & \varepsilon_z(Y) & -\varepsilon_y(Y) \\ -\varepsilon_z(Y) & 1 & \varepsilon_x(Y) \\ \varepsilon_y(Y) & -\varepsilon_x(Y) & 1 \end{pmatrix} \tag{12.8}$$

$$\mathbf{R}(z) = \begin{pmatrix} 1 & \varepsilon_z(Z) & -\varepsilon_y(Z) \\ -\varepsilon_z(Z) & 1 & \varepsilon_x(Z) \\ \varepsilon_y(Z) & -\varepsilon_x(Z) & 1 \end{pmatrix} \tag{12.9}$$

Coordinates of point $\mathbf{P}$ in carriage system $\mathbf{P}_2 = (X_{p2}, Y_{p2}, Z_{p2})^T$ and in ram system $\mathbf{P}_3 = (X_{p3}, Y_{p3}, Z_{p3})^T$ are related by

$$\mathbf{P}_1 = \mathbf{Y} + \mathbf{R}^{-1}(y)\mathbf{P}_2 \tag{12.10}$$

and

$$\mathbf{P}_2 = \mathbf{Z} + \mathbf{R}^{-1}(z)\mathbf{P}_3 \tag{12.11}$$

Equations 12.10 and 12.11 can also be written in homogeneous form

$$
\begin{pmatrix} X_{p1} \\ Y_{p1} \\ Z_{p1} \\ 1 \end{pmatrix}
=
\begin{pmatrix}
1 & -\varepsilon_z(Y) & \varepsilon_y(Y) & \delta_x(Y) - Y\alpha_{xy} \\
\varepsilon_z(Y) & 1 & -\varepsilon_x(Y) & Y + \delta_y(Y) \\
-\varepsilon_y(Y) & \varepsilon_x(Y) & 1 & \delta_z(Y) \\
0 & 0 & 0 & 1
\end{pmatrix}
\begin{pmatrix} X_{p2} \\ Y_{p2} \\ Z_{p2} \\ 1 \end{pmatrix}
\tag{12.12}
$$

and

$$
\begin{pmatrix} X_{p2} \\ Y_{p2} \\ Z_{p2} \\ 1 \end{pmatrix}
=
\begin{pmatrix}
1 & -\varepsilon_z(Z) & \varepsilon_y(Z) & \delta_x(Z) - Z\alpha_{xz} \\
\varepsilon_z(Z) & 1 & -\varepsilon_x(Z) & \delta_y(Z) - Z\alpha_{yz} \\
-\varepsilon_y(Z) & \varepsilon_x(Z) & 1 & Z + \delta_z(Z) \\
0 & 0 & 0 & 1
\end{pmatrix}
\begin{pmatrix} X_{p3} \\ Y_{p3} \\ Z_{p3} \\ 1 \end{pmatrix}
\tag{12.13}
$$

Applying Equation 12.4 in succession using Equations 12.10 and 12.11 gives

$$\mathbf{P} = \mathbf{X} + \mathbf{R}^{-1}(x)\left\{\mathbf{Y} + \mathbf{R}^{-1}(y)[\mathbf{Z} + \mathbf{R}^{-1}(z)\mathbf{P}_3]\right\} \tag{12.14}$$

Applying Equation 12.5 in succession using Equations 12.12 and 12.13 gives

$$
\begin{pmatrix} X_p \\ Y_p \\ Z_p \\ 1 \end{pmatrix}
=
\begin{pmatrix}
1 & -\varepsilon_z(X) & \varepsilon_y(X) & X+\delta_x(X) \\
\varepsilon_z(X) & 1 & -\varepsilon_x(X) & \delta_y(X) \\
-\varepsilon_y(X) & \varepsilon_x(X) & 1 & \delta_z(X) \\
0 & 0 & 0 & 1
\end{pmatrix}
\begin{pmatrix}
1 & -\varepsilon_z(Y) & \varepsilon_y(Y) & \delta_x(Y) - Y\alpha_{xy} \\
\varepsilon_z(Y) & 1 & -\varepsilon_x(Y) & Y + \delta_y(Y) \\
-\varepsilon_y(Y) & \varepsilon_x(Y) & 1 & \delta_z(Y) \\
0 & 0 & 0 & 1
\end{pmatrix}
$$

$$
\begin{pmatrix}
1 & -\varepsilon_z(Z) & \varepsilon_y(Z) & \delta_x(Z) - Z\alpha_{xz} \\
\varepsilon_z(Z) & 1 & -\varepsilon_x(Z) & \delta_y(Z) - Z\alpha_{yz} \\
-\varepsilon_y(Z) & \varepsilon_x(Z) & 1 & Z + \delta_z(Z) \\
0 & 0 & 0 & 1
\end{pmatrix}
\begin{pmatrix} X_{p3} \\ Y_{p3} \\ Z_{p3} \\ 1 \end{pmatrix}
\tag{12.15}
$$

After expanding Equation 12.15, the position of point $\mathbf{P}$ in the table system can be fully determined. In the ideal case, that is, in the case of a perfect machine, there should be $X_p = X + X_{p3}$, $Y_p = Y + Y_{p3}$, and $Z_p = Z + Z_{p3}$. The differences $\Delta X = X_p - X - X_{p3}$,

$\Delta Y = Y_p - Y - Y_{p3}$, and $\Delta Z = Z_p - Z - Z_{p3}$ are the position errors of the probe tip caused by CMM error motions. From Equation 12.15, the following can be calculated:

$$\Delta X = \delta_x(X) + \delta_x(Y) + \delta_x(Z) - Y\alpha_{xy} - Z\alpha_{zx} - Y\varepsilon_z(X)$$

$$+Z[\varepsilon_y(X) + \varepsilon_y(Y)] - Y_{p3}[\varepsilon_z(X) + \varepsilon_z(Y) + \varepsilon_z(Z)] \qquad (12.16)$$

$$+Z_{p3}[\varepsilon_y(X) + \varepsilon_y(Y) + \varepsilon_y(Z)]$$

$$\Delta Y = \delta_y(X) + \delta_y(Y) + \delta_y(Z) - Z\alpha_{yz} - Z[\varepsilon_x(X) + \varepsilon_x(Y)]$$

$$+X_{p3}[\varepsilon_z(X) + \varepsilon_z(Y) + \varepsilon_z(Z)] \qquad (12.17)$$

$$-Z_{p3}[\varepsilon_x(X) + \varepsilon_x(Y) + \varepsilon_x(Z)]$$

$$\Delta Z = \delta_z(X) + \delta_z(Y) + \delta_z(Z) + Y\varepsilon_x(X)$$

$$- X_{p3}[\varepsilon_y(X) + \varepsilon_y(Y) + \varepsilon_y(Z)] \qquad (12.18)$$

$$+ Y_{p3}[\varepsilon_x(X) + \varepsilon_x(Y) + \varepsilon_x(Z)]$$

where ΔX, ΔY, and ΔZ are error motions of the probe tip, respectively, in x, y, and z directions in the table coordinate system. This implies that probe measurements will have errors equaling $-\Delta X$, $-\Delta Y$, and $-\Delta Z$, respectively. In order to compensate the readings, corrections equal to ΔX, ΔY, and ΔZ should be added to the measured coordinates. Note particularly the role of probe offsets and how they are multiplied by the angular errors of the machine axes.

12.3.2 GEOMETRIC MODELS FOR OTHER TYPES OF MACHINES

For establishing the geometric model, only the relative motions among the components are important. In this aspect, 3D machines are classified into four basic groups: **FXYZ**, **XFYZ**, **YXFZ**, and **ZYXF**, as shown in Figures 12.8a through 12.8d, respectively. The machine shown in Figure 12.7 belongs to the first group where the table is fixed, the bridge moves in x direction with reference to the table, the carriage moves in y direction with reference to the bridge, and the ram moves in the z direction with reference to the carriage. As an example, Equation 12.14 represents an **FXYZ** type machine.

Figure 12.8b showing the CMM with fixed bridge and moving table is an **XFYZ** type machine. In this type, the workpiece moves with the table in the x direction, the carriage moves in y direction with reference to the fixed bridge, and the ram moves in z direction with reference to the carriage.

For an **XFYZ** type CMM, the workpiece coordinate system **OXYZ** is set on the moving table, coordinate system $O_1X_1Y_1Z_1$ is set on the fixed bridge, and carriage system $O_2X_2Y_2Z_2$ and ram system $O_3X_3Y_3Z_3$ are the same as in an **FXYZ** type CMM. They are shown in Figure 12.8b.

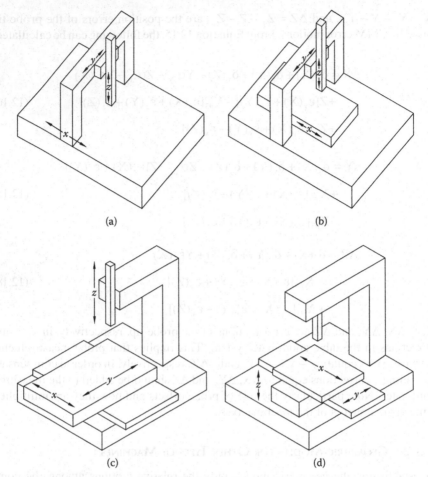

(a)   (b)

(c)   (d)

FIGURE 12.8 Four types of machines. (Redrawn from Zhang et al., *Ann CIRP*, 34(1):445, 1988.)

To obtain the coordinates of point **P** in the fixed bridge system, multiply **P** by $\mathbf{R}^{-1}(x)$ and add displacement **X**, resulting in

$$\mathbf{P}_1 = \mathbf{X} + \mathbf{R}^{-1}(x)\mathbf{P} \tag{12.19}$$

$\mathbf{P}_1$, $\mathbf{P}_2$, and $\mathbf{P}_3$ are related by the same equations as in **FXYZ** type machines.

$$\mathbf{P}_1 = \mathbf{Y} + \mathbf{R}^{-1}(y)\mathbf{P}_2 \tag{12.20}$$

$$\mathbf{P}_2 = \mathbf{Z} + \mathbf{R}^{-1}(z)\mathbf{P}_3 \tag{12.21}$$

Hence,

$$\mathbf{P} = \mathbf{R}(x)\left\{-\mathbf{X} + \mathbf{Y} + \mathbf{R}^{-1}(y)\left[\mathbf{Z} + \mathbf{R}^{-1}(z)\mathbf{P}_3\right]\right\} \tag{12.22}$$

Equation 12.22 can be also written in homogeneous form

$$
\begin{pmatrix} X_p \\ Y_p \\ Z_p \\ 1 \end{pmatrix} = \begin{pmatrix} 1 & \varepsilon_z(X) & -\varepsilon_y(X) & 0 \\ -\varepsilon_z(X) & 1 & \varepsilon_x(X) & 0 \\ \varepsilon_y(X) & -\varepsilon_x(X) & 1 & 0 \\ 0 & 0 & 0 & 1 \end{pmatrix} \begin{pmatrix} 1 & -\varepsilon_z(Y) & \varepsilon_y(Y) & -X-\delta_x(X)+\delta_x(Y)-Y\alpha_{xy} \\ \varepsilon_z(Y) & 1 & -\varepsilon_x(Y) & -\delta_y(X)+Y+\delta_y(Y) \\ -\varepsilon_y(Y) & \varepsilon_x(Y) & 1 & -\delta_z(X)+\delta_z(Y) \\ 0 & 0 & 0 & 1 \end{pmatrix}
$$

$$
\begin{pmatrix} 1 & -\varepsilon_z(Z) & \varepsilon_y(Z) & \delta_x(Z)-Z\alpha_{xz} \\ \varepsilon_z(Z) & 1 & -\varepsilon_x(Z) & \delta_y(Z)-Z\alpha_{yz} \\ -\varepsilon_y(Z) & \varepsilon_x(Z) & 1 & Z+\delta_z(Z) \\ 0 & 0 & 0 & 1 \end{pmatrix} \begin{pmatrix} X_{p3} \\ Y_{p3} \\ Z_{p3} \\ 1 \end{pmatrix} \tag{12.23}
$$

After expanding Equation 12.23, the position errors of the probe tip caused by error motions of the CMM can be obtained:

$$
\Delta X = -\delta_x(X)+\delta_x(Y)+\delta_x(Z)-Y\alpha_{xy}-Z\alpha_{xz}+Y\varepsilon_z(X)
$$

$$
+Z\left[-\varepsilon_y(X)+\varepsilon_y(Y)\right]-Y_{p3}\left[-\varepsilon_z(X)+\varepsilon_z(Y)+\varepsilon_z(Z)\right] \tag{12.24}
$$

$$
+Z_{p3}\left[-\varepsilon_y(X)+\varepsilon_y(Y)+\varepsilon_y(Z)\right]
$$

$$
\Delta Y = -\delta_y(X)+\delta_y(Y)+\delta_y(Z)-Z\alpha_{yz}+X\varepsilon_z(X)
$$

$$
-Z\left[-\varepsilon_x(X)+\varepsilon_x(Y)\right]+X_{p3}\left[-\varepsilon_z(X)+\varepsilon_z(Y)+\varepsilon_z(Z)\right] \tag{12.25}
$$

$$
-Z_{p3}\left[-\varepsilon_x(X)+\varepsilon_x(Y)+\varepsilon_x(Z)\right]
$$

$$
\Delta Z = -\delta_z(X)+\delta_z(Y)+\delta_z(Z)-X\varepsilon_y(X)-Y\varepsilon_x(X)
$$

$$
-X_{p3}\left[-\varepsilon_y(X)+\varepsilon_y(Y)+\varepsilon_y(Z)\right] \tag{12.26}
$$

$$
+Y_{p3}\left[-\varepsilon_x(X)+\varepsilon_x(Y)+\varepsilon_x(Z)\right]
$$

In **YXFZ** type machine, the table moves in both x and y directions while the ram moves in z direction, the workpiece coordinate system **OXYZ** is set on the table movable in y direction, coordinate system $O_1X_1Y_1Z_1$ is set on the table movable in x direction, system $O_2X_2Y_2Z_2$ is set on the machine frame, and system $O_3X_3Y_3Z_3$ is set on the ram, as shown in Figure 12.8c. The coordinate transformation relation between **P** and P_1 is given by

$$
R^{-1}(y)P + Y = P_1 \tag{12.27}
$$

The coordinate transformation relation between P_1 and P_2 is given by

$$
R^{-1}(x)P_1 + X = P_2 \tag{12.28}
$$

The coordinate transformation relation between P_2 and P_3 is the same as in **FXYZ** and **XFYZ** type CMMs given by Equation 12.21. Hence,

$$P = R(y)\{-Y + R(x)[-X + Z + R^{-1}(z)P_3]\} \tag{12.29}$$

Equation 12.29 can be also written in homogeneous form

$$
\begin{pmatrix} X_p \\ Y_p \\ Z_p \\ 1 \end{pmatrix} =
\begin{pmatrix}
1 & \varepsilon_z(Y) & -\varepsilon_y(Y) & 0 \\
-\varepsilon_z(Y) & 1 & \varepsilon_x(Y) & 0 \\
\varepsilon_y(Y) & -\varepsilon_x(Y) & 1 & 0 \\
0 & 0 & 0 & 1
\end{pmatrix}
\begin{pmatrix}
1 & \varepsilon_z(X) & -\varepsilon_y(X) & -\delta_x(Y) + Y\alpha_{xy} \\
-\varepsilon_z(X) & 1 & \varepsilon_x(X) & -Y - \delta_y(Y) \\
\varepsilon_y(X) & -\varepsilon_x(X) & 1 & -\delta_z(Y) \\
0 & 0 & 0 & 1
\end{pmatrix}
$$

$$
\begin{pmatrix}
1 & -\varepsilon_z(Z) & \varepsilon_y(Z) & -x - \delta_x(X) + \delta_x(Z) - Z\alpha_{xz} \\
\varepsilon_z(Z) & 1 & -\varepsilon_x(Z) & -\delta_y(X) + \delta_y(Z) - Z\alpha_{yz} \\
-\varepsilon_y(Z) & \varepsilon_x(Z) & 1 & -\delta_z(X) + Z + \delta_z(Z) \\
0 & 0 & 0 & 1
\end{pmatrix}
\begin{pmatrix} X_{p3} \\ Y_{p3} \\ Z_{p3} \\ 1 \end{pmatrix}
\tag{12.30}
$$

After expanding Equation 12.30, the position error of the probe tip caused by error motions of the CMM can be obtained:

$$\Delta X = -\delta_x(X) - \delta_x(Y) + \delta_x(Z) + Y\alpha_{xy} - Z\alpha_{xz} - Y\varepsilon_z(Y)$$

$$-Z\big[\varepsilon_y(X) + \varepsilon_y(Y)\big] - Y_{p3}\big[-\varepsilon_z(X) - \varepsilon_z(Y) + \varepsilon_z(Z)\big] \tag{12.31}$$

$$+Z_{p3}\big[-\varepsilon_y(X) - \varepsilon_y(Y) + \varepsilon_y(Z)\big]$$

$$\Delta Y = -\delta_y(X) - \delta_y(Y) + \delta_y(Z) - Z\alpha_{yz} + Z\big[\varepsilon_x(X) + \varepsilon_x(Y)\big]$$

$$+X\big[\varepsilon_z(X) + \varepsilon_z(Y)\big] + X_{p3}\big[-\varepsilon_z(X) - \varepsilon_z(Y) + \varepsilon_z(Z)\big] \tag{12.32}$$

$$-Z_{p3}\big[-\varepsilon_x(X) - \varepsilon_x(Y) + \varepsilon_x(Z)\big]$$

$$\Delta Z = -\delta_z(X) - \delta_z(Y) + \delta_z(Z) + Y\varepsilon_x(Y) - X\big[\varepsilon_y(X) + \varepsilon_y(Y)\big]$$

$$-X_{p3}\big[-\varepsilon_y(X) - \varepsilon_y(Y) + \varepsilon_y(Z)\big] + Y_{p3}\big[-\varepsilon_x(X) - \varepsilon_x(Y) + \varepsilon_x(Z)\big] \tag{12.33}$$

The geometrical model of **XYZF** and other type machines will not be discussed in detail since the procedure to develop the equations would be the same.

12.3.3 VERIFICATION FOR QUASI-RIGID BODY ASSUMPTION MODEL

All the above discussions are based on a quasi-rigid body model. The CMM should have high stiffness. There are several ways to check if the machine meets quasi-rigid body model requirement.

1. To check the variation of flatness of the table or straightness of the guide-ways: To meet the quasi-rigid body model requirement, the table of the moving bridge type CMM shown in Figure 12.5 should have high enough stiffness. Otherwise the force deformation caused by the deadweight of the machine components will cause the table to bend. If the guideways of the

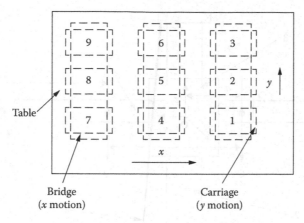

FIGURE 12.9 Test for quasi-rigid body assumption. (Redrawn from Zhang et al., *Ann CIRP*, 34(1):445, 1985.)

gantry machine are not stiff enough, the force deformation will not be a function of only one argument. Based on this understanding, the fitness of the CMM can be checked by measuring the changes of the table flatness while the machine bridge and carriage are moved to nine different positions, as shown in Figure 12.9. If the variation of the table flatness in these nine positions is less than a certain threshold, for example, one-fifth of the allowed measurement uncertainty, the CMM can be considered fitting a quasi-rigid body model.

2. To check the variation of angular error motion: In a quasi-rigid body machine, both positioning errors and straightness error motions depend on the line along which these errors are measured. For example, both $\delta_x(X)$ and $\delta_z(X)$ might be different when they are measured along lines with different y coordinate. However, the angular error motions $\varepsilon_x(X)$, $\varepsilon_y(X)$, and $\varepsilon_z(X)$ should be independent from the positions of lines along which they are measured. If the variations of the angular error motions multiplied by the maximum y travel are less than one-fifth of the allowed measurement uncertainty, the CMM can be considered fitting a quasi-rigid body model.

3. To check the results of error compensation: When a CMM meets the quasi-rigid body model requirement and its systematic errors are dominant, significant accuracy enhancement may be achieved through error compensation. Departure from the quasi-rigid body model could be the major reason for unsatisfactory results from error compensation if the machine calibration has been carried out accurately and uncertainties are relatively small.

12.3.4 GEOMETRIC MODEL OF A MACHINE THAT DOES NOT OBEY THE AXIS INDEPENDENCE ASSUMPTION

It is more difficult to derive the geometric model of a machine that does not obey the axis independence assumption (see also Section 12.2.2) since the errors of a given axis change with the position of another axis. This requires the measurement of the individual errors by moving other axes as well (Hocken et al. 1977).

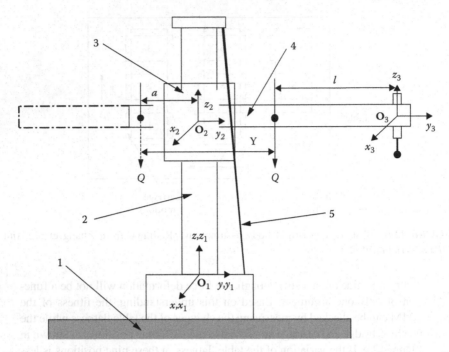

FIGURE 12.10 Example of a quasi-rigid body machine without axis independence. (1) table; (2) column; (3) carriage; (4) arm; (5) reinforcement.

In Figure 12.10, the diagram of a CMM with horizontal arm, which is a typical example of such a machine, is shown. Four coordinate systems, $\mathbf{OXYZ}$, $\mathbf{O_1 X_1 Y_1 Z_1}$, $\mathbf{O_2 X_2 Y_2 Z_2}$, and $\mathbf{O_3 X_3 Y_3 Z_3}$, are set on table (1), column (2), carriage (3), and arm (4), respectively.

The deformation of column (2) depends not only on the position of carriage (3) but also on the position of horizontal arm (4). It means the deformation of column (2) is a function of both Y and Z displacements of the probe and the CMM does not meet the requirement of axis independence.

First, the CMM is treated as a quasi-rigid body machine, which belongs to $\mathbf{FXZY}$ type machine. $\mathbf{FXZY}$ type machine is a modified version of $\mathbf{FXYZ}$ type machine. The only difference between the two is that an $\mathbf{FXZY}$ has z as the second motion and y motion as the third. For an $\mathbf{FXZY}$ type machine, Equations 12.6 and 12.7 should be written as

$$\mathbf{O_1 O_2} = \mathbf{Z} = \begin{pmatrix} \delta_x(Z) - Z\alpha_{xz} \\ \delta_y(Z) \\ Z + \delta_z(Z) \end{pmatrix} \tag{12.34}$$

$$\mathbf{O_2 O_3} = \mathbf{Y} = \begin{pmatrix} \delta_x(Y) - Y\alpha_{xy} \\ Y + \delta_y(Y) \\ \delta_z(Y) - Y\alpha_{yz} \end{pmatrix} \tag{12.35}$$

Equation 12.14 should be changed to

$$P = X + R^{-1}(x)\left\{Z + R^{-1}(z)\left[Y + R^{-1}(y)P_3\right]\right\} \tag{12.36}$$

Now the effects of force deformation are discussed. Suppose the deadweight of arm (4) equals Q. Its center of gravity is located at a distance a on the left of column (2) while arm (4) is at its initial position $(Y = 0)$. When arm (4) has a displacement Y, the torque acting on column (2) equals

$$M = Q(Y - a) \tag{12.37}$$

The displacement of point O_2 in y direction equals

$$\delta Y_2 = \frac{MZ^2}{2EJ} = \frac{Q(Y - a)Z^2}{2EJ} \tag{12.38}$$

The angular deformation at point O_2 equals

$$\delta\theta_2 = \frac{MZ}{EJ} = \frac{Q(Y - a)Z}{EJ} \tag{12.39}$$

A sag in z direction is caused at point O_3 due to angular deformation $\delta\theta_2$ given by

$$\delta Z_3 = \delta\theta_2 \cdot (l + Y - a) = \frac{Q(Y - a)(l + Y - a)Z}{EJ} \tag{12.40}$$

where E and J are the Young's Modulus (Pa) and moment of inertia (m⁴) of column (2), respectively; l is the distance between point O_3 (m), where the probe is mounted, and the gravity center of arm (4); Y, Z are the displacement of arm (4) and carriage (3) in y and z directions (m), respectively.

In Equation 12.40, the deformation of arm (4) under its own weight is not considered since this deformation is a function of Y displacement and it is included in the quasi-rigid body model. For increasing the stiffness of column (2), rigidity reinforcing rod (5) is used. It makes the calculation of the deformations quite complicated. However, the deformations of point O_3 in both y and z directions are functions of both Y and Z displacements of the probe and can be expressed as $\delta_y(Y, Z)$ and $\delta_z(Y, Z)$. $\delta_y(Y, Z) = \delta Y_2$ and $\delta_z(Y, Z) = \delta Z_3$ and the role of rod (5) is to enlarge the effective moment of inertia J.

It is difficult to theoretically calculate $\delta_y(Y, Z) = \delta Y_2$ and $\delta_z(Y, Z) = \delta Z_3$ accurately. However, Equations 12.38 and 12.40 show the forms of their relations with displacements Y and Z. In practice, the results obtained from machine calibration include always both errors caused by geometric factors and force deformations. It is necessary to separate them to obtain a complete error map within the whole measuring volume. Usually, angular error motions are measured first. Angular error motions caused by geometric factors should be independent from the line along which the errors are measured. By evaluating the angular error motions along several parallel lines, both angular error motions caused by geometric factors and by force deformation shown by Equation 12.39 can be determined separately.

Then, the linear error motions are measured. Linear error motions caused by geometric factors are functions of a single argument plus Abbe errors, which are given by the Abbe offsets multiplied by angular error motions. By evaluating $\delta_y(Y, Z)$ and $\delta_z(Y, Z)$ along several lines, the unknown values of Q, a, l, and J in Equations 12.37 through 12.40 and functional relationship of $\delta_y(Y, Z)$ and $\delta_z(Y, Z)$ with y and z can be determined. In this way, a complete error map within the whole measuring volume can be determined, and it gives the whole error model of the machine.

12.3.5   GEOMETRIC MODEL OF COORDINATE MEASURING MACHINES WITH A ROTARY TABLE

1. Application fields: A rotary table is an important CMM accessory. It is mainly used for measuring bodies of rotation or parts with features equally spaced on a circle, such as gears, polyhedrons, and parts with hole patterns, as shown in Figure 12.11.

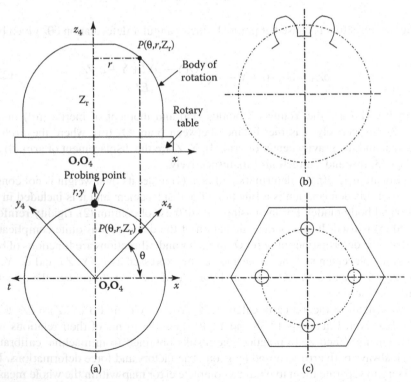

FIGURE 12.11 Features measured with a rotary table. (a) Body of rotation. (b) Gear and (c) Polyhedron with equally spaced holes.

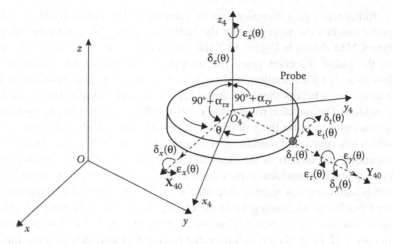

FIGURE 12.12 Geometric model of a rotary table.

2. Coordinate systems: The workpiece is mounted on the rotary table and rotates with it. For establishing its mathematical model, a fifth coordinate system $O_4X_4Y_4Z_4$ is set on the rotary table, as shown in Figure 12.12. Usually, its axis of rotation is selected as axis z_4. The point of intersection of z_4 axis with **OXY** plane is selected as the origin of the coordinate system O_4 and $O_4X_4Y_4Z_4$ rotates with the rotary table. The initial positions of x_4 and y_4 are expressed as X_{40} and Y_{40}, respectively. O_4X_{40} is parallel to plane **OXZ**.

3. Individual errors: Since the rotary table rotates, all its error motions are expressed as functions of rotated angle θ. There is a total of six error motions: linear errors $\delta_x(\theta)$, $\delta_y(\theta)$, and $\delta_z(\theta)$ of the origin of coordinate system O_4 and angular errors of axis z_4 about three mutually perpendicular axes $\varepsilon_x(\theta)$, $\varepsilon_y(\theta)$, and $\varepsilon_z(\theta)$. $\varepsilon_z(\theta)$ is the error of rotated angle θ while $\varepsilon_x(\theta)$ and $\varepsilon_y(\theta)$ are tilt motions of the rotational axis about two other axes x_4 and y_4.

   However, in many cases, especially when measuring bodies of rotation, it is more convenient to express the error motions in another way. For a body of rotation, all its cross sections perpendicular to the axis of rotation are circles. It is more convenient to measure the body of rotation in a cylindrical coordinate system (θ, r, Z_r), where θ is the argument angle, r is the radius, and Z_r is the height of point **P**, respectively, as shown in Figure 12.11a. For bodies of rotation, only error motions in the radial direction cause errors in measurement. It is reasonable to decompose the error motions in the plane perpendicular to the axis of rotation into radial error motion $\delta_r(\theta)$ and tangential error motion $\delta_t(\theta)$. The tangential error motion $\delta_t(\theta)$ has negligible effect for measuring bodies of rotation. The radial and tangential directions do not change with rotation.

Either the x or y direction can be selected as the radial direction. The probe touches the workpiece in the radial direction. For a moving bridge type CMM shown in Figure 12.7, the y direction is preferred to be selected as the radial direction since the weight of component for y motion is less than that for x motion, and for y motion, there is only Abbe offset in z direction, whereas there are Abbe offsets in both y and z directions for x motion. The radial error motion $\delta_r(\theta)$ can be calibrated easily by using an indicator to sense the y deviation of a well-centered reference mandrel, which has very small roundness error and rotates with the rotary table. The mounting eccentricity of the mandrel can be eliminated easily by data processing and its roundness error can be eliminated by using reversal or multistep techniques as needed (see also Chapter 13, "Reversal" Techniques for Coordinate Measuring Machine Calibration). The tangential direction is 90° anticlockwise rotated about axis z_4 against radial direction, as shown in Figure 12.12. $\delta_t(\theta)$ can be calibrated by using an indicator to sense the X deviation of the reference mandrel.

It is also reasonable to decompose the tilt error motions into angular error motion about the tangential axis $\varepsilon_t(\theta)$ and angular error motion about the radial axis $\varepsilon_r(\theta)$. For point $P(\theta, r, Z_r)$, angular error motion about the tangential axis causes error motion in the radial direction equal to $Z_r\varepsilon_t(\theta)$ and error motion in the z direction equal to $-r\varepsilon_t(\theta)$. Angular error motion about the radial axis causes error motion in the tangential direction equal to $-Z_r\varepsilon_r(\theta)$.

Besides linear and angular error motions, there are also squareness errors between the axis of rotation and x and y axes of the CMM. Squareness error between O_4Z_4 and y axis, α_{ry}, causes an error motion $-Z_r\alpha_{ry}$ in the y direction and an error motion $Y_{p4}\alpha_{ry}$ in the z direction. Squareness error between O_4Z_4 and x axis, α_{rx}, causes an error motion $-Z_r\alpha_{rx}$ in x direction and an error motion $X_{p4}\alpha_{rx}$ in z direction, where X_{p4} and Y_{p4} are Cartesian coordinates of probing point in the $O_4X_4Y_4Z_4$ system.

4. Geometric model of a CMM with rotary table for measuring bodies of rotation: In the case where the $+y$ direction of the CMM is selected as the radial direction and $-x$ direction is taken as the tangential direction, the equivalent total error motion of the probe for measuring a body of rotation at point (θ, r, Z_r) can be written as

$$\Delta\theta = -\varepsilon_z(\theta) - [\Delta X + \delta_t(\theta) + Z_r\alpha_{rx} - Z_r\varepsilon_r(\theta)]/r \qquad (12.41)$$

$$\Delta r = \Delta Y - \delta_r(\theta) + Z_r[\alpha_{ry} - \varepsilon_t(\theta)] \qquad (12.42)$$

$$\Delta Z_r = \Delta Z - \delta_z(\theta) - r[\alpha_{ry} - \varepsilon_t(\theta)] \qquad (12.43)$$

where ΔX, ΔY, and ΔZ are the x, y, and z components of CMM error motion, which can be calculated from Equations 12.16 through 12.18, 12.24 through 12.26, or 12.31 through 12.33 in accordance with the machine type.

The error motions caused by the rotary table are those of the workpiece. Signs of the error motions are reversed when they are converted to error motions of the probe. When probing is along the $+x$ axis of the machine and $+y$ direction is taken as the tangential direction, then Equations 12.41 through 12.43 should be changed to

$$\Delta\theta = -\varepsilon_z(\theta) + [\Delta Y - \delta_t(\theta) + Z_r\alpha_{ry} + Z_r\varepsilon_r(\theta)]/r \tag{12.44}$$

$$\Delta r = \Delta X - \delta_r(\theta) + Z_r[\alpha_{rx} - \varepsilon_t(\theta)] \tag{12.45}$$

$$\Delta Z_r = \Delta Z - \delta_z(\theta) - r[\alpha_{rx} - \varepsilon_t(\theta)] \tag{12.46}$$

5. Geometric model of a CMM with rotary table for measuring equally spaced features: There are three types of measurements: (1) Measurement of the dimension and form errors of individual features; (2) measurement of pitch errors of equally spaced features; (3) measurement of position errors of features with reference to other elements of the part.

In the first case when the dimension and form errors of a particular feature are measured, such as the profile error of a tooth, the dimension, and the form error of a hole, generally speaking, the rotary table is kept at a fixed position; that is, there is no rotation of the rotary table in such measurements. So, the geometric model of such measurements is just the model of the CMM itself. Errors of the rotary table need not to be taken into account.

In case of measuring the pitch errors of equally spaced features, such as adjacent and cumulative pitch errors of teeth, spacing errors of the holes, the probe repeats almost the same motion for measuring every individual feature. The rotation of the rotary table enables the probe to measure all the features in sequence. In this case, only the errors of the rotary table should be taken into account. When the y direction is taken as the radial direction, the mathematical model of the measurement can be expressed as

$$\Delta\theta = -\varepsilon_z(\theta) - [\delta_t(\theta) + Z_r\alpha_{rx} - Z_r\varepsilon_r(\theta)]/r \tag{12.47}$$

When x direction is taken as the radial direction, angular spacing error can be expressed as

$$\Delta\theta = -\varepsilon_z(\theta) + [-\delta_t(\theta) + Z_r\alpha_{ry} + Z_r\varepsilon_r(\theta)]/r \tag{12.48}$$

For converting them to linear spacing errors, it is enough to multiply $\Delta\theta$ by r of the characteristic point of the feature. In case of measuring the spacing errors of a series of holes, the center of the hole defined in accordance with certain criterion serves as its characteristic point.

Both the CMM errors and those of the rotary table are involved while the positions of features with reference to other elements of the part are measured. Some parameters, such as the tooth thickness and tooth direction, look like parameters of the individual feature, but they are related to the whole part in nature. Tooth direction is evaluated with reference to the axis of the gear. Tooth thickness is evaluated at the dividing circle, which is related with other elements of the gear.

To derive the general geometric model of a CMM with a rotary table, the following coordinate transformation should be carried out:

$$\dot{\mathbf{P}} = \mathbf{R}^{-1}(\theta)\mathbf{P}_4 + \mathbf{O}_4 + \delta\mathbf{O}_4 \qquad (12.49)$$

where $\mathbf{P}_4 = (X_{p4}, Y_{p4}, Z_{p4})^T$ is the probe tip vector in the rotary table system, $\mathbf{O}_4 = (X_{O4}, Y_{O4}, Z_{O4})^T$ is the vector of origin of the rotary table system in $\mathbf{OXYZ}$ system, $\mathbf{R}^{-1}(\theta)$ is the inverse rotational matrix of the rotary table, and $\delta\mathbf{O}_4$ is the vector representing linear error motions of point $\mathbf{O}_4$.

$$\mathbf{R}^{-1}(\theta) = \begin{pmatrix} \cos(\theta + \varepsilon_z(\theta)) & -\sin(\theta + \varepsilon_z(\theta)) & \varepsilon_y(\theta) \\ \sin(\theta + \varepsilon_z(\theta)) & \cos(\theta + \varepsilon_z(\theta)) & -\varepsilon_x(\theta) \\ -\varepsilon_y(\theta)\cos(\theta) + \varepsilon_x(\theta)\sin(\theta) & \varepsilon_y(\theta)\sin(\theta) + \varepsilon_x(\theta)\cos(\theta) & 1 \end{pmatrix} \quad (12.50)$$

An important distinguishing feature of inverse rotational matrix $\mathbf{R}^{-1}(\theta)$ is that since θ is not an infinitesimal angle, $\cos\theta$ cannot be simplified as 1, and $\sin\theta$ cannot be replaced by θ.

Equation 12.49 can also be written in homogeneous form as

$$\begin{pmatrix} X_p \\ Y_p \\ Z_p \\ 1 \end{pmatrix} = \begin{pmatrix} \cos(\theta + \varepsilon_z(\theta)) & -\sin(\theta + \varepsilon_z(\theta)) & \varepsilon_y(\theta) & X_{O4} + \delta_x(\theta) - Z_{p4}\alpha_{rx} \\ \sin(\theta + \varepsilon_z(\theta)) & \cos(\theta + \varepsilon_z(\theta)) & -\varepsilon_x(\theta) & Y_{O4} + \delta_y(\theta) - Z_{p4}\alpha_{ry} \\ -\varepsilon_y(\theta)\cos(\theta) + \varepsilon_x(\theta)\sin(\theta) & \varepsilon_y(\theta)\sin(\theta) + \varepsilon_x(\theta)\cos(\theta) & 1 & Z_{O4} + \delta_z(\theta) \\ 0 & 0 & 0 & 1 \end{pmatrix} \begin{pmatrix} X_{p4} \\ Y_{p4} \\ Z_{p4} \\ 1 \end{pmatrix}$$

$$(12.51)$$

For a perfect CMM with a perfect rotary table

$$\begin{pmatrix} X_p - \Delta X \\ Y_p - \Delta Y \\ Z_p - \Delta Z \\ 1 \end{pmatrix} = \begin{pmatrix} \cos\theta & -\sin\theta & 0 & X_{O4} \\ \sin\theta & \cos\theta & 0 & Y_{O4} \\ 0 & 0 & 1 & Z_{O4} \\ 0 & 0 & 0 & 1 \end{pmatrix} \begin{pmatrix} X'_{p4} \\ Y'_{p4} \\ Z'_{p4} \\ 1 \end{pmatrix} \qquad (12.52)$$

where $(X'_{p4}, Y'_{p4}, Z'_{p4})^T$ are coordinates of the probe tip in the rotary table coordinate system in the ideal case. The error motions caused by both error motions of the CMM and rotary table can be determined as

$$\begin{pmatrix} \Delta X_4 \\ \Delta Y_4 \\ \Delta Z_4 \end{pmatrix} = \begin{pmatrix} X_{p4} \\ Y_{p4} \\ Z_{p4} \end{pmatrix} - \begin{pmatrix} X'_{p4} \\ Y'_{p4} \\ Z'_{p4} \end{pmatrix} \qquad (12.53)$$

Generally, Z_{O4} and all error motions are small. By ignoring all the second-order errors and substituting Equations 12.51 and 12.52 into 12.53, the following equations are obtained:

$$\Delta X_4 = [\Delta X - \delta_x(\theta) + Z_{p4}\alpha_{rx}]\cos\theta + [\Delta Y - \delta_y(\theta) + Z_{p4}\alpha_{ry}]\sin\theta$$
$$- (X_p - X_{O4})\sin\theta\varepsilon_z(\theta) + (Y_p - Y_{O4})\cos\theta\varepsilon_z(\theta) \quad (12.54)$$
$$+ Z_p[-\cos\theta\varepsilon_y(\theta) + \sin\theta\varepsilon_x(\theta)]$$

$$\Delta Y_4 = [-\Delta X + \delta_x(\theta) - Z_{p4}\alpha_{rx}]\sin\theta + [\Delta Y - \delta_y(\theta) + Z_{p4}\alpha_{ry}]\cos\theta$$
$$- (X_p - X_{O4})\cos\theta\varepsilon_z(\theta) - (Y_p - Y_{O4})\sin\theta\varepsilon_z(\theta) \quad (12.55)$$
$$+ Z_p[\cos\theta\varepsilon_x(\theta) + \sin\theta\varepsilon_y(\theta)]$$

$$\Delta Z_4 = [\cos\theta\varepsilon_y(\theta) + \sin\theta\varepsilon_x(\theta) - \alpha_{rx}](X_p - X_{O4})$$
$$+ [-\cos\theta\varepsilon_x(\theta) + \sin\theta\varepsilon_y(\theta) - \alpha_{ry}](Y_p - Y_{O4}) - \delta_z(\theta) + \Delta Z \quad (12.56)$$

12.3.6 THERMAL MODEL OF THE COORDINATE MEASURING MACHINE

In the case where there are no temperature gradients in the ambient space and all the materials of a CMM's components are homogeneous, that is, their thermal coefficients of expansion are the same, the machine components suffer only thermal expansions and no bending or torsion when their temperature departs from 20°C. Such thermal deformation is called *simple thermal deformation*. In this case, all angular and straightness error motions, as well as squareness errors, do not change with temperature. Only positioning errors $\delta_x(X)$, $\delta_y(Y)$, and $\delta_z(Z)$ change with temperature. Simple error compensation may be used for $\delta_x(X)$, $\delta_y(Y)$, $\delta_z(Z)$ and the thermal expansion/contraction of the workpiece where

$$\Delta_u(u) = \delta_u(u) + uk_u(t_u - 20) - uk_w(t_w - 20) \quad (12.57)$$

where u can be x, y, or z and k_u and t_u are thermal coefficient of expansion and temperature of u scale, which could be x, y, z scales ((2), (4), (6) in Figure 12.7), respectively. k_w and t_w are those of the workpiece.

Since all compensation values are linearly proportional to displacements X, Y, and Z, such thermal error compensation is called first-order compensation. The factors influencing the accuracy of error compensation for simple thermal deformation include the following:

1. Uncertainties of the temperature measuring instruments
2. Uncertainties of nominal thermal expansion coefficients
3. Temperature gradients of the machine components and workpiece
4. Departure from the simple thermal deformation model

Temperatures are usually measured on the surface of components. However, thermal expansions of the machine and workpiece are determined by their body temperatures. To reduce these errors, the machine components should be made of high-conductive materials and actions for accelerating the heat transfer should be taken. The workpiece should be moved to the machine prior to the measurement to have adequate soak out time to allow it to equilibrate to a uniform temperature over its whole volume.

Usually, the values of thermal coefficients of expansion are stored in the computer in advance. The real effective thermal coefficient of expansion of the workpiece probably will be different from piece to piece. The effective thermal coefficients of expansion of the machine components are difficult to determine since a variety of materials with different thermal coefficients of expansion may be involved. It is recommended to calibrate the effective thermal coefficients of expansion of each scale individually, on site, with the machine in its working state.

The effective thermal coefficient of a scale can be defined as the variation of the displacement errors $\delta_x(X)$, $\delta_y(Y)$, or $\delta_z(Z)$ per unit change of temperature. Figure 12.13 shows the variation of the positioning error versus temperature. The average slope of curve gives the effective thermal coefficient of expansion of the corresponding scale.

Error compensation introduced by Equation 12.57 is based on the assumption there is only thermal expansion and no bending or torsion. In case there are significant bending and torsion deformations, Equation 12.57 may not provide satisfactory thermal error compensation results.

The assumption of simple thermal deformation should be verified by evaluating angular error motions at different temperatures. For simple thermal deformation, angular error motions should not change with the variation of temperature. For checking the fitness of using simple thermal deformation, all nine angular error

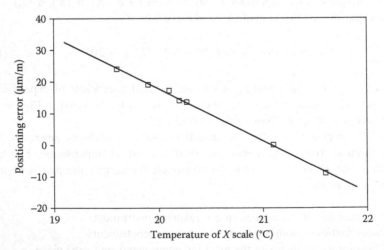

FIGURE 12.13 Determination of the thermal coefficient of the scale. (Redrawn from Zhang, G. et al., *Ann CIRP*, 34(1):445, 1985.)

motions should be checked at different temperatures, but this is usually economically impractical.

In reality, there are always temperature gradients. The heat generated in the machine can be removed only when the temperature of the heat removal media is lower than that of the machine component, and this unavoidably causes certain temperature gradients. To make the temperature gradient as small as possible, the materials from which the machine is built should be highly conductive thermally and the thermal capacity of heat removal media should be quite large. For this purpose, aluminum, which is highly conductive, is used for building some CMMs; although its thermal coefficient of expansion is large, simple thermal deformation is much easier to compensate. An oil shower is used for some precision machines to increase the thermal capacity of heat removal media (see Chapter 10, Temperature Fundamentals).

Besides temperature gradients, the nonuniform thermal behavior of the construction materials is another important factor causing the machine component to bend or twist. The bending of a glass scale mounted on a case as shown in Figure 12.14a is an example. Since the scale's glass and the material of the case have different thermal coefficients of expansion, the scale may bend when their temperatures vary. One of the solutions is using a metallic scale instead of a glass one and another solution is fixing the scale only at its one end, as shown in Figure 12.14b. Some manufacturers even suggest only ferrous metals should be used as the construction materials.

In case of having temperature gradient $\gamma(°C/m)$, the beam bends, as shown in Figure 12.15. Suppose the thickness of the beam is H, the radius of curvature after bending equals R, the total bending angle equals θ, then

$$R\theta = l \tag{12.58}$$

$$(R + H)\theta = l(1 + \gamma H\alpha) \tag{12.59}$$

Hence, the bending angle

$$\theta = l\gamma\alpha \tag{12.60}$$

For a beam supported at two end points A and B, its central point C has a deflection δ_1 with reference to line AB and

$$\delta_1 = R\left(1 - \cos\frac{\theta}{2}\right) = \frac{R\theta^2}{8} = \frac{l\theta}{8} = \frac{l^2\gamma\alpha}{8} \tag{12.61}$$

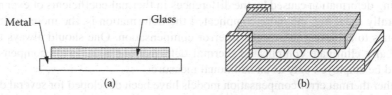

Metal — Glass

(a)   (b)

FIGURE 12.14 Mounting of the scale. (a) A glass scale mounted in a metallic case (b) A floating mount.

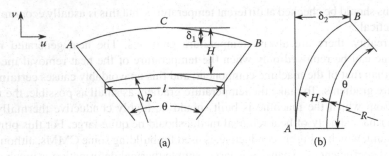

FIGURE 12.15 Bending of the beam. (a) A beam supported at two ends. (b) A cantilever beam.

For a cantilever beam fixed at point A, the deflection of point B with reference to point A

$$\delta_2 = \frac{l\theta}{2} = \frac{l^2 \gamma \alpha}{2} \qquad (12.62)$$

Thermal error compensation can be introduced in accordance with Equations 12.61 or 12.62. In both cases, the deflection curve is a second-order curve and the total deflection is proportional to l^2. This type of thermal error compensation can be called *second-order thermal error compensation*.

Besides compensation for linear deflections, error compensation should be made for angular errors also. In case of complicated thermal deformations, the angular error motions vary with temperatures. Thermal error compensation should be introduced for all angular error motion values in corresponding equations such as Equations 12.16 through 12.18, 12.24 through 12.26, and 12.31 through 12.33.

Above and beyond the factors influencing the accuracy of error compensation for simple thermal deformation, there are a lot of other uncertain factors in the case of complicated thermal deformations. All the above equations are derived under the assumption that there is only a constant temperature gradient in the v direction and no temperature gradients in the other directions. Once there is a temperature change, it is hard to make a case that such an assumption holds. More than simple bending could be produced by a difference in the thermal coefficients of expansion. Thermal coefficients of expansion do not change linearly with temperature, but in a more complicated manner. Another type of model is required for compensation of the bending deformation caused by the differences in thermal coefficients of expansion. Generally speaking, the more complicated the deformation is, the more difficult it becomes to achieve good results in error compensation. One should always try to physically eliminate complicated thermal deformations first. Error compensation should be employed only as an additional measure.

Other thermal error compensation models have been developed for several different applications. Santolaria et al. (Santolaria et al. 2009) shows one such realization for articulated-arm CMMs (see also Chapter 10, Temperature Fundamentals).

12.4 APPLICATION PROBLEMS OF ERROR COMPENSATION TECHNIQUES IN COORDINATE MEASURING MACHINES

12.4.1 PREREQUIREMENTS

Error compensation is a powerful and essential technique for enhancing the accuracy of CMMs. However, it does not mean this technique can be used indiscriminately to solve all problems. There are certain prerequirements for applying error compensation techniques.

1. A suitable model is available: The feasibility of compensating certain errors first depends on the existence of a suitable mathematical model. Without a model, the errors cannot be compensated effectively or cannot be justified economically.

   The most popular methods used for error compensation are the ones for compensating the geometric errors of quasi-rigid body model machines and error compensation for simple thermal deformations in the form of thermal expansion or contraction. However, it is best to eliminate other types of errors by reasonable design, careful manufacturing, and strict control of environmental conditions (Blaedel 1980; Slocum 1992a; Luttrell 2007) before using compensation.

   For example, increasing both motion and probing speed of a CMM in order to enhance the efficiency of measurement is one of the important tendencies in CMM development. For this purpose, the weight and cross section of the motion components should be reduced. Both increasing motion speed and acceleration and also decreasing the cross sections of the machine components tend to increase dynamic errors. Dynamic errors become the bottleneck for further increasing the speed. Compensation for dynamic errors is desirable. Many efforts have been devoted to this hot topic and some achievements have been obtained (see also Section 12.5). Nonetheless, dynamic error compensation techniques are still under study. The major problem is a lack of suitable models, which would be able to correctly and simply relate the dynamic errors with the structural and operational parameters.

   Approximations introduced by the model constitute one of the major factors limiting the accuracy of error compensation.

2. The major errors can be compensated: Due to the limitations in the chosen mathematical model and calibration techniques, not all errors of a CMM can be compensated. For example, the bending and torsion errors caused by thermal effects cannot be compensated by using a simple thermal deformation model. Due to several reasons, the non-real-time error compensation type is most popularly used but it only compensates for systematic errors. Error compensation techniques work well only where systematic errors dominate. Prudent measures in machine design and manufacture and in environment control should be taken for reducing other errors, including probing and dynamic errors along with errors caused by complicated

thermal deformation. Only after that can the accuracy of CMMs be enhanced by error compensation.

Short-range periodic errors are also difficult to compensate. The major difficulties include not only the relocation of the zero point (home) of the CMM but also the sampling interval during machine calibration. It is especially true when physical artifacts are used for calibration. Usually, the spacings between adjacent reference elements are quite large and such artifacts are unable to calibrate periodic errors.

Short-period errors are difficult to compensate even when using a laser interferometer as the major calibration device. First, the calibration work would be quite time consuming. Second, the inaccuracy of the home position could make error compensation meaningless. Usually, short-period errors are checked by laser interferometer at the manufacturer's site. Proper measures, such as improving the moiré signal of optical gratings and improving the subdivision circuits, should be taken to eliminate such short-period errors. Software error compensation is powerful but it cannot do everything, and one should not expect it to do everything.

In addition, the machine should have good long-term stability. Error compensation would lose its effectiveness if the machine errors change quickly with time.

3. The CMM should have a very repeatable zero point (home) since all the error compensation is carried out in the machine coordinate system.
4. The data processing system must be able to process the compensation data and to introduce error compensation in time and have enough internal memory to store all the related data. In the case of hardware error compensation, the compensation device should have good dynamic response to allow real-time compensation.
5. Compensation must be justified economically. The expenses in making error compensation include labor and material for machine calibration and those of developing the related model and software. In case of hardware error compensation, they also include the cost of the compensation device and the cost for machine modification. The cost–benefit ratio for error compensation should be better than other means for achieving similar results.

12.4.2 Verification and Troubleshooting

The results of error compensation should be verified by measuring several lines to see if the measurement uncertainties have been reduced significantly. This test may be conducted by using a laser interferometer, gage blocks, step gages, or other artifacts. To meet the verification objective, the lines measured during verifications should not be the same as those used for machine calibration. The effectiveness of error compensation should be checked over the whole measuring volume. Certainly, to measure a large number of lines with different orientations could take a long time. As a common practice, four body diagonals *AG*, *CE*, *BH*, and *DF* shown in Figure 12.16 and three lines parallel to the axes are often used for verification purposes (ISO 2009a). This is also done on machine tools (ISO 2002b; ASME 2005a).

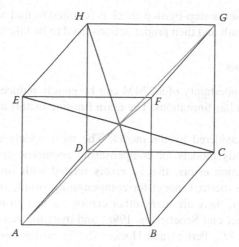

FIGURE 12.16 Four body diagonals of a CMM.

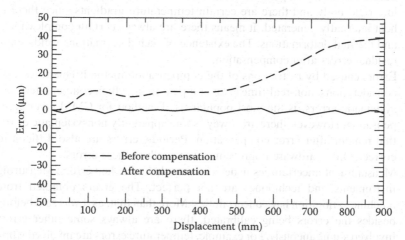

FIGURE 12.17 Reduction of error after compensation. (Redrawn from Zhang et al., *Ann CIRP*, 34(1):445, 1985.)

For testing the effect of thermal error compensation, the machine should be measured at different temperatures.

Figure 12.17 shows the result of error compensation for diagonal line *AG* of a particular machine (Zhang et al. 1985). The dotted line shows the displacement error before error compensation and the solid line shows that after compensation. The displacement error was reduced from 41 µm to 1.5 µm. The errors were measured by a laser interferometer.

Verification along four space diagonals cannot be a full check for the effectiveness of error compensation. However, in most cases, it can give a general picture of the machine errors since all 21 geometric errors, as well as thermal errors, do affect these lines. It is not unusual for the verification to not give the expected results.

In unsatisfactory cases, a step-by-step check is required to find out the cause of such an unsatisfactory result and then proper actions need to be taken to improve it.

12.4.3 LIMITATIONS

The measurement uncertainty of a CMM can be greatly reduced by applying error compensation, but it has limitations. The main limiting factors are as follows.

1. Errors not considered by the model: The most widely used models are quasi-rigid body models for compensating geometric errors and part of force deformation errors, that is, errors related with simple force deformation, and a thermal model for compensating simple thermal deformations. However, there are many other errors, such as probing (Estler et al. 1996; van Vliet and Schellekens 1998) and dynamic errors (Weekers and Schellekens 1997; Pereira and Hocken 2007), which cannot be compensated by the mentioned models.

2. Errors caused by imperfections of the model: Actually, there is no absolute rigid body and there are certain temperature gradients since there is heat internally generated. It means there are always certain complex force and thermal deformations. The existence of such deformations yields some residual errors after compensation.

3. Errors caused by restrictions of the compensation method: Due to several considerations, non-real-time error compensation, which compensates only systematic errors, is the most popular used method for CMM error compensation. However, there are always some apparently nonsystematic errors that remain after error compensation. Periodic errors are also difficult to correct. Only hardware improvements can reduce these errors.

4. Measurement uncertainties in determining the individual errors: Measuring instruments and techniques are not perfect. The errors obtained from machine calibration have uncertainties. More than that, as it was said before, besides the errors being calibrated, there are always some other errors involved simultaneously. For example, temperature errors are involved while displacement errors $\delta_x(X)$, $\delta_y(Y)$, and $\delta_z(Z)$ are calibrated. Systematic errors are always accompanied by uncertainties. Averaging only reduces but cannot eliminate the uncertainties.

   Error compensation should be made for every point within the working volume of a CMM. However, it is impossible to calibrate all errors of a machine at every point. The errors of all points, which were not calibrated directly, are derived by interpolation that also introduces some additional errors. This statement is especially true when artifacts with relatively large spacing between adjacent reference elements are used for machine calibration and the machine has periodic errors.

5. Errors caused by data processing and compensation software: Digitizing errors, interpolation errors, fitting errors, and algorithm errors are some examples of software errors.

6. Errors of compensation devices, including their dynamic errors.

12.5 TRENDS IN ERROR COMPENSATION DEVELOPMENT

The study on error compensation of CMMs started in the 1970s and was put into practical application in the 1980s. Significant progress has been achieved during the ensuing decades. Error compensation has become an indispensable component of modern direct computer-controlled (DCC) CMMs. However, there are still a lot of problems to be solved. Error compensation is still a fertile research topic. The future trends in error compensation development can be summarized as follows:

1. Means and methodologies for quick and accurate extraction of individual errors: The main purpose of error compensation is enhancing the accuracy of CMMs economically. The achievable accuracy of error compensation depends on the fitness of model and on the accuracy of error extraction to an even larger degree. Here, the term *error extraction* is used instead of *error measurement* or *calibration* since, in most cases, the parameter directly measured is not just the desired error but also it is in combination with some other errors. This situation is true especially when the CMM is calibrated by using certain artifact types.

    In terms of economy, the expenses of measuring means and labor for calibration and extraction of errors play a critical role. The error model of CMMs is the same for the same type of machine. However, the errors should be calibrated and extracted for each individual machine on the user's site. It is also desirable for the user of a CMM to check the machine periodically and update the compensation files. Most expenses of error compensation are related to the costs of measuring devices including the artifacts used and labor force for error calibration and extraction. The measuring instruments and artifacts used for calibration should ideally be of low cost, accurate, easy to use, stable, robust, and lightweight.

    It is not strange that the research in this direction attracts the interest of a large number of scientists and engineers all over the world. There are significant disagreements in terms of what are the best means and methodologies for error calibration and extraction, if it is better to use the parametric method or other complex methods of machine calibration, if it is better to use instrumentation or artifacts to carry out the machine calibration, and what is the best instrument or artifact for machine calibration and the application fields of a particular instrument or artifact. Such disagreements are not harmful but, to the contrary, help people to improve the means and methodologies for quick and accurate extraction of errors. Research in these areas will continue as one of the major directions in the coming decades.
2. Accuracy of error calibration and compensation: The accuracy requirements for measuring instruments and CMMs continually increase. Coordinate accuracy at the nanometer level has been put on the agenda. To enhance the accuracy of a CMM, efforts should be made in all aspects, to make the design more reasonable, to improve the manufacturing accuracy, to get better control of the working environment, and to compensate the errors effectively. The achievable accuracy of error compensation

depends on fitness of the model, modeling technique, and accuracy of error extraction.

3. Dynamic error compensation: The production tempo increases every day. To follow this tendency, CMMs with higher operational speeds should be developed. CMM structures become lighter, and more noncontact probes are used to measure parts on the fly. Compensation for dynamic errors is an urgent need nowadays. Although some achievements have been obtained in this field, they are still far behind the needs (Weekers and Schellekens 1997; Pereira and Hocken 2007). There is no applicable model that can quite relate the dynamic error with structural and operational parameters of a CMM. Real-time error compensation that is becoming possible should be used in combination with non-real-time to better compensate the errors. Means for measuring errors dynamically should also be further developed (Kroll 2003).

4. Error compensation for quasi-rigid body CMMs: From one side, with the increase of operational speed, CMM structures become lighter and lighter. At the same time, CMM accuracy requirements become ever more demanding. Complicated force deformations, which are negligible for low-accuracy machines, become essential for high-accuracy ones. In Figure 12.10, an example of quasi-rigid body model is given where the force deformation is treated separately. However, in the case shown in Figure 12.10, only the deformation of column (2) is a function of two arguments. What happens if the deformations of more than one component are functions of two or more arguments? In such cases, it would be difficult to separate the deformation errors from geometric ones by measuring several parallel lines. A more complicated model is required.

5. Error compensation for CMMs with complicated thermal deformation: Temperature gradients, which yield to convoluted thermal deformations, are unavoidable. For construction purposes, materials with different thermal coefficients of expansion are often used, and this leads to complicated thermal deformations. The problem is to determine if these complicated thermal deformations are negligible. They may be negligible for commonly used machines but become dominant for high-accuracy ones. As it was said before, due to the difficulties of compensation for complicated thermal deformations, it is reasonable to put more strict requirements on environment control and machine structure design rather than performing error compensation.

The case shown in Figure 12.15 is the simplest one illustrating complicated thermal deformation with a constant temperature gradient in v direction. The real case can be much more complicated than that. Temperature gradients exist in all directions and they are not constant. Moreover, there are nonuniform materials with different thermal coefficients of expansion used with different forms of bending. It is difficult to derive the thermomechanical model in such cases. Experimental approaches associated with theoretical studies are often preferred. Using a neural network is one of the possible approaches for modeling. In research, one of the key problems is determining the minimum number of sensors needed and where they should be mounted in order to fully understand the temperature field. Another question is how to separate different types of thermal errors from geometric errors. Generally speaking, the research on thermal errors lags behind the research on geometric errors. One of the major reasons is that it is difficult to simulate the thermal environment in operation. Another reason is that it requires the joint effort of experts from different fields.

6. Real-time error compensation: Until now, the dominant form for compensating geometric errors of CMMs has been non-real-time error compensation, since economy is one of the major considerations for applying error compensation. However, it does not mean that real-time error compensation will not find its application in the future. Non-real-time error compensation compensates for systematic errors, whereas real-time compensates both systematic and nonsystematic errors, including periodic errors, which are difficult to compensate by non-real-time means.

Another distinguishing feature of real-time error compensation is that there is no such strict requirement on the mathematical model. Due to the identity, many errors are identical for the measuring and compensation loops. Dynamic errors, errors caused by complicated force and thermal deformations, which are difficult to compensate by non-real-time error compensation, might be compensated to a large degree by real-time error compensation without a perfect model. For example, using the comparator method, there are two probes working in parallel. One measures the workpiece, another measures the reference. Due to the identity of these two measurements, many errors can be compensated without a mathematical model.

The key problems are reducing the cost of online error measuring devices and making them small and compatible with the CMM.

7. Compensation for probing errors: After the systematic errors of the machine structure have been compensated, probing errors contribute a significant part to the total error budget (see Chapter 6, Probing Systems for Coordinate Measuring Machines). Optical noncontact probes possess many advantages, such as no contact, no measuring force, and the possibility of measuring with high speed although they have their own difficulties (see Chapter 7, Multisensor Coordinate Metrology). The application of optical noncontact probes is an important tendency in CMM development. However, most optical noncontact probes have relative low accuracy. Their measurement uncertainty often is several tens of micrometers. Probing errors are often the dominant errors in the total error budget. On the other hand, there are many other errors in optical probes to compensate, such as scale error and nonlinearity errors of video probes, errors of triangulation probes caused by inclination of measured surface, dependence of its reading on the curvature of measured objects. Another large problem is that most manufactured parts have specular or partially specular surfaces. Simply put, the study on probing error compensation is an urgent need for enhancing the accuracy of CMMs.

8. Error compensation for accessories: For enhancing the functional ability of CMMs, a variety of accessories are used. Motorized heads equip most CMMs. Rotary tables and probe autochangers are other examples which are indispensable for measuring certain parts. All these devices may introduce significant contributions to the total error budget. To study the error compensation techniques for these accessories is another essential task in the future (see also Chapter 6, Probing Systems for Coordinate Measuring Machines).

9. Error compensation for nonorthogonal measuring systems: The majority of currently used CMMs are Cartesian. Non-Cartesian measuring systems will find wider applications in large-scale measurements and on shop floors due to their flexibility and lower cost (see also Chapter 17, Non-Cartesian Coordinate Measuring Systems). Non-Cartesian measuring systems should be developed with associate error compensation techniques (ASME 2004, 2006a; Santolaria et al. 2008). The key techniques are error calibration, extraction, and modeling techniques.

10. Development of devices for real-time error compensation: The dominant form of error compensation currently used is software error compensation. It is of low cost, and there is no trouble with dynamic servo driving. However, with the development of probes with a large sensing area, such as video probes and multifunction CMMs, which can be used as measuring machines and machining devices for example, lithographic machines as well, it would not be enough just to correct the measured data by software. Compensation devices are needed for correcting the error motions of the machine in real time. There are two ways for making compensation motion, as shown in Figures 12.2 and 12.3, respectively. The key problems are resolution and accuracy of the compensation device, its dynamic response, compactness, and cost.

11. Improving the dynamic performance of compensation software and devices: In most cases the speed of the compensation software is not a problem at the current time. However, with the development of high-speed CMMs, real-time error compensation, including the dynamic response of the compensation software and especially that of compensation devices, will become a major hurdle. Efforts should be made in improving the data processing mode and algorithm of compensation software along with approaches and performance of compensation devices.

12. Error compensation becomes an essential idea in machine design: Error compensation was first used for refurbishing existing CMMs to enhance their accuracies (Hocken et al. 1977; Zhang et al. 1985). Nowadays, error compensation is applied almost in all DCC CMMs as a measure to assure their accuracy. Error compensation should become an essential idea and basic principle in all machine types including CMMs.

In CMM design, a detailed and in-depth analysis of all the errors and their effects on the final performance of the machine should be conducted (error budget analysis). Error compensation should be used as an essential measure for enhancing its accuracy economically. A detail analysis should be made to determine which errors can be compensated and if that can be justified economically. Great attention should be paid to factors, such as no repeatability errors, short-period errors, low rigidity, nonuniform thermal behavior, yielding to errors that cannot be compensated, and others which cannot be compensated economically. Error compensation should be made not only on the machine level but also on the element and component levels.

Another important detail is allowing more room for error compensation at the design stage. As it was mentioned above, real-time compensation is able to compensate a large number of errors if there are two probes, one measures the workpiece and the other measures the reference. However, that is almost impossible on current machines. In the design of future machines, the possibilities which can be offered by error compensation should be taken into account. Error compensation should become an essential design principle.

13 "Reversal" Techniques for Coordinate Measuring Machine Calibration

Robert J. Hocken

CONTENTS

In this chapter, the term *reversal techniques* is used to cover a wide range of self-calibration techniques that are used or could be used on coordinate measuring machines (CMMs). These are a subset of a wider set of self-calibration techniques described in the literature. A good review of the subject, which was used extensively in this chapter, can be found in the study by Evans, Hocken, and Estler (1996) and the references therein.

13.1 CLASSIC REVERSALS

These reversals require a repositioning of the instrument or the artifact between at least two measurements. The simplest form is the level reversal shown in Figure 13.1. Suppose that the level is imperfect due to a piece of dirt attached to one foot (a common problem in masonry). When the level is in one position, it measures the angle of the surface, *s*, plus the angle of the level. When rotated 180° and put at the same place on the surface, it measures the angle of the surface minus the angle of the level. Splitting the difference gives the correct angle of the surface.

If one watches a good mason laying a brick wall, level reversal is seen constantly, as it is hard to keep small pieces of mortar from adhering to the surface of the level. When initially leveling a measurement machine with either a bubble level or an electronic level, level reversal is recommended.*

---

* Similar procedures are used in surveying where the telescope of a theodolite is turned 180° about a horizontal axis and then 180° about a vertical axis and resighted on a target. If the angles are averaged, height of standard errors (squareness) and collimation errors can be removed. The procedure is commonly called "face left, face right."

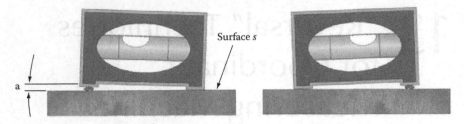

FIGURE 13.1   Level reversal.

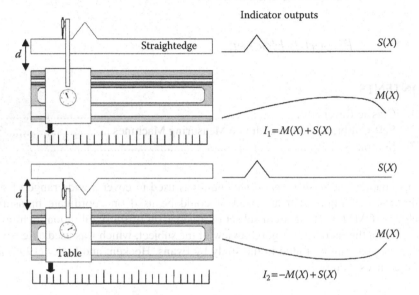

FIGURE 13.2   Straightness reversal.

Another classic reversal useful for CMM calibration is called straightedge reversal.* This procedure is used to measure the straightness of motion of a moving carriage and is illustrated in Figure 13.2. First, a calibrated indicator is used to measure the changing distance between the carriage and a straightedge aligned with the carriage. Next, the straightedge is rotated 180° about the measurement line on its face and is remeasured in the "reverse" position. According to Evans et al., if one calls the machine slide straightness $M(x)$ and the straightedge straightness $S(x)$, then the indicator readings in the two positions are given by the equations in Figure 13.2. $M(x)$ and $S(x)$ can then be calculated by taking half the sum and differences of I_1 and I_2. After the calculation of $M(x)$ and $S(x)$, a straight-line term should be removed, as it is the result of the lack of perfect alignment. Traditionally, the straight line was

---

* The oldest reference found on straightedge reversal is from 1909 (Cobleigh 1909). Here the reversal technique was used to make a straightedge by continually reversing and removing material until lines traced along the straightedge in the reversed and forward directions are the same.

removed graphically by zeroing the end points (or by graphically estimating a central line), but a least squares line is now preferred.

Note that the straightedge must remain stable throughout the process, the reversal must be such that the measurement line is at the same distance from the table, that is no offset, and there should be no movement of the straightedge along the axis direction, x, in the figure (movement of the straightedge in that direction is commonly called "shear"). Errors caused by position changes will depend on the out-of-straightness of the straightedge (for shear) and the angular error of the moving table (for offset). The reversal procedure works best if both errors mentioned are slowly varying functions of the position.

For machines that have continuous motion rotary axes, another classic reversal is quite useful. It is universally called the Donaldson reversal after its originator (Donaldson 1972). It is precisely analogous to the straightedge reversal but measures radial motion (ASME 2010) of a rotating axis and allows the use of an imperfect master ball just as the straightedge reversal uses an imperfect straightedge. This procedure is shown schematically in Figure 13.3. First, the ball out-of-roundness is measured in one position, and then the ball is rotated 180° with respect to the rotary table and is remeasured with the indicator on the opposite side of center. Let $X(\theta)$ be the radial motion of the rotary axis and $R(\theta)$ be the out-of-roundness of the master ball. Then, the equivalent equations to the straightedge case are represented in Figure 13.3, with similar sum and difference solutions for $R(\theta)$ and $X(\theta)$.

Eccentricity should be removed from the solutions. Further, if the rotation angle is not exactly 180°, errors due to master ball deviations from roundness can be seen, as is the case with shear in a straightedge. Results from this method can be quite accurate (Marsh 2009).

Another reversal that is useful for rotary axes is due to Estler (Estler 1986; Salsbury 2003b). It is called the Estler Face Motion Reversal and is shown diagrammatically in Figure 13.4. This method allows the separation of tilt and axial motions from out-of-flatness of the axis face. In the first setup, two indicators

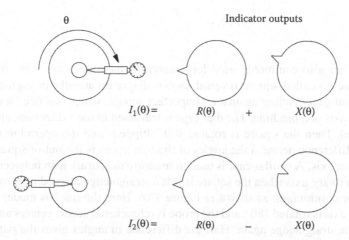

$$\theta \qquad\qquad\qquad\qquad \text{Indicator outputs}$$

$$I_1(\theta) = \qquad R(\theta) \quad + \quad X(\theta)$$

$$I_2(\theta) = \qquad R(\theta) \quad - \quad X(\theta)$$

FIGURE 13.3 Donaldson reversal.

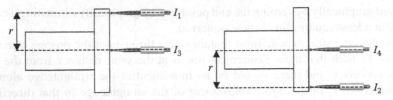

FIGURE 13.4 Estler reversal.

are used, one on the rotation axis and the other a distance r off the axis. The spindle is rotated 360° and the indicator readings and angles are recorded. In the second setup, the part is rotated 180° and the second indicator is repositioned to follow it. Again, a rotation and a measurement are performed. The first indicator is not really needed for the second measurement, but if left on the center, it improves knowledge of the axial motion. The data are analyzed by first averaging the results of the on-axis indicator, call it $A(\theta)$ following Evans, Hocken, and Estler (1996). Then, one can calculate the tilt error motion of the spindle $T(\theta)$ and the flatness of the part $P(\theta)$, where

$$I_3 = I_4 = A(\theta)$$

$$A(\theta) = \frac{I_3 + I_4}{2}$$

(13.1)

$$I_1 = P(\theta) + rT(\theta) + A(\theta)$$
$$I_2 = P(\theta) - rT(\theta) + A(\theta)$$

(13.2)

$$P(\theta) = \frac{I_1 + I_2}{2} - \frac{I_3 + I_4}{2}$$

$$T(\theta) = \frac{I_1 - I_2}{2r}$$

(13.3)

Reversals are also commonly used for squareness measurements. The simplest of these is the so-called square reversal shown diagrammatically in Figure 13.5. It shows a dial-gage reading against an imperfect square, which has one face aligned with the y axis of a machine. The dial gage is traversed in the x direction, and a trace is recorded. Then the square is rotated 180° (flipped) and the operation repeated. Half the difference between the angles of the two traces is the out-of-squareness of the machine axis. A similar case is used to measure the parallelism between a linear axis and a rotary axis when the square is just a straightedge aligned on a rotary axis (spindle or positioning), as shown in Figure 13.6. Here the trace is made, and then the rotary axis is rotated 180°, and the probe is rebracketed across centers and is used to trace the straightedge again. Half the difference in angles gives the parallelism. Many other examples are in the B5.57 standard (ASME 1998).

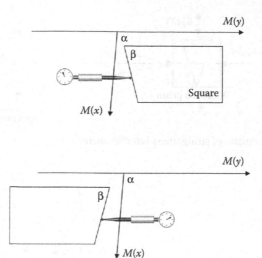

FIGURE 13.5   Squareness reversal.

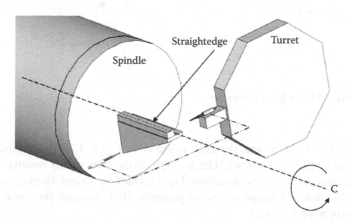

FIGURE 13.6   Parallelism reversal.

Straightness interferometers were introduced by Hewlett-Packard in the 1970s and are now produced by several companies. Typically, they use an angled reference mirror and a Wollaston prism, as shown in Figure 13.7. Lateral movement of the two optical elements with respect to each other (y direction in the figure) will cause changes in the optical path lengths of the two beams and therefore a change in the optical count.

If the mirrors of the straightness reflector are not perfectly flat, the deviations from flatness yield incorrect results as the optics move laterally. This error may be removed by another reversal technique introduced by Bryan and Carter (1989a, b). The data for an axis are taken first with the straightness mirror in one position.

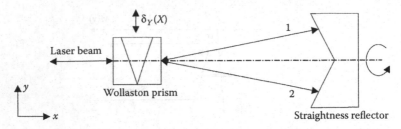

FIGURE 13.7   Schematic of straightness interferometer.

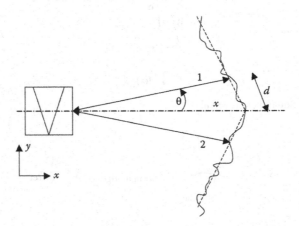

FIGURE 13.8   Mirror flatness effects.

Then, it is rotated 180° around the axis, as shown in Figure 13.8 and the straightness is measured at the same points. The average of the two results removes the flatness error. Interested readers are referred to the study by Evans, Hocken, and Estler (1996) for the detailed equations and alignments. (For example, the same positions on the mirror must be used.)

Note that all these classic reversal techniques are used to measure parametric machine errors and thus are useful for manufacturers as well as machine users.

Reversal techniques may also be used for determining the properties of the probe tip on measuring machines. Sometimes, the whole tip is mapped for use in ultra accuracy CMMs (see Chapter 16, Typical Applications and Küng, Meli, and Thalmann 2007), whereas in others, it is only desired to measure the probe diameter along one line, as shown in Figure 13.9. This technique was developed by Physikalisch-Technische Bundesanstalt (PTB; Evans, Hocken, and Estler 1996). Note that from the three measurements shown in the figure, the diameters of the probe tips can be calculated with simple algebra.

For CMMs, however, there exists a very different class of "reversal" like techniques that are usually referred to as "self-calibration." They are described in Section 13.2.

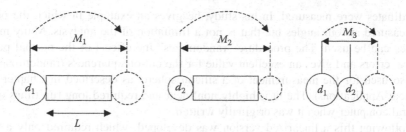

FIGURE 13.9   PTB probe diameter measurement.

13.2   SELF-CALIBRATION ON COORDINATE MEASURING MACHINES

On machines with multiple degrees of freedom, uncalibrated artifacts may be used for self-calibration. The concept is to measure the artifact in a number of carefully chosen positions and from the measurement results deduce the machine errors and the artifact errors (Sartori and Zhang 1995; Belforte et al. 1987; Balsamo et al. 1996). This should not be confused with measuring machine errors with a calibrated artifact. These measurements can only be used to find geometric errors as the results are independent of the chosen length scale (metric), which must be defined by the machine user. Ideally, such measurements could yield complete geometric information about the machine, but practical artifacts have limited numbers of measurement points, and the time of measurement is also constrained. Therefore, errors with a higher spatial frequency than the point density can be missed or erroneously interpreted due to aliasing. Most modern CMMs, however, have slowly varying (in space) geometric and scale errors and are thus ideally suited for this type of self-calibration. There are two very different methods.

In the first method, the general procedure is as follows: An artifact, usually a grid plate or a ball plate, is measured in at least two positions in two dimensions (2D) and more in three dimensions (3D). Ball bars, space frames, and even plane surfaces have also been used (Clément, Bourdet, and Weill 1981; Jouy and Clément 1986). Next, in any case, a model of machine errors is developed. This model may be very simple (Hocken and Borchardt 1979) containing only a few dominant terms or quite complex requiring a set of basis functions and perhaps some assumptions about the form of the machine errors (Raugh 1985, 1997). After multiple measurements, a least squares fit for the artifact coordinates and the machine error parameters is performed on the measurement data. Some experiments have fit the data to get representations of the normal parametric errors (see also Chapter 12, Error Compensation of Coordinate Measuring Machines; Fu 2000), others go directly to an X, Y error map. Often, it is not mentioned that the machine would need a different error map for different probe offsets. The earliest form of this type of calibration was found by Reeve (1974). He used a very simple model for the machine errors because he was concerned with the calibration of reference ball plates rather than the machine; his machine model contained only out-of-squareness. In his model, the plate was rotated to many different angles about the z axis of a measuring machine, and the X, Y center coordinates were measured. In his study, he gives an example in which the plate is measured at four angles but that is not a limitation of the analysis. Many more angles can be used. The procedure "randomizes" the effects of the normal parametric errors and gives an excellent value for the out-of-squareness (randomization is also used for the measurement of a silicon sphere, as described in Chapter 16, Typical Applications). The fit is highly nonlinear and required long run times on a central computer when it was originally written.

Following this a linearized version was developed, which required only a 90° rotation, although the rotation had to be within a small angle of 90° (Hocken and Borchardt 1979). This method used a machine model that contained out-of-squareness and a "metric error," which was the difference between the scales of the x and y axes. In normal use, one axis was assumed to have an accurate scale. Besides these two errors, this method calculated the difference between the actual rotation and 90° as well as any positional offset caused by the mechanical rotation. How this method works can be seen from an oversimplified example (Evans, Hocken, and Estler 1996). In that study, a ball bar, as shown in Figure 13.10, was rotated exactly 90° with no linear motion of the rotation point. Call the measured position of the second ball (the ball not at the origin) X_1, Y_1 in the first position and X_2, Y_2 in the second position. If α is the out-of-squareness and γ is the incremental scale error in the x axis, then

$$X_1 = (1+\gamma)X + \alpha Y$$
$$Y_1 = Y$$
$$X_2 = (1+\gamma)Y - \alpha X$$
$$Y_2 = -X$$

(13.4)

and the solutions for the machine squareness and scale difference are

$$\gamma = \frac{(X_2 - Y_1)Y_1 - (X_1 + Y_2)Y_2}{Y_1^2 + Y_2^2}$$
$$\alpha = \frac{X_1 Y_1 + X_2 Y_2}{Y_1^2 + Y_2^2}$$

(13.5)

A further advance in this procedure was developed by Raugh (1985) for the calibration of the stages on direct writing e-beam machines at Hewlett-Packard. In his model, the errors for each axis were parameterized, and the least squares fit was made for the coefficients of the fit parameters as well as the squareness and scale differences. Again a fairly large program was required. He continued to try and develop a complete theory (Raugh 1997) but has yet to complete this work. He did discover, however, that the procedures of Reeve and Hocken (Raugh 1997) would miss certain rotationally symmetric errors unless a finite translation was added to the procedure.

The above procedures were in 2D and the techniques were extended to 3D by measuring a number of planes. See the study by Kruth, VanHerck, and Jonge (1994) and references therein.

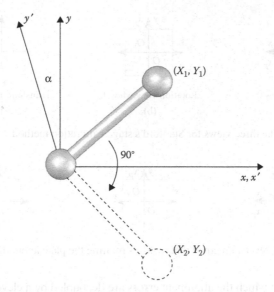

FIGURE 13.10   Idealized ball bar rotation.

This method of multiple measurements of the same artifact and least squares fitting was also used on rotary tables (Kunzmann et al. 1989). The abstract of his patent states the following:

> The four components of the deviations of a rotary table from an ideal axis of rotation can be obtained by a single measuring method wherein a test body (4) having a plurality of well-defined measurement points is placed on the rotary table (1) and the positions of said measuring points are then determined by means of a coordinate-measuring instrument (3), for each of various angular positions of the rotary table, there being thus a set of ascertained measurement-point coordinates for each of the angular positions of the rotary table. From the sets of measurement-point coordinates, the travel deviations f.sub.a, f.sub.r and f.sub.t of the axis of rotation, and the angle-position deviation p.sub.w are then determined by calculation.

In American nomenclature (ASME 2010), the deviations are called axial, radial, and tilt motions (f_a, f_r, f_t, respectively) as well as angular displacement accuracy (p_w).

Nearly independent of these developments for measuring machines (even the Raugh model that was checked by using the e-beam machine as a scanning electron microscope), efforts were ongoing in the microelectronics world to improve the accuracy of mask making by self-calibration (Ye 1996; Ye et al. 1997). In this stage calibration method introduced by Stanford University, three views are chosen (Figure 13.11a through c): (1) the initial view; (2) the rotation view, where the plate is rotated counterclockwise 90° about the center of the sample arrays; and (3) the translation view, where the plate is translated along +x by one sample site interval a.

Because of the existence of the alignment errors, the equations for each view couple the stage errors with the alignment errors. Consequently, solving for all these unknown errors involves a large computation (Ye et al. 1997). The beauty of Stanford's method

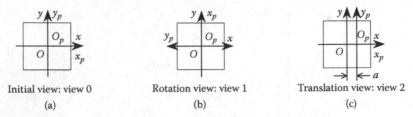

FIGURE 13.11   The three views for Stanford's stage calibration method.

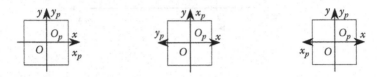

FIGURE 13.12   A reversal method that requires rotating the plate about the y axis.

is its algorithm, in which the alignment errors are decoupled by a clever mathematical treatment. Furthermore, by applying the discrete Fourier transform (DFT) algorithm, the discrete coefficients of the stage error map are further decoupled orthogonally, and therefore the calibration result is less sensitive to random measurement errors.

Another similar technique uses plates that can be accessed from both sides. It is in some sense similar to the straightedge reversal in that a rotation out of the plane of measurement is required. The three views required are shown in Figure 13.12.

This method yielded values for the parametric errors, which were compared to those obtained by traditional calibration with good agreement (Fu 2000). An extended algorithm that provided a 3D solution to self-calibration for testing CMMs was presented in 2006 (Dang, Yoo, and Kim 2006). Using a 3D grid artifact of steel balls, four separate sets of measurements were taken in different positions of the artifact. Then, the errors related to the artifact were identified using algebraic manipulation, hence a complete 3D error map of the machine systematics was constructed. The results were checked by experiment. Accuracy was limited by the measurement repeatability of the CMM tested.

13.3   SUMMARY

In this chapter, many of the reversal techniques common in machine calibration were described. Several of the methods described, such as square reversal, are already part of American national standards for machine tools (B5.57) and are in wide use. As pointed out in a review by Evans, Hocken, and Estler (1996), such tests are also common in optical metrology. The interested reader is referred to that study for many non-CMM applications.

14 Measurement Uncertainty for Coordinate Measuring Systems

Paulo H. Pereira and Robert J. Hocken

CONTENTS

14.1 BACKGROUND

Measurement uncertainty is the uncertainty of results of a measurement on a coordinate measuring system (CMS) computed according to the *Guide to the Expression of Uncertainty in Measurement* (GUM; JCGM 2008a). It is considered task-specific when a certain feature is measured using a defined inspection strategy. The state of the art in estimating this measurement uncertainty was discussed, in some detail, in a CIRP* Keynote paper in 2001 (Wilhelm, Hocken, and Schwenke 2001). This chapter briefly discusses major points in that paper and then includes some technical developments that have occurred since that time.

In the Keynote paper, the authors divided CMS uncertainties into five main categories: hardware, workpiece, sampling strategy, fitting and evaluation algorithms, and extrinsic factors. They then discussed how each of these categories influenced the task-specific uncertainty. Many of their comments have already been covered in this book in Chapter 9, Performance Evaluation. The interested reader is referred to the

---

* College International pour la Recherche en Productique—The International Academy for Production Engineering (www.cirp.net).

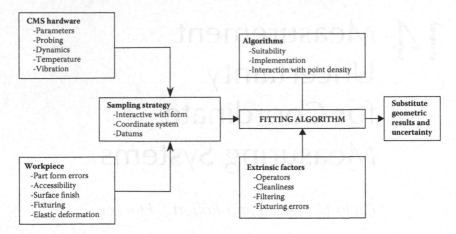

FIGURE 14.1 Error components that lead to uncertainties. (Adapted from Wilhelm, R. G., R. Hocken, and H. Schwenke, *Ann CIRP*, 50/2:553, 2001.)

Keynote paper for the details, but a summary block diagram is given in Figure 14.1 (Wilhelm, Hocken, and Schwenke 2001).

The authors then introduced uncertainty models that were available in the literature at that time. These included sensitivity analysis, expert judgment, experimental methods using calibrated objects, computer simulation, statistical estimations from measurement history, and hybrid methods. Sensitivity analysis was to be used following the procedures of GUM, "in cases where a clear, analytical solution can be formulated for the measurand and the function of measurement parameters." Expert judgment was defined, as in the GUM, which defines it as essentially a value judgment determined by the experience of an expert. Experimental methods using calibrated objects are fairly clear and very similar to the substitution method discussed by Phillips in Chapter 9, Performance Evaluation. Computer simulation was a broader category and was divided into numerous subcategories such as the virtual coordinate measuring machine (CMM), simulation by constraints, the expert CMM, and Monte Carlo simulation. Figure 14.2, taken from the Keynote paper, illustrates the concept of the virtual CMM. The virtual CMM includes the machine hardware, the sampling strategy, the evaluation software, and factors such as probing, geometry, and the environment.

The computer simulations are all now generally lumped into the Monte Carlo simulation described in Chapter 9, Performance Evaluation, and commercial software packages are available, which do a fairly reasonable job of performing these simulations with the proper inputs. The last two methods, statistical estimations from measurement history and hybrid methods, are still also in use. Statistical estimations are generally used when large numbers of nominally similar featured parts are measured over long periods of time.

As stated in the Keynote paper, there is no one common point of view as to how to best predict this task-specific uncertainty. Clearly, the constraints of requiring more uncertainty statements according to the new definition of traceability (which is "property of the result of a measurement or the value of the standard whereby it can be related to stated preferences using national or international standards through an

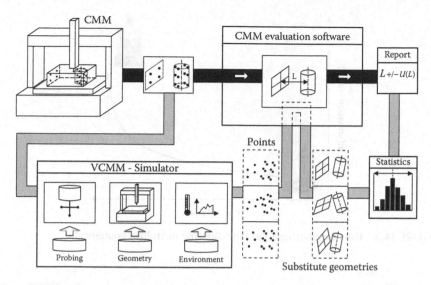

FIGURE 14.2   The virtual CMM concept. (Adapted from Wilhelm, R. G., R. Hocken, and H. Schwenke, *Ann CIRP*, 50/2:553, 2001.)

unbroken chain of comparisons all having stated uncertainties") make establishing agreement on developing task-specific uncertainties more important. There is currently some discussion (Sprauel et al. 2003; Reznik and Dabke 2004) such that not even national measurement institutes and other prestigious laboratories completely agree on the values (Jusko, Salsbury, and Kunzmann 1999).

Ultimately, the objective of estimating uncertainty is to make sure that the measurement results are reliable. On planning a metrology system, an uncertainty analysis does help the metrologist to choose the correct configuration for the intended application (Kunzmann et al. 2005; see also example in Section 14.2.1). The value in this case consists in planning the metrology portion together with the manufacturing process and making the proper investments as to obtain a working solution from the beginning of the intended application.

Since measuring is an integral part of the manufacturing process, larger uncertainty values cause the apparent process capability to be lower than the actual. This is because one cannot determine the true value of a measurand due to the ever-present measurement uncertainty. Thus, in order to obtain the proper level of performance of a manufacturing cell, one must plan the accompanying metrology system, taking into account how much variation (measurement uncertainty) it will add to the results (Kunzmann et al. 2005; Pereira 2005).

As part of planning the metrology aspects of a new manufacturing cell, a metrologist has to consider which gage or other measuring equipment to use. Simply put, there are many choices depending on the application. The qualitative chart in Figure 14.3 illustrates the concept that, in general, simpler gages may cost less initially but will not deliver a lot of information. On the other hand, a more complex piece of equipment like a CMM will be more flexible and capable of measuring more features while demanding a higher initial investment. There are several aspects to

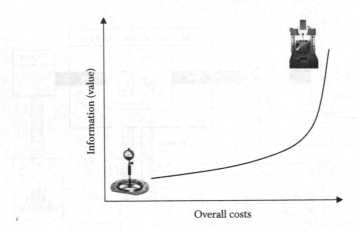

FIGURE 14.3   Cost-versus-benefit ratio for choosing metrology equipment.

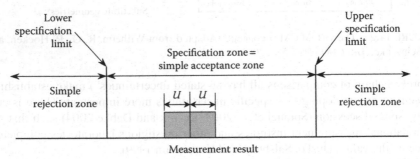

FIGURE 14.4   Simple acceptance zone. (Adapted from ASME B89.7.3.1-2001, Guidelines for Decision Rules: *Considering Measurement Uncertainty in Determining Conformance to Specifications*, New York: American Society of Mechanical Engineers, 2001).

consider besides uncertainty when choosing a gage, including tolerances to be measured, quantity of parts to be checked, size and type of parts, type of data collection, required training, and so on (see also Chapter 15, Application Considerations).

Another change that has really emphasized the need for proper estimation of CMM uncertainty is the development of new decision rules. In the end, the outcome of a measurement gives the user information that is used to make a decision about the manufacturing process and the workpieces produced.* In the past, metrologists simply planned for a metrology process that would have 10 times better accuracy than the feature being measured (10:1 gagemaker's rule; Qberg and Jones 1920; Radford 1922; Rolt 1929; ASME 2001b), and then no one would worry about measurement errors other than properly maintaining and using the gages. With that requirement met, a simple acceptance rule by which parts are considered conforming whenever they fall within the specification limits, shown by Figure 14.4 (ASME 2001b), was considered adequate for many decades.

---

* The use of proper statistical process control techniques (Juran and Godfrey 1999; Montgomery 2009) must be also considered but is not covered here.

The B89.7.4.1 Technical Report (ASME 2005b) defines the measurement capability index, C_m, which is analogous to other capability indices, as

$$C_m \equiv \frac{T}{2U} \tag{14.1}$$

where

T is the tolerance zone

U is the expanded measurement uncertainty (defined in detail later in the text)

This is a useful index to characterize the quality of a measuring system. A 10:1 ratio means that $C_m = 10$. The higher the measurement capability index, the better the measuring system quality for the specific application.

With increasing costs in achieving the 10:1 ratio, it becomes more important to users to better understand the uncertainty in order to manage costs properly and make informed decisions about their parts. There are always risks and costs involved that the user needs to consider when making decisions on how to deal with results close (within uncertainty) to the specification limits (ASME 2005b). One option to reduce the risks is to use the stringent acceptance zone (Figure 14.5) by which parts are considered conforming only when they fall within the specification range minus the uncertainty. This has the advantage of reducing the risks to the customer, as no questionable measurements are considered conforming. However, this option increases the producer's costs as the rejection rate increases.

There is also the possibility of having a relaxed acceptance zone (Figure 14.6) by which all measurements falling within the uncertainty added to the specification limits are deemed acceptable. This reduces the cost to the producer but increases the customer's risks.

There is some controversy in decision rules considering uncertainties to prove conformance or nonconformance of parts in the case of commercial transactions (ASME 2006b). The ISO 14253-1 Standard (ISO 1998) states that, by default, sellers should use the stringent acceptance rule when proving conformance, whereas buyers should use the relaxed acceptance rule. This can create undesirable situations in certain cases. An example of this is a company that buys and then resells products that would be penalized twice by this default rule. Certainly, significantly reducing

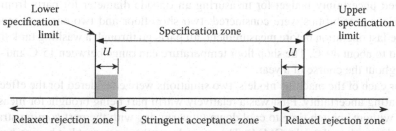

FIGURE 14.5 Stringent acceptance zone—relaxed acceptance zone. (Adapted from ASME B89.7.3.1-2001, Guidelines for Decision Rules: *Considering Measurement Uncertainty in Determining Conformance to Specifications*, New York: American Society of Mechanical Engineers, 2001.)

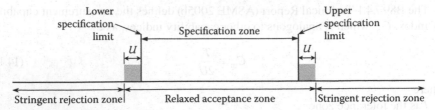

FIGURE 14.6 Relaxed acceptance zone—stringent rejection zone. (Adapted from ASME B89.7.3.1-2001, Guidelines for Decision Rules: *Considering Measurement Uncertainty in Determining Conformance to Specifications*, New York: American Society of Mechanical Engineers, 2001.)

measurement uncertainties and having well-controlled manufacturing processes would all but eliminate the questions of measurements falling within the uncertainty range from the specification limits. The ISO 14253-1 Standard does state that other arrangements could be made between seller and buyer. The Technical Specification ISO/TS 14253-4 (ISO 2010b) elaborates on this issue.

14.2   EXAMPLES

Much of this discussion can be quantified by examining worked examples. This is done in Sections 14.2.1 and 14.2.2.

14.2.1   METROLOGY PLANNING FOR A NEW MANUFACTURING CELL

Consider the generic example of planning a cell to make cylindrical parts from Ø50 to Ø700 mm. The cell output was to be around 200,000 parts a year and each had several features, including size and location, to be measured. Not all parts, but a sizable number, needed to be measured by the system. Because of such conditions, manual gaging processes were excluded from the possibilities. One decision was that the metrology solution would be very close to the cell to minimize response time and not depend on a metrology laboratory located far away. All that narrowed the choices to having a CMM in the cell.

Even then, there were some choices of machine models that could be installed directly on the shop floor or enclosed for temperature control. Table 14.1 shows the detailed uncertainty budget for measuring an outside diameter for parts from this cell. Four possibilities were considered: two shop floor and two enclosed models. As the last operation before measurement, all parts go through a washing tank that is heated to about 49°C. The shop floor temperature can range between 15°C and 45°C throughout the course of a year.

For each of the machine models, two situations were considered for the effects of estimating uncertainty. First was a relatively warm part being brought for measurement without enough time to cool down (a warm day with the part coming straight from the wash tank to the CMM). The second was the same part but having had the time to cool down (a cooler day).

Four main uncertainty contributors (Salsbury 1995; ISO 1999b; Chapter 19, Performance Evaluation) were considered: (1) machine uncertainty, (2) sampling

TABLE 14.1
Parameters for Metrology Planning Example

Description	Details	Observation
Feature	Outer diameter 500 ± 0.1 mm	
Part temperature range	Warm condition = 8°C; cool condition = 4°C	
Average part temperature	Warm condition = 35°C; cool condition = 24°C	
Part material	Steel	
Sampling strategy	11 hits equally spaced	
Scale temperature range for shop CMMs	4°C	During longest measurement cycle
Average scale temperature for shop CMMs	Warm condition = 35°C; cool condition = 24°C	
Scale temperature range for enclosed CMMs	1°C (20 ± 2°C)	
Average scale temperature for enclosed CMMs	20°C	
Uncertainty of temperature measurement device	± 0.1°C	
Shop CMM 1 specs	$MPE_E = 1.7 + L/286$ μm	Temperature ≤ 22°C
	$MPE_E = 2.4 + L/195$ μm	Temperature ≤ 40°C
	$MPE_P = 1.7$ μm	
Shop CMM 1 scale	Coefficient of thermal expansion = 1.8 μm/m/°C	
Shop CMM 2 specs	$MPE_E = 1.5 + L/300$ μm	Temperature ≤ 22°C
	$MPE_E = 1.5 + L/150$ μm	Temperature ≤ 40°C
	$MPE_P = 1.7$ μm	
Shop CMM 2 scale	Coefficient of thermal expansion = 0.1 μm/m/°C	
Laboratory CMM 1 specs	$MPE_E = 1.7 + L/300$ μm	
	$MPE_P = 1.8$ μm	
Laboratory CMM 1 scale	Coefficient of thermal expansion = 6.5 μm/m/°C	
Laboratory CMM 2 specs	$MPE_E = 2.2 + L/333$ μm	
	$MPE_P = 2.2$ μm	
Laboratory CMM 2 scale	Coefficient of thermal expansion = 6.5 μm/m/°C	
Probe tip qualification error for all four CMMs	2 μm	All have comparable probing systems

uncertainty, (3) thermally-induced uncertainty, and (4) datum uncertainty. Most cases usually have only a handful of contributors of importance to the overall uncertainty. The combined uncertainty, u_c, is calculated as the root mean square of all individual uncertainty components. Considering the sources to be independent, the uncertainty Equation 9.3 becomes

$$u_c = \sqrt{u_M^2 + u_{sampling}^2 + u_T^2 + u_{datum}^2} \qquad (14.2)$$

The expanded uncertainty, U, is obtained by multiplying the standard uncertainty by a coverage factor as per

$$U = k \times u_c \qquad (14.3)$$

where

u_M is the machine uncertainty.

$u_{sampling}$ is the sampling uncertainty.

u_T is the temperature induced uncertainty.

u_{datum} is the datum uncertainty.

k is the coverage factor. Conventionally, a coverage factor of 2 is chosen, which results in a confidence level of 95% for the uncertainty estimate.

14.2.1.1 Machine Uncertainty

The machine uncertainty comes from the geometric errors of the CMM at any point in its work volume. Most of the CMM models available in the market today use one form or another of specification according to published standards (ASME 1997, 2008; ISO 2009a) that can be used for calculating the machine uncertainty per Equation 14.4, or similar,

$$u_M = MPE_E = a + b \times L \qquad (14.4)$$

where

MPE_E is the maximum permissible error as supplied ("a" and "b") by the manufacturers

L is the length being evaluated (or largest distance to a datum, whichever is larger)

The calculation for machine uncertainty can be refined by replacing the specifications for the actual results from the last time they were checked. This is more realistic for the cases when the machine is in place and will almost always be smaller. For comparisons of new machines, the best way is to use the specifications, which is a more conservative approach.

14.2.1.2 Sampling Uncertainty

Sampling uncertainty is due to measuring a limited number of points from an infinite number of possible points that constitute a feature. To quantify this contributor, a statistical sampling approach is used. The standard deviation of a sample mean is the standard deviation of the population divided by the square root of the sample size, n. The more the number of points there are in a sample, the smaller the associated uncertainty. Uncertainty in measuring a feature is the result of the point coordinate uncertainty taking into consideration the number of points measured as per

$$u_{\text{sampling}} = \frac{\sigma_{\text{pc}}}{\sqrt{n-x}} \tag{14.5}$$

where

σ_{pc} is the point coordinate error

n is the number of points probed to define the feature

x is the minimum number of points required to define a given feature (e.g., $x = 2$ for a line, $x = 3$ for plane or circle, $x = 4$ for sphere, and $x = 5$ for cylinder)

In general, the styli used for measurement are not the same as the one used for probing performance tests like ISO 10360-5 (ISO 2010a) or B89 (ASME 1997) probing test. In those cases, it is necessary to add to the specified performance test some error due to the different stylus being used for probing the specific feature, as per Equation 14.6. This value can come from standardized tests executed with the stylus used for the measurement or from the probe qualification routine.

$$\sigma_{\text{pc}} = \sqrt{\sigma_{\text{MPEp}}^2 + \sigma_{\text{stylus}}^2} \tag{14.6}$$

where

σ_{pc} is the point coordinate error

σ_{MPEp} is the probing error

σ_{stylus} is the additional probing error due to the stylus

14.2.1.3 Uncertainty Due to Temperature

The temperature of a measuring environment, machine, and part should ideally be 20°C (refer to Chapter 10, Temperature Fundamentals; ISO 2002a). If this condition were satisfied, then there would be no uncertainty due to temperature effects. However, in practice, this does not happen, hence uncertainty due to thermal effects is always present to some extent. In fact, many times, temperature is the most significant uncertainty contributor.

Uncertainty of nominal expansion of a part or scale is due to (1) uncertainty in expansion coefficient, (2) uncertainty in temperature measuring device, and (3) variation of temperature from its mean value due to lack of control and/or inadequate soaking of parts. Using Equation 14.7, total thermal uncertainty, u_T, can be calculated as (ISO 2003d)

$$u_T = \sqrt{\text{UNE}_S^2 + \text{UNE}_W^2 + \text{LUTM}_S^2 + \text{LUTM}_W^2} \tag{14.7}$$

where

UNE is the uncertainty of nominal thermal expansion

LUTM is the length uncertainty due to temperature measurement

UNE is given by Equation 14.8

$$\text{UNE}_X = \sqrt{L^2 \times u_{\alpha_X}^2 \times (T_{X\text{avg}} - 20)^2} \tag{14.8}$$

where
   X = S or W, respectively, for scale or workpiece
   L is the length being considered
   u_{α_X} is the uncertainty associated with coefficients of expansion given by Equation 14.9

$$u_{\alpha_X} = C_D \times 0.1 \times \alpha_X \qquad (14.9)$$

where
   C_D is the coefficient tied to the temperature distribution (used to transform limits of variation into standard deviation). The most used are 0.5 for normal, 0.6 for rectangular, and 0.7 for U-shaped distributions (ISO 1999b).
   α_X is the coefficient of thermal expansion for the scale or workpiece (usually from published values or from a calibration certificate).
   The 0.1 factor comes from the approximation that both calibrated and published coefficients of thermal expansion have a ±10% variation.
   LUTM$_X$ is given by Equation 14.10.

$$\text{LUTM}_X = \sqrt{\left(L^2 \times \alpha_X^2 \times u_{TX1}^2\right) + \left(L^2 \times \alpha_X^2 \times u_{TX2}^2\right)} \qquad (14.10)$$

where
   u_{TX1} is the uncertainty due to variation of temperature of the machine scale or workpiece. It can be calculated by $u_{TX1} = (T_{X\max} - T_{X\min}) \times C_D$.
   u_{TX2} is the uncertainty component due to uncertainty of the temperature-measuring device (used for measuring scale or workpiece temperature) itself, which can be obtained from calibration certificates.

14.2.1.4   Datum Uncertainty

Some features are associated with datums, which themselves are also measured and hence have associated uncertainties. If a feature is evaluated with respect to a datum (e.g., true position and profile), then its feature uncertainty is also affected by datum uncertainties.

Even though datums are features, not all of them equally influence uncertainty. The primary datum controls three degrees of freedom (out of six degrees of freedom of a rigid body), the secondary datum controls two degrees of freedom, and the tertiary datum controls one degree of freedom. Therefore, Equation 14.11 expresses datum uncertainty.

$$u_{\text{datum}} = \sqrt{\left(\frac{3}{6} \times u_{\text{primary}}\right)^2 + \left(\frac{2}{6} \times u_{\text{secondary}}\right)^2 + \left(\frac{1}{6} \times u_{\text{tertiary}}\right)^2} \qquad (14.11)$$

where the uncertainty for each datum is its own feature uncertainty (Equation 14.5).

TABLE 14.2

Uncertainty Contributors for Outer Diameter Measurement on Shop CMMs with the Sample Part at 35°C

Source	Formula	Result (µm)	Observation
Machine	Shop CMM $1 = 2.4 + \dfrac{500}{195}$	4.96	
	Shop CMM $2 = 1.5 + \dfrac{500}{150}$	4.83	
Sampling	$\dfrac{\sqrt{1.7^2 + 2^2}}{\sqrt{11-3}}$	0.93	Same specification for both models
Thermal			
UNE_S			
Shop CMM 1	$\sqrt{0.5^2 \times (0.6 \times 0.1 \times 1.8)^2 \times (35-20)^2}$	0.81	With rectangular distribution
Shop CMM 2	$\sqrt{0.5^2 \times (0.6 \times 0.1 \times 0.1)^2 \times (35-20)^2}$	0.045	
UNE_W	$\sqrt{0.5^2 \times (0.6 \times 0.1 \times 11.5)^2 \times (35-20)^2}$	5.18	With rectangular distribution
$LUTM_S$			
Shop CMM 1	$\sqrt{(0.5^2 \times 1.8^2 \times (4 \times 0.6)^2) + (0.5^2 \times 1.8^2 \times 0.1^2)}$	2.16	With rectangular distribution
Shop CMM 2	$\sqrt{(0.5^2 \times 0.1^2 \times (4 \times 0.6)^2) + (0.5^2 \times 0.1^2 \times 0.1^2)}$	0.12	
$LUTM_W$	$\sqrt{(0.5^2 \times 11.5^2 \times (8 \times 0.6)^2) + (0.5^2 \times 11.5^2 \times 0.1^2)}$	27.61	With rectangular distribution
u_T (Shop CMM 1)	$\sqrt{0.81^2 + 5.18^2 + 2.16^2 + 27.61^2}$	28.19	
u_T (Shop CMM 2)	$\sqrt{0.045^2 + 5.18^2 + 0.12^2 + 27.61^2}$	28.09	
Datum	Not applicable to size features	0	
Combined uncertainty	Shop CMM $1 = \sqrt{4.96^2 + 0.93^2 + 28.19^2 + 0^2}$	28.63	
	Shop CMM $2 = \sqrt{4.83^2 + 0.93^2 + 28.09^2 + 0^2}$	28.52	
Expanded uncertainty	Shop CMM $1 = 2 \times u_c = 2 \times 28.63$	57.26	
	Shop CMM $2 = 2 \times u_c = 2 \times 28.52$	57.04	

Table 14.2 shows the application of Equations 14.2 through 14.11 together with the details from Table 14.1 to the metrology planning example, for the condition of an outside diameter of a warm part measured on the mentioned shop CMMs.

Similarly, Table 14.3 shows the application of the same equations, together with the details from Table 14.1, for the condition of an outside diameter of a cooler part measured on the shop CMMs.

TABLE 14.3

Uncertainty Contributors for Outer Diameter Measurement on Shop CMMs with the Sample Part at 24°C

Source	Formula	Result (μm)	Observation
Machine	Shop CMM $1 = 1.7 + \dfrac{500}{286}$	3.45	
	Shop CMM $2 = 1.5 + \dfrac{500}{300}$	3.17	
Sampling	$\dfrac{\sqrt{1.7^2 + 2^2}}{\sqrt{11-3}}$ for both shop CMMs	0.93	Same specification for both models
Thermal			
UNE$_S$			
Shop CMM 1	$\sqrt{0.5^2 \times (0.6 \times 0.1 \times 1.8)^2 \times (24-20)^2}$	0.22	With rectangular distribution
Shop CMM 2	$\sqrt{0.5^2 \times (0.6 \times 0.1 \times 0.1)^2 \times (24-20)^2}$	0.012	
UNE$_W$	$\sqrt{0.5^2 \times (0.6 \times 0.1 \times 11.5)^2 \times (24-20)^2}$	1.38	With rectangular distribution
LUTM$_S$			
Shop CMM 1	$\sqrt{(0.5^2 \times 1.8^2 \times (4 \times 0.6)^2) + (0.5^2 \times 1.8^2 \times 0.1^2)}$	2.16	With rectangular distribution
Shop CMM 2	$\sqrt{(0.5^2 \times 0.1^2 \times (4 \times 0.6)^2) + (0.5^2 \times 0.1^2 \times 0.1^2)}$	0.12	
LUTM$_W$	$\sqrt{(0.5^2 \times 11.5^2 \times (4 \times 0.6)^2) + (0.5^2 \times 11.5^2 \times 0.1^2)}$	13.81	With rectangular distribution
u_T (Shop CMM 1)	$\sqrt{0.22^2 + 1.38^2 + 2.16^2 + 13.81^2}$	14.05	
u_T (Shop CMM 2)	$\sqrt{0.012^2 + 1.38^2 + 0.12^2 + 13.81^2}$	13.88	
Datum	Not applicable to size features	0	
Combined uncertainty	Shop CMM $1 = \sqrt{3.45^2 + 0.93^2 + 14.05^2 + 0^2}$	14.50	
	Shop CMM $2 = \sqrt{3.17^2 + 0.93^2 + 13.88^2 + 0^2}$	14.27	
Expanded uncertainty	Shop CMM $1 = 2 \times u_c = 2 \times 14.50$	29.00	
	Shop CMM $2 = 2 \times u_c = 2 \times 14.27$	28.54	

Likewise, Table 14.4 shows the application of the same equations, together with the details from Table 14.1, for the condition of the outside diameter of a warm part measured on the enclosed (laboratory) CMMs.

Equally, Table 14.5 shows the application of the same equations, together with the details from Table 14.1, for the condition of the outside diameter of a cooler part measured on the enclosed (laboratory) CMMs.

TABLE 14.4

Uncertainty Contributors for Outer Diameter Measurement on Laboratory CMMs with the Sample Part at 35°C

Source	Formula	Result (µm)	Observation
Machine	Laboratory CMM $1 = 1.7 + \dfrac{500}{300}$	3.37	
	Laboratory CMM $2 = 2.2 + \dfrac{500}{333}$	3.70	
Sampling	Laboratory CMM $1 = \dfrac{\sqrt{1.8^2 + 2^2}}{\sqrt{11-3}}$	0.95	
	Laboratory CMM $2 = \dfrac{\sqrt{2.2^2 + 2^2}}{\sqrt{11-3}}$	1.05	
Thermal			
UNE_S	$\sqrt{0.5^2 \times (0.6 \times 0.1 \times 6.5)^2 \times (20-20)^2}$ (same scale coefficients of thermal expansion)	0	With rectangular distribution
UNE_W	$\sqrt{0.5^2 \times (0.6 \times 0.1 \times 11.5)^2 \times (35-20)^2}$	5.18	With rectangular distribution
$LUTM_S$	$\sqrt{(0.5^2 \times 6.5^2 \times (1 \times 0.6)^2) + (0.5^2 \times 6.5^2 \times 0.1^2)}$ (same scale coefficients of thermal expansion)	1.98	With rectangular distribution
$LUTM_W$	$\sqrt{(0.5^2 \times 11.5^2 \times (8 \times 0.6)^2) + (0.5^2 \times 11.5^2 \times 0.1^2)}$	27.61	With rectangular distribution
u_T	$\sqrt{0^2 + 5.18^2 + 1.98^2 + 27.61^2}$	28.16	
Datum	Not applicable to size features	0	
Combined uncertainty	Laboratory CMM $1 = \sqrt{3.37^2 + 0.95^2 + 28.16^2 + 0^2}$	28.38	
	Laboratory CMM $2 = \sqrt{3.7^2 + 1.05^2 + 28.16^2 + 0^2}$	28.42	
Expanded uncertainty	Laboratory CMM $1 = 2 \times u_c = 2 \times 28.38$	56.75	
	Laboratory CMM $2 = 2 \times u_c = 2 \times 28.42$	56.84	

Table 14.6 shows the compilation of all the uncertainty results from the different possibilities, along with their ratios to the considered feature tolerance (Table 14.1). These results allow several interesting inferences as follows:

- Similar machines from different manufacturers have the same tolerance-to-uncertainty ratios. Thus, this parameter alone is not enough to differentiate the brands.
- The temperature uncertainty completely dominates the overall budgets for this example. It is very clear that warm parts cannot be checked with reasonable uncertainty levels.
- A shop CMM with temperature-stabilized parts would work, but guaranteeing the parts are thermally stable can be a significant issue.

TABLE 14.5

Uncertainty Contributors for Outer Diameter Measurement on Laboratory CMMs with the Sample Part at 24°C

Source	Formula	Result (μm)	Observation
Machine	Laboratory CMM $1 = 1.7 + \dfrac{500}{300}$	3.37	
	Laboratory CMM $2 = 2.2 + \dfrac{500}{333}$	3.70	
Sampling	Laboratory CMM $1 = \dfrac{\sqrt{1.8^2 + 2^2}}{\sqrt{11-3}}$	0.95	
	Laboratory CMM $2 = \dfrac{\sqrt{2.2^2 + 2^2}}{\sqrt{11-3}}$	1.05	
Thermal			
UNE_S	$\sqrt{0.5^2 \times (0.6 \times 0.1 \times 6.5)^2 \times (20-20)^2}$ (same scale coefficients of thermal expansion)	0	With rectangular distribution
UNE_W	$\sqrt{0.5^2 \times (0.6 \times 0.1 \times 11.5)^2 \times (24-20)^2}$	1.38	With rectangular distribution
$LUTM_S$	$\sqrt{(0.5^2 \times 6.5^2 \times (1 \times 0.6)^2) + (0.5^2 \times 6.5^2 \times 0.1^2)}$ (same scale coefficients of thermal expansion)	1.98	With rectangular distribution
$LUTM_W$	$\sqrt{(0.5^2 \times 11.5^2 \times (4 \times 0.6)^2) + (0.5^2 \times 11.5^2 \times 0.1^2)}$	13.81	With rectangular distribution
u_T	$\sqrt{0^2 + 1.38^2 + 1.98^2 + 13.81^2}$	14.02	
Datum	Not applicable to size features	0	
Combined uncertainty	Laboratory CMM $1 = \sqrt{3.37^2 + 0.95^2 + 14.02^2 + 0^2}$	14.45	
	Laboratory CMM $2 = \sqrt{3.7^2 + 1.05^2 + 14.02^2 + 0^2}$	14.54	
Expanded uncertainty	Laboratory CMM $1 = 2 \times u_c = 2 \times 14.45$	28.90	
	Laboratory CMM $2 = 2 \times u_c = 2 \times 14.54$	29.08	

For this particular example, the best solution would be to employ one of the enclosed machine models considered with a way to guarantee part temperature stabilization prior to measurement. However, other aspects also need to be considered besides uncertainty before choosing the right metrology system for an application (see also Chapters 15, Application Considerations, and 19, Financial Evaluations).

TABLE 14.6
Uncertainty Summary Values

CMM	Workpiece at 35°C		Workpiece at 24°C	
	Expanded Uncertainty (μm)	Tolerance-to-Uncertainty Ratio	Expanded Uncertainty (μm)	Tolerance-to-Uncertainty Ratio
Shop CMM 1	57.26	3.5	29.00	6.9
Shop CMM 2	57.04	3.5	28.54	7.0
Enclosed CMM 1	56.75	3.5	28.90	6.9
Enclosed CMM 2	56.84	3.5	29.08	6.9

14.2.2 OTHER EXAMPLES

There are several other examples in the literature. For example, ISO/TS 14253-2 (ISO 1999b) shows an uncertainty budget for certifying a ring gage; ASME B89.7.3.2 (ASME 2007) outlines an uncertainty budget for using a caliper on the shop floor; ISO/TR 16015 (ISO 2003d) exemplifies the thermal uncertainties for a length measurement. Salsbury (Salsbury 1995) details a few examples of measuring some features with a CMM. Drescher (Drescher 2004) details an uncertainty budget for measuring airfoil cooling holes using a five-axis CMM with dual sensors. The ASME B89.1.13 Standard (ASME 2001a) has an uncertainty budget for calibrating micrometers; ASME B89.1.9 (ASME 2002a) lists possible uncertainty sources when calibrating gage blocks by mechanical comparison. Doiron and Stoup (Doiron and Stoup 1997) describe uncertainties for calibrating several artifacts, including gage wires and ring gages.

14.3 CONCLUSION

It is clear there has been considerable progress in the past several years in developing methodologies to estimate task-specific uncertainties, particularly for CMMs used in the contact mode acquiring points on an individual basis. However, there are some challenges that need to be addressed to improve these estimates and the ease of their computation.

First, several of the models require input data that are not commonly measured as part of machine acceptance or interim testing. The utility of these models could be improved by developing standards for machine acceptance and testing that allowed rapid and accurate measurement of the required inputs or by having the uncertainty estimation methods that use inputs from tests performed routinely as per available standards like B89 (ASME 1997, 2008) and ISO 10360 series ISO 2000b, 2000c, 2009a, 2010a).

Second, although many of the models were tested in laboratory environments, it is unclear how they would handle the thermal environments common on the factory floor, where temperatures can change by up to 20°C in a single day* and operators usually neglect or do not know how to utilize part temperature sensors even when they are available. The complexity of thermal effects is well known (Bryan et al. 1966; Weckenmann and Heinrichowski 1985; Sartori et al. 1989) and an ISO Technical Report (ISO 2003d) exists to help compute their uncertainty, but the methods are not universally understood even in the case of constant temperatures that differ from 20°C. Cases in which temperatures are changing are difficult to treat due to the varying time constants of machine components (rams, scales, etc.), parts, fixtures, and probes (Kruth, Van den Bergh, and VanHerck 2001).

Third, none of the models appear to successfully address the issues of the interaction of the sampling strategy with possible part form error, which has many ramifications. Even in the case of circles and cylinders (Weckenmann and Heinrichowski 1985; Hocken, Raja, and Babu 1993; Weckenmann, Knauer, and Kunzmann, 1998), the situation is complex (and difficult to quantify by real measurement due to time and sampling size constraints), but in the case of free-form surfaces, it is probably worse (Edgeworth and Wilhelm 1996, 1999b; Savio and De Chiffre 2001, 2002), and magnification effects due to small datums with form error make matters more challenging.

Fourth, there is a large class of coordinate measuring equipment that has only been partially addressed in the literature. This includes large-volume metrology systems, such as laser trackers (ASME 2006a), photogrammetry systems and stereo triangulation (Clarke et al. 2001), as well as vision-based CMMs (Kim and McKeown 1996; Hansen and De Chiffre 1997; Fu 2000), articulated-arm CMMs (ASME 2004), scanning CMMs (Pereira and Hocken 2007), and laser triangulation-based machines, among others.

Fifth, extrinsic factors are nearly impossible to take into full account without an expert actually examining the measurement situation, conducting the equivalent of a gage repeatability and reproducibility (GR&R) study and using the data and personal judgment to estimate a type B uncertainty.

In addition, for uncertainty estimation to become widespread, it needs to be made more automated and simpler such that nonexperts can use it. One possible way would be to have the uncertainty estimation tool as a module tied to the CMM off-line programming software.

ACKNOWLEDGMENTS

Special thanks to Brett A. Hill of Caterpillar Inc. for kindly reviewing the text and the many helpful suggestions. Also, many thanks to Robert Wilhelm, Robert Hocken, and Heinrich Schwenke, who wrote the original paper on which the introduction to this chapter was based (Wilhelm, Hocken, and Schwenke 2001).

---

* The average daily temperature change for mail code (ZIP) 61630 (Central Illinois, USA) according to www.weather.com, for the five warmest months of the year, is about 13°C. A factory in that location, unless it has temperature control, basically follows the external temperature.

15 Application Considerations

Paulo H. Pereira and Dean E. Beutel

CONTENTS

A successful installation of a coordinate measuring machine* (CMM) in a manu-facturing facility is a delicate mix of the proper equipment with the parts it will measure, the environment in which it will function, the people who will operate it, those who will maintain it, and those who will use its results. Careful consideration and planning of the installation, programming, fixturing, data output formatting for the application, and operating personnel will yield higher productivity and quality. Failure to consider one or more of the key elements will lead to difficulty in operat-ing the CMM, as well as ineffective use of resources.

The acquisition of a CMM, or any capital asset for that matter, is a significant investment for any company. The CMM is the most significant development for dimensional quality control since the development of gage blocks. CMMs can inspect nearly every part in the shop, generally faster and more accurately than many other methods. But other measuring devices, such as fixed gages, comparators, functional gages, electronic or air gages, optical projectors, gear checkers, roundness-measuring devices, or even manual devices (micrometers, calipers, and height stands, to name a few), need to be carefully considered for the application during the planning process.

The selection of the optimal measuring equipment includes, at a high level, the consideration of people aspects such as safety, quality aspects as in the tolerances involved, velocity aspects as in needed throughput, and initial and lifetime costs. Other typical aspects to consider are projected uptime and reliability, as well as part size, environmental factors, variation in the parts to be inspected, and required data output (Krejci 1992).

In cases where the measuring task is performed to control manufacturing pro-cesses, the relationship between the inspection equipment and the operation it will support must be considered as carefully as the selection of the production machines. The fit of the measuring equipment to the process is the most important factor to consider. The proper measuring device will provide measurement data necessary and adequate to control the manufacturing process, troubleshoot problems that may arise, maintain production rates, provide first-piece inspections of setups, and adapt to changing production needs (Kunzmann et al. 2005).

---

* CMM is used generically in this chapter to include all measuring machines capable of measuring coordinates such as traditional Cartesian models (see also Chapter 4, Cartesian Coordinate Measuring Machines), as well as other types such as AACMMs and laser trackers (see also Chapter 17, Non-Cartesian Coordinate Measuring Systems).

Users must also consider the timely introduction of new measurement technologies. It is common to find companies that buy and install a CMM and simply assume it can be in use for a very long time, that is, usually more than 10 years. This can be very inefficient as technologies continuously evolve and better solutions for the applications become available. By staying up to date with measurement technologies, users can benefit from accuracy, throughput, and reliability improvements to their operations. One possible target is to maintain the average age of major inspection equipment at or below five years, but this is a decision requiring careful analysis of the specific situation considering technical and business aspects. By replacing 10% of its metrology assets every year (every piece of equipment is replaced at about 10 years of age), a factory can achieve the five-year average.

This chapter covers the primary considerations when selecting a CMM for a particular measuring application. These considerations constitute a starting point for preparing a spreadsheet (Table 15.1) to compare different offerings. Each company will usually have to consider several other unique factors of importance to better evaluate CMMs for the specific application. Each purchaser will find certain features more important than others and assign weighting factors accordingly. Such a spreadsheet needs to include all aspects that the purchaser deems important for a successful installation. This list must be carefully created before contacting possible vendors to avoid useless iterations just to clarify or add specific requirements. These aspects would normally consist of the following:

People aspects
    Safety and ergonomics
    Operator interfaces and programming language
    Training requirements
Quality aspects
    Measurement volume and machine configuration
    Accuracy and repeatability (uncertainty)
    Probing method
    Programming alternatives
    Data output
Velocity aspects
    Speed or throughput (takt time)
    Part handling and fixtures
    Location (layout)
Cost aspects
    Initial costs
    Environmental requirements
    Warranty/maintenance
    Lifetime costs

Each of these aspects is considered individually in this chapter. The evaluation factors recommended are presented in three sections consisting of hardware

TABLE 15.1

Example Portion of an Evaluation Spreadsheet[a]

Requirement	Vendor A	Vendor B	Vendor C
Measuring volume (see Section 15.1.1)			
x-axis 4000 mm	4000	4070	4000
y-axis 2000 mm	2100	2100	2000
z-axis 1500 mm	1750	1600	1500
Weight requirements	comply	comply	comply
Accuracy[b] (see Section 15.1.2)			
ISO 10360-2 (ISO 2009a) $MPE_E = 5 + 5L/1000$ μm	$5 + 5L/1000$	$5 + 5L/1000$	$4 + 5L/1000$
ISO 10360-5 (ISO 2010a) $MPE_P = 4.0$ μm	5.0	4.5	4.0
Repeatability according to ASME B89.4.1-1997 (ASME 1997) = 2.5 μm	2.5	2.0	2.5
Multiple tip test as per ISO 10360-5 (ISO 2010a)			
Short stylus, 50 mm			
$MPE_{AL} = 6.5$ μm	comply	6.0 μm	Exception
$MPE_{AS} = 4.0$ μm	comply	4.0 μm	4.0 μm
$MPE_{AF} = 8.0$ μm	comply	7.0 μm	8.0 μm
Long stylus, 200 mm			
$MPE_{AL} = 8.0$ μm	comply	8.0 μm	7.0 μm
$MPE_{AS} = 5.0$ μm	comply	4.5 μm	5.0 μm
$MPE_{AF} = 9.5$ μm	comply	9.0 μm	9.5 μm
Measurement uncertainty (see Section 15.1.3)			
Target tolerance-to-uncertainty ratio = 10:1[c]			
Provide uncertainty budgets for the following tolerances as per NIST TN1297 (Taylor and Kuyatt 1994) or GUM (JCGM 2008a)			
0.2 mm true position part # 123–4567	7:1	8:1	11:1
0.25 mm flatness part # 234–5678	6:1	8:1	9:1
0.06 mm roundness part # 345–6789	4:1	5:1	7:1

	15'45"	14'18"	29'16"
Speed (see Section 15.1.4)			
Test part cycle time (part to part)			
Probing System (see Section 15.1.5)			
Automatic stylus changer with "n" garages	comply	comply	comply
Software requirements (see Section 15.2)			
Off-line programming with capability to estimate uncertainty values	comply	exception	comply
Ability to evaluate gears as per attached specification	exception	exception	comply
Interface with CAD packages "A," "B," and "C"	comply	comply	comply
98% uptime guarantee—to be monitored monthly. Each noncompliant month to extend warranty by 2 months[d]	comply	comply	comply

[a] Requirements are in the left column and vendor offerings are to the right. Actual spreadsheets have many other items.

[b] Other performance parameters may be needed depending on the system configuration. For example, if a rotary table is used, then performance tests as per ISO 10360-3 (ISO 2000b) are to be conducted. See also Chapter 9, Performance Evaluation.

[c] Desired level based on specific conditions. See also Chapters 9, Performance Evaluation, and 14, Measurement Uncertainty for Coordinate Measuring Systems.

[d] This is just an example as, along with all other details, it needs to be negotiated between buyer and seller.

capability, software capability, and other considerations. The list of requirements is just a starting point. Once negotiations start between buyer and seller, both must be ready to interact in order to find the best solution for their specific needs.

15.1 HARDWARE CAPABILITY

With the broad spectrum of CMM hardware to choose from, the wide range of capabilities, and the diversity of options regarding performance characteristics, there are important fundamentals a purchaser should consider when making a selection.

15.1.1 MEASURING VOLUME, MACHINE CONFIGURATION, AND PART WEIGHT

The first aspect to consider in selecting a CMM is the measuring volume required to measure the variety of parts involved and how the machine is to deal with their overall characteristics. Although this is usually a rather straightforward matter, it is often complicated by the trade-off involving machine configurations that are most adaptable for the workpieces to be measured, the desired speed, the required accuracies, the needed floor space, and the cost of the system.

15.1.1.1 Sizing Recommendations

The primary consideration in determining the physical size of the measuring volume is the range of dimensions of the parts to be measured. In almost all applications, the workpiece should be contained in the measuring volume. Although this sounds obvious, there are instances involving parts with one exceptionally long dimension where the purchaser may consider restaging the part to limit the required machine volume. There are potential safety risks and loss of productivity and accuracy when considering this approach that must carefully examined. When evaluating cost versus measuring volume, it is often enticing to believe that a smaller machine at a slightly lower price may be more cost effective. This frequently proves to be false economy and can lead to higher operation costs if the parts do not quite fit in the volume along with fixtures and other needed accessories. In addition to its useful life, an adequately sized CMM will provide an attractive return on investment.

The user has several recommendations regarding how to size a machine. First, the machine should be able to access the workpiece datum system and all related features fully without repositioning it in its fixture. If a part is repositioned between inspections using different datum systems, throughput and accuracy would be affected.

The probe and stylus configuration also affect machine sizing. Many parts do require long stylus configurations. This is most common when measuring parts with deep features that can be accessed only by reaching through the external faces (Figure 15.1). Extra space is required for probe and stylus articulators to allow unobstructed movement around the part and/or fixture.

The CMM must also be sized to accommodate fixtures that are used to hold the workpiece during inspection. Such fixturing may significantly increase the volume needed for the measurement task as the probe must travel around it, as exemplified in Figure 15.2.

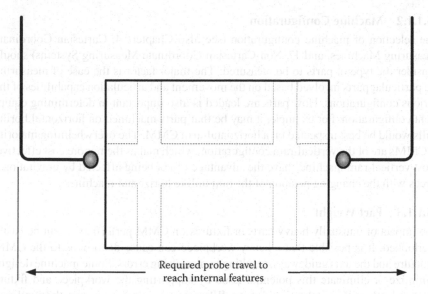

Required probe travel to
reach internal features

FIGURE 15.1 Reaching internal features often requires longer travel and thus greater machine volume.

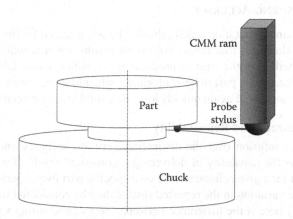

CMM ram

Part

Probe
stylus

Chuck

FIGURE 15.2 An example of the effect of fixturing on the required machine volume. The part is mounted on a chuck for measurement, but the ram needs to clear it for taking points on the bottom face.

CMMs with probe changers may need to allow additional volume to accommodate them. Most models on the market require a probe changer to be inside the machine working volume to allow proper access to it. Some models offer retractable tool changers that move out of the way during part measurement, which could save space.

15.1.1.2 Machine Configuration

The selection of machine configuration (see also Chapters 4, Cartesian Coordinate Measuring Machines, and 17, Non-Cartesian Coordinate Measuring Systems) should consider the type of parts to be measured. The major factor is the ease of measuring the particular parts involved based on the movement and articulation capabilities of the various configurations. How parts are loaded is also important in determining equipment configuration. For example, it may be that parts machined on horizontal boring mills would be best inspected on a horizontal-arm CMM. The overwhelming majority of CMMs are of the vertical-ram configuration, which makes them more cost effective. Most vertical-ram machines have the advantage of not being affected by gravitational forces with the changing position of the ram unlike horizontal machines.

15.1.1.3 Part Weight

The impact of unusually heavy parts or fixtures on CMM performance can be easily overlooked. It is possible that a heavy workpiece can cause distortions to the CMM structure and the axis guideways, resulting in measuring errors. Some machine designs minimize or eliminate this potential issue by supporting the workpiece and fixture independently of the machine guideways. When part weight is a concern, the purchaser should include specifications and tests to assure that it does not adversely affect CMM measuring accuracy (ASME 1997; see also Chapter 9, Performance Evaluations).

15.1.2 MEASURING ACCURACY

CMMs soon gain a reputation among those who are affected by the measurements they provide. Machines that report consistent results when repeatedly inspecting the same part will earn the trust of production personnel. Those CMMs with poor repeatability that show a part to be conforming when first measured and then out of tolerance when measured again quickly lose the confidence of production personnel.

15.1.2.1 Accuracy and Repeatability

There is often confusion over the distinction between accuracy and repeatability. Repeatability is the capability of delivering a consistent result. If a CMM reports the same result for a given characteristic on a specific part over a period of time, it is repeatable. The variation in the reported results may be considered the repeatability error. Accuracy here is the difference between the mean resulting value given by a CMM and the calibrated value as determined by some better measurement method.

A valid comparison of accuracies between different CMMs should be based on a common standard,* for example, B89 (ASME 1997, 2004, 2006a, 2008) or International Organization for Standardization (ISO) 10360 series (ISO 2000a, 2000b, 2000c, 2001b, 2009a, 2010a). Even then, users should be cautious when very high accuracy or unusual applications are involved because standardized tests may not uncover all errors of a machine (Pereira and Beutel 2003). The performance tests themselves have uncertainties that are complex to evaluate (Salsbury 2003a; ISO 2006) and could sometimes be as large as the specification values themselves.

---

* See also Chapter 9, Performance Evaluation.

Another option is a study of actual workpieces with the CMMs being compared. Such a workpiece study is a gage repeatability and reproducibility (Gage R&R, or GR&R) procedure (AIAG 2002). Conducting comparisons in this manner used to be a common practice when standardized procedures were not as accepted or developed. However, asking suppliers to measure multiple parts is an expensive and time-consuming proposition, particularly when they are large, complex and/or expensive. Moreover, several steps can be taken to artificially improve GR&R results of a CMM (slower speeds, point averaging, etc.), which would invalidate the results.

15.1.2.2 Uncertainty Budget

Ultimately, users should require CMM suppliers to provide an uncertainty budget (Taylor and Kuyatt 1994; ANSI/NCSL 1997; JCGM 2008a) for measuring their most demanding parts with their proposed machines under certain conditions. However, such budgets are complex for CMM measurements and could lead to more confusion in picking a suitable machine if not properly done (refer also to Section 15.1.3 and Chapter 14, Measurement Uncertainty for Coordinate Measuring Systems).

Besides helping in choosing a CMM, a correct uncertainty budget can identify other areas of the inspection process that require attention. For example, part temperature could be the largest contributor to measurement uncertainty. In this case, instead of investing more in a higher-accuracy machine, or its room, to reduce uncertainty, it would be much more efficient to control the temperature of the parts to guarantee that they are measured in a more thermally stable condition.

It has been known for a long time that measuring equipment is to be more accurate than the manufacturing process requirements. The classic gagemakers' rule states that the gage tolerance should be one-tenth of the product tolerance (Qberg and Jones 1920; Radford 1922; Rolt 1929; ASME 2001b). For example, to measure a tolerance of 0.010 mm, a gage with an uncertainty of 0.001 mm is required. In some cases, this ratio cannot be achieved. Depending on the application, a 4:1 tolerance-to-uncertainty ratio can be acceptable, but it depends on business and technical reasons (ASME 2001b).

During the evaluation of production processes, there is variation of the gage, which is inherently part of the variation observed in the process. The variation of the process combined with that of the gage is observed. It is possible to presume a manufacturing process to have large variation while it is actually caused by the measuring equipment (Kunzmann et al. 2005).

The impact that measuring variation has on the ability to control a manufacturing process must be considered. The measured results are an observed process variation that consists of the actual process variation augmented by that from the measurement. Assuming normal statistical distributions, the arithmetic average of the means of the distributions is the mean of the total process. The variances can be added by the square root of the sum of the squares to obtain a variation for the total process. The standard deviation is the square root of the sum of the individual variances (AIAG 2002).

$$\sigma_t = \sqrt{\sigma_p^2 + \sigma_m^2}$$

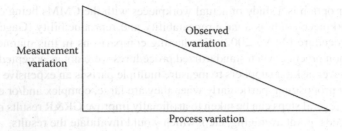

FIGURE 15.3   The observed variation may be viewed as the hypotenuse of a right triangle with the two sides representing process variation and measurement variation.

This relationship is represented as a right triangle with the process variation as one leg, the measurement variation as the other, and the hypotenuse being the "observed variation." This representation facilitates understanding of the relationship between measuring accuracy and its effect on observed process variation. Figure 15.3 shows a representation of this relationship (not to scale). It must be emphasized that both the equation above and its representation are valid only when the process and gage variations follow normal (Gaussian) distributions. Measurement variation directly affects process capability metrics such as C_p and C_{pk} (Kunzmann et al. 2005; see also Chapter 14, Measurement Uncertainty for Coordinate Measuring Systems).

If a CMM is used to support a process that has a variation approaching or exceeding the engineering functional tolerances, it becomes necessary to measure every part (sorting). The CMM in this situation must be very accurate and repeatable as each inspection yields a pass/fail decision. If the feature is critical, the prudent decision may be to set the control limits tighter than the engineering requirements to prevent the acceptance of nonconforming parts (stringent acceptance; ASME 2001b). This approach minimizes the risk of accepting nonconforming parts at the expense of increased probability of rejecting conforming ones. The tightening of the control limits can be reduced through the use of more accurate and repeatable CMMs. Uncertainty estimates and decision rules must be applied to decide when to accept or reject a part (ISO 1998, 1999b; ASME 2001b; Bachmann et al. 2004; see also Section 15.1.3).

For example, a measurement length can be reported as 100 mm but, without knowing how the measurement was taken, it is impossible to know how close that is to the true value.* If the reading was taken using a measuring tape, one could estimate an uncertainty of about 2 mm. With a CMM under proper conditions, the result could still be 100 mm but with an uncertainty of 0.01 mm or better. So, for a tolerance of ±0.020 mm, a workpiece with a length of 99 mm would be conforming in the first case and nonconforming in the second. The application, and thus the tolerance, dictates the measurement method to be used, and uncertainty is an indication of the measurement quality.

---

* By definition (JCGM 2008b), the true value of a measurand is unknown.

15.1.3 TRACEABILITY AND UNCERTAINTY FOR PART MEASUREMENTS*

The Joint Committee for Guides in Metrology–International Vocabulary of Metrology (JCGM 2008c) and ISO 10012 (ISO 2003a) imply that all part measurements should have an uncertainty estimate per the *Guide to the Expression of Uncertainty in Measurement* (GUM; ANSI/NCSL 1997; JCGM 2008a) to have traceability to the unit of length. However, both are ambiguous enough that there is much discussion about the topic (Doiron 2004), and even an American Society of Mechanical Engineers (ASME) technical report about it has been published (ASME 2006b). Industry, in general, cannot simply create uncertainty statements for the millions of measurements performed each day. There is still debate with one line of thought insisting that, to claim traceability, one must have more than the paper trail of calibrations of the measuring instruments required by surpassed military standards (Department of Defense 1988†). It is necessary to have uncertainty statements for every single measurement as per this line of thought. On the other hand, many claim traceability by certifying their gages only having uncertainty estimates while keeping an adequate tolerance-to-accuracy ratio for the actual part measurements the gages perform.

15.1.4 SPEED OR THROUGHPUT

Throughput is a difficult characteristic of a CMM to quantify. The ASME B89.4.7 standard (ASME 1997) states that each user should develop a unique test for throughput. The CMM throughput should be compared to the frequency of inspections required to control the process or, in general, to meet the inspection demands of the application, that is, correct takt (cycle) time while meeting the needed measurement uncertainty. There is a wide variety of measurement tasks with their unique requirements such as high and low volume combined with high and low mix of components that need to be considered to properly configure the system.

CMM manufacturers use different parameters for evaluating throughput. The parameters are typically speed, acceleration/deceleration, and touches per minute. If speed by itself is high, it does not necessarily mean higher throughput depending on the application (for example for small parts).

There are a few helpful guidelines for the purchaser to evaluate throughput effectively. First, examine the time to measure a single point from the probe-approach distance. CMMs spend most of their time by making touches that are close together. Second, if very large components are to be measured, the traverse rates (speed and acceleration values) of CMMs need to be examined. Third, select a workpiece most representative of those that will be inspected on the machine and have the prospective CMM suppliers measure it while recording the time required to do so. This needs to consider the correct program, with all the necessary points and moves included.

---

* For a complete discussion, refer to Chapter 14, Measurement Uncertainty for Coordinate Measuring Systems.
† Officially cancelled in 1995 in favor of civilian documents ISO 10012 (ISO 2003a) and ANSI/NCSL Z540-1 (ANSI/NCSL 1994).

Another option is that certain CMM programming languages have the ability to properly simulate inspection routines considering cycle time. A common measurement strategy (carefully mimicking normal use) should be used to assure that different measuring times are the result of CMM performance only. The results of this test will provide the best indication of inspection time throughput. Finally, do not ignore accuracy and repeatability. Speeding up a CMM to achieve a higher throughput often degrades measuring accuracy. A correct uncertainty budget must also account for speed of measurement.

15.1.5 PROBE TYPES*

The selection of the probe for a particular measuring task is often as important as the selection of the CMM itself (Bobo 1999), since they have a large influence on accuracy and, for smaller systems, on cost. Many of the key challenges in effective coordinate measuring are found in the workpiece-sensing area. Understanding how to apply probes and other workpiece-sensing devices is critical to the success of CMM implementation.

15.1.5.1 Contact Probes

The effectiveness of CMMs increased greatly when hard probes were replaced with electronic touch-trigger probes. Touch-trigger probes have been well proven for measuring production parts, but they lose some effectiveness in tight quarters as they depend on the machine reaching a constant velocity at the point of contact, which is more difficult when the distances are limited.

Users of some touch-trigger probe models must learn how to adjust their contact force as per the manufacturer's recommendations. Improper adjustments can lead to either measuring variations or excessive false triggers. Touch-trigger probes using strain gage or similar technologies eliminate the need for adjustment.

An alternative to the touch-trigger probe is the scanning (analog) probe. Such probes allow points to be taken while the stylus remains in contact with the workpiece. These probes are useful when measuring a large number of points in a limited area or when workpiece configuration limits the approach distance. Often, scanning probes can hold styli much heavier than their touch-trigger counterparts but are slower for single-point probing.

15.1.5.2 Non-Contact Sensors

Contacting probes are not the solution to all CMM measuring tasks. For example, locating the edges of extremely thin sheet metal components is difficult with a spherical probe tip. Artwork on circuit boards is also difficult because mechanical probes cannot contact the edges of the conductive tracks. Non-contact probes are best for these tasks[†] (Jalkio 1999).

Vision probes are the most common non-contact probes. A vision probe is simply a precision camera that is mounted to the CMM ram. The software analyzes the

---

* See also Chapter 6, Probing Systems for Coordinate Measuring Machines.
† See also Chapter 7, Multisensor Coordinate Metrology.

image on the screen, dividing it into picture elements (pixels), and then determines the offset of features in the plane of the camera from a precalibrated probe center-line. The position of features along the camera axis is determined by focus, providing the capability to measure the position of an object in all three coordinates.

Laser-triangulation probes are useful when measuring large numbers of points along relatively gentle contours. These probes work by reflecting a light beam, usually a laser, on the workpiece and focusing the diffused scatter of light onto a photodetector. Parts such as aircraft wing panels are typical applications for laser probes. The contours of parts such as rubber seals and bellows are also effectively measured with laser probes since the problem of distortion under the pressure of a contact probe is eliminated. Still, the operator must be aware that these probes will not perform well under certain conditions. If the surface is polished, there will be insufficient diffused light returning to the photodetectors, and hence, it is difficult to take a reading by the system. The system is limited by the spot size of a laser (approximately 0.2 mm diameter).

15.2 SOFTWARE CAPABILITY*

The capability of the software is often as important as, and sometimes more important than the capability of the hardware for a successful CMM application.

15.2.1 PROGRAMMING ROUTINES

One of the most significant advantages of some CMM models is their capability to perform fully-automatic measurements. Automatic inspection has several advantages over manual inspection such as speed, better repeatability, and being less prone to operator errors. Some different programming approaches are available:

- Manual versus direct computer-controlled (DCC)
- Self-teach programming
- Programming by part family (parametric programming)
- Computer-aided design (CAD)-based programming (off-line programming)

15.2.1.1 Advantages of Programmable CMMs

Automatic inspection is faster and more accurate at contacting the probing target than a manually-operated CMM. Cost of measuring parts with a fully automatic CMM is usually less than half that of a manual one (see Chapter 19, Financial Evaluations).

Programmed or automated inspections are more consistent than those conducted manually. This is partly due to the effects of local part variation (flatness and waviness) on the inspection. Programmed runs will contact the same target on the part usually within 0.005–0.05 mm, minimizing the effects of local variation on alignment and part measurement. With manual operation, the target points are generally contacted within 5 mm, possibly changing the readings. Manual operation

---

* See also Chapter 8, Coordinate Measuring System Algorithms and Filters.

also results in varying contact speeds and forces, which can also affect measuring accuracy (see Chapter 9, Performance Evaluation).

Manual CMM inspection is a tedious job requiring continuous concentration and focus. The nature of the task requires that the operator perform each step in sequence, and several steps must be performed before a result is generated. Manual CMM inspection can be physically demanding. Some parts require that the operator assume difficult postures to see the probe taking the measurement on the workpiece without disturbing the setup. Often, operating a manual CMM requires a higher skill than operating an automatic (DCC) model. Repeatability and throughput suffer as well. On the other hand, there is a higher initial cost in achieving automatic CMM operation.

An articulated arm CMM (AACMM) is an example of a CMM that depends on the operator to move it since there are no motors (Figure 15.4). Its software can perform data analysis but cannot control its movements (see also Chapter 17, Non-Cartesian Coordinate Measuring Systems).

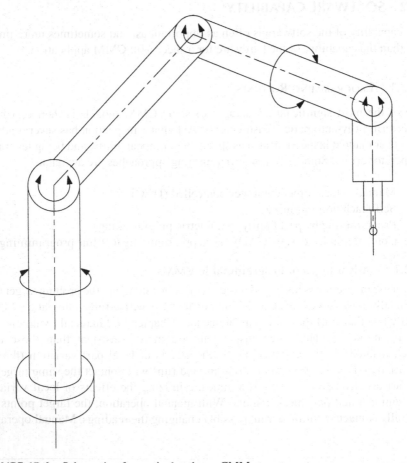

FIGURE 15.4 Schematic of an articulated arm CMM.

Cartesian CMMs are also offered in manual mode. The axes are constrained but have no means of moving by themselves. These machines have the advantage of not requiring complex programs for the movements but need a more skilled operator to move them around and collect data.

15.2.1.2   Teach Programming

Under self-teach, the system records each step the operator makes for an inspection task. The recording of these steps becomes an automatic program that can be used to inspect future parts. However, self-teaching a program is not as simple as inspecting. The operator must record nonprobing moves in the self-teach phase to permit the machine to move from feature to feature without collisions during the automatic cycle. Furthermore, the inspector doing the programming must document the probes and setup used in sufficient detail to enable another inspector to repeat the inspection.

15.2.1.3   Parametric Programming

An astute inspector/programmer can reduce the programming effort considerably by developing programs around similar (or a family of) parts. In some cases, the family may be differentiated by the inclusion or omission of a few features. Such a family may consist of a group of pulleys, gears or shafts, which have basic similarity other than sizes and a few features that are different from one specific part in the family to another. Even in those cases, the core program is used to inspect the common features while individual branches take care of unique ones.

It is not uncommon to have a family of parts that is identical in configuration but varied in size. The only difference on the drawing is often the nominal values of the features. For this case, only one program can serve multiple parts. The operator only chooses the correct part number and the program retrieves the proper nominal dimensions that are fed into the program variables.

It takes more time initially to develop a parametric program, but that investment is repaid many times over during its usage. Using parametric programs can reduce the total number of programs by more than 90% while the number of lines of code can be reduced by more than 50% in typical cases when compared with individual programs. This makes a parametric program much easier to maintain than the tens, or even hundreds, of individual programs it replaces.

15.2.1.4   Off-Line Programming

CAD-based programming provides several analysis and productivity advantages for CMM programming. The use of CAD systems offers an opportunity for inspection of more complex features and potentially economic operation of a DCC CMM for a lot size of one. By interfacing a CAD system with a CMM, it is possible to inspect complex contours that are normally inspected on an optical comparator or with contour templates.

One of the greatest advantages of CAD-based programming is pictorial reporting. Pictorial reporting can save time in decision making because an operator can look at the CAD rendering of a part with the measured deviations included. Pictorial rendering relays part deviations without having to use numerical results so that people from different disciplines can more readily understand them (see also Section 15.2.4).

15.2.2 OPERATOR INTERFACES

Due to the variation in machine applications and operation modes, CMM manufacturers have developed a wide range of operator interfaces. Some are elegantly simple to understand and use. Pictures are used to represent the various functions required, and software checks are performed to assure that the inspector stays on the proper inspection sequence. Other interfaces are as complex as CAD systems and may bewilder all but the most proficient operators. There are interfaces that are custom designed for special applications. Many users choose to develop their own interface functions. An effective operator interface improves the capability of the CMM.

Many CMMs use semiautomatic cycles during the inspection. These cycles provide for automatic inspection of a single feature or a number of features with minimum input. These routines are very successful in the development of a program to inspect bolt patterns, for example. CMM programming languages offer standard routines for measuring bolt patterns, roundness of cylinders, and many other common measuring needs.

15.2.3 DATA EVALUATION SYSTEMS

To use CMMs effectively, it is important to understand that all they measure are points in space as defined by coordinate axes. All that comes from the machine is a series of X, Y, Z data relative to a coordinate system, which could be absolute or relative. Coordinates are of little use by themselves. Manufacturers need to know the size of bores, their relative orientation to the datums, how they align to each other, where they are relative to the mounting surfaces, and other relevant details. The coordinate data must be converted into the geometric definition of the workpiece.

Nearly every CMM manufacturer provides the basic algorithms for fitting circles, planes, cones, cylinders, and ellipses to the acquired coordinate points. High-level systems also offer the user the choice of fitting technique to be used on features, that is, the traditional least squares, maximum inscribed feature, minimum circumscribed feature, or others. CMM manufacturers also offer data analysis systems for complex contours. All these features can add considerable capability to the measuring system but require more highly skilled programming personnel.

Algorithms can misrepresent the results (Moroni, Polini, and Rasella 2003), especially for more complex geometries or extreme cases. There are techniques (Shakarji 2002) and even standards (ASME 2000; ISO 2001b) that can be applied for software testing. In general, mature CMM languages have stable algorithms behind them, often certified by a national measurement institute. Nonetheless, users should still be aware of possible software flaws and require the vendor to prove adequacy if there is a reason to do so (see also Chapter 8, Coordinate Measuring System Algorithms and Filters).

15.2.4 DATA OUTPUT FORMATS

The most sophisticated CMM is useless unless the people who need the data understand the results. CMMs can generate reams of extremely detailed information on any workpiece inspected. The reaction to many reports from a CMMs is "How can one understand all those numbers?"

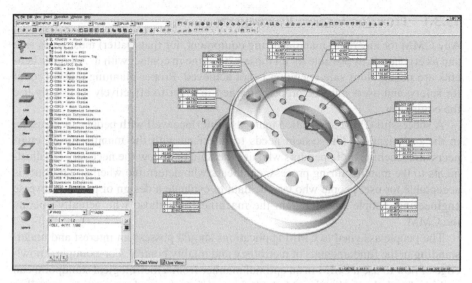

FIGURE 15.5   Pictorial report of measurements on a part. (Courtesy of Hexagon Metrology, Inc.)

Typically, reports may be created with two main objectives: (1) to demonstrate dimensions that are in conformance to design requirements and (2) to provide relevant feedback for manufacturing process adjustments. Depending on the usage of the information, the programmer must adapt the reporting in order for the user to be more efficient when interpreting the results. The specific output format depends on the intended use and audience for the reports. The output of a CMM is information, and information is beneficial only when it is transmitted to and understood by a decision maker.

A common printed output lists a short description of each dimension on a line followed by the measured dimension, the tolerance, the deviation, and the extent of any out-of-tolerance condition. It is often useful to flag conditions that are of interest (typically those that indicate a process is out of control). Pictorial reports are available for understanding measuring results, as exemplified by Figure 15.5. A more complete pictorial may also include depictions of actual deviations as compared to nominal surfaces.

15.2.5   DATA INTERFACES

It is likely a printed page may not be a fully-acceptable output medium for all applications. Many companies use central databases to store inspection data for statistical process control purposes. Reverse engineering and product analysis often require that measuring results or data points be transmitted to a CAD workstation. Automatic transmission of data makes the ability of given software to seamlessly communicate with others through several formats very desirable.

15.3   OTHER CONSIDERATIONS

Beyond the hardware and software considerations, the purchaser of a CMM needs to plan for other aspects of the project, like part handling, probing, fixturing, environmental conditions, layout, sampling strategy, and the "people" aspects of the project.

15.3.1 People or Training Requirements

Any CMM (or any other manufacturing equipment, for that matter) must be as safe and as ergonomic as possible for all those who come in contact with it. All other benefits are negligible if safe operation is not achieved. Proper planning must account for safety, and users must be properly trained in order to effectively and safely use the equipment.

A successful CMM application must also interface well with people, or its users, in the organization. The system needs to interface with the manufacturing engineers and machine tool operators who use the data to review the necessity of adjustments to the manufacturing process, engineers who determine whether a product is acceptable for use, people who use the data to evaluate a design or perform reverse engineering, people who program the machines, and those who actually operate the CMM.

The people assigned to CMM applications should possess an interest and understanding in the fundamentals of metrology and quality control. For example, knowledge of how to use a CMM in an optimal manner from a metrological standpoint can make or break the application. Methods for part fixturing and preparation, as well as temperature considerations, are critical. Also, understanding the specific inspection requirements is paramount. Users need to know how to read part drawings and to visualize three-dimensional features of parts.

A training plan must be in place in case people need to be trained. Proper training must be completed by the time a new system is installed. Usually, equipment vendors offer a variety of training classes, but there are many other sources ranging from community colleges to independent consultants.

15.3.2 Part Handling, Probing, and Fixturing

Fixture design and probe selection can affect data reliability and throughput of the CMM in important ways. Poorly-engineered methods for delivering parts to the CMM and for placing them in the measuring volume can debilitate an otherwise effective CMM application. This problem can often be treated quite adequately by the selection and position of a proper crane, for example. Portable CMMs (e.g., laser trackers and AACMMs) usually do not have a fixed location, making it more critical that all manufacturer's recommendations be followed to obtain proper readings. Safety and ergonomic aspects of part handling must be properly addressed as well.

Probe management involves selection of the probe types (see Section 15.1.5), stylus sizes, probe qualification, extensions and approach angles to obtain the most accurate and repeatable measurements. When extensions (or long styli) are required, they must be rigid yet as light as possible to allow the least impact to uncertainty. An important point is that the number of connections must be minimized when devising a certain stylus configuration to minimize deflections.

Probe changers allow automated, unattended routines. Effective probe articulation and probe-changing systems facilitate access to difficult inspection locations and avoid the need for part setup changes that can lengthen inspection time.

Similarly, proper selection of fixtures can improve inspection access to the part features. Besides optimizing access, orienting and fixturing the part can reduce point-to-point movement distances and reduce or eliminate probe changes to shorten inspection time. Improper fixturing not only complicates inspection routines but it leads to accuracy and repeatability problems by allowing slight shifting or deformation of the part. This is especially true of parts with thin-walled features. These fragile parts must be properly fixtured so that probing forces, especially with scanning probe heads, do not deform them, causing improper measurements or even damage to the surfaces.

Certain parts may also change when released from their manufacturing fixtures that normally use higher clamping forces. The manufacturing processes (material removal, welding, heat treatment, among others) also induce or release internal stresses of the material, which may manifest themselves when in a free state or if improperly held during inspection. Such factors must be considered during the design phase of the measurement process.

15.3.3 Environmental Conditions

There are numerous important issues that need to be addressed to prevent environmental factors from degrading CMM measuring accuracy.* It makes no sense to invest in a CMM with a low measuring uncertainty, only to place it in an incorrect environment, preventing it from delivering its best results.

15.3.3.1 Part Cleaning

Physical factors related to the part can affect the quality of data obtained from the CMM. The most prevalent of these are burrs, chips, and oil films. If parts are not carefully cleaned, dust and dirt can accumulate on the stylus tip, compromising the performance and reliability of the CMM readings. Occasions when variations in measured results are caused by foreign particle contamination are numerous, but most can easily be avoided. Usually, issues due to part cleanliness are not repeatable. It is recommended that a proper wash and blow-off cycle be provided after the machining operation and before the measuring process. An alternative is to manually clean the parts.

15.3.3.2 Temperature

The major environmental factor affecting measurements is temperature. Chapter 10, Temperature Fundamentals is dedicated exclusively to the fundamentals of temperature. In addition, Chapter 11, Environmental Control covers temperature-controlled rooms for CMM installations.

Since all dimensions and tolerances are specified at 20°C (ISO 2002a), the dimensional measurements must be made at this temperature or properly compensated to it. Most CMM users do not pay enough attention to temperature control unless problems with measuring results are observed.

---

* See also Chapter 11, Environmental Control.

The calculations for determining the effects of temperature are rather straightforward. The parameters to be considered are (1) the size of the features to be measured, (2) the coefficient of temperature expansion of the material used to make the parts, (3) the maximum difference in temperature between 20°C and that of the part or measuring environment, and (4) the level of uncertainty required of the measured results.

If the part and the CMM are at the same temperature (other than 20°C), the measuring error will be due to the difference between the machine temperature coefficient and that of the part. When measuring a 500-mm aluminum part on a CMM with steel scales in a temperature environment just 1°C from standard reference temperature of 20°C, there will be a measuring error of approximately 0.006 mm. Such an analysis is based on assuming stable temperature conditions. If the temperature is changing, which is normally the case, the measuring results will have larger errors because of the CMM and part distortions that occur when temperature gradients exist, as well as differences between the measuring scales and workpiece temperatures.

When considering temperature, the problems that can be caused by heat flow from radiant energy should also be considered. The most obvious problems take place when sunlight is shining through a window and striking the CMM, the part to be measured, or simply allowing sunlight into the room. Another equally-obvious condition that must be avoided is having air blowing directly onto the CMM. This happens often and can cause significant data inconsistency, which is difficult to pinpoint.

Automatic temperature compensation is available on most CMMs, particularly for the ones intended for shop floor applications. A CMM with this capability provides measurement data that are converted to 20°C standard reference temperature. Sensors monitor the temperature of the CMM scales and of the part to be measured, and software automatically calculates the dimensional measurement at the standard temperature, factoring in coefficients of expansion. This procedure does have influence on the overall measurement uncertainty (see Chapter 14, Measurement Uncertainty for Coordinate Measuring Systems; ISO 2003d).

The temperature compensation program includes a table for selecting coefficients of temperature expansion for most standard materials. The operator can enter others for particular materials, as needed. Temperature compensation software, though, cannot adjust for rapid temperature changes or gradients during the measurement cycle. When the temperature does change rapidly, the data are usually flagged, so the operator knows the readings were taken under conditions that do not yield maximum accuracy.

Specific tests are outlined in the B89 standards (ASME 1973, 1997) that users can employ to evaluate the thermal errors of a given machine (also refer to Chapter 10, Temperature Fundamentals). Temperature compensation is usually ineffective for intricate geometry parts as the mathematical models are not detailed enough to account for the complex thermal gradients and distortions. In those cases, the best approach is to ensure the parts are properly acclimatized prior to their measurement.

CMMs constructed of the same material have more predictable thermal distortions and therefore are somewhat simpler to compensate. Enclosures can buffer the CMM against rapid temperature changes, provide for air circulation to protect it from spatial temperature gradients, and filter out dirt and oil mist, which can contaminate electronics and bearing way surfaces.

15.3.3.3 Vibration

Usually, to some degree, vibrations negatively impact all CMMs. Vibration has become a more frequent problem with more CMMs being moved to the shop environment. Chapter 11, Environmental Control includes a more in-depth discussion on isolation equipment and theory behind how it protects the measuring equipment from the effects of externally-induced vibrations. Obviously, a punch press, compressor, fork truck traffic, or other heavy machinery near a CMM usually cause problems. Less obvious to detect is low-amplitude vibration that may fall within a CMM resonant frequency range, affecting its operation. CMM manufacturers specify amplitude and frequency ranges that are permissible for their machines. A typical example would be a maximum of one micrometer peak-to-peak amplitude at 5–50 Hz.

Automatic part loaders that are part of a CMM system installation can be a source of vibration. To avoid this, the CMM needs to be isolated from the part-handling system. The CMM manufacturer involved in the project can offer support to avoid costly mistakes and redesigns.

15.3.3.4 Humidity

Compared with other environmental considerations, humidity is a relatively minor problem. To prevent oxidation or rust on gage blocks and other metrological devices, the accepted practice is to maintain the relative humidity in the inspection environment at about 40%–45%. Humidity does change the index of refraction of air, which affects laser interferometers (used for error mapping the machines and as scales for certain models), but compensation techniques are readily known and available.

15.3.3.5 Electrical Power

Problems with the quality of an electrical power supply for a CMM tend to be easily addressed with an isolation transformer or shielding and suppression devices. Such problems usually surface quickly and the remedies are inexpensive. The user should have some knowledge about the quality of the available power supply and avoid using lines that are also used for welding or similar equipment with large capacitors in their circuits. The typical allowable variation is ±10% of the specified voltage, but users need to heed the manufacturer's specific recommendations. Computer equipment supporting the CMM must also be adequately protected from power surges.

15.3.3.6 Compressed Air

Many CMMs use precision air bearings that have specific compressed air requirements. Therefore, the purchaser of a CMM must ensure that the compressed air available at the installation meets, or can be made to meet, these requirements. Problems with water and oil in the compressed air are the most common but easily avoidable. In addition, attention is required if the available air pressure is low or fluctuates too much. Also, provisions must be made to avoid sudden loss of air pressure that can damage the bearings or guideways in the CMM. Most CMMs in the market automatically stop operations if air pressure falls below a threshold to minimize damage.

15.3.4 Sampling Strategy

When planning the metrology process, one must think of how many parts there will be and how they will be checked. The first will determine the resources needed and the second will establish how long the measurements will take.

15.3.4.1 Part Sampling (Inspection Frequency)

The number of parts to be checked is a function of how critical they are to the end product, how stable the manufacturing process is, and how the end customers would deal with nonconforming items. In general, high cost or high-risk components (luxury items, airplanes, medical devices) have small margins for errors. In those cases, the producer must make sure defective items are not shipped. The producer's cost is also a factor, as zero defects can be costly if manufacturing processes are not properly designed and maintained.

When devising the part-sampling strategy, quality planning personnel must consider all these factors to make sure they achieve the right balance between costs and outgoing quality (ASME 1999; Juran and Godfrey 1999; ANSI/ASQ 2003a, 2003b). Once quality plans are defined, one can determine the expected CMM burden based on production levels. Contingencies for unusual events, like machine downtime or a need for checking more parts than planned, have to be accounted for when estimating the burden. The total burden per plan needs to be such that it balances the risks of having the system overburdened against the costs of overcapacity. Usually, it is cheaper to have some extra measuring capacity than stopping several machine tools or incurring the risk of poor quality otherwise. A reasonable burden level during planning stages for a CMM is about 60%, which is a reasonable compromise between costs for the system and the benefits it brings to the operation. However, that is a decision specific to each situation considering the costs of not checking parts versus those of having an asset with less than full utilization.

15.3.4.2 Point Sampling*

Measurement results mean very little if they are incorrect. A simple example is the inspection of a large diameter with a roundness callout on a CMM when few points are collected over a small area. This generates large errors in the readings (Hocken, Raja, and Babu 1993; Phillips et al. 1998). Most CMMs still rely entirely on the operator and/or programmer to decide how many points to take, where to take them, and how to collect and analyze the data. Normally, without uncertainty analysis (see also Section 15.1.3), extensive testing or prior information, or simulation, it is difficult to have a clear idea of how the measurement strategy really affects the results (Krejci 1991; Weckenmann, Knauer, and Kunzmann 1998; Phillips et al. 2003).

Hard gages and most manual gages usually have the advantage of mimicking the mating part when used to measure a feature. CMMs have to mathematically simulate that through analysis of properly collected coordinates (see also Section 15.2.3).

A well-trained and experienced metrologist has the working knowledge to make the proper decisions and, if not, tests can be conducted to determine an

---

* See also Chapter 5, Operating a Coordinate Measuring Machine.

acceptable way to perform the measurement. However, experienced metrologists are difficult to find and the experimental process takes time. In the future, CMMs will be intelligent enough to make those decisions based on uncertainty estimates and to plan the moves around the part (Spitz 1999; Zhang et al. 2002; Chen, Wang, and Yang 2004; Srinivasan, Acharya, and Anand 2004; Wu, Liu, and Zhang 2004), and even to adapt to changing process conditions (Edgeworth and Wilhelm 1999a).

15.3.5 LAYOUT OR WORKFLOW

When planning where to situate a CMM within a manufacturing facility, it is very important to consider the workflow of parts going through it. It is critical to have the CMM in a location where the workflow does not need to be disrupted to route parts through it. One must consider the inspection plan (see Section 15.3.4) to calculate how many parts will be measured. Also, possible future changes to the manufacturing process have to be considered.

On the other hand, with the normal constant change in products and processes, it may prove difficult to have the machine always at an optimal location. There are costs involved in moving a machine and potential detriment to equipment life and accuracy. Planners have to consider inspection as an integral part of their manufacturing process when designing (or revising) their layout and factory flow.

15.3.6 WARRANTY AND MAINTENANCE

The more expensive a CMM is, the more critical it is to one's operations. Thus, the more important it becomes to have well-defined warranty clauses. It is paramount to negotiate the warranty details—and all others—before signing a purchasing agreement. Vendors may push a buyer to sign a contract before everything is ironed out, with the argument of meeting a certain delivery deadline. Users must resist it, as proper planning and a clear contract will save time, money, and grief during the whole project and subsequent machine use.*

A user can require a minimum uptime for the duration of the warranty period (usually 95% or higher) with a financial penalty to the supplier in case that is not met. The user then needs to diligently record the uptime. One possible arrangement would be to hold a portion of the payment until the end of the period, which may increase initial costs sensibly, but it guarantees the supplier will have a clear stake in maintaining the agreed-upon uptime. Another possibility is to extend the warranty by a negotiated time if the minimum uptime is not achieved on a monthly basis.

Maintenance of the CMM after installation is also a key aspect. No one can afford a system that is not reliable because of the consequence to production schedules and quality levels. During the initial period, usually the supplier takes care of the machine as part of the warranty. However, the user is the one who needs to perform

---

* "A drop of sweat spent in a drill is a drop of blood saved in a battle."—Chinese proverb (Yuan 2011)

the periodic tasks for properly maintaining the machine.* A well-kept CMM, very likely, will be more reliable and better maintain its accuracies for a longer time.

As part of maintenance, the user must establish and follow a certification schedule in order to guarantee proper traceability (see Section 15.1.3 and Chapters 2, The International Standard of Length, and 9, Performance Evaluation). The supplier should provide annual certifications during the warranty period and would be able to do so after that for a cost. It is also highly recommended for users to perform interim (simpler) checks on a regular basis to identify possible problems between certifications.

Availability of replacement parts is particularly challenging when the supplier is overseas and does not have a strong local presence.† In that case, the user needs to consider, for example, making a condition of purchase that the vendor have certain spare parts readily available.

15.3.7 INFORMATION SOURCES

In selecting a CMM, there are many factors that go beyond the performance and capability considerations covered in this chapter. Such factors as reliability, uptime, cost to repair, calibration frequency and cost, as well as the viability of the manufacturer to continually provide prompt service, technology upgrades and follow-up training, are all important. To evaluate these less-tangible factors, along with all others, it is necessary to gather inputs from a broad range of users. This fact-finding effort is time consuming but worthwhile.

Within some corporations, there is a CMM coordination group that has representatives from several divisions meet to compare experiences, as well as to learn about improved techniques and new products. Individual professionals can also join external users' groups. Most of the larger CMM manufacturers sponsor users' forums with events throughout their market areas. There are also online forums usually with open access to all who are interested.‡ These can be very beneficial for exchanging information. It is important, though, to pursue multiple sources of information. Other options include attending trade shows and seminars and simply contacting colleagues, other users, and manufacturers asking for information.

15.3.8 METROLOGY SERVICES OUTSOURCING

When considering outsourcing metrology services, the total costs and risks must be evaluated and not only the individual inspection cost per unit (Juran and Godfrey 1999; Smith 2002). For some situations, it can be the right solution, but there are many possible traps. Normally, managers only see the apparent cost reduction, ignoring the side costs mainly by giving up control of the measurement process. Also, most of the

---

* Like checking and replacing air filters, maintaining temperature control, cleanliness, and lubrication, as applicable.
† Usually a representative, or even a supplier subsidiary, does not have the business volumes to justify having a comprehensive stock of replacement parts.
‡ See, for example, http://www.cmmtalk.com/welcome.htm; http://www.yourcmm.com.

costs still remain if the work is to be conducted in house by a service provider. This is because the service receiver usually still pays for energy, floor space, and capital costs. In the case when parts are routed to an outside metrology house, there is the additional cost and time to transport the parts back and forth. Moreover, a critical piece for a successful metrology operation is product and process knowledge. It is certainly more difficult to train a supplier and then maintain the knowledge about all the unique details of one's products, components, and processes.

15.4 SYSTEM COST

All details pertaining to equipment acquisition, installation, and usage affect initial as well as ongoing costs. It is important to carefully consider all requirements, not only present but future, to make the correct choices. Many times, it is necessary to compromise but that has to be done wisely not to limit the system usefulness.

One can compare buying a CMM to buying any other tool: it needs to match the need. Acquiring the CMM requires a detailed analysis to match the need with the solution, which includes the machine itself and all the other details discussed in this chapter. The overall cost of the system needs to be correctly established based on the needs of the application.

Establishing and maintaining a functional metrology department can be challenging. Shifting production schedules and processes can make the inspection load extremely variable. Traditionally, companies try to keep their metrology resources at a bare minimum, assuming metrology is all appraisal—ignoring the value in reducing internal and external failures (see also Chapter 19, Financial Evaluations). This should not be the case since there is value added to the parts when they are measured, as knowledge about them provides feedback to production and guarantees they conform to specifications, reducing costs due to internal failures. Moreover, better-quality products delivered to customers reduce external failure costs. The key is to control the overall cost of poor quality (Juran and Godfrey 1999).

15.5 SUMMARY

Although the requirements for each CMM application are different, there is a common thread for consideration that makes the difference between a successful and unsuccessful installation. More often than not, CMMs are either underutilized or do not meet expectations. To avoid this situation, the buyer needs to very carefully review the application requirements and evaluate the different alternatives against the requirements. A spreadsheet is an effective approach to use the selection process (refer back to Table 15.1).

Certainly, the buyer needs to check, recheck, verify, and understand all the information provided by the CMM salespeople before populating the comparative spreadsheet with the details. This pertains especially to accuracy and uncertainty, measuring speed, match to the application, and programming routines. Proper planning, resources, and time are essential so that all the necessary steps are taken to assure a successful choice and installation yielding many years of fulfilling expectations. This chapter has reviewed the main considerations to achieve a successful installation.

ACKNOWLEDGMENTS

Special thanks to Jeff Donovan, Caterpillar Inc., and Dr. Sheila E. Rowe, Assistant Professor, North Carolina A & T State University, for kindly reviewing the text and for helpful suggestions. Also, many thanks to Jay Nilsson who wrote the original Application Considerations chapter on which this one was based (Nilsson 1995).

16 Typical Applications

Wolfgang Knapp

CONTENTS

To provide a broad view of coordinate measuring system (CMS) capabilities, 12 case studies have been selected from six countries, involving several types of CMSs from seven manufacturers. These case studies, with companies and authors mentioned, range from the most typical (Sections 16.2 and 16.5) to those being used for research and gage calibration (Sections 16.9 to 16.12). Categories discussed include systems for checking large parts (Sections 16.3 and 16.4), medical applications (Section 16.8) and CMS with hybrid and optical probing (Sections 16.1 and 16.7) and computed tomography (Section 16.6).

16.1 CASE STUDY 1: LOCATION OF COOLING HOLES IN HOLLOW, CAST, AND TURBINE AIRFOILS

United Technologies Corporation – Pratt & Whitney/
East Hartford, Connecticut Dr. Joseph Drescher/
Manufacturing Technology Division, Pratt & Whitney

Cooling holes are used to maintain strength in turbine airfoils at the high temperatures required for optimal engine performance. Accurate location of the holes is critical to proper airflow and uniform cooling. There may be hundreds of holes in a single airfoil and each hole is considered a separate feature, thus requiring dimensional verification.

16.1.1 Airfoil Inspection Equipment

Dual-sensor five-axis coordinate measuring machines (CMMs) have been used for many years in this application. The current generation of CMM (Figure 16.1) used in this case study was built by Optical Gaging Products (OGP), Inc., a division of Quality Vision International of Rochester, New York. A cast crossway forms a bridge with granite uprights that is fixed to a granite base. The machines have linear motor drives for the horizontal axes and a circulating ball screw drive on the vertical axis, all with single optical tape scale per axis. Linear way bearings are rolling elements. Rotary axes have direct current (DC) motors with worm drive and optical encoders. The work volume in the *x*, *y*, and *z* axes is 610 × 660 × 400 mm. Motion control used is personal computer (PC) based and proprietary.

The CMMs are enclosed and maintained at 21.5°C ± 0.5°C within the shop, which is nominally maintained at 24°C ± 3°C. This is accomplished with a 2-kW alternating current (AC) unit mounted external to the enclosure and using shop-chilled water in the condenser heat exchanger. Parts are loaded manually on CMMs through a sliding door arrangement on the front of the enclosure. Cycle times are typically between 5 and 10 minutes. The chosen set point for air temperature within the enclosure is a compromise necessitated by the frequent opening of the door. Uncertainty due to thermal expansion and that due to differential thermal expansion

FIGURE 16.1 The Optical Gaging Products SmartScope Quest 650CH configured with dual rotary table, touch-trigger probes, and dual magnification optics. (Courtesy of Pratt & Whitney—Turbine Module Center.)

are increased compared to a standard 20°C control. But the stability of the machine structure is improved with a smaller differential to the shop air temperature.

Machines employ both touch-trigger and video probes (dual sensor). The touch-trigger probe is used to establish the part frame of reference based on a kinematic arrangement of points and normal vectors on the contoured surface of the airfoil. The video camera probe is necessary for quick and automatic measurement because of the hole size, which is typically 300 μm in diameter. Working distance (WD) on these CMMs may be adjusted by changing the final objective of the optical column. A 150-mm WD works well for a broad range of parts. A tilt–rotary axis configuration is required in this application so that each hole may be viewed and the position measured as defined in a plane perpendicular to the nominally cylindrical hole centerline (the camera has to look straight down the hole).

16.1.2 WORKPIECES AND FIXTURES

The CMMs are dedicated to measuring the location of cooling holes in turbine air-foils. Fixtures are designed to hold the part so that all nest points that define the datum planes can be accessed by the touch-trigger probe. In addition, all holes to be checked must be visible to the video probe. The fixtures must locate the parts within approximately 2 mm of a known, nominal position in all 6 degrees of free-dom (DOF). Probe deviations at the nest points are recorded and the actual location of the part is estimated using the first-order approximation to the exact six-DOF fit. A second set of probe deviations is then recorded in the corrected coordinate system and a second correction applied. Iteration is required because the probe points on the part surface are, in general, within three-dimensional (3-D) and nonsymmetrical surface profiles. Therefore, the linearization of the six-DOF fit is well justified. In practice, two sets of probe data are sufficient to reduce associated errors to a small percentage of the total measurement accuracy.

Holes are often designed, produced, and inspected at high angles to the surface normal such that the optical axis of the video has a low angle to the part surface. Parts are metallic and usually quite reflective. The contrast required to recognize and define a hole in the video image analysis is obtained in most situations with two features of these CMMs: The first is a programmable ring light, orthogonal to the optical axis, having a diameter approximately equal to the optical WD. The included angle between outer ring and optical axis usually gives sufficient diffusely reflected light for detection. When it is not sufficient due to specularity or due to obstructions in the light path, the rate of sampling and discharge of the video charge-coupled device (CCD) array may be slowed. The image is integrated. Contrast increases as does settling time. The software provides several integration settings.

16.1.3 QUALITY ASSURANCE

The accurate location of cooling holes in rotating blades and static vanes is critical to the performance and life of the turbine module. As such, the conformance of these features must be guaranteed. For the development of a new part or a new cooling hole configuration, an inspection program is created. It is verified by correlation with results from an independent inspection method. The production engineer, using the feedback from the CMM, targets the drilling process to achieve true position of all holes within 50% of the true position drawing tolerance. Allowing for measurement system variation, the goal is then to achieve less than 75% of tolerance in an initial production run incorporating 100% inspection. When this is satisfied, the statistical process control is initiated, with periodic sampling, tracking, and application of appropriate rules. In addition, inspections are made at the start of a new production run when a new tool setup has occurred.

16.1.4 DATA HANDLING

Inspection programs are stored centrally and accessed directly through the CMM controller at the time when a part is inspected. The program is automatically deleted from the controller on completion. This system allows efficient and reliable version control and generally reduces the possibility that the wrong program is used.

Results of the inspection program are automatically sent to a local printer so the operator can quickly judge the conformance of the part. The results are also filed to a central database through a network connection.

16.1.5 CALIBRATION AND REVERIFICATION

Verification of CMM performance takes place at two time intervals and two levels of complexity. The primary calibration is on a yearly cycle. This interval is adjusted annually, if necessary, following an analysis of all calibration results within an equipment classification. Probing error and size measurement error are checked following International Organization for Standardization (ISO) 10360-5:2010 (ISO 2010a) and ISO 10360-2:2009 (ISO 2009a) standards, respectively. The five-axis error is measured with a modified version of the four-axis error test described in the standards American Society of Mechanical Engineers (ASME) B89 (ASME 1997) and ISO 10360-3:2000 (ISO 2000b). These tests use only the touch-trigger probe. Tests for the additional errors associated with the video/optical system and the referencing of video probe to touch probe do not exist in ASME or ISO standards currently. For this purpose, a reference material artifact, or check standard, is employed. This artifact represents an airfoil. It has a coordinate system defined by six points on the surface and six normal vectors at those points. It has holes of the same size as typical cooling holes. The orientation of the holes requires full exercise of the five-axis system for measurement. Although not metrologically traceable, this check standard has been well characterized in terms of hole location. Multiple measurements on different machines have been made to reduce nonsystematic components of uncertainty. Reversal techniques have been used to eliminate many machine-related systematic sources of uncertainty. The results of the check standard measurement incorporate errors of the total measuring system and are therefore a significant portion of the system's measurement uncertainty for any specific application.

On a weekly interval, a simple test artifact is measured. This artifact is neither traceable nor is the uncertainty well established for the positions of holes. The locations of holes are simply measured relative to the positions of the same holes established at the completion of the original calibration. This artifact is also used to quickly verify accuracy after scheduled maintenance, repairs, or other occurrences.

When results of any calibration or reverification test exceed established limits, the CMM manufacturer's setup procedures are employed as necessary. These procedures incorporate a two-dimensional (2-D) glass grid standard, a laser interferometer, a rotary axis calibration artifact, field-of-view standards, and test sphere standards.

16.1.6 MEASUREMENT UNCERTAINTY

The best estimate of measurement uncertainty for the general application of measuring cooling hole positions on turbine airfoils comes from the check standard measurement and the uncertainty in the calibration of the check standard. The uncertainty in the calibration of the check standard (location of holes) varies according to individual hole. The worst case has a standard uncertainty of 0.005 mm in either direction orthogonal to the hole centerline (x and y). The standard deviation of errors from a typical check standard measurement is 0.004 mm in either direction. This gives a standard uncertainty in each orthogonal measurement direction of $(0.005^2 + 0.004^2)^{1/2}$ = 0.0064 mm. The measurand in this application is true position or, effectively, radial deviation. Independent, normal distributions with zero mean for uncertainty in the x and y directions combine to give a Rayleigh distribution for uncertainty in radial location. For $u(x) = u(y)$, the Rayleigh mean is (Morse and Voelcker 1996)

$$\mu_r = u(x)\sqrt{\pi/2} = 0.0064\sqrt{\pi/2} = 0.008 \text{ mm}$$

The square root of Rayleigh variance gives the combined standard uncertainty as

$$\sigma_r = u(x)\sqrt{2 - \pi/2} = 0.0064\sqrt{2 - \pi/2} = 0.004 \text{ mm}$$

Adding two standard uncertainties to the mean conservatively estimates the expanded uncertainty for around 95% confidence at 0.016 mm. More quantitatively, the estimate for expanded uncertainty can be obtained by evaluating the cumulative distribution function at 95% as

$$P(r) = 0.95 = 1 - e^{-r^2/2u^2(x)} = 1 - e^{-r^2/2(0.0064)^2}$$

and a simple Monte Carlo analysis will agree with this result. The radial deviation is calculated from two populations of normal x and y deviations, each having zero mean and a standard deviation of 0.0064 mm. If 100,000 points are generated, 95,000 of them should be less than or equal to 0.0157 mm. This is the stated uncertainty.

16.1.7 SUMMARY

The current generation of CMMs equipped with video probe has greatly improved the quality of measurements at Pratt & Whitney in the application of turbine airfoil cooling holes. These CMMs reduced the cycle time of measurements and enabled more frequent measurements for less cost. Most importantly, measurement accuracy has improved by a factor of five. This allows the production engineer an added safety margin in tooling, drilling, and other process variables. And it provides the design engineer with improved overall process capability.

16.2   CASE STUDY 2: GEAR MEASUREMENT WITH COORDINATE MEASURING MACHINES ON THE SHOP FLOOR

Caterpillar Inc.
Paulo H. Pereira/Chief Metrologist – Integrated Manufacturing
Operations Division – East Peoria, Illinois

The component manufacturing area of the Caterpillar Inc., IMOD—East Peoria facility makes a wide range of high-torque gears for utilization on several earthmoving products. It produces over 400,000 gears a year of different types ranging in size from $\emptyset$ 50 to 1,300 mm.

In order to provide more timely feedback to the manufacturing process, it was necessary to change the condition of taking the parts from a cell to a laboratory 100 m away to complete quality checks. Reduction in feedback time on component quality is a critical element of improving overall manufacturing process efficiency and component quality. The three key areas that were improved with this installation were as follows: (1) reduction in setup time, (2) faster quality verification after tool changes, and (3) reduction in rework/scrap due to the local availability of a CMM. It was decided to place a CMM within a gear-manufacturing cell to achieve these goals. This was the first of such type of CMMs and there are plans for replicating the same concept to other gear-manufacturing cells.

16.2.1   Coordinate Measuring Machine Equipment

The measuring equipment is a moving bridge CMM model Global 7107 manufactured by Hexagon Metrology, North Kingstown, RI running on QUINDOS 7 software. It is installed in a controlled environment that is held to 20°C ± 2°C. It utilizes an LSP-X5 fixed scanning probe head. The enclosure is small, just big enough to house the machine with space around it for servicing. Floor vibration tests indicated the need for air springs vibration isolation, which was installed.

16.2.2   Workpieces and Fixtures

This machine supports manufacture of both internal and external parallel tooth gears up to $\emptyset$ 600 mm. It utilizes a simple fixture that holds the parts magnetically. This is a single fixture that does not require changeover when a new part number is to be manufactured by the cell. The fixture only requires the users to center the parts within ±5 mm. There is a conveyor system that transports the parts into the machine working volume.

Machine tool operators (this CMM does not have a dedicated operator) do not have to enter the CMM enclosure; rather they just position the parts on a pallet and then press a button that starts the system, moving the part into a station with temperature control. To stabilize the temperature of parts, air at 20°C is blasted on them at 5 m/s. While the part is being acclimatized (the shop environment is not cooled during the warm months but heated during colder ones), the operator enters the part data onto a computer screen and allows the system to take over. A sensor monitors the part temperature and once the part reaches a threshold temperature (within 15°C–25°C), the part is ready to be moved to the CMM if it is available since there are two loading stations for better utilization.

Figure 16.2 shows the system with the two loading and cooling stations. The machine itself is inside the enclosure. One can also see in the figure the computer on which users input data about the parts to be measured and retrieve results. Figure 16.3 depicts a gear being measured on the CMM. The pallet carrying the gears uses magnets to hold them and is lifted by three pneumatic cylinders before the measurement commences.

FIGURE 16.2   Gear cell CMM. (Courtesy of Caterpillar Inc.)

FIGURE 16.3   Gear measurement. (Courtesy of Caterpillar Inc.)

16.2.3 Organization

The quality department is in charge of all quality within the facility. All of incoming, in-process, and outgoing quality checks are the responsibility of this department. The inspectors can hold up production and stop the assembly line in case quality issues are identified. After corrective action is identified and implemented for the issues found, production resumes. In the case of this machine, which does not have dedicated operators, quality personnel are called when potential quality issues arise.

16.2.4 Data Handling

Measurement data from the parts are stored in individual reports in a network location for easy access and are also sent to a statistical process control (SPC) database for the calculation of performance data. Reports and SPC data are used by operators, quality personnel, planners, processors, and engineers in assessing what needs to be done if an issue is raised.

16.2.5 Periodic Inspection of Coordinate Measuring Machines[*]

Machine performance is verified every week immediately after probe calibration by quality department personnel. Multiple tip test and probing test as per ISO 10360-5 (ISO 2010a) and repeatability test as per B89 (ASME 1997) are performed and tracked every week. Volumetric ball bar tests as per B89 are performed and tracked monthly. Should problems be detected the machine is taken out of use and the issues are immediately addressed. Complete annual certifications are also performed on the machine.

16.2.6 Measurement Uncertainty[†]

The main reasons for choosing a CMM inside an enclosure versus a floor model were issues regarding daily temperature variations and part temperature stability. In Central Illinois, daily temperature gradients average around 12°C during 5 months of the year, which is beyond the maximum allowable variations of currently available floor CMM models. On top of that, the lack of part temperature control increases measurement uncertainties to unacceptable levels. By estimating uncertainties, it was possible to make the right decision in this case study.

Table 16.1 shows a summary of the uncertainty calculations performed for one part measured by this CMM before deciding which model to acquire. Tolerance to uncertainty ratios (TURs) were evaluated for two features considered critical on a given part number, measurement over pins, and internal diameter. These were some of the tightest tolerances that this system was expected to handle.

---

[*] Refer also to Chapter 9, Performance Evaluation.
[†] Refer also to Chapter 14, Measurement Uncertainty for Coordinate Measuring Systems.

TABLE 16.1

Summary of Estimated Uncertainty and TUR Values

Part Features		Shop CMMs		Enclosed CMMs		
		Brand 2	Brand 2	Brand 1	Brand 2	
Measurement over	40°C	31.6	31.3	24.1	24.1	μm
pins Ø 132 mm	TUR	*1.6*	*1.6*	*2.1*	*2.1*	
	28°C	22.6	22.5	15.8	15.8	μm
	TUR	2.2	2.2	3.2	3.2	
	25°C	12.0	12.0	11.4	11.4	μm
	TUR	4.2	4.2	4.4	4.4	
Internal diameter	40°C	17.9	17.7	13.7	13.7	μm
Ø 74.7 mm	TUR	*1.5*	*1.5*	*1.9*	*1.9*	
	28°C	12.8	12.8	9.1	9.1	μm
	TUR	2.0	2.0	2.9	2.9	
	25°C	6.9	7.0	6.6	6.6	μm
	TUR	3.7	3.7	3.9	3.9	

Note: roman type = unacceptable values
italic type = acceptable values

Both enclosed and shop-floor CMM models of two different vendors were considered. Three part temperature conditions were evaluated: (1) warmest at 40°C, (2) average at 28°C, and (3) cool at 25°C. The warmest situation occurs because finished gears out of the cell go through a parts washer that is set to approximately 49°C before making it available for inspection. The average temperature situation for parts happens when the environment is not too warm and the parts have enough time after passing through the washer to cool. Maintaining the environment at a temperature around 20°C–25°C and allowing enough time for the parts to cool after passing through the washer would bring the parts temperature to around 25°C. Machine temperatures for uncertainty calculations for the shop-floor models were considered according to the shop-floor conditions; machine temperatures for the enclosed models were considered according to the performance of the enclosure. For this evaluation, TURs ≥ 4 were considered acceptable.

The uncertainty estimates showed that the best solution was to enclose the machine, mainly due to part temperature control and also because the CMMs themselves have less error when they are enclosed.

16.2.7 SUMMARY

The measurement system described in this case study has been in operation successfully for over 3 years. It proved the power of using uncertainty estimates when planning such a system. Having inspection capabilities closer to the manufacturing unit is also beneficial due to quicker response times and consequent gains in efficiency.

16.3 CASE STUDY 3: APPLICATIONS OF LARGE COORDINATE MEASURING MACHINES IN INDUSTRIES

China Precision Engineering Institute for Aircraft (CPEI)/China
Professor Guoxiong Zhang/Tianjin University, China

China Precision Engineering Institute for Aircraft (CPEI) has more than 30 years of experience in CMM manufacturing. It has put 68 different types of more than 700 CMMs into the market. Its products are widely used in aerospace, aviation, nuclear, shipbuilding, vehicle, textile, and machine-building industries.

16.3.1 COORDINATE MEASURING MACHINE EQUIPMENT

Recently, a large gantry CMM with x, y, and z travels 6000, 3500, and 2500 mm, respectively, was built in the CPEI. The uncertainty of measurement in space is $E = 8.0 + 8.0L/1000$ µm, where L is the traveled distance in millimeters. The repeatability of the machine is 8 µm. It is a computer numerical control (CNC) machine with dual synchronized x motions. Special care was taken for the compensation of gantry deformation, which might be one of the major error sources for this kind of machine. Figure 16.4 is a photograph of the CMM.

An SP25 laser scanning probe made by Renishaw, Ltd., United Kingdom, in conjunction with a PH10MQ motorized head is used for data sampling. The probe can work in both scanning and touch-trigger modes; this ability of the probe makes it very convenient for measuring sculptured surfaces, such as aircraft wings, turbine blades, and 3-D prismatic parts.

FIGURE 16.4 LM603525 gantry CMM. (Courtesy of China Precision Engineering Institute for Aircraft.)

16.3.2 WORKPIECES AND THEIR HANDLING

In Figures 16.5 through 16.7, some workpieces checked by this CMM are illustrated. Figure 16.5 is an airplane. One of the airplane wings measured on this machine has a length of 16.67 m, height of 4.2 m, and width of 1 m. Its weight is 3.5 t. The forms of the airplane wings and the hole positions on them are checked. In Figure 16.6, the propeller of a vessel is shown. The form errors of the sculptured surface of the propeller are measured. In Figure 16.7 a steam turbine is shown. It is required to measure the form errors of the sculptured turbine blades and their relative positions. Other requirements are raised by engineering machine industry and mining machine industry, vehicle industry, and many others. The CMMs are used for both reverse

FIGURE 16.5 Workpiece measured—airplane.

FIGURE 16.6 Workpiece measured—propeller.

FIGURE 16.7 Workpiece measured—turbine.

engineering and quality control. The measurement is especially important for study purposes in the product development stage. It is estimated that about 15 such large CMMs per year are required in China. It is obvious that China's domestic CMM industry cannot meet this requirement.

Most parts measured on this CMM are directly mounted on the table without fixturing. For the parts that cannot be mounted on the machine table stably, special fixtures are needed.

16.3.3 DATA HANDLING

The scanning probe used in this machine is suitable for a diverse range of applications, materials, and surfaces. The scanning probe is a miniature measuring machine by itself that can acquire several hundred surface points per second, enabling the measurement of form as well as size and position of a part. The scanning probe can also be used to acquire discrete points in a way similar to touch-trigger probes. It can provide a fast way to capture form and profile data from prismatic or complex components.

Rational DMIS*, a software that has passed the Physikalisch-Technische Bundesanstalt (PTB) software test, is used for data handling. The basic functions of the software include

- Automatic measurement of basic geometric features such as point, line, plane, circle, cylinder, cone, slot, ball, ellipse, and paraboloid, and calculation and output of relevant parameters
- Calculations for intersection, symmetry, projection, distance, and angle of the geometry elements
- Position tolerance analysis including parallelism, squareness, coaxiality, and symmetry
- Coordinate transformation

---

* Dimensional Measuring Interface Standard (ISO 2003c)

- Probe operation module
- Teach mode programming and off-line programming
- Inverting function
- Statistics function

A program for system diagnosis and error compensation is included.

16.3.4 PERIODIC INSPECTION OF MEASURING MACHINE CAPABILITY

The workpieces measured on this machine are critical, and they are all checked. The accuracy of the CMM is inspected once a year by using a laser interferometer.

16.3.5 ORGANIZATION

The CMM is under the control of the quality assurance (QA) engineer who is responsible for programming, fixture design, and environment and utility monitoring. For large parts, environment conditions are essential for ensuring the required accuracy of measurement. The required temperature is 20°C ± 1°C. The variation in temperature should be no larger than 0.5°C/h and the thermal gradient in space 0.5°C/m. Relative humidity should be within 60% ± 5%. Vibrations of the machine foundation should be within the following ranges: frequency range 0.2–2.0 Hz, vibration displacement $<5 \times 10^{-6}$ m; frequency range 2–20 Hz, vibration speed $<50 \times 10^{-6}$ m/s; and frequency range 20–200 Hz, vibration acceleration $<5 \times 10^{-3}$ m/s^2. Utility air supply minimum operating pressure should be $\geq 1 \times 10^6$ Pa, and air consumption should be ≥ 0.3 m^3/min.

16.3.6 SUMMARY

There is an increasing demand for large-scale part measurement. Usually these parts are critical and costly, so they all must be checked throughly and completely. In most cases they are also complicated in form. Both efficiency and environment conditions are of great importance for these parts. Some special problems such as bending and thermal deformations, dual synchronized drive, and many others should be carefully considered in the design of such machines. China is just at the starting stage of building such large CMMs.

16.4 CASE STUDY 4: LEITZ PMM-F INSPECTS WIND ENERGY PLANT COMPONENTS

Manfred Frank – free journalist
Sabine Eisel – Hexagon Metrology GmbH – Germany
Geert van Landuyt – TVL Toeleveringsbedrijf van Landuyt NV
Juergen Engelhardt – formerly with Hexagon Metrology GmbH – Germany

Only 50 µm is the tolerance on the axle bores of the 4-t transmission case used in a wind turbine. The Belgian manufacturer TVL Toeleveringsbedrijf van Landuyt NV achieves reproducible measurement of the complex and nearly inaccessible geometry in about 30 minutes on a Leitz precision CMM made by Hexagon Metrology.

16.4.1 Coordinate Measuring Machine Equipment

The CMM PMM-F 30.20.16 is 4.5 m long, 4.36 m wide, and 5.09 m tall, with a measurement volume of $3000 \times 2000 \times 1600$ mm. A special foundation was poured to provide a stable, vibration-free base for the 28-t machine. The U-formed body made of gray granite does not absorb as much humidity as black granite, increasing overall stability of the measurement results. The PMM-F at TVL offers high reproducibility and a proved measurement uncertainty of 2.5 µm. When it comes to high-speed scanning, the Leitz PMM-F is in its turf. In addition, an integrated collision protection system prevents damage to the ram.

Another special feature of PMM-F is the dynamic tool changer rack. As a rule, complex inspections of large parts require several styli changes during the measurement process, because no one stylus can reach all measurement points. However, a conventional tool rack at the end of the measurement volume restricts the range of the styli system enormously, especially with particularly long styli configurations. Too much active measurement range is lost by a rigid tool rack. The newly developed dynamic tool rack swings itself out of the measurement volume after a styli change is completed (Figure 16.8). The complete area is then available for the ensuing styli positions. The exchanged stylus immediately starts with the measurement process.

Every Leitz CMM made by Hexagon Metrology is calibrated, and its measurement results and reproducibility are tested. Each CMM is delivered completely preassembled, with the y-axis dismounted and later remounted at the site. After final assembly, the geometric errors of the CMM are corrected again using a laser measuring device. All errors stemming from almost two dozen sources, such as rotations, translations, and errors of perpendicularity, are corrected. Test measurements are carried out to confirm compliance with the stated specifications. Thanks to the highly linear force/deflection characteristics of the Leitz probe, extremely reproducible probing results ensue.

FIGURE 16.8 Automatically articulating tool changer rack. The probes do not reduce the measurement volume.

16.4.2 WORKPIECES AND FIXTURES

The TVL produces transmission housings (Figure 16.9) for 5-MW wind energy turbines (Figure 16.10) with a length of up to 2500 mm and bores having diameters of up to 2000 mm. As an example, the tolerance for a 650-mm bore is only ±25 μm and must be measured verifiably on every housing. This means that drilling is toleranced at a mere 0.008% of nominal size.

The transmission housings are transported using a crane to the Leitz PMM-F site. The housings are measured in a room with the same temperature conditions as on the shop floor (Figure 16.11). This eliminates the usual waiting times associated with parts adjusting to measuring temperature.

After the CMM performs a fully automated position-recognition routine and the temperature sensor on the measurement object is activated, the measurement procedure can begin. The temperature sensor measures the current gearbox housing temperature, and the measurement program later integrates the value during data analysis. The longitudinal elongation of the measurement object is compensated for in the final calculations based on the readouts collected by the temperature sensor. The collection of measurement data takes about 30 minutes, and the data contains some measurement points found at hard-to-reach locations on the measurement object.

16.4.3 ORGANIZATION

When TVL started the production of gearbox housings for wind energy turbines, around 100 housings were completed in 2000; today, they have reached an annual figure of 900 with further growth anticipated. In the past, TVL could not measure

FIGURE 16.9 High-speed scanning deep inside the transmission housing.

FIGURE 16.10   Wind energy plant. (Courtesy of www.pixelio.de, Marco Barnebeck.)

FIGURE 16.11   The Leitz PMM-F for measurement of transmission cases used in wind turbines.

the big transmission housings internally. The company was working together with a metrology service provider in Germany that had the necessary measurement equipment. As production quantities grew, transportation to the service provider became a costly endeavor. Thus, TVL decided in favor of its own CMM.

The investment was planned carefully and completed between TVL and Hexagon Metrology, as solutions had to be developed for TVL's quality process like measuring strategy, measuring uncertainty, scanning of elements versus single point probing, and the consideration of special issues. In addition, the development of the dynamic tool changer rack was the outcome of that close cooperation between both companies.

16.4.4 Data Handling

The machine-independent 3-D measurement software QUINDOS involved in data handling runs on all TVL equipment and provides a common user environment. The only difference lies in the specific hardware in use at TVL, such as touch-trigger or measurement probes. Broader advantages of universally applying the software are the interchangeability of measurement programs between different gages as well as the use of common statistical files or reports. This not only minimizes the costs for the software maintenance at TVL for their special measurement programs but also reduces the costs for additional training of their staff since the same releases are always installed on all equipment simultaneously.

The QUINDOS basic license provides all the commands that are required for the inspection of standard parts, such as engine blocks and gearboxes. It also includes features like form and true position evaluation according to ISO, free programmable plots, generation of probing points, and more. In addition to the basic package, there are about 35 optional packages. They cover almost all the measurement tasks performed and special geometries manufactured in today's industrial world.

16.4.5 Periodic Inspection of Coordinate Measuring Machines

The Leitz PMM-F at TVL is typically recalibrated once a year.

16.4.6 Measurement Uncertainty

According to ISO 10360-2 (ISO 2009a), the volumetric length measurement error of the Leitz PMM-F is $E_{0,\text{MPE}} = 2.8 + L/400$ µm and the volumetric probing error $P_{\text{FTU, MPE}}$ (ISO 2010a) equals 2.4 µm. The scanning probing error MPE_{THP} amounts to 3.5 µm at 68 seconds, according to ISO 10360-4 (ISO 2000c). The maximum positioning speed is 400 mm/s and the maximum acceleration 2500 mm/s^2.

16.4.7 Summary

The Leitz PMM-F at TVL offers high reproducibility and a proved measurement uncertainty of 2.5 µm. Precision, speed, support and, last but not the least, price made Leitz CMM the ideal machine for TVL. Since November 2005, every single 4 t of transmission housing produced at TVL has been measured on a high-precision Leitz PMM-F CMM.

16.5 CASE STUDY 5: UNTENDED AUTOMATION CELLS

Peer Inc./Waukegan, Illinois
Jim Salsbury/Mitutoyo America Corporation/Aurora, Illinois

In support of untended machining cells, Peer, Inc., Waukegan, Illinois, utilizes CMMs on the shop floor for both real-time feedback control and 100% inspection of tightly toleranced machined components.

16.5.1 COORDINATE MEASURING MACHINE EQUIPMENT

The automation cells at Peer, Inc. include machining centers, CMMs, and robots for moving workpieces between stations. Each cell contains two CMMs, both Mitutoyo Crysta-Apex C CMMs equipped with Renishaw TP200 probing systems. Crysta-Apex C CMM is a moving bridge, air bearing CMM. Both CMMs are dedicated to a single automation cell. All the equipment in the cell, including the CMMs, is located on the shop floor with no special environmental isolation.

16.5.2 WORKPIECES AND FIXTURES

Automation cells are for the manufacture of thrust blocks, which are found in various hydraulic gear pumps. Some of the critical dimensions on the thrust blocks have tolerances as small as 0.01 mm. One of the CMMs is used for rapid inspection of critical dimensions, and the results are used in a real-time feedback loop to keep the manufacturing process within its control limits. The other CMM is used for off-line inspection of all dimensions. Due to the tight tolerances, 100% inspection is used to ensure zero defects.

Dedicated fixtures are used due to the automated nature of the CMM inspection process at Peer, Inc. (Figure 16.12). The fixtures are loaded and off-loaded by the robotic arm. A compressed-air blow-out station is also part of the automated cell and is used to prepare the workpieces for measurement on the CMMs. The combination of fixturing, robotics, and CMM programming allows for untended operation of the CMM (Figure 16.13). The cells run 24 hours a day, 7 days a week. Operators oversee all aspects of the process, but the machining and measurement stations are designed to function without operators.

In addition to providing fully automated inspection, the flexibility of the CMM allows Peer, Inc. to reconfigure the inspection process for new or different workpieces. The entire cell, including machining and measurement, is designed for both automation and flexibility.

16.5.3 ORGANIZATION

The director of manufacturing at Peer, Inc. is responsible for the management of the CMMs and the automation cells. The measurement process is automated and runs untended. Operators on the shop floor monitor the multiple automation cells.

FIGURE 16.12 The CMM inspection process within the automation cell. (Courtesy of Mitutoyo America Corporation.)

FIGURE 16.13 An automation cell at Peer, Inc. (Courtesy of Mitutoyo America Corporation.)

16.5.4 Data Handling

The CMM data is managed by Measurlink software, a data management and process control software product from Mitutoyo. If the CMM results indicate that a workpiece is out of tolerance, the robot marks the workpiece and moves it to a separate area. If too many workpieces begin to fail, the untended system shuts down until the problem can be investigated and resolved manually.

16.5.5 PERIODIC INSPECTION OF COORDINATE MEASURING MACHINES

The Mitutoyo America Corporation periodically calibrates the Crysta-Apex C CMMs at Peer, Inc. The calibration is done in accordance to ISO 10360-2 (ISO 2009a). The CMMs dedicated to a particular manufacturing process, such as the CMMs at Peer, Inc., are also often checked periodically by using some type of "golden" workpiece. The golden, or master, workpiece is often a normal workpiece that has been clearly marked, for example, sometimes painted gold. This same workpiece is measured on the CMM at some periodic interval, such as daily, weekly, or monthly. The results are checked to ensure no significant changes are occurring to the measurement process. If a normal workpiece is not available, or if the workpiece is not stable enough to be a long-term master, then a workpiece like the master can be made.

16.5.6 MEASUREMENT UNCERTAINTY

When CMMs are used for controlling manufacturing processes, the most important quality issue with a CMM is that the variability of the results is reduced as much as possible. The repeatability of the CMM process is critical to properly support manufacturing. Long-term repeatability studies, in combination with periodic calibration of the CMM, ensure that the measurement uncertainty is adequate for the application.

16.5.7 SUMMARY

The automation cells at Peer, Inc. showcase the combined speed, flexibility, accuracy, and programmability of CMMs. There may be more accurate or faster inspections solutions, but CMMs can provide a sound economical solution to a variety of shop-floor inspection challenges.

16.6 CASE STUDY 6: GEOMETRY MEASUREMENT OF CAST ALUMINUM CYLINDER HEADS USING INDUSTRIAL COMPUTED TOMOGRAPHY

Physikalisch-Technische Bundesanstalt PTB,
Braunschweig und Berlin, Germany
Dr. Markus Bartscher/PTB, Braunschweig, Germany
Dr. Uwe Hilpert/Wenzel Volumetrik, Singen/Hohentwiel, Germany
Dr. Dirk Fiedler/Nemak Wernigerode, Germany

The PTB, the National Metrology Institute of Germany, works on the traceability of geometry measurements. For several years, PTB—together with partners from industry—has been analyzing and optimizing the measurement capability of industrial computed tomography (CT) using X-rays. Nemak Wernigerode, an automotive part foundry in Germany, and PTB performed studies on the measurement capabilities of CT.

16.6.1 COORDINATE MEASURING MACHINE EQUIPMENT

The measurement equipment is located in the QA department of the Nemak Wernigerode aluminum foundry. The CMM equipment consists of an industrial CT scanner using a 450-kV X-ray tube and an X-ray line detector (fan-beam CT system; Figure 16.14). This CT measuring system is manufactured by Bio-Imaging Research, Inc. (now Varian Security & Inspection Products), Lincolnshire, Illinois, and is located in a climate-controlled area within a lead enclosure. The CT system is provided with integrated computing for measurement control and for surface extraction from CT volume raw data. Postprocessing of the extracted workpiece surface data and volume visualization of the CT volume data are carried out by several computer-aided design (CAD) workstations. Reference measurements have been performed with tactile CMMs of PTB.

16.6.2 WORKPIECE AND MEASURING TASK

The workpieces to be measured are raw cast automotive aluminum cylinder heads made in the production facility of Nemak Wernigerode. Several measurement tasks are performed.

One prominent task is actual–nominal value comparison between actual state (as measured by the CT system) and nominal geometry as given by CAD models. Therefore, parts of the geometry (e.g., the water and oil jackets) have to be segmented from the whole part geometry. This is done regularly with a production part approval process (PPAP). The classical process chain consists of consecutive slicing and tactile measuring of parts. This process is quite slow, is destructive, and is inaccurate as reference geometries are lost due to the slicing process. It was, therefore,

FIGURE 16.14 Industrial 450-kV CT system applied for geometry measurements of cast aluminum cylinder heads.

strongly desired to substitute integral parts of this process by the nondestructive measurement technique CT. An additional measurement task to be performed with the CT system is nondestructive testing, like the detection of shrinkage cavities and other inhomogeneities caused by the casting process.

16.6.3 DATA HANDLING

The CT system collects, during every revolution of the workpiece (about 1 minute) on the rotary table, X-ray projections of the part. As the tomographic reconstruction of data starts immediately after the first rotation, the irradiated slice of the part of thickness 1 mm is reconstructed just after the respective rotation. The whole part is measured by consecutive rotations and by the repeated measuring of spatially shifted slices. The slice increment is typically 0.5 mm. For the visualization of workpiece data, dedicated volume visualization software is used. This software also enables measurement of regular geometries (cylinders, spheres, planes, etc.) and defect analysis of the part under study (see Figure 16.15).

For actual–nominal value comparison, surface data must be extracted from the volume data by a threshold process. The surface data given in STL-format* representation are analyzed by means of CAD inspection software. The analysis of part geometry is usually carried out by graphical representation of the deviation (deviation topology; see Figure 16.16) or the point-by-point analysis of absolute deviation values.

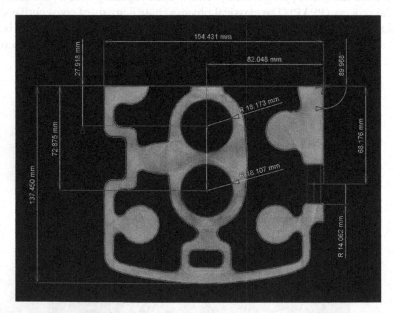

FIGURE 16.15 Geometry determination of a cast aluminum cylinder head segment within single slice, measured by CT.

---

* The STL is a file format native to stereolithography CAD software.

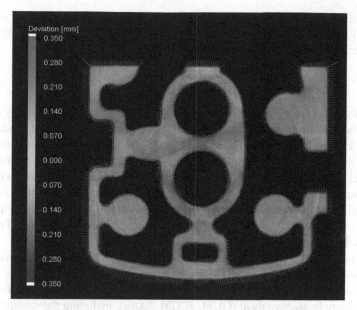

FIGURE 16.16 Actual nominal data comparison of cast aluminum cylinder head segment within single slice. Whiskers indicate the deviation from the CAD model.

16.6.4 ACCURACY ENHANCEMENTS

To enhance the accuracy of measurements, several actions have been taken. First, the parameterization of the measurement task has been optimized during a 2-year research project of PTB and Nemak Wernigerode. Additionally, calibrated reference bodies are measured simultaneously with the workpiece under study. As reference bodies, a ball bar with ceramic balls on a carbon fiber rod and two adequate aluminum rings are applied. From the reference body data, scaling factors and threshold values for the surface extraction process are deduced. These parameters are applied to the CT-measured workpiece data.

The need for geometry-preserving thresholds is a common problem that occurs when extracting surfaces from CT volume data. For this purpose, two rings consisting of the same aluminum alloy as the measured workpiece and with complementary inner and outer radii are used. In the surface extraction process, the threshold is varied and the measured radii of the two rings are compared. As the complementary radii of the rings are recovered, this threshold value is used to reproduce the geometry of the workpiece. Also, modern analysis software packages already have adaptive threshold processes implemented in them, which can adapt to the local geometry of a workpiece.

16.6.5 PERIODIC INSPECTION OF MEASURING MACHINE CAPABILITY

A periodic inspection schedule for CT systems performing geometry control did not exist when the system was set up. Jointly, PTB and Nemak Wernigerode developed a periodic inspection schedule. It consists of the measurement of calibrated ball bars (ceramic balls and carbon fiber rod), calibrated aluminum rings, and a calibrated aluminum cylinder head segment. The measurements are evaluated with respect to size and form deviations, and also with respect to nominal–actual value differences in comparison with a "valid" machine condition.

16.6.6 MEASUREMENT UNCERTAINTY STUDY

The PTB and Nemak Wernigerode performed a measurement uncertainty study on a real production part: a cylinder head of size $85 \times 195 \times 360$ mm made from a common cast aluminum alloy. In all 25 CT measurements were performed during a 9-month period, with a replacement of the X-ray tube after the tenth measurement. The voxel size (size of the volumetric cell assessed by the CT measurement) of the CT measurement was $0.37 \times 0.37 \times 0.5$ mm. The cylinder head was then calibrated at 103 points in the outer region by tactile CMM. Finally, the part was opened by milling at selected points. In this way, 64 additional points were calibrated by tactile CMM in the inner region of the part.

The final result of the measurement uncertainty study was a probability statement for the single point measurement uncertainty of extracted surface points over the assessed sampling. The data were evaluated according to the *Guide to the Expression of Uncertainty in Measurement* (GUM; JCGM 2008a), including the statistical spread of CT results, influence of workpiece properties (especially roughness), and influence of uncorrected systematic errors. Of all measured points (inner and outer region of the part), 93% had a measurement uncertainty below the smallest voxel edge length (0.37 mm), that is, subvoxel precision was achieved (Bartscher et al. 2007). Due to the fact that decisions on surface point positions are made by averaging over several surface points, typical measurement uncertainties of 0.2–0.3 mm can be achieved in practice.

16.6.7 SUMMARY

The work of PTB and Nemak Wernigerode presented in this case study enables Nemak Wernigerode to carry out complete measurement tasks entirely by means of the CT measurement system. CT measurements now encompass essential sections of the PPAP for new cylinder heads and the geometry process control of foundry output. Now, CT has become an integral part of the QA system of the foundry. Main customers of Nemak Wernigerode have accepted the use of CT as a verified coordinate measurement technology.

16.7 CASE STUDY 7: NONCONTACT MEASUREMENT OF SCULPTURED SURFACE OF ROTATION

Tianjin University (TJ)/China
Professor Zhang Guoxiong/Tianjin University, China

Tianjin University (TJ) is a leading university in the field of instrumentation and metrology in China. The State Key Laboratory of Precision Measuring Technology is established there, and a large number of research projects in the field have been accomplished in the university.

16.7.1 Coordinate Measuring Machine Equipment

Surfaces of rotation consist of a large portion of sculptured surfaces. The measurement of surfaces of rotation can be realized during the continuous rotation of a workpiece mounted on a rotary table while the probe moves along the generatrix of a surface step by step. The distinguished features of this method are simplicity of probe motion, high reliability, and efficiency, which are especially important for measuring fragile parts. Optical noncontact probes possess some advantages, such as no measuring force, no deformation, and high speed in comparison with contact probes.

A CIOTA1203DH moving bridge CMM made by the 303 Research Institute, Beijing, China, is used as the main equipment. The x, y, and z travels are 1320, 970, and 610 mm, respectively. The uncertainty of measurement along axial lines is $G = 3.5 + 3.5L/1000$ µm; the uncertainty of measurement in space is $M = 4.0 + 4.0L/1000$ µm, where L is the traveled distance in millimeters. The rotary table is made by Yantai Machine Tool Accessory Works, Shandong, China. The radial error motion measured at a height of 300 mm above the mounting surface does not exceed 4 µm, and the face error motion measured at a distance of 400 mm from the axis does not exceed 4 µm. The probe used is an LK031 laser triangulation probe made by Keyence Company, Japan, which has a resolution of 1 µm and a measuring range of ±5 mm with a nominal distance of measurement of 30 mm. Such a large nominal distance and measuring range provide convenience in alignment, and the safety problem of collision avoidance can be solved relatively easily. For getting accurate results in measurements, it is required to make the laser beam normal to the measured surface. For this purpose, the laser triangulation probe is mounted on a PH10M motorized head, made by Renishaw, Ltd. Figure 16.17 shows a picture of

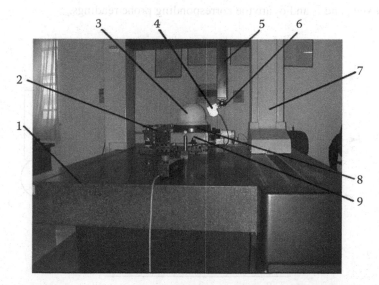

FIGURE 16.17 Measuring surface of revolution on a CIOTA1203DH CMM. (Courtesy of Tianjin University.)

the system where numbers indicate the components: 1—table, 2—auto-centering device, 3—workpiece, 4—laser triangulation probe, 5—ram of the CMM, 6—motorized head, 7—bridge, 8—mounting disc, and 9—rotary table.

After the workpiece (3) has been centered by the auto-centering device (2), the measurement starts. The ram (5) carries the motorized head (6) and probe (4) to the desired position, and the motorized head (6) positions the laser beam approximately perpendicular to the workpiece surface. A circular section is measured when the rotary table (9) rotates 360°. After that, ram (5) carries the motorized head (6) and probe (4) to the next position along the generatrix of the surface, and the motorized head (6) rotates to the desired angular position in case of necessity. In this way, the whole sculptured surface can be measured section by section.

Measurement of the surface of rotation with a rotary table in cylindrical coordinate system gives high accuracy in form measurements, such as roundness and concentricity. The error motions of the CMM components, except the rotary table and probe, have almost no effect on the measuring accuracy. However, the probe gives only the variation of the radius. For obtaining the absolute value of the diameter, the location of the axis of rotary table x_0 in the CMM coordinate system and the effective arm length R of the motorized head should be calibrated. The diagram for calibrating the axis position and effective arm length is shown in Figure 16.18.

In Figure 16.18, the gage (2), which might be a gage block or a ring gage, with known dimension L is mounted on the rotary table (1). The gage (2) is measured by the probe (5) from both sides by moving the ram (3) and rotating the motorized head (4), as shown in Figure 16.18. It is obvious that

$$x_2 - x_1 = 2R + \delta_1 + \delta_2 + L$$

where x_1 and x_2 are CMM readings when the ram is on the left and right sides, respectively; and δ_1 and δ_2 are the corresponding probe readings.

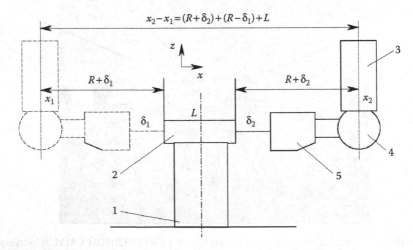

FIGURE 16.18 Diagram showing axis position and arm length calibration.

For determining the location of the axis of rotary table x_0 in the CMM coordinate system, a second set of measurements is taken after rotating the table (1) 180°; the corresponding readings x_1', x_2', δ_1', and δ_2' of the CMM and the probe are obtained. From these two measurements, both x_0 and R are determined:

$$x_0 = [(x_2 - R - \delta_2) + (x_1 + R + \delta_1) + (x_2' - R - \delta_2') + (x_1' + R + \delta_1')]/4$$

Hence

$$x_0 = [(x_2 - \delta_2) + (x_1 + \delta_1) + (x_2' - \delta_2') + (x_1' + \delta_1')]/4$$

$$R = [(x_2 - x_1) + (x_2' - x_1') - (\delta_1 + \delta_2 + \delta_1' + \delta_2') - 2L]/4$$

16.7.2 Workpieces and Their Handling

Workpiece (3) is mounted on the rotary table (9) (Figure 16.17) by a loading device. The loading device is able to mount the workpiece with an eccentricity no higher than 5 mm which is not accurate enough for measurement. Auto-centering device (2) is used to reduce the eccentricity to less than 0.1 mm. During the auto-centering process, the triangulation probe (4) measures a cylindrical surface of the workpiece (3) while the rotary table (9) rotates 360°. The maximum radius, r_{max}, and the minimum radius, r_{min}, of the section and the angle at which r_{max} is reached are recorded. After that, rotary table (9) rotates again until the maximum radius point faces the pushing rod of the auto-centering device (2). The rod pushes the mounting disc (8) until the reading of the triangulation probe reaches $(r_{max} + r_{min})/2$. Following this, the rotary table (9) rotates a whole circle again to check if the mounting eccentricity is lower than 0.1 mm. In most cases, the eccentricity can be reduced to less than 0.1 mm after one or two runs.

The measuring uncertainty of the machine is less than 0.02 mm.

16.7.3 Data Handling

The flowchart of data processing software is shown in Figure 16.19. Five sets of data are sampled at each measured point. They are CMM readings x_i and z_i; angular reading of motorized head α_i; probe reading δ_i; and the rotated angle of rotary table φ_i. Based on these sampled data and the effective radius of the probing system R, coordinate of the rotary table axis x_0 obtained from calibrating the coordinates of the measured point Q_i (X_i, φ_i, Z_i) can be calculated. X_i is the polar radius, φ_i is the argument, and Z_i is the height of point Q_i in the cylindrical system, respectively.

All the obtained coordinates in the cylindrical system are preprocessed to eliminate noise and coarser errors. All these cylindrical coordinates are transformed from the machine system to the workpiece system. The key point is finding the axis of rotation of the workpiece. For a surface of rotation, the geometry on a plane perpendicular to the z axis should be a circle. For the purpose of finding the axis of rotation of the workpiece, the least-squares center of the circle at each cross section perpendicular to the z axis is first determined. Then, the least-squares average line of

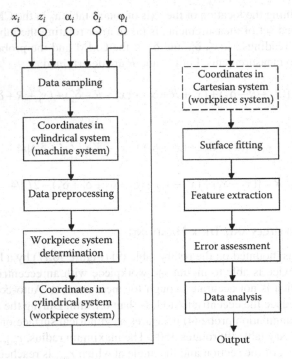

FIGURE 16.19   Data processing flowchart.

the centers at sections with different z is calculated and this line is taken as the z axis of the workpiece system, and all coordinates are transformed from the machine system to the workpiece system.

The measured data in cylindrical coordinate system might be transformed to Cartesian coordinate system if needed. Quite often, a surface of rotation consists of several elemental curved surfaces, such as spheres, cones, cylinders, and rings. It is required to know the specific features of these elemental curved surfaces, for example, radius of a sphere, its center position, its vertex point, and half angle and radius at a certain height of a cone. In this case, surface fitting and feature extraction are required.

16.7.4   Periodic Inspection of Measuring Machine Capability

The workpieces measured on this machine are critical ones, which means all must be checked. In a Cartesian CMM, the distance between two points is determined by relative positions of these two points in the workpiece coordinate system. The home position of the CMM has no effect on measured results. However, in a CMM with a rotary table all the measurements are carried out in a cylindrical system; any error in the home position of the CMM has a direct effect on the position of the rotary table axis. The home position error will be introduced twice to the diameter measured. To eliminate these errors, a ring gage and an automatic routine for calibrating the rotary axis position x_0 and arm length R are provided and the machine is inspected every day or after going back to the home position before running the measurements. Another, more complicated artifact similar to the form of workpieces is used to check the CMM accuracy every month.

16.7.5 ORGANIZATION

The workpieces measured on this machine are critical and fragile. The machine can be operated only by qualified workers by typing in a unique password. The software can be modified only by high-ranking engineers by inputting another personal identification number.

16.7.6 SUMMARY

A special-purpose CMM with a rotary table and a noncontact probe has been developed for measuring the sculptured surface of rotation. The measurement is realized during the continuous rotation of the workpiece mounted on the rotary table while the probe moves along the generatrix of the surface step by step. This method possesses lots of advantages such as simplicity of probe motion, high reliability and efficiency, and high accuracy in form error measurements. However, for obtaining the absolute diameters of the part and for correct contouring along the generatrix, the effective arm length of the motorized head and the position of axis of rotation should be calibrated. A ring gage and an automatic routine for calibrating the rotary axis position x_0 and arm length R are developed to inspect the machine every day or after going back to the home position. An auto-centering device is also developed. The measuring uncertainty of the machine is less than 0.02 mm.

16.8 CASE STUDY 8: MEASURING PRECISION MEDICAL PROSTHESES

Grace Medical, Inc./Memphis, Tennessee
Tony Prescott/President, Diane Holton/Quality Engineer
Quality Vision International, Inc., Rochester, New York
Fred Mason/Vice President

Grace Medical, Inc., Memphis, Tennessee, manufactures devices that are used in the disciplines of ophthalmology (diagnosis and treatment of disorders of the eye) and otolaryngology (treatment of the ear, nose, and throat).

16.8.1 COORDINATE MEASURING MACHINE EQUIPMENT

Grace Medical achieves the necessary accuracy by using a SmartScope Flash 200 benchtop 3-Dl optical CMM system (Figure 16.20) developed by OGP. The CMM has a measuring volume of $x = 200$ mm, $y = 200$ mm, and $z = 150$ mm, with areal and linear uncertainties of $E_2 = (1.8 + 6L/1000)$ μm in the xy directions and $E_1 = (3.5 + 6L/1000)$ μm in the z direction, respectively (where L is stage travel in millimeters).

FIGURE 16.20 The Optical Gaging Products SmartScope Flash 200 CMM in use at Grace Medical. (Courtesy of Grace Medical, Inc.)

The CMM is fitted with a MicroTheta rotary (MTR) table with 2 arcsecond positioning resolution. The MTR table is a "set angle" device. In the Grace Medical application, it is used to rotate a single part being measured in order to present other surfaces of the part to the video measuring system optics. Angles are not measured in this process. Instead, the rotary table reduces part handling and minimizes the number of fixtures necessary for measuring more surfaces of these small, cylindrical parts. Since the parts are rotationally symmetric, the absolute angular position is not critical.

The primary measuring technique is optical, whereby magnified images of the parts captured by a camera are analyzed by computer software algorithms to locate multiple edge and surface data points at a time. With high-accuracy closed-loop positioning, the dimensional and angular relationships among all data points collected within the measurement volume of the CMM are retained in the system's metrology software. Multiple programmable light-emitting diode (LED) illuminators optimize image intensity and contrast for best measurement repeatability and reproducibility.

16.8.2 Workpieces and Fixtures

Typically, 10 parts per run are measured in an automated inspection routine. During inspection, the CMM measures critical dimensions on parts arranged in a holding fixture. Blind holes (250 μm in diameter × 750 μm deep) are measured in the z axis to the nearest 5.0 μm. The MTR indexer and MeasureMind 3D MultiSensor metrology software with comprehensive rotary axis calibration routines add a fourth axis measuring capability for measuring precision parts made on Swiss-style turning machines. The rotary table enables the CMM to automatically rotate parts within a measurement routine to easily measure anything from simple geometric forms to complex free-form shapes without user interaction.

16.8.3 Organization

The Grace Medical facility is ISO 13485 (ISO 2003b) certified, which is a quality management standard for medical devices. The SmartScope Flash 200 CMM is currently used in a number of company operations, including product development, engineering, production, process control, QA, final inspection, and quality training.

16.8.4 Data Handling

The system automatically generates inspection reports that are used as part of device history records. Grace Medical has also used the SmartScope Flash 200 CMM to perform fully automated lights-out qualification of components. Since part measurement routines are developed for each part and they operate identically for all such parts loaded on a fixture, parts are loaded on a fixture; the measurement sequence is initiated; and the measurement, data transfer, and collection are completely automatically.

16.8.5 Periodic Inspection of Coordinate Measuring Machines

Since it is used in a production environment, the OGP CMM needs to be viewed and evaluated as a very sophisticated automated production device and not merely as another piece of inspection equipment. Cycle times, fixturing for multiple parts to optimize uptime, and quick changeover of inspection routines and tooling are important considerations.

Grace Medical uses master pieces prior to the inspection of each lot of parts. They do, according to an internal quality procedure, qualify a part routine prior to use. This is achieved by performing a gage R&R study (AIAG 2002) on the most critical characteristic of the part, determined by the quality engineering (QE) and product development (PDE) departments and approved by the QA department. The SmartScope Flash measurement system is calibrated semiannually, and any measurement uncertainty is covered by calibration certification, which refers to traceability to National Institute of Standards and Technology (NIST) certificate 821/253792-94, and so on.

16.8.6 Measurement Uncertainty

Use of the SmartScope Flash 200 CMM may result in changes in the production flow that will need analysis, explanation, and training to ensure the most positive impact on productivity and efficiency. Grace Medical constantly assesses and validates its machine routines, and the OGP CMM is no exception. Because of process improvements attributable to the SmartScope Flash CMM, 100% inspection is no longer required for many parts and processes.

In the manufacture of MTR, the following is the OGP certification process: The measuring accuracy of each MTR is verified using a calibrated nine-sided optical polygon and a Taylor–Hobson autocollimator. The polygon is mounted on the face plate of the MTR, and then the autocollimator is used to compare the rotational position of the MTR with the rotational position of the optical polygon. Readings are taken at each polygon face in both clockwise (CW) and counterclockwise (CCW) rotations. Three sets of CW and CCW readings are taken. The data are recorded and analyzed and, if within specification, a certificate is issued. The uncertainty of the measuring data is evaluated to confirm the measuring process.

16.8.7 SUMMARY

The OGP SmartScope Flash CMM fits the measuring needs of a broad range of materials and geometries encountered in precision medical prosthetic device manufacture. This CMM's noncontact optical measuring technique makes it equally suitable for measurement of precision-molded flash-free silicone parts as well as titanium implants with their complex shaft geometries on surfaces as small as 0.125 mm in diameter and tolerances of ±0.0125 mm (parts; see Figure 16.21). By providing immediate, accurate measurements of dimensions such as these, the SmartScope Flash 200 CMM helps to speed up the production of devices critical to patient health, while improving manufacturing quality control.

16.9 CASE STUDY 9: PRECISION COORDINATE MEASURING MACHINE FOR CALIBRATION OF LENGTH STANDARDS AND MASTER GAGES

Mitutoyo America Corporation, Aurora, Illinois
Jim Salsbury/Mitutoyo America Corporation

To meet growing needs for high accuracy and high throughput calibrations of various length standards and other master gages, the calibration laboratory of Mitutoyo America Corporation decided to implement a high-precision CMM.

16.9.1 COORDINATE MEASURING MACHINE EQUIPMENT

In 2003, a Mitutoyo Legex 910 CMM was installed in the calibration laboratory of Mitutoyo America located in Plymouth, Michigan. This same machine was moved in 2006 to the primary calibration laboratory of Mitutoyo America located in Elk Grove

FIGURE 16.21 Precision prosthetic devices on a United States 1 ¢ coin to show scale: The part on the right is made out of titanium and has a diameter of 0.2 mm between the ribs. The other part is made of two pieces. The metal part is laser-welded gold wire of 0.125 mm diameter. Both these parts are middle ear prostheses. (Courtesy of Grace Medical, Inc.)

Village, Illinois. The measuring volume of Legex 910 is $900 \times 1000 \times 600$ mm, and the accuracy specification, in accordance with ISO 10360-2 (ISO 2009a), is $E_{0,\text{MPE}} = (0.35 + L/1000)$ µm. The machine is a fixed bridge design with a cast iron base and a central drive on the moving table to reduce angular error motion. The Legex also incorporates air-damped spring isolators to minimize the impact of external vibrations. An air dryer and cleaner is utilized to maintain a stable temperature ($\pm 0.1°C$) of the air supply to the machine air bearings. The Legex is equipped with crystallized-glass scales with a resolution of 0.01 µm and an ultralow linear coefficient of thermal expansion (CTE) of $0.01 \times 10^{-6}/°C$ to further reduce thermal issues.

The Legex supports a variety of probing systems. For the most accurate measurements of 3-D length standards, the Mitutoyo MPP-300 probe is utilized. The MPP-300 is a high-accuracy analog head with a resolution of 0.01 µm. The Legex can also be used with the Mitutoyo QVP Vision probe, which is a noncontact optical probing system. The QVP Vision probe can be interchanged with the MPP-300 for use in measuring glass line scales and other 2-D glass standards. The QVP attaches to a Renishaw PH10 motorized probe head. This probe head can therefore be utilized with a large variety of touch-trigger probes when the flexibility of an articulating probing system is needed.

The accuracy specifications of the Legex are valid over a temperature range of $20°C \pm 2°C$. The calibration laboratory of Mitutoyo America is controlled to $20.0°C \pm 0.5°C$ across the entire laboratory (Figure 16.22). The Legex is equipped with a built-in temperature compensation system, and this system is used for all measurements.

16.9.2 WORKPIECES AND FIXTURES

The primary measurement task for the Legex 910 CMM is the calibration of Mitutoyo step gages (Figure 16.23), many of which are used by the field service department of

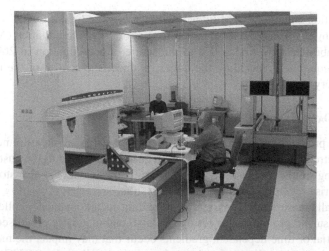

FIGURE 16.22 Original installation of the Legex 910 in the calibration laboratory of Mitutoyo America. (Courtesy of Mitutoyo America Corporation.)

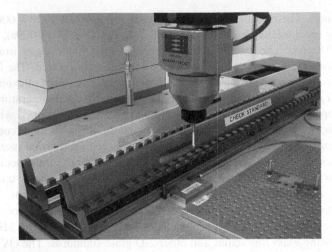

FIGURE 16.23 Typical setup for measuring step gages on the Legex 910. (Courtesy of Mitutoyo America Corporation.)

Mitutoyo America for onsite calibration of CMMs. Over 200 of these types of step gages are calibrated each year. A large variety of other gages are also calibrated on the Legex, including glass line scales, ball bars, ring gages, ball plates, and other specialized gages.

In order to achieve a measurement uncertainty less than the accuracy specification of the CMM, a "sweet spot" located in the middle of the moving table is carefully calibrated. Whenever possible and always when the lowest uncertainty is needed, the gage to be measured is positioned within this sweet spot.

16.9.3 Organization

The calibration laboratory of Mitutoyo America is accredited ISO/IEC 17025 (ISO/IEC 2005) by the American Association for Laboratory Accreditation (A2LA). The Legex capabilities were first added to the scope of accreditation in 2003. All the laboratory equipment, including the Legex, is under the control of the manager of the laboratory.

16.9.4 Data Handling

To improve productivity and eliminate transcription errors, all measurement data are moved electronically to the final calibration certificate issued to the customer. This is done using simple spreadsheet tools. All original raw data are also stored in case questions arise regarding the final reported results.

During all measurement runs, temperature data are automatically collected by the CMM measuring program. The results are also automatically processed to ensure that no spatial or time-dependent gradients occur that may impact the measurement results. Various preset limits for thermal variations, all under 0.05°C, have been established as part of the uncertainty analysis.

16.9.5 PERIODIC INSPECTION OF COORDINATE MEASURING MACHINES

Due to the high accuracy of the length calibrations made on the Legex, a methodology that incorporates five levels of periodic calibrations and inspections is used.

1. Calibration to ISO 10360-2 by the manufacturer
2. Calibration of the y axis to a tighter linear specification
3. Calibration of the workpiece temperature sensors
4. Calibration of the probe stylus tip diameter
5. Inspection of an independently calibrated check standard

The Legex is calibrated by the field service department of Mitutoyo America to the accuracy specifications of the CMM in accordance with ISO 10360-2 (ISO 2009a). The specification for $E_{0, \text{MPE}} = (0.35 + L/1000)$ μm is of particular importance. Although this calibration, which is accredited by A2LA as ISO/IEC 17025 (ISO/IEC 2005), is usually done annually, it may be more frequently performed if needed. During the calibration, particular attention is paid to the results along a line parallel to the moving table axis, which is the y axis of the CMM. In order to achieve the desired uncertainty for length measurements, the sweet spot along this axis has a reduced specification of $E_{0, \text{MPE}} = (0.2 + 0.4L/1000)$ μm. Due to the low uncertainty, the calibration is done using special gage blocks that are made of the same material as the Legex scales with an ultralow CTE.

Once the CMM performance is verified, a separate calibration is done on the workpiece temperature sensors. Instead of using traditional thermometry calibration methods, the temperature sensors are calibrated in situ using a method of differential length measurements (Salsbury and Inloes 2006). Two calibrated 500-mm gage blocks are placed one next to the other, and their lengths are measured on the Legex. The workpiece temperature sensors are placed in contact with the gage blocks. One gage block is made of steel, and the other is made of the low-thermal-expansion material. Any error in the workpiece temperature sensors will impact the measurement result of the steel gage block by over 100 times than the gage block made of the low-thermal-expansion material. This method can therefore be used to calibrate the temperature sensors in the same manner in which the sensors are regularly used.

For many measurements, only one stylus is used on the Legex. A special calibration of the diameter of this stylus tip is done using a 25-mm steel gage block. The normal calibration sphere is not used as the accuracy of the diameter of the sphere is not adequate. The quality of the stylus tip calibration, as well as the linear accuracy of the Legex, is monitored through the use of a check standard. A step gage was calibrated externally at NIST, and this step gage is measured daily. If the results of this check ever exceed the best uncertainty level of the Legex, then work stops; the problem is investigated and resolved prior to continuing the calibrations.

16.9.6 MEASUREMENT UNCERTAINTY

Since the Legex is used for ISO/IEC 17025 (ISO/IEC 2005)-accredited calibration work, a formal expression of measurement uncertainty in accordance with the GUM

TABLE 16.2

Uncertainty Analysis for the Calibration of a 1000-mm Steel Step Gage on the Legex 910

Source of Uncertainty	Standard Uncertainty (μm)
Linear accuracy of the y axis	0.30
Calibration uncertainty of linear accuracy	0.13
Stylus tip calibration	0.05
Temperature sensor calibration	0.25
Thermal gradients	0.14
Uncertainty in thermal expansion	0.14

(JCGM 2008a) is required. The uncertainty also has to pass routine technical assessments and proficiency testing. The best uncertainty reported on the A2LA scope of accreditation is $0.25 + 0.7L/1000$ μm, using a coverage factor of $k = 2$. A summary of a typical uncertainty analysis is shown in Table 16.2.

16.9.7 SUMMARY

The Mitutoyo Legex 910 CMM was installed in the calibration laboratory of Mitutoyo America to provide a highly accurate, productive, and flexible measurement solution for the calibration of various length standards and other master gages. By combining the Legex accuracy, a good environment, i.e. laboratory air temperature control at 20 °C ± 0.5 °C and air supply temperature within ± 0.1 °C, and some novel procedures, the experiences of Mitutoyo America have demonstrated that a CMM can be used for world-class level calibrations. At the time of this writing, a new laboratory facility is being planned that will improve on the temperature control of the laboratory environment. With this new facility, it is expected that the measurement uncertainty can be further reduced to meet the future calibration needs of Mitutoyo America.

16.10 CASE STUDY 10: MEASUREMENT OF AN IMAGE SLICER MIRROR ARRAY, USING THE ISARA ULTRAPRECISION COORDINATE MEASURING MACHINE

IBS Precision Engineering, The Netherlands
ir. [Dipl.Ing.] I. Widdershoven, dr. ir. [Dr. Dipl.Ing.] H.A.M. Spaan,
dr. ir. [Dr. Dipl.Ing.] M.A.A. Morel Engineering

The Isara ultraprecision 3D CMM, which is manufactured by IBS Precision Engineering, the Netherlands, is capable of performing ultraprecision measurements with a "volumetric uncertainty" of less than 50 nm on complex workpieces.

16.10.1 Coordinate Measuring Machine Equipment

The company IBS Precision Engineering is the manufacturer of the Isara 3D CMM, an ultraprecision 3-D CMM developed in cooperation with Philips Applied Technologies, Eindhoven, the Netherlands. The Isara 3D CMM (see Figure 16.24) features a measuring volume of $100 \times 100 \times 40$ mm. The 3-D measuring uncertainty is defined as the expanded uncertainty with a coverage factor of two (ISO 1998), thus specifying the uncertainty of the 3-D position of a single measurement point within the measurement volume. This volumetric uncertainty is less than 50 nm within the complete volume, including contributions from the probe system. The Isara 3D CMM features a moving product table whose 3-D position is measured by three laser interferometers, which are mounted in a stationary metrology frame. The metrology frame also provides a mount for the probe system. As the probe tip is located at the virtual intersection point of the three laser beams, the Isara 3D CMM conforms to the "Abbe principle" in three dimensions.

Touch-probe systems used on the Isara 3D CMM typically have a spherical tip with a diameter of 0.5 mm or less. The ultraprecision touch probe features an elastically suspended stylus; the deflection of the stylus during probing is measured by the internal measurement system of the probe. Probing forces are very low (typically less than 1 mN), which makes the system suitable for ultraprecision measurement of small and accurate features without damage to the (optical quality) surfaces of the workpiece.

Critical parts in the metrology loop of the CMM are made of low-expansion materials to minimize the effect of changes in environmental air temperature. The metrology frame, which holds the laser interferometers and the probe system, is

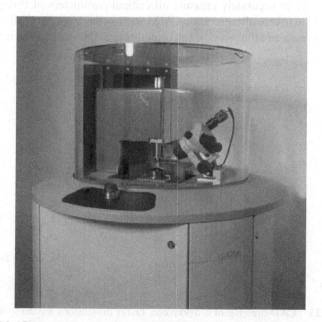

FIGURE 16.24   Photograph of the Isara 3D CMM.

entirely made from invar. The product table is a monolithic Zerodur part, whose three sides serve as measurement mirrors for the interferometers. An environmental temperature stability of 0.1°C/h is recommended, so that thermal expansion of the workpiece itself, for example, is reduced to a minimum. The CMM is placed in a temperature-controlled laboratory; the workpiece is allowed to adjust to the room temperature for at least one hour. As the base frame of the machine features passive air dampers, additional vibration isolation is not necessary.

The Isara 3D CMM is capable of performing discrete point measurements as well as scanning measurements. The metrology software allows the operator to import a CAD model of the workpiece, to program automated measuring sequences, and to compare the measurement data to the nominal model.

16.10.2 WORKPIECES AND FIXTURES

The typical range of workpieces for the Isara 3D CMM includes spherical and aspheric optics, free forms, small parts, and so on. In this case study, the workpiece under investigation is a prototype image slicer mirror array. This image slicer mirror is an optical component, used in the mid-infrared imager (MIRI) and spectrometer for the James Webb Telescope. The workpiece was manufactured by the Cranfield University Precision Engineering Centre, United Kingdom.

The image slicer consists of several spherical mirror segments. The critical parameters to be measured are the form accuracy of segments (sphericity) and the physical 3-D offsets between the virtual center points of the spherical segments and several reference surfaces (see Figures 16.25 and 16.26). The tolerances on the x and y offsets of the center points are specified as ±0.02 mm.

The only way to accurately measure all critical parameters of this image slicer mirror array is to use an ultraprecision CMM such as the Isara 3D CMM; this allows the operator to measure the position and orientation of the vertical reference surfaces

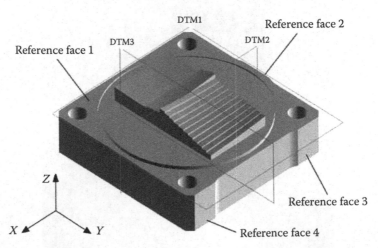

FIGURE 16.25   CAD drawing of a workpiece. Outer dimensions are 30 × 30 × 11.5 mm. (Courtesy of UK Astronomy Technology Centre.)

FIGURE 16.26   Photograph showing a close-up of the Isara probe and the image slicer on the Isara work.

as well as the position and form accuracy of the spherical segments in a single orientation of the workpiece, thus providing the required data on the positions of the virtual center points with respect to the reference surfaces. Other measurement strategies, such as optical measurements, may provide an accurate evaluation of the sphericity of mirror segments and may even be able to determine the relative positions of the segment center points with respect to each other, although they are not capable of determining the positions of these center points with respect to the vertical reference surfaces. The low measuring uncertainty of the Isara 3D CMM is required to meet the accuracy demands of the product, as demonstrated in the uncertainty analysis provided in Section 16.10.4.

At the start of the measurement, a CAD model of the workpiece is imported into the coordinate metrology software (Figure 16.27). The exact location and orientation of the product within the measuring volume is determined by performing alignment measurements. The product coordinate system is then adjusted to coincide with the CAD model. After alignment, it is possible to generate probing paths for automated measurements.

16.10.3   Data Handling

All spherical mirror segments have a radius of 162.5 mm. The image slicer mirror array features 14 spherical segments, which are 12 mm long and 1 mm wide; each segment is measured by taking 50 measurement points on the surface. One segment, the fifteenth one, is 12 × 8 mm; a total of 110 measurement points are performed on the surface of this segment. In addition, the positions of the vertical reference planes are measured. By using the reference planes for the alignment of measurement data with respect to the CAD model, the product coordinate system is aligned to the actual product so that all measurements and calculations can be directly presented in

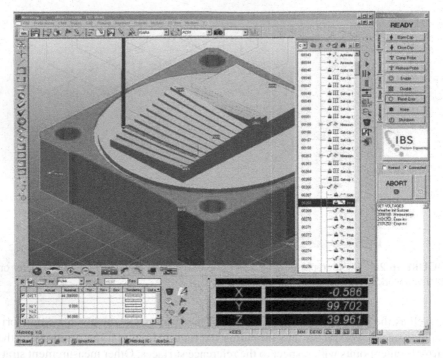

FIGURE 16.27   Screenshot of the metrology software.

product coordinates. As per the definition, the z axis is oriented square to reference plane 1 and the x axis is oriented parallel to the reference plane described by the two reference planes 3 and 4.

For each of the segments, a best-fitting sphere is calculated. The x, y, and z offsets of the sphere center points with respect to the reference planes are then calculated and compared with the nominal values.

16.10.4   MEASUREMENT UNCERTAINTY

The 3-D measurement uncertainty of the Isara 3D CMM is determined by various influencing error sources. From a thorough analysis of the measurement system, it has been determined that the volumetric measuring uncertainty is less than 50 nm (2σ), including the uncertainty of laser interferometers, calibration errors, and measuring uncertainty of the probe. A measuring uncertainty is present for all point measurements and will influence the measuring uncertainty of the sphere fit needed for the workpiece described in Section 16.10.2. The calculations presented here determine how this measuring uncertainty affects the resulting uncertainty in the lateral position of the sphere centers (x and y directions according to the coordinate system in Figure 16.25). The simulation results are limited to the x and y offsets, as the tolerances of these parameters (±20 μm) are the most critical.

The uncertainty calculation is performed by a Monte Carlo simulation, an experimental mathematical approach that calculates solutions of a physical process by simulating the behavior of a process and averaging over a large number of simulations. The resulting uncertainties for the parameters of the sphere fits are considerably larger than the volumetric measuring uncertainty of a single point measurement performed by the Isara 3D CMM. The image slicer array features mirror segments, which are only a small fraction of the total sphere. This causes the results of the least-squares fit to be very sensitive to small deviations in measured data. The simulation shows that for 14 narrow mirror segments, the estimated uncertainty of the x offset is ≤ 2.3 μm, the uncertainty of the y offset is approximately 5.6 μm. The uncertainty of the y offset is larger than the uncertainty of the x offset because the y coordinate is defined along the width of the slices. As the segments are only 1 mm wide, the resulting uncertainty of the y coordinate of the best-fit sphere is higher; small deviations in the point measurements may cause a relatively large change in the resulting y offset. For the fifteenth segment, which is considerably wider in the y direction, the uncertainty of the x offset is only 0.2 μm and the uncertainty of the y offset is only 0.6 μm.

As a guideline, the uncertainty of the measurement process (2σ) should generally be less than 10% of the tolerance band of the features that are measured. The measurement uncertainty for the most critical parameters, the x and y offsets, should therefore be less than 4 μm. For the x offset, the Isara measurements meet this specification. The uncertainty of the y offset for mirror segments 1–14 is slightly larger, but still only 14% of the tolerance band.

16.10.5 SUMMARY

The case study presented here shows an example of a complex product with demanding tolerances; inspection of this product requires measurement equipment with high measuring accuracy.

In this case, the product features small spherical mirror segments; as only a limited amount of measurement data can be taken from these small segments, measurement deviations at the nanometer level already influence the estimation of the sphere center location at the micrometer level. By means of Monte Carlo analysis, it is determined that the measurement uncertainty of the Isara ultraprecision 3D CMM is sufficient to reach an uncertainty $\leq 14\%$ of the tolerance band of the feature parameters that are measured.

16.11 CASE STUDY 11: ULTRAPRECISION MICRO-COORDINATE MEASURING MACHINE FOR SMALL WORKPIECES

Felix Meli, Alain Küng, Ruedi Thalmann,
Federal Office of Metrology METAS, Switzerland

The METAS micro-CMM is a laser interferometer-based 3-D measuring machine with an innovative low-force touch probe. Due to its consequent metrology concept, the corrections introduced for the reference frame, and the probe sphericity deviation, it exhibits unprecedented 3-D measurement accuracy. It is used for high-precision measurement of small workpieces in both point-to-point and scanning modes.

16.11.1 COORDINATE MEASURING MACHINE EQUIPMENT

The ultraprecision CMM motion stage was developed at Philips CFT (Ruijl and van Eijk 2003). Its vacuum preloaded air bearing stage is driven by Lorenz actuators and its motion is controlled by an interferometric position measurement. The working volume of the stage is 90 × 90 × 38 mm. During measurement, the probe head stands still while the stage moves the workpiece in all directions around the probe. The workpiece sits in a Zerodur cube corner with three perpendicular flat mirrors on the outside that form the reference coordinate system. The displacement of all axes is measured without Abbe offset since all three interferometer beams point to the center of the probing sphere. The metrology frame that holds the interferometers and the probe head is completely decoupled from forces acting on the stage.

An innovative 3-D touch probe supporting exchangeable probes with sphere diameters from 0.1 to 1.0 mm and probing forces less than 0.5 mN was developed (Meli et al. 2003; Küng, Meli, and Thalmann 2007). A parallel kinematics structure was chosen for minimizing the moving mass and for having isotropic properties of the probe head. Based on three parallelograms with flexure hinges (Figure 16.28), this

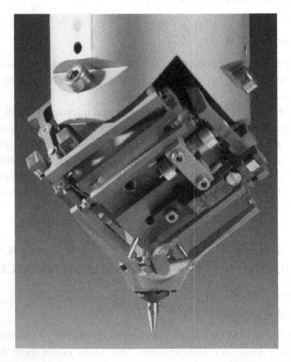

FIGURE 16.28 Photograph of the new three-dimensional touch-probe head.

structure leaves the probing sphere exactly three DOF. Whereas all rotational movements are blocked, the translational motion is separated into its xyz components,
which are measured by three inductive sensors. Due to the special orientation of the
probe head coordinate system all axes are identical with respect to gravity, which
results in an equal probing force in all directions. The main part of the structure is
manufactured out of a single piece of aluminum using electro-discharge machining.
The flexure hinges have a thickness of only 60 μm, which results in an isotropic stiffness of 20 mN/mm. The measurement range is ±0.2 mm, whereas the mechanical
limits allow a tip deflection of ±0.5 mm in all directions. The effective moving mass
is only 7 g. The magnetic holding of the probing element allows easy tip replacement
and cleaning.

16.11.2 WORKPIECES AND FIXTURES

The workpieces measured on the micro-CMM have sizes of a few millimeters to a
few centimeters and features down to 100 μm. The applications come typically from
fields like microsystems technology, microoptics, medical industry, micromechani-
cal industry, and automotive industry, which make use of parts such as injection
valves, lenses, optical apertures, small gages, precision spheres, fiber connectors,
and high frequency connectors.

Although very small measuring forces are used, small lightweight objects
will not remain in place only by gravity. Usually, these workpieces cannot be
clamped because they would be considerably deformed by even very small clamping forces. Mostly they are glued to the base plate by cyanoacrylate glue, which
can be dissolved easily using organic solvents once the measurement process is
complete.

16.11.3 ORGANIZATION

The Swiss National Metrology Institute METAS ensures the consistency and traceability of measurements done in Switzerland. It conducts research and development
to realize and disseminate the national measurement standards according to the
International System of Units (SI). It is responsible for conformity assessment and
type approval of measuring instruments in legally regulated applications. It provides
calibration and measurement services for accredited laboratories and industries,
both in Switzerland and across borders. It runs a quality system according to ISO
17025 (ISO/IEC 2005).

16.11.4 DATA HANDLING

The low-level machine controller software was developed at METAS. It controls
the procedure for point-to-point and scanning data acquisition and takes into
account all machine-specific, geometrical, and probe sphere corrections. The high-
level software QUINDOS (Hexagon Metrology PTS GmbH, Wetzlar, Germany) is
used to provide easy programming of measurement sequences and complex data
evaluations.

16.11.5 Calibration and Correction of the Micro-Coordinate Measuring Machine

Three built-in laser interferometers with calibrated laser frequency are the basis for all calibrations and for the traceability of all measurements to the definition of the meter. The orientation of the laser beams is carefully adjusted to guarantee negligible Abbe and cosine errors. The refractive index of air is determined by continuously monitoring pressure, temperature, humidity, and carbon dioxide concentration. Due to the rigorous metrology design of the CMM, only a few components have direct influence on the measurement uncertainty and therefore need a careful characterization for the subsequent corrections. Furthermore, these components can be characterized independently because there is no correlation among their influences.

The most important component is the reference mirror system made from Zerodur. The flatness of the three mirrors was determined using an additional Zerodur flat and a reversal and averaging method. This Zerodur flat has a gold-covered side with flatness better than 15 nm. It was probed in various configurations using a one-dimensional (1-D) capacitive probe. Additional measurements using ball bars and a ball plate were performed to determine the perpendicularity between the three mirrors.

For multidirectional probing, the size and shape of the probing element, usually a ruby sphere, becomes a major contributor to uncertainty. A 3-D probe shape calibration and correction procedure was implemented. The probe spheres, with diameters as low as 70 μm, are calibrated on a reference sphere with known sphericity deviation, taking into account its shape and orientation.

The sphericity and also the absolute diameter of the 1-mm calibration spheres were calibrated on the micro-CMM itself using an error separation method: In this "three-sphere calibration method," each sphere is used once as the probe measuring the 3-D shape of the two others (Küng, Meli, and Thalmann 2007). In fact, for a good error separation, each sphere pair has to be measured additionally in several rotational orientations. After some computational steps, the absolute diameter and the local radius deviations can be determined for all three spheres. This method is independent of any external reference and is traceable to the micro-CMM metrology system itself. Figure 16.29 shows the magnified sphericity deviation of one of the 1-mm reference ruby spheres. The exact diameter is 1.000859 mm and the peak-to-valley form deviation is 33 nm (sphericity). The flexible magnetic probe holder of this probe head is essential for the application of this method.

The probe shape correction was also implemented for the scanning mode. In scanning, the accurate calibration of the probe head transducer is even more important. Besides the deflection sensitivity, the orientation of the probe coordinate system also must be known very well. All these parameters are determined using an automatic calibration procedure and the aforementioned reference sphere.

16.11.6 Performance Verification

The final performance of the micro-CMM was verified with the help of artifacts such as gage blocks, big and small precision spheres, double ball bars, and also an invar ball plate (Küng and Meli 2007). The determined 2-D and 3-D deviations

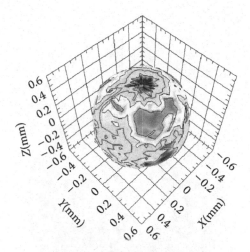

FIGURE 16.29 Three-dimensional representation of the calibrated reference sphere sphericity deviation (peak-to-valley: 33 nm).

(k = 2) using the ball plate and the ball bars were the same: 27 nm + $2 \cdot 10^{-7} \cdot L$ (where L is the measured lenght in mm). Because each sphere center was determined from several points, this corresponds to a unidirectional uncertainty. For multidirectional probing, the remaining probe shape error after correction gives an additional contribution of up to 20 nm.

The performance verification of the scanning mode was particularly interesting. Roundness measurements on a 1-mm ruby sphere have shown an average repeatability of 5 nm in each point. Surprisingly, this value is not higher than that for point-to-point probing. The performance according to the ISO 10360-4 (ISO 2000c) scanning acceptance test, as shown in Figure 16.30, using a 0.3-mm probe on a 1-mm sphere was investigated. The acquisition time for the four profiles was only 95 seconds. The determined unfiltered sphericity of 87 nm seems reasonable knowing that the value includes the sphere form error (about 53 nm), sphere surface roughness (about 15 nm), and probe shape calibration error (about 20 nm). One should keep in mind that this ISO test is meant for conventional CMMs where the reference sphere is considered to be perfect with negligible form deviation.

To obtain a quantitative value for scanning accuracy, we compared micro-CMM results with measurements made on a Taylor–Hobson TR73 high-precision roundness-measuring machine. Figure 16.31 shows three roundness measurements made on a 4-mm ruby sphere, using the micro-CMM in the scanning mode. In order to test the probe shape correction for various directions, the sphere was rotated by 120° after each measurement. The average local standard deviation of the three measurements was 5 nm, and the maximum difference between any two profiles was 20 nm. The measured least-squares roundness deviation (filtered at 50 upr) differs by only 3 nm from the TR73 value. The maximum profile difference was less than 15 nm. The advantage of the micro-CMM measurement is that it provides at the same time the absolute diameter of the measured sphere also.

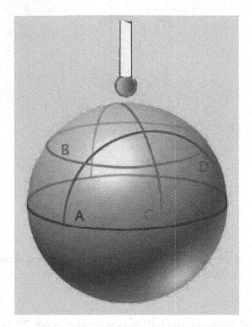

FIGURE 16.30 The four profiles to be measured for the ISO 10360-4 scanning acceptance test.

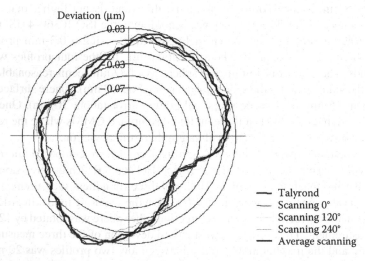

FIGURE 16.31 Comparison of micro-CMM scanning and high-accuracy roundness measurement.

16.11.7   Measurement Uncertainty

The major contributions to micro-CMM measurement uncertainty are mirror flatness, mirror angles, the probe shape, and probing repeatability. This results in an uncertainty for a single point-to-point distance of $27 + 2 \cdot 10^{-7} \cdot L$ for unidirectional probing and 34 nm $+ 2 \cdot 10^{-7} \cdot L$ for points obtained using multidirectional probing.

For a specific measurement task, the uncertainty can be higher (or lower) because it is always related to the task and depends, besides on CMM accuracy, strongly on the geometry and size of the device to be measured, the applied measurement strategy, and point distribution (see also Chapter 14, Measurement Uncertainty for Coordinate Measuring Systems). Typical workpiece influences are form deviation and surface roughness. Temperature deviations and thermal drift both influence the CMM and the device under test. In fact, the scanning mode of the micro-CMM often helps to lower the uncertainty because it reduces the measurement time and acquires many more data points, which lead to better fitting results.

16.11.8   Summary

The combination of a new 3-D touch probe with small exchangeable probing elements and low probing forces together with an ultraprecision CMM exhibits single point repeatability of typically 5 nm. With the help of an error separation method the absolute diameter and the sphericity deviation of small calibration spheres can be determined, and thus it serves to calibrate the probing spheres and to map their form deviation. The implementation of scanning speeds up the measurement procedures with no loss in accuracy compared to a point-to-point procedure and enables METAS to offer affordable traceable 3-D measurement services on small parts with uncertainties in the range of 50 nm within the whole CMM working volume.

16.12   CASE STUDY 12: MEASUREMENTS OF LARGE SILICON SPHERES USING A COORDINATE MEASURING MACHINE

John R. Stoup – National Institute of Standards and
Technology, Gaithersburg, Maryland

16.12.1   Introduction

The NIST M48 CMM was manufactured by Moore Special Tool in Bridgeport, Connecticut. The instrument is used by NIST to support U.S. industrial requirements for very-high-accuracy measurements of 1-D, 2-D, and 3-D artifacts. The M48 has been painstakingly developed and tested for over 20 years and now provides artifact measurements with uncertainty levels of less than 100 nm.

16.12.2 Coordinate Measuring Machine Equipment

The NIST M48 CMM (Figure 16.32) is currently operated as a three-axis machine that can incorporate a rotary fourth axis under specific circumstances. The motion of the x and y axes of the machine are accomplished using precision lead screws coupled with constant force springs to eliminate backlash in the motion. The table and carriage motions are restricted using twin V-ways and hundreds of precision cylindrical roller bearings. A kinematically supported granite surface plate was added to the machine table to assure rigid body behavior during table motion. The machine design takes advantage of the high moving mass of the machine and the mechanical averaging of the bearings to provide extremely smooth and repeatable motion. The z axis of the machine consists of a ceramic ram and constant force air bearings in a configuration that provides stable translation. The standard tactile probe is an induction-type probe constructed using a three-parallelogram design connected with thin flexures to allow for live sensing along each of the three axes simultaneously, and it can provide repeatability as low as 7 nm.

The machine is housed in an advanced thermally controlled laboratory that supplies 300 room air changes/hour to eliminate thermal gradients around the machine structure. The machine is vibration isolated and all secondary heat sources are removed from the room to maintain a thermal stability of 10 mK around the machine.

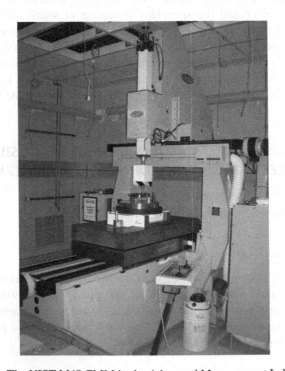

FIGURE 16.32 The NIST M48 CMM in the Advanced Measurement Laboratory, NIST.

16.12.3   Measurement Achievement

The machine was used to measure the average diameter of two precision, silicon spheres of nominal diameter near 93.6 mm. A measurement technique was devised that took advantage of the specific strengths of the machine and the artifacts while restricting the influences derived from the machine's few weaknesses. The standard uncertainty of these average diameter measurements was calculated as less than 20 nm.

16.12.4   Measurement Method

The geometry and surface characteristics of silicon spheres allowed for very accurate determination of the sphere average diameter. However, the flexibility built into the CMM becomes a distinct disadvantage when trying to achieve absolute accuracy at a level of several nanometers. The final measurement plan naturally separated into two distinct paths of focus: (1) Strengths of the machine, such as smooth mechanical motions, stable room environment, and repeatability characteristics, were exploited as much as possible. Historical measurement data has proved that the M48 performs very consistently and is very repeatable over long periods of time. The machine also behaves very predictably as a result of internal heating effects during these 12–24-hour measurement runs. (2) The influences of any unaddressed and multiaxial errors on the measurement results are minimized. The M48 always excelled during 1-D comparison-style operations where the flexibility of the machine is severely restricted. The silicon sphere measurements needed to approach this ideal as closely as possible.

Classical comparison measurement theory emphasizes that the most robust measurement design is a direct comparison of like artifacts of similar geometry and material. Since large, master silicon spheres were not available or practical, precision stainless steel spheres with exceptional geometry and surface finish characteristics were selected as master artifacts. Three master spheres were selected and then calibrated along specific, marked diameters using interferometric techniques. Measuring along a specific diameter for each master sphere removes most of the effects of sphere roundness from the overall measurement uncertainty.

The stainless steel master spheres were selected to be large so that they could be rigidly and kinematically mounted on the CMM work surface using existing magnetic mounting fixtures and still allow unobstructed access to the equators of the spheres. Using master spheres of three different sizes also facilitates some sampling of the high-frequency unmapped CMM positioning errors during the silicon sphere comparison process.

The selection of these master sizes also provided the benefit of reducing the required accuracy of subsequent temperature measurement. The ratio between average master sphere size and CTE closely matched the ratio between diameter and CTE of the silicon spheres, which resulted in nearly identical thermal expansion response. This fortunate situation dramatically reduced the importance of artifact temperature measurement and allowed the temperature sensors to be placed in more convenient locations around the artifact setup. The important thermal uncertainty component thus became a problem of thermal gradient control and was easily managed.

FIGURE 16.33 The silicon sphere measurement setup using the rotary table.

The key component to the success of these measurements was the ability to perform what amounted to a 3-D measurement while maintaining the restriction of only 1-D motion. Measuring along a single axis of both the machine and the probe minimized many complex error components including local machine and probe squareness, probe bidirectional errors, and sphericity error of stylus tip, as well as other smaller machine motion errors not corrected in the machine error map.

A novel technique was devised to facilitate the requirement of 1-D data collection, which is shown in Figure 16.33. The probe deflection sphere, three stainless steel master spheres, and silicon sphere were mounted on a precision rotary table interfaced with the CMM machine controller. The silicon sphere was mounted directly central to the rotation axis of the table. This assured that if the table were rotated to predetermined angular positions, the two-point diameter measurements of the silicon sphere were still performed directly through the central coordinates of the sphere. The mounting technique was designed so that the sphere could be removed and replaced without damage to the delicate surface. The three master spheres were positioned in different locations around the rotary table such that the calibrated diameters were located parallel along the machine's x axis.

16.12.5 MEASUREMENT PROCEDURE

The measurement procedure combined redundancy and repeatability into a thermal drift, eliminating data collection design. The three stainless steel master sphere locations were determined and a two-point diameter measurement was repeatedly collected for each along the CMM's x axis. The silicon sphere location was then determined and a two-point diameter was also repeatedly measured along the machine's x axis. The silicon sphere was then rotated at 15° intervals and subsequent two-point diameter measurements were repeatedly collected along the same

probe and machine's x axis at these new radial positions. Following each 15° interval silicon sphere diameter measurement, the rotary table was moved to the original
position and the three master sphere two-point diameters were again measured.
This process was performed through a 180° rotation of the silicon sphere then performed in the reverse direction to provide a thermal drift elimination technique to
the data set.

The silicon sphere data collected during any single measurement sequence effectively resulted in a 13-point average diameter around a single cross-sectional plane
of the sphere. The repetitive measurements of the master spheres provided a good
averaging of the master sphere values and brought any probe repeatability and drift
concerns along with any machine repositioning inaccuracy into the random fluctuations of the data set. The measurement process was repeated along many other cross-
sectional planes to sample the silicon sphere geometry. This required the sphere to
be manually rotated in its kinematic mount.

The final, very important advantage of the rotary table setup was for sampling
any uncorrected CMM positioning errors. Restricting the positioning of all measurement artifacts onto the rotary table surface allowed the easy repositioning of the
whole rotary table assembly on the CMM table. The rotary table was moved to nine
locations around the CMM work surface thereby sampling any position-dependent
systematic errors into the random fluctuations of the data set. The silicon sphere
was also rotated in the mount through the three orthogonal axes in addition to other
random positions and measured with the rotary table at a fixed position. These tests
isolated effects of silicon sphere geometry from machine positioning errors.

16.12.6 SUMMARY

Each master sphere was measured a total of 13 times along a specific calibrated
diameter, whereas the silicon sphere was measured once along 12 different diameters located on a plane through the center of the sphere. The rotary table positioning
makes it possible for all these measurements to be 1-D measurements performed
along the same machine axis. The average standard deviation for the three master
spheres gives a good indication of the repeatability of the two-point diameter measurement process. This value was about 15 nm.

The silicon sphere average diameter was calculated from among 150–200 individual two-point diameters. The standard deviation from these data was about
30 nm for each silicon sphere. This indicates that the sphericity of the artifacts was
very good and that the average results had a very low statistical uncertainty. The
silicon sphere 3-D form was also measured using an X-ray optics calibration interferometer developed at NIST to verify that the form results were not skewed due to
lobing effects on the spheres' surfaces. These data, combined with a very detailed
uncertainty statement, yielded the anticipated <20 nm final standard measurement
uncertainty for the process.

Performance of the NIST M48 CMM in the average diameter measurements of
two exceptionally fine 93.6-mm silicon spheres was impressive. The unprecedented
quality of the artifact's geometry and surface finish allowed for a rigorous analysis
of some very small and difficult-to-isolate measurement uncertainty sources in the

operation of the M48 CMM. This analysis combined with the restricted design of the measurement methods and the inherent accuracy of the M48 produced measurement results with very low measurement uncertainty.

ACKNOWLEDGMENTS

The author extends special thanks to all the contributors who took the time to write each of the excellent examples presented in this chapter about the applications of different CMSs.

17 Non-Cartesian Coordinate Measuring Systems

Guoxiong Zhang

CONTENTS

17.1 INTRODUCTION

Most coordinate measuring machines (CMMs) used in production and research are based on the Cartesian coordinate system. They have three mutually perpendicular guideways and scales to build a Cartesian coordinate system.* All the coordinates of the characteristic points of a workpiece are measured in this coordinate system. People are accustomed to the orthogonal coordinate system in their daily lives, as a result of which this system is easier to use. Most workpiece dimensions are given in Cartesian coordinate systems. Cartesian CMMs ensure relatively high accuracy in measurements. The displacement accuracy of an orthogonal CMM is mainly ensured by scales and the coordinate accuracy by guideways in combination with the scales. These are the main reasons why Cartesian CMMs have gained wide acceptance. When using cylindrical or spherical coordinate systems, the errors in angular measurements cause much larger uncertainties when the point measured is far from the rotational axis.

However, with the development of science and production technology, there are increasing demands to develop non-Cartesian coordinate measuring systems, mainly for the following reasons (Zhang et al. 1999):

- More and more large engineering projects require accurate measurements. Large components such as those in airplanes, large mining trucks or similar cannot be measured by traditional CMMs because of the difficulty and high cost of building such large CMMs and also because of the impossibility of putting these objects on a typical measuring machine (see also Chapter 16, Typical Applications). Therefore, it is required to build measuring systems, usually a non-Cartesian CMS, on the site where an object is mounted and used.
- A Cartesian CMM must have precise guideways and scales longer than the workpieces measured. This makes an orthogonal CMM costly and large.

---

* See also Chapter 4, Cartesian Coordinate Measuring Machines.

Low-cost and portable coordinate measuring devices are often preferred on shop floors.

- The long guideways make the movable components of CMMs large and heavy. With the increase in production tempo, there is a demand to develop lightweight, flexible coordinate measuring systems and non-Cartesian measuring systems that can better meet this requirement in many cases.

There are several approaches to realize measurements in non-Cartesian coordinate systems: First is using a set of articulated arms. The second approach is based on triangulation. The third is measuring a part in a spherical coordinate system. The fourth approach is based on measuring the distances of a characteristic point from several reference points. All these types of systems are discussed in this chapter, which is named "Non-Cartesian Coordinate Measuring Systems," since in many cases such systems are composed of several distributed components rather than solid machines.

Non-Cartesian coordinate measuring systems are still under development. Generally speaking, most non-Cartesian coordinate measuring systems have lower accuracy than Cartesian ones. Data processing algorithms are more complicated. Novel calibration techniques are also needed.

17.2 ARTICULATED ARM COORDINATE MEASURING MACHINES

The articulated arm coordinate measuring machine consists of several articulated arms equipped with angular encoders, used to measure the coordinates of an object in space by reading the rotated angles of the articulated arms from angular encoders.

17.2.1 WORKING PRINCIPLE

For the sake of simplicity, this discussion on AACMMs is started from the case of measurements on a plane by a single arm. A probe mounted on the end of an articulated arm, which rotates about an axis, can measure the points located on the circumference of a circle with radius equal to the length of the arm. An articulated arm CMM (AACMM) with two linkage arms is theoretically able to measure all the points located on a plane between two concentric circles with radii $l_1 + l_2$ and $|l_1 - l_2|$, as shown in Figure 17.1, where l_1 and l_2 are the lengths of the first and second linkage arms, respectively. Here, the word "theoretically" is used since this conclusion is true only when all the linkage arms are geometric lines and all the joints can be represented by geometric points. The actual measurable zone is smaller than that.

Optical encoders are built into the joints and these encoders measure the rotated angles θ_1 and θ_2 of linkage arms (1) and (2), respectively. Coordinates of the probing point $P(X, Y)$ can be fully determined by knowing $\theta_1, \theta_2, l_1,$ and l_2. For determining the functional relations between X, Y and $\theta_1, \theta_2, l_1, l_2$, two coordinate systems OXY and $O_1X_1Y_1$ with origins at their joints are set. Whereas OXY is fixed, $O_1X_1Y_1$ rotates with linkage arm (1). At the initial position, $O_1X_1Y_1$ is parallel to OXY. The following vector equation can be derived:

$$\mathbf{P} = \mathbf{R}_1^{-1}(\mathbf{L}_1 + \mathbf{R}_2^{-1}\mathbf{L}_2) \qquad (17.1)$$

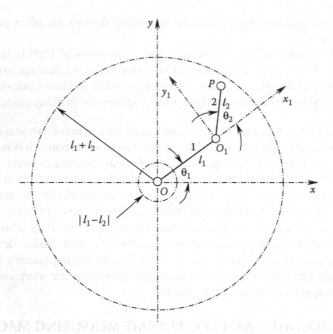

FIGURE 17.1 Articulated arms.

where $\mathbf{P} = (X, Y)^T$ is the position vectors of probing point $\mathbf{P}$ in $\mathbf{OXY}$ system; T indicates transposition of the vector; $\mathbf{R}_1 = \begin{pmatrix} \cos\theta_1 & \sin\theta_1 \\ -\sin\theta_1 & \cos\theta_1 \end{pmatrix}$ and $\mathbf{R}_2 = \begin{pmatrix} \cos\theta_2 & \sin\theta_2 \\ -\sin\theta_2 & \cos\theta_2 \end{pmatrix}$ are rotational matrices of the first and second linkage arms, respectively. Superscript -1 indicates the inversion of a matrix. $\mathbf{L}_1$ and $\mathbf{L}_2$ are the vectorial representations of articulated arms.

Equation 17.1 can also be written in the homogeneous form:

$$\begin{pmatrix} X \\ Y \\ 1 \end{pmatrix} = \begin{pmatrix} \cos\theta_1 & -\sin\theta_1 & 0 \\ \sin\theta_1 & \cos\theta_1 & 0 \\ 0 & 0 & 1 \end{pmatrix} \begin{pmatrix} \cos\theta_2 & -\sin\theta_2 & l_1\cos\theta_1' \\ \sin\theta_2 & \cos\theta_2 & l_1\sin\theta_1' \\ 0 & 0 & 1 \end{pmatrix} \begin{pmatrix} l_2\cos\theta_2' \\ l_2\sin\theta_2' \\ 1 \end{pmatrix}$$

$$= \begin{pmatrix} l_1\cos(\theta_1 + \theta_1') + l_2\cos(\theta_1 + \theta_2 + \theta_2') \\ l_1\sin(\theta_1 + \theta_1') + l_2\sin(\theta_1 + \theta_2 + \theta_2') \\ 1 \end{pmatrix}$$

(17.2)

Hence

$$X = l_1\cos(\theta_1 + \theta_1') + l_2\cos(\theta_1 + \theta_2 + \theta_2') \qquad (17.3)$$

$$Y = l_1\sin(\theta_1 + \theta_1') + l_2\sin(\theta_1 + \theta_2 + \theta_2') \qquad (17.4)$$

where θ_1' and θ_2' are angles formed by linkage arms (1) and (2) with axes $\mathbf{OX}$ and $\mathbf{O}_1\mathbf{X}_1$, respectively.

In order to measure the whole area within a certain range and to make AACMMs more flexible, usually no fewer than three linkage arms are used. For a CMM with n linkage arms, the position of probing point $\mathbf{P}$ can be determined as

$$\mathbf{P} = \mathbf{R}_1^{-1}(\mathbf{L}_1 + \mathbf{R}_2^{-1}(\mathbf{L}_2 + \ldots + \mathbf{R}_i^{-1}(\mathbf{L}_i + \ldots + \mathbf{R}_n^{-1}(\mathbf{L}_n))\ldots)) \tag{17.5}$$

or

$$
\begin{pmatrix} X \\ Y \\ 1 \end{pmatrix} =
\begin{pmatrix} \cos\theta_1 & -\sin\theta_1 & 0 \\ \sin\theta_1 & \cos\theta_1 & 0 \\ 0 & 0 & 1 \end{pmatrix}
\ldots
\begin{pmatrix} \cos\theta_i & -\sin\theta_i & l_{i-1}\cos\theta'_{i-1} \\ \sin\theta_i & \cos\theta_i & l_{i-1}\sin\theta'_{i-1} \\ 0 & 0 & 1 \end{pmatrix}
\ldots
$$
$$
\begin{pmatrix} \cos\theta_n & -\sin\theta_n & l_{n-1}\cos\theta'_{n-1} \\ \sin\theta_n & \cos\theta_n & l_{n-1}\sin\theta'_{n-1} \\ 0 & 0 & 1 \end{pmatrix}
\begin{pmatrix} l_n\cos\theta'_n \\ l_n\sin\theta'_n \\ 1 \end{pmatrix} \tag{17.6}
$$

$$X = \sum_{i=1}^{n} l_i \cos\left(\sum_{k=1}^{i} \theta_k + \theta'_i\right) \tag{17.7}$$

$$Y = \sum_{i=1}^{n} l_i \sin\left(\sum_{k=1}^{i} \theta_k + \theta'_i\right) \tag{17.8}$$

In three-dimensional (3D) measurements, Equation 17.5 is still effective; however, both the rotational matrices $\mathbf{R}_i$ and the vectors $\mathbf{L}_i$ ($i = 1$ to n) are much more complicated. In principle, each linkage arm can rotate about all the three coordinate axes. In Figure 17.2, a two-linkage-arm system rotating about two mutually perpendicular directions is shown. The angles φ_i and θ_i ($i = 1$ and 2) are angles of rotation of the ith

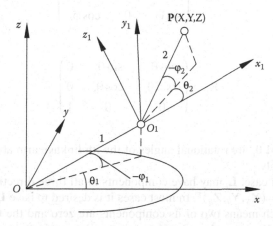

FIGURE 17.2 A two-linkage-arm system rotating about two axes.

linkage arm about the y and z axes, respectively. Angles φ_1 and φ_2 shown in Figure 17.2 are negative ones since positive φ_i goes down. The ith linkage arm has coordinates $\mathbf{L}_i = (l_i \quad 0 \quad 0)^T$ at its initial position and rotates first about its z axis and then about its y axis. In this case, rotational matrix $\mathbf{R}_i$ is defined as

$$
\begin{aligned}
\mathbf{R}_i &= \begin{pmatrix} \cos\theta_i & \sin\theta_i & 0 \\ -\sin\theta_i & \cos\theta_i & 0 \\ 0 & 0 & 1 \end{pmatrix} \begin{pmatrix} \cos\varphi_i & 0 & -\sin\varphi_i \\ 0 & 1 & 0 \\ \sin\varphi_i & 0 & \cos\varphi_i \end{pmatrix} \\
&= \begin{pmatrix} \cos\theta_i \cos\varphi_i & \sin\theta_i & -\cos\theta_i \sin\varphi_i \\ -\sin\theta_i \cos\varphi_i & \cos\theta_i & \sin\theta_i \sin\varphi_i \\ \sin\varphi_i & 0 & \cos\varphi_i \end{pmatrix}
\end{aligned}
\tag{17.9}
$$

The rotational matrices $\mathbf{R}_i$ depend not only on the angles of rotation about the three axes but also on the order of rotation. Although in principle it is possible to build a gimbal joint, which can rotate about all the three axes, it is difficult to build encoders that can measure angles about all three or even two axes accurately. In most cases, each linkage arm rotates about only one axis. The rotational matrices of linkage arm $\mathbf{L}_i$ about x, y, and z axes can be expressed as

$$
\mathbf{R}_{ix} = \begin{pmatrix} 1 & 0 & 0 \\ 0 & \cos\psi_i & \sin\psi_i \\ 0 & -\sin\psi_i & \cos\psi_i \end{pmatrix}
\tag{17.10}
$$

$$
\mathbf{R}_{iy} = \begin{pmatrix} \cos\varphi_i & 0 & -\sin\varphi_i \\ 0 & 1 & 0 \\ \sin\varphi_i & 0 & \cos\varphi_i \end{pmatrix}
\tag{17.11}
$$

$$
\mathbf{R}_{iz} = \begin{pmatrix} \cos\theta_i & \sin\theta_i & 0 \\ -\sin\theta_i & \cos\theta_i & 0 \\ 0 & 0 & 1 \end{pmatrix}
\tag{17.12}
$$

where ψ_i, φ_i, and θ_i are rotational angles of the ith linkage arm about the x, y, and z axes, respectively.

In the general case, $\mathbf{L}_i$ may have components in all three directions and may be expressed as $\mathbf{L}_i = (X_{li}, Y_{li}, Z_{li})^T$. In most cases it is desired to have $\mathbf{L}_i$ parallel to one of the axes, which means two of its components are zero and the third one equals the arm length. However, due to construction and manufacturing limitations and alignment errors all the components may be nonzero. Two examples are shown in Figure 17.3.

After substituting all the corresponding L_i and R_i into Equation 17.5, the coordinates of probing point $P(X, Y, Z)$ can be fully determined.

17.2.2 CONSTRUCTION REALIZATION

An AACMM employs a series of rotating components around generally perpendicular axes. A typical construction realization of an AACMM is shown in Figure 17.4. The unit is portable and can be mounted at the desired place through its base ((4) in the figure). A hard, touch-trigger, or scanning probe ((5) in the figure) is mounted

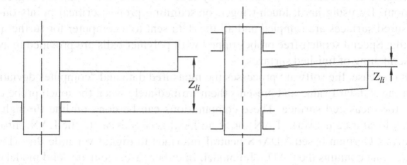

FIGURE 17.3 Offsets caused by construction consideration (left) and manufacturing errors (right).

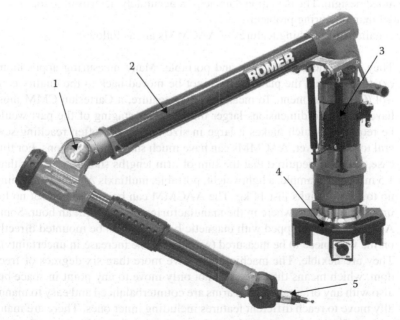

FIGURE 17.4 An example of articulated arm coordinate measuring machine. (Courtesy of Hexagon Metrology, Inc.)

at the end of the last arm. The AACMM should have the desired number of joints to ensure the required flexibility and accessibility to different features. Usually, the maximum number of joints used is seven. High-precision optical encoders are built into all joints (1) to measure the rotation of one arm with reference to another. Due to the difficulty in mounting the optical encoders measuring the rotated angles about several mutually perpendicular axes, most joints offer only rotation about one axis. All the linkage arms (2) are made of lightweight materials such as a carbon fiber composite or an aluminum alloy. They are counterbalanced inside or outside the joints as shown by (3) in the figure to make the arm easy to operate. Thermal elements are fixed inside rods (2) to measure their temperatures and compensate for thermal errors. The rotation of each linkage arm about a single axis provides one degree of freedom. As a result, the system may have more than six degrees of freedom. By using hard, touch-trigger, or scanning probes, critical points on the measured surfaces are sampled and all the data sent to a computer for further processing. Special scratch-free probes tipped with polymer balls are provided to avoid the scratching of finished surfaces.

In one case, the software processes the measured data and "computes deviations from the standard value and presents them immediately when the touch probe contacts the measured surface. These computations can be done on-line through the plant's local area network (LAN) or, if no local access exists to the LAN, through the arm's German-based VDAFS neutral interface to digital versatile disc (DVD) storage that contains the CAD solid model. In either case, requisite solid model data are always available at any measuring location" (http://www.faro.com/content.aspx? ct=&content=news&item=151, accessed on April 02, 2011; CAD stands for computer-aided design). The deviation data help in accurately determining the source of potential manufacturing problems.

The main distinguishing features of AACMMs are as follows:

- They are lightweight, small, and portable. Many measuring applications existing right on the plant floor cannot be moved back to the quality control (QC) department. To measure a 3-m feature, a Cartesian CMM must have travel and dimensions larger than 3 m (or staging of the part would be required), which makes it large in size and heavy, often reaching several tons. However, AACMMs can have much smaller dimensions. For this case, it is only required that the sum of arm lengths should be larger than 1.5 m. As an example, a lightweight, portable, multiaxis AACMM reaching up to 3.7 m weighs just 14 kg. The AACMM can be moved and set up for immediate use anywhere in the manufacturing plant within an hour. Some AACMMs are equipped with magnetic bases and can be mounted directly on the workpiece to be measured (with a possible increase in uncertainty).
- They are flexible. The machine may have more than six degrees of freedom, which means the probe can not only move to any point in space but also with any orientation. The arms are counterbalanced and easy to manually move to reach different features including inner ones. There are many such applications, particularly on the production line, without easy access using conventional CMMs. These problems can be solved by AACMMs.

- They are of low cost. Long high-grade guideways and scales are essential for Cartesian CMMs. These elements as well as the massive structure of the machines make them very expensive, and they often cost several hundred thousand dollars. AACMMs have only some small precision joints and encoders. The typical cost of an AACMM having several meters of measuring range is only about U.S. $20,000 to $50,000.

A major problem of an AACMM is that, generally speaking, its measuring accuracy is much lower than that of a conventional Cartesian CMM. The measuring accuracy of an AACMM is mainly governed by the accuracies of its encoders. The encoders built into the joints have to be small. Small encoders are subject to larger errors than are large encoders. For converting angle measurement error to coordinate measurement error the former is multiplied by the distance between the measured point and the joint, and this leads to a large error in measurement when the measured point is far from the joint. Most AACMMs have measurement uncertainties on the order of several tens of micrometers, although some smaller units have achieved an accuracy of ±5 μm within a measuring range of 600 mm according to manufacturer's literature.

The second problem is that AACMMs should be well calibrated since it is difficult to ensure accuracy of both the length of the linkage arm and its initial angle with reference to the coordinate axis during manufacturing. Careful calibration enhances the accuracy of the AACMM significantly. Calibration is carried out by measuring certain accurately calibrated artifacts or by comparison with more accurate instruments, and it is the duty of the manufacturer.

The third is that the AACMM is difficult to subject to computer control because of the following reasons: The algorithms for computer control are much more complicated than Cartesian ones, there are infinite solutions for inverse kinematics, and it is much more complicated to solve the collision avoidance problem with computer control than with Cartesian machines. For these reasons, AACMMs are used as low-cost manual machines and their accuracies as well as output are highly dependent on the skills of the operator.

17.2.3 Accuracy Analysis

The main error sources of AACMMs are discussed in Sections 17.2.3.1 through 17.2.3.7.

17.2.3.1 Errors of the Angular Encoders

Small-size angular encoders are subject to large errors. As shown by Equation 17.5, the position of a measured point is, first of all, determined by the rotational matrices $\mathbf{R}_i$. Errors of angular encoders change the rotational matrices shown in Equations 17.10 through 17.12 to

$$\mathbf{R}_{ix} = \begin{pmatrix} 1 & 0 & 0 \\ 0 & \cos(\psi_i + \Delta\psi_i) & \sin(\psi_i + \Delta\psi_i) \\ 0 & -\sin(\psi_i + \Delta\psi_i) & \cos(\psi_i + \Delta\psi_i) \end{pmatrix} \qquad (17.13)$$

$$\mathbf{R}_{iy} = \begin{pmatrix} \cos(\varphi_i + \Delta\varphi_i) & 0 & -\sin(\varphi_i + \Delta\varphi_i) \\ 0 & 1 & 0 \\ \sin(\varphi_i + \Delta\varphi_i) & 0 & \cos(\varphi_i + \Delta\varphi_i) \end{pmatrix} \qquad (17.14)$$

$$\mathbf{R}_{iz} = \begin{pmatrix} \cos(\theta_i + \Delta\theta_i) & \sin(\theta_i + \Delta\theta_i) & 0 \\ -\sin(\theta_i + \Delta\theta_i) & \cos(\theta_i + \Delta\theta_i) & 0 \\ 0 & 0 & 1 \end{pmatrix} \qquad (17.15)$$

where $\Delta\theta_i$, $\Delta\varphi_i$, and $\Delta\psi_i$ are errors of the angular encoders measuring θ, φ, and ψ, respectively.

17.2.3.2 Squareness Errors

In an ideal machine, the rotational axes should be mutually perpendicular. When they are not mutually perpendicular, some errors in measurement occur. Suppose the first rotational axis is the z axis and the second linkage arm rotates about the x axis, then the squareness error between these axes, α_{xz}, causes an additional rotation, α_{xz}, about the y axis, as shown in Figure 17.5.

The rotational matrix of the second arm $\mathbf{R}_{2x}$, which has both squareness error and error of the encoder, can be expressed as

$$\mathbf{R}_{2x} = \begin{pmatrix} 1 & 0 & -\alpha_{xz} \\ 0 & \cos(\psi_i + \Delta\psi_i) & \sin(\psi_i + \Delta\psi_i) \\ \alpha_{xz} & -\sin(\psi_i + \Delta\psi_i) & \cos(\psi_i + \Delta\psi_i) \end{pmatrix} \qquad (17.16)$$

17.2.3.3 Error Motions of the Articulated Arm

Not only errors of angular encoders and squareness error but also angular error motions of the articulated arm cause the rotational matrices $\mathbf{R}_i$ to change. For an arm that nominally rotates about the z axis due to error of encoder $\Delta\theta_i$ and angular error motions about the x and y axes, $\varepsilon_x(\theta)$ and $\varepsilon_y(\theta)$, rotational matrices $\mathbf{R}_{iz}$ take

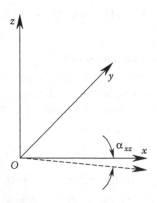

FIGURE 17.5 Squareness error.

the following form (one must specify whether the errors are measured in the fixed or the rotating sensitive direction since the matrices would be different for each case):

$$\mathbf{R}_{iz} = \begin{pmatrix} \cos\theta_i & \sin\theta_i & 0 \\ -\sin\theta_i & \cos\theta_i & 0 \\ 0 & 0 & 1 \end{pmatrix} \begin{pmatrix} 1 & \Delta\theta_i & -\varepsilon_y(\theta) + \alpha_{xz} \\ -\Delta\theta_i & 1 & \varepsilon_x(\theta) + \alpha_{yz} \\ \varepsilon_y(\theta) - \alpha_{xz} & -\varepsilon_x(\theta) - \alpha_{yz} & 1 \end{pmatrix} \quad (17.17)$$

Linear error motions of the linkage arm also cause error in measurement. However, they have no influence on rotational matrices. They change the form of $\mathbf{L}_i$. For a linkage arm with error motions $\delta_{ix}(\theta)$, $\delta_{iy}(\theta)$, and $\delta_{iz}(\theta)$ in x, y, and z directions, respectively, the effective linkage arm takes the form

$$\mathbf{L}_i = (\mathbf{X}_{li} + \delta_x(\theta) + \Delta\mathbf{X}_{ti}, \mathbf{Y}_{li} + \delta_y(\theta) + \Delta\mathbf{Y}_{ti}, \mathbf{Z}_{li} + \delta_z(\theta) + \Delta\mathbf{Z}_{ti})^{\mathrm{T}} \quad (17.18)$$

where $\Delta\mathbf{X}_{ti}$, $\Delta\mathbf{Y}_{ti}$, $\Delta\mathbf{Z}_{ti}$ are changes of arm lengths caused by temperature variations, which are explained in 17.2.3.4.

17.2.3.4 Thermal Errors

The AACMMs are operated manually. The operator holds one of the linkage arms (usually the one closest to the probe) and moves it by hand. Both heat from the operator and environmental sources cause the temperatures of rods (linkage arms) to change. The expansion of a rod depends on both the temperature variation and the thermal coefficient of expansion of the rod, and this equals

$$\Delta u_{ti} = k_i u_{li}(t_i - 20) \quad (17.19)$$

where u_{li} is the offset of the ith rod in u direction, u can be x, y, or z directions, and k_i and t_i are the thermal coefficient of expansion and temperatures of the ith rod, respectively.

Due to thermal errors, the effective linkage arm $\mathbf{L}_i$ changes to

$$\mathbf{L}_i = (\mathbf{X}_{li} + \Delta\mathbf{X}_{ti}, \mathbf{Y}_{li} + \Delta\mathbf{Y}_{ti}, \mathbf{Z}_{li} + \Delta\mathbf{Z}_{ti})^{\mathrm{T}} \quad (17.20)$$

For reducing thermal errors, the following measures should be taken: A thermal isolation layer should be put outside the rod where the operator handles it, environmental temperature should be controlled well, the rods should be made of materials having low thermal coefficients of expansion, and thermal elements should be fixed inside the rod, as described for Figure 17.4, to introduce thermal error compensation for accurate AACMMs.

17.2.3.5 Elastic Deformations of Rods

Elastic deformations can be caused by dead weights of rods, manipulating forces from the operator, dynamic forces caused by accelerations, and measuring force. For reducing the deformation caused by the dead weights of rods, they should be made of materials with high Young's modulus-to-specific weight ratio, such as carbon fiber composites, and in the form of tubes, which have high rigidity-to-weight ratio. For reducing the manipulating forces, all the joints should have low friction and all the arms should be counterbalanced. For reducing the dynamic forces, the arms should be moved gently. Generally speaking, measuring force should not be a big problem in measurement. However, an AACMM usually has much lower rigidity than the conventional Cartesian CMM and the measuring force may produce significant torque on the first linkage arm.

In general, the first linkage arms, that is, the arms closest to the stand or the arms on the shoulder, suffer the largest deformations since all the dead weights and dynamic forces of successive linkage arms act on the first linkage arms and the (force acting) lever arms are quite long. From another side, arms in the elbow have much lower stiffness than arms in the shoulder. The bending deformations occurring on these arms may also be significant. The bending deformations of all the preceding arms cause the successive arms to rotate, additional angular error motions are introduced into rotational matrices of the latter, and the deformations also cause the successive arms to move linearly. Their effects are the same as the linear error motions of the linkage arm itself. The bending effect on the bent arm itself is quite complicated.

17.2.3.6   Probing Error

Probing error has a one-to-one relation with the final measurement error (refer to Chapter 6, Probing Systems for Coordinate Measuring Machines).

17.2.3.7   Calibration Errors

All parameters of $\mathbf{L}_i$ and error motions are obtained from calibration. Calibration errors introduce additional errors in the expressions of $\mathbf{L}_i$ and rotational matrices.

To ensure that articulated CMMs perform with high accuracy, attention should be paid to ensuring accuracies of angular encoders and joints and to thermal and force deformations. The machine should be carefully calibrated to determine $\mathbf{L}_i$ and linear and angular offsets accurately. The first linkage arms (the arms closest to the stand) contribute the most in the total error budget. Special attention should be paid to these arms.

17.2.4   Performance Evaluation of Articulated Arm Coordinate Measuring Machines

The performance of AACMMs should be evaluated in accordance with the American national standard B.89.4.22-2004 "Methods for Performance Evaluation of Articulated Arm Coordinate Measuring Machines" (ASME 2004). The American standard will be briefly summarized here.

This standard provides a glossary of terms and classifications for machines. It also provides a specification form, which includes environmental parameters with acceptable ranges to be specified by the AACMM supplier. Since these machines are often used in very different environments, it also allows the environmental specifications to be deferred.

The standard also requires that the supplier provide detailed mounting requirements as inadequate mounting is a common source of error in these devices. Tests for ensuring the rigidity of the mount are delineated. Provided as well are recommendations regarding mounting orientation, as vertical mounting (with the base on the floor, or rigid mount) is very different from horizontal mounting.

The performance part of this specification consists of three main parts: (1) effective diameter test, (2) single-point articulation performance test, and (3) volumetric performance test. The purpose of each test is described here. For detailed procedures and position recommendations, the reader is referred to the standard.

The tests are to be performed in an environment meeting the requirements outlined by the equipment manufacturer. Due to their portability, AACMMs are often used in less-than-ideal environments. The specific conditions must be considered, and if they are worse than what are required for the validity of equipment performance then the specifications need to be derated accordingly. Due to the nature of AACMMs, the skill of the operator plays an important role in determining their achievable accuracy. Less experienced users usually cannot achieve the best possible accuracy initially. The performance tests can be used as a metric for users in training.

17.2.4.1 Effective Diameter Performance Test (B89.4.22, Section 5.2)

In this test, the diameter of a specified master sphere is measured three times in a position near the midpoint of the arm's extension. The sampling strategy is specified. The average value of the diameter and its standard deviation are reported. Changes in the articulation of the arm for the three measurements are minimized.

17.2.4.2 Single-Point Articulation Performance Test (B89.4.22, Section 5.3)

This test delineates variations in position measurement by examining the range of values measured by the arm at a single point with different articulations. Three points are chosen: (1) one close to the arm, (2) one near maximum arm extension, and (3) a third somewhere in the middle of the arm's range. At each point, a target (like a trihedral seat for a hard probe or a ball for a touch probe) is selected. The location of the target is measured 10 times at each of the three positions with different arm articulations. For each point, the maximum deviation from the average position and a quantity related to twice the standard deviation of the differences are reported as

$$2s_{\text{SPAT}} = 2\sqrt{\frac{\sum \delta_i^2}{(n-1)}} \tag{17.21}$$

17.2.4.3 Volumetric Performance Test (B89.4.22, Section 5.4)

The volumetric performance test is conducted by measuring the length of properly fixtured ball bars of short and long lengths in 20 different locations in the arm's work zone. Short and long lengths are specified in terms of the arm's radial lengths and positions, and orientations of the ball bar are specified. The ball bar lengths are corrected for temperature. Measurements are repeated at each position, and a procedure for handling outliers is given. Deviations from the calibrated ball bar lengths are reported plus a quantity related to twice the standard deviation.

The standard also provides a section on test equipment and 10 appendices ranging from a user's guide to explanation of the statistics utilized.

17.3 TRIANGULATION SYSTEMS

In triangulation systems, the position of a point is determined by measuring its direction angles from two or more reference points. Although it originated from surveying, triangulation finds wide applications in large-scale metrology along with shop-floor measurements.

17.3.1 THEODOLITE SYSTEMS

Theodolite systems, consisting two or more theodolites, are used for large-scale on-site measurements.

17.3.1.1 Working Principle

The working principle of a theodolite system is shown in Figure 17.6 (Allan 1988). Theodolites are basically sighting devices and require operators to aim them at the target point. The target can be a reticle or a small circle on a plate kinematically supported on the surface measured or a laser spot. The plate with reticle or circle is

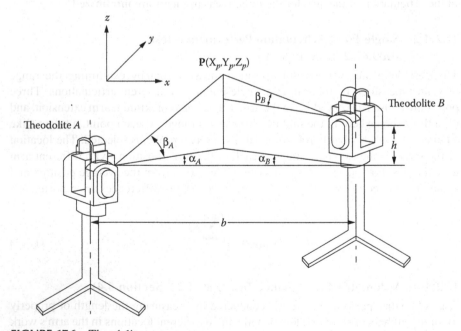

FIGURE 17.6   Theodolite system.

inconvenient to use not only because the plate has to be moved manually but also because the distance between the center of the reticle or circle to the surface measured is not a constant. It depends on the curvature of the surface being measured. It is much more convenient to use an auxiliary laser source, which projects a spot on and then scans the surface to be measured. Two angles, horizontal angle α and zenith angle β, are produced from each theodolite head. The coordinates of point **P** are determined by

$$X_p = b \frac{\cos \alpha_A \sin \alpha_B}{\sin(\alpha_A + \alpha_B)} \tag{17.22}$$

$$Y_p = b \frac{\sin \alpha_A \sin \alpha_B}{\sin(\alpha_A + \alpha_B)} \tag{17.23}$$

$$Z_p = b \frac{\sin \alpha_B \tan \beta_A}{\sin(\alpha_A + \alpha_B)} = h + b \frac{\sin \alpha_A \tan \beta_B}{\sin(\alpha_A + \alpha_B)} \tag{17.24}$$

where b is the length of the baseline, that is, the horizontal distance between the optical centers of two theodolite heads; h is their height offset; subscripts A and B indicate the readings from theodolite heads A and B, respectively. Both b and h are calibrated against a reference bar mounted at different places with different orientations or by other methods.

Theodolite systems possess many distinguishing features. They can be used for measuring large engineering objects up to several hundred meters. The only restriction is the beam reach. They are noncontact and portable. They can be moved to the site of the engineering object. A coordinate system is set on-site. In modern theodolites, charge-coupled device (CCD) cameras are incorporated for real-time image acquisition. The motions of theodolite heads are motorized and automatic target recognition is incorporated into the telescope. The theodolite heads are able to track the target automatically. The operator sets certain threshold values for the deviation of the aiming points of theodolites from a target spot. The automatic control system is intended to minimize this deviation (Wendt and Zumbrunn 1993). Data are collected only if the deviation is lower than this threshold. The theodolites are also able to send the sampled data wirelessly or through cable to a host computer. A team from the National Institute of Standards and Technology (NIST) developed this technique with computer assistance to measure spherical liquefied natural gas (LNG) tanks and won an Industrial Research magazine award (IR100) for one of the best inventions of 1980 (Estler et al. 2002). Three theodolites were used to provide redundancy of measurement.

The measuring system consisting of two theodolites has limited visible area, as shown schematically in Figure 17.7. Since both theodolites A and B have limited view angles, only the area covered by both their beams is visible. The actual measurable area is smaller than this value, since some features are hidden by the front surface of the object and become invisible. Only the area between C and D in the figure is actually measurable. For expanding the measurable area, up to eight theodolites are used in some systems.

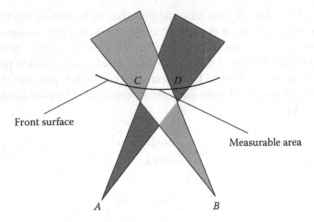

FIGURE 17.7   Measurable area.

17.3.1.2 Accuracy Analysis

Since there are many factors affecting the accuracy of a theodolite system, generally speaking, it has lower accuracy than Cartesian CMMs. It is very difficult to get accuracies better than $(1.0 \text{ to } 1.5) \times 10^{-5}$. The main factors affecting the accuracy of a theodolite system are as follows:

Angle measurement uncertainty: To determine the coordinates of a point, four angles α_A, β_A, α_B, and β_B are measured. Although with the development of the angle measurement technique it is possible to obtain subarcsecond resolution in individual angle measurement, the total measurement uncertainty contributed by angle measurement is limited by refraction and may reach $(0.5 \text{ to } 1.0) \times 10^{-5}$ under the right circumstances.

Squareness errors and offset errors among rotational axes and the optical axis: Equations 17.22 through 17.24 are derived under the assumption that all three axes, collimating axis, vertical axis, and armature axis, are mutually perpendicular (Figure 17.8). For example, the squareness error between collimating and vertical axes causes errors in measuring angles on a horizontal plane. The squareness error between armature and vertical axes introduces an additional rotational matrix in the coordinate transformation. These errors should be carefully checked and adjusted during the assembly of theodolite heads. The practice of reversing the theodolites (face left–face right), resighting, and reporting the average allows assessment and some compensation for collimation error and squareness error (height of standards). Besides, all three axes should intersect at a point. The offsets of these axes introduce additional errors. These errors should be checked during the assembly and alignment of the theodolites by combined tests.

Mounting accuracy of theodolites: Not only should all three axes of a theodolite be mutually perpendicular but also should the corresponding axes of two theodolites be parallel. Theodolites are moved to the site before measurement. For convenience of motion, there can be wheels underneath the

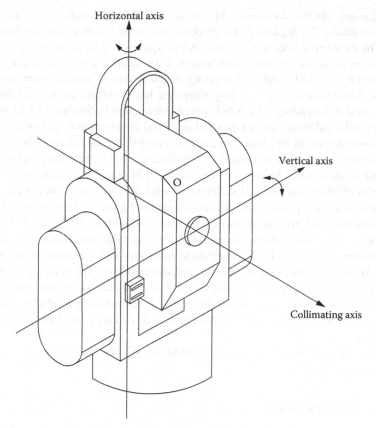

FIGURE 17.8 Three axes of the theodolite head.

theodolites to be moved. During the measurement, the theodolites should be mounted stably on kinematic supports and be well aligned with respect to gravity. The curvature of Earth may be ignored when the spacing between two theodolites is less than 10 m. Otherwise, the correction for Earth's curvature should be introduced to make the vertical axes of two theodolites parallel. In some precise electronic theodolites, corrections can be made automatically when inclinations are less than several minutes.

Aiming errors: Features are measured point by point. A small round spot projected by a laser on the surface measured is often used as the target, and it gives good accuracy in comparison with many other forms of targets. The aiming accuracy also depends on the telescope system. A good telescope system offers 2 arcseconds of collimation accuracy. Another important detail is the mutual alignment of the telescope systems of two theodolites. For this purpose, the reticle of one telescope is used as the target of the other telescope and vice versa.

Calibration errors: As shown in Equations 17.22 through 17.24, all the coordinates X_p, Y_p, and Z_p are proportional to the length of the baseline, b. The measuring accuracy of a theodolite system to a large degree depends on the accuracy of baseline calibration. If the accuracy of the baseline calibration is 1×10^{-5}, then the accuracy of triangulation measurement cannot be better than 1×10^{-5}. A long reference bar is subject to thermal deformation and bending. The NIST system mentioned in Section 17.3.1.1 used specially calibrated invar tapes and achieved an uncertainty in baselines of about one part in 10^6. Tapes are easier to carry than long reference bars. The aiming error would contribute to significant error when a short reference bar is used and a large-scale measurement is conducted. Reference bars with different lengths and different mounts are often used to conduct an overall accuracy test of the theodolite system. Figure 17.9 shows one of the recommended arrangements of bars and theodolites (Zhang et al. 1999). Figure 17.10 shows the arrangement of reference bars and reticle plates.

Environmental errors: Both temperature variations and vibrations may cause relative positions of the theodolites and dimensions of the object to change. Tripods are particularly sensitive to gradients caused by the operator's body heat. It should be noted that vertical temperature gradients, which are common, cause bending of the light, which creates a first-order angular error.

Special attention should be paid to calibration, aiming, and alignment during measurement.

17.3.2 PHOTOGRAMMETRY

A photogrammetric system is a kind of noncontact measuring system based on triangulation using multiple cameras or the same camera in multiple positions (Mikhail, Bethel, and McGlone 2001). It is also called "stereovision" (Minoru and Akira 1986).

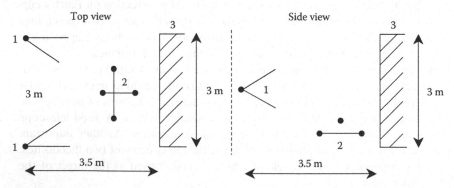

FIGURE 17.9 Overall performance test of a theodolite system; it shows arrangement of the bars (2), theodolites (1), and the measuring area (3).

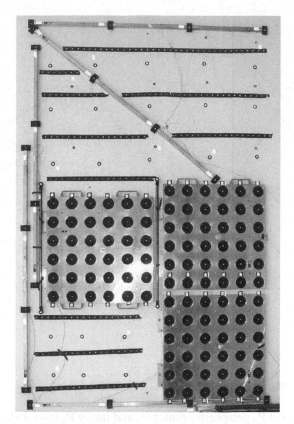

FIGURE 17.10 Overall performance test of a theodolite system—arrangement of the reference bars and plates.

17.3.2.1 Working Principle

In a photogrammetric system the positions of characteristic points of the measured object are determined by merging the images formed by several cameras shooting from different angles. For ultraprecision work, film and plate cameras were used earlier, but CCD cameras are now becoming the norm. It is possible to take pictures using several cameras and merge their images to extract the contours of certain features. However, in most cases a laser spot or spots are projected on the measured surface to form a characteristic point or points. In Figure 17.11, O_j is the perspective center of the jth camera and $\mathbf{P}_{ij}$ is the image of point $\mathbf{P}_i$ on the image plane of the jth camera. To determine the position of point $\mathbf{P}_i$ in space, its images should appear in at least two cameras and the angle θ_{jk} formed by the lines connecting point $\mathbf{P}_i$ and perspective centers O_j and O_k of the two cameras should not be too small. Otherwise, the position of point $\mathbf{P}_i$ in space cannot be determined accurately.

Usually the collinearity model (Tsai 1987), which is also called the "pinhole model," is used for deriving the functional relationship between position of point $\mathbf{P}_i$ and position of its image $\mathbf{P}_{ij}$ on the image plane of the jth camera. In the collinearity

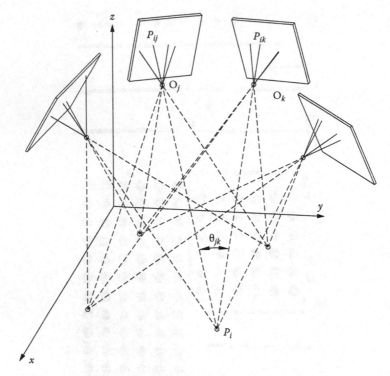

FIGURE 17.11　Working principle of a photogrammetric system.

model, object point $\mathbf{P}_i$, perspective center O_j, and image $\mathbf{P}_{ij}$ are collinear. The following equations can be derived:

$$\begin{bmatrix} X_c \\ Y_c \\ Z_c \end{bmatrix} = \mathbf{R} \cdot \begin{bmatrix} X_w \\ Y_w \\ Z_w \end{bmatrix} + \mathbf{T} \tag{17.25}$$

or

$$X_u = mX_c = f\frac{X_c}{Z_c} \tag{17.26}$$

$$Y_u = mY_c = f\frac{Y_c}{Z_c} \tag{17.27}$$

where

$$\mathbf{R} = \begin{pmatrix} r_1 & r_2 & r_3 \\ r_4 & r_5 & r_6 \\ r_7 & r_8 & r_9 \end{pmatrix}$$
rotational matrix of the camera, which depends on the mounting posture of the camera;

$$\mathbf{T} = \begin{bmatrix} T_x \\ T_y \\ T_z \end{bmatrix} \text{—translation matrix of the camera, which depends on the position}$$

of the camera

X_w, Y_w, Z_w—coordinates of the measured point in the world (workpiece) coordinate system

X_c, Y_c, Z_c—coordinates of the measured point in the camera coordinate system

X_u, Y_u—coordinates of the undistorted image on the image plane

m—magnification factor of the image

f—camera constant

The lenses of each camera also have imperfections. In most cases, the main distortion of the image is radial distortion (spherical aberration). In this case, the coordinates of the distorted image X_d, Y_d are related to X_u, Y_u by the following equations:

$$X_{di} = X_{ui}(1 + kq_i^2) \tag{17.28}$$

$$Y_{di} = Y_{ui}(1 + kq_i^2) \tag{17.29}$$

where k is the distortion coefficient and $q_i^2 = X_{di}^2 + Y_{di}^2$.

After expanding Equation 17.25, the following linear equations are obtained:

$$X_c = r_1 X_w + r_2 Y_w + r_3 Z_w + T_x \tag{17.30}$$

$$Y_c = r_4 X_w + r_5 Y_w + r_6 Z_w + T_y \tag{17.31}$$

$$Z_c = r_7 X_w + r_8 Y_w + r_9 Z_w + T_z \tag{17.32}$$

and

$$X_{di} = X_{ui}(1 + kq_i^2) = (1 + kq_i^2)f \frac{r_1 X_{wi} + r_2 Y_{wi} + r_3 Z_{wi} + T_x}{r_7 X_{wi} + r_8 Y_{wi} + r_9 Z_{wi} + T_z} \tag{17.33}$$

$$Y_{di} = Y_{ui}(1 + kq_i^2) = (1 + kq_i^2)f \frac{r_4 X_{wi} + r_5 Y_{wi} + r_6 Z_{wi} + T_y}{r_7 X_{wi} + r_8 Y_{wi} + r_9 Z_{wi} + T_z} \tag{17.34}$$

All the unknown parameters of the camera in Equations 17.25 through 17.34, such as f, k, r_h ($h = 1$ to 9), T_x, T_y, T_z can be determined through camera calibration, except the magnification factor m, which requires an external scale in the image. Equations 17.33 and 17.34 also show that there are three unknowns, X_{wi}, Y_{wi}, and Z_{wi} but only two equations. It is impossible to determine the position of the measured point from its image in a fixed camera. It can also be seen from Figure 17.11 that all points located on the line

P_iO_j have the same image at P_{ij}. To determine the position of point P_i, it is required to obtain at least two images taken by two cameras or a camera at two different positions.

In the second case, the relative positions of the camera before and after movement should be known. In fact, the camera obtaining images at two different positions is equivalent to two cameras. The measured point is located at the intersection of two lines passing through its image points in two cameras and corresponding perspective centers. The angle between these two lines should not be too small. Otherwise, a small error in image position might cause a significant error in target position measurement.

Photogrammetric systems are used for shop-floor measurement, although efficiency is a problem. Besides, there is often a large number of points on the workpieces, especially for sculptured surfaces, that need to be measured. To enhance the efficiency of measurement, a light stripe or other pattern is often projected on the surface, as shown in Figure 17.12. In the figure, a system with three cameras applicable for measuring high-reflective surfaces is shown (Zheng 2003). A polarizer is mounted in front of the laser and three analyzers are mounted in front of the cameras. The light stripe projected on the surface is linearly polarized. The light reflected from a specular surface is also linearly polarized, whereas diffuse light forms a polarization ellipsoid. The polarization directions of the analyzers are perpendicular to that of the polarizer. The specular light is eliminated to a large degree and only the diffuse light reaches the three CCD cameras.

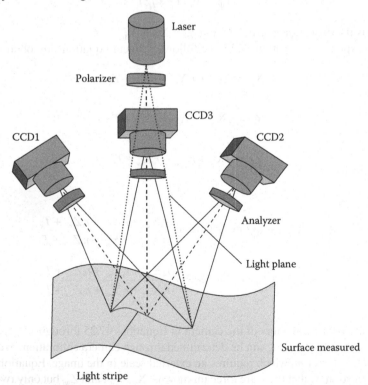

FIGURE 17.12 Photogrammetric system with three charge-coupled device (CCD) cameras and a light stripe.

17.3.2.2 Camera Calibration

Camera calibration includes the calibration of camera's intrinsic parameters, such as camera constant f; distortion coefficient k; coordinates of the image center C_x, C_y; and extrinsic parameters $\mathbf{T}$ and $\mathbf{R}$, that is, their position and orientation in the world coordinate system. A photogrammetric system is used as a portable device for measuring the dimensional features of an object. After the cameras are mounted on the site, they should be carefully calibrated using a calibrated artifact. A popular artifact is a grid plate, but for large objects targets on tapes are used. The grid plate can be calibrated with an accuracy of better than 1 μm and mounted at different heights, as shown in Figure 17.13.

Suppose the first calibration is carried out while the grid plate is located at position F_2 and $Z_w = 0$ is set at this position. The images of n points $\mathbf{P}_i(X_i, Y_i, Z_i)$ are captured on the image plane. The direct readings are the pixel numbers of these image points (X_{fi}, Y_{fi}). Their coordinates in the image coordinate system in millimeters (X_{di}, Y_{di}) can be written as

$$X_{di} = (X_{fi} - C_x)d'_x s_x^{-1} \tag{17.35}$$

$$Y_{di} = (Y_{fi} - C_y)d_y \tag{17.36}$$

where $d_x = d_x N_{cx}/N_{fx}$; d_x and d_y are spacings of the CCD element in x and y directions, respectively; N_{cx} is the total number of pixels in x direction; N_{fx} is the total number of sampled points in x direction; d'_x is the interval between sampled points in x direction; C_x and C_y are expressed in pixels; and s_x is the uncertainty factor caused by the variation in video scanning frequency.

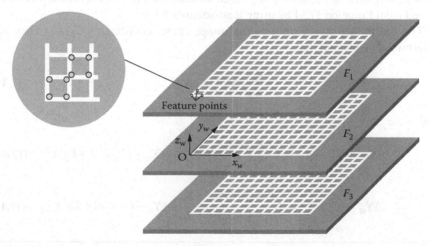

FIGURE 17.13 Grid plate for camera calibration.

In the first step, $s_x = 1$, $k = 0$, and $C_x = C_y = 0$ are taken. From Equations 17.33 and 17.34 the following vector equation is obtained*:

$$\left(X_{wi}Y_{di} \quad Y_{wi}Y_{di} \quad Y_{di} \quad -X_{wi}X_{di} \quad -Y_{wi}X_{di} \right) \begin{pmatrix} \dfrac{r_1}{T_y} \\[4pt] \dfrac{r_2}{T_y} \\[4pt] \dfrac{T_x}{T_y} \\[4pt] \dfrac{r_4}{T_y} \\[4pt] \dfrac{r_5}{T_y} \end{pmatrix} = X_{di} \qquad (17.37)$$

In the equation, X_{wi} and Y_{wi} are known, and X_{di} and Y_{di} are obtained from Equations 17.35 and 17.36 based on measured X_{fi} and Y_{fi}. By measuring $n \geq 5$ points, the unknown values r_1/T_y, r_2/T_y, T_x/T_y, r_4/T_y, and r_5/T_y are determined by a least-squares algorithm. There are nine components in matrix $\mathbf{R}$, but only three independent ones. There are six orthogonality constraints for r_h ($h = 1$ to 9). By adding these orthogonality constraints, all r_h and T_x, T_y can be determined.

In the second step, the following equation is derived from Equations 17.31, 17.34, and 17.36:

$$\left(Y_{ci} \quad -(Y_{fi} - C_y)d_y \right) \begin{pmatrix} f \\ T_z \end{pmatrix} = (Y_{fi} - C_y)(r_7 X_{wi} + r_8 Y_{wi})d_y \qquad (17.38)$$

By using the data obtained in the first step and taking $C_y = 0$, f and T_z are determined.

The third step is determining the distortion coefficient k. For this purpose, image points distributed in a relatively large area should be used. The coefficient k is determined from Equation 17.34 by using least-squares fitting.

The fourth step is determining the image center coordinates C_x, C_y. They are determined by minimizing

$$\sum_{i=1}^{n} (\Delta X_{ui}^2 + \Delta Y_{ui}^2) \qquad (17.39)$$

and

$$\Delta X_{ui} = X'_{ui} - X_{ui} = f \frac{r_1 X_{wi} + r_2 Y_{wi} + T_x}{r_7 X_{wi} + r_8 Y_{wi} + T_z} - (X_{fi} - C_x)d'_x (1 + kq_i^2)^{-1} \qquad (17.40)$$

$$\Delta Y_{ui} = Y'_{ui} - Y_{ui} = f \frac{r_4 X_{wi} + r_5 Y_{wi} + T_y}{r_7 X_{wi} + r_8 Y_{wi} + T_z} - (Y_{fi} - C_y)d_y (1 + kq_i^2)^{-1} \qquad (17.41)$$

---

* A row vector times a column vector is a dot product.

The final step is determining the uncertainty factor s_x. In the first four steps, it is assumed that $s_x = 1$ and that $X_{di} = (X_{fi} - C_x)d'_x$. Actually these assumptions are not correct: $s_x \neq 1$ and $(X_{fi} - C_x)d'_x = X_{di}s_x = X'_{di}$. For determining s_x, the artifact is moved to positions F_1 and F_3, as shown in Figure 17.13. Since in these positions $Z_w \neq 0$, Equation 17.37 changes to

$$\left(X_{wi}Y_{di} \quad Y_{wi}Y_{di} \quad Z_{wi}Y_{di} \quad Y_{di} \quad -X_{wi}X'_{di} \quad -Y_{wi}X'_{di} \quad -Z_{wi}X'_{di} \right) \begin{pmatrix} s_x r_1 T_y^{-1} \\ s_x r_2 T_y^{-1} \\ s_x r_3 T_y^{-1} \\ s_x T_x T_y^{-1} \\ r_4 T_y^{-1} \\ r_5 T_y^{-1} \\ r_6 T_y^{-1} \end{pmatrix} = X'_{di} \quad (17.42)$$

All parameters except s_x in Equation 17.42 are known, and s_x is the least-squares solution.

17.3.2.3 Image Matching

The position of a feature point in space cannot be determined just by its image in one camera. Image matching is an important step for determining the position of a feature point in space. If there is only one spot on the measured surface, the image matching is simple. However, a light stripe, as shown in Figure 17.12, is often used for enhancing measurement efficiency. In this case and when measuring the contours directly, image matching is critical.

To improve the accuracy and reliability of image matching, both geometric constraints and gray-level similarity constraints are used. Geometric constraints include the following three constraints: (1) Uniqueness: Each point has a unique image point on the image plane. (2) Projection constraint: Each point on the image plane corresponds to different points in space. However, they all lie on the line connecting the perspective center of camera and the image point. (3) Light stripe constraint: The point in space must lie on the light stripe, which generates the image.

In Figure 17.14, a three-camera system is shown. In this figure, O_{ci} ($i = 1$ to 3) is the perspective center of camera i. Its image plane is F_i. The image of light stripe on image plane F_i is l_i. The projection of point $\mathbf{P}$ on F_i is $\mathbf{P}_i$, which is the intersection point of PO_{ci} with F_i. The images $\mathbf{P}_1$, $\mathbf{P}_2$, and $\mathbf{P}_3$ are called *homologous* images. The plane formed by $PO_{ci}O_{cj}$ is called the *epipolar* plane S_{ij}. Intersection of the epipolar plane S_{ij} with plane F_i forms an epipolar line L_{ij}. The lines L_{ij} and L_{ji} are called *conjugation epipolar* lines. The intersection point of the optical centerline $O_{ci}O_{cj}$ with image plane F_i is E_{ij}, and it is called the *epipole*.

The homologous image $\mathbf{P}_2$ of image $\mathbf{P}_1$ on plane F_2 must be on the epipolar line L_{21}. From another side, it should be on the image of the light stripe l_2. So it must be located at the intersection point $\mathbf{P}_2$ of L_{21} and l_2, as shown in Figure 17.14. The homologous image $\mathbf{P}_3$ of image $\mathbf{P}_1$ on plane F_3 must be on the epipolar line L_{31}. From

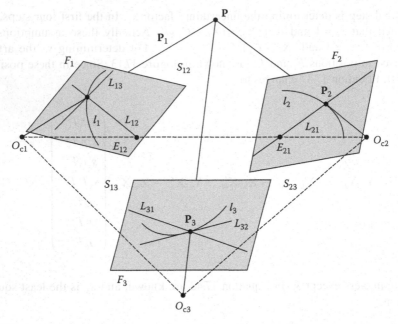

FIGURE 17.14 Image matching.

another side, it should be on the image of light stripe l_3. So it must be located at the intersection point $\mathbf{P}_3$ of L_{31} and l_3. In the meantime, it should also be located on the epipolar line L_{32}. Since all the images are subject to certain errors for improving the accuracy and reliability of image matching, the gray-level similarity constraint is applied simultaneously. The gray-level similarity of points $\mathbf{P}_1$, $\mathbf{P}_2$, and $\mathbf{P}_3$ is evaluated according to the following equations. First the average $\overline{f_i(u_i, v_i)}$ and the standard deviation $\sigma(f_i)$ of the gray-level $f_i(u_i, v_i)$ in a neighborhood area of point $\mathbf{P}_i$ with $(2N+1) \times (2M+1)$ pixels are calculated.

$$\overline{f_i(u_i, v_i)} = \frac{\sum\limits_{n=-N}^{N} \sum\limits_{m=-M}^{M} f_i(u_{i+n}, v_{i+m})}{(2N+1)(2M+1)} \tag{17.43}$$

$$\sigma(f_i) = \sqrt{\frac{\sum\limits_{n=-N}^{N} \sum\limits_{m=-M}^{M} f_i^2(u_{i+n}, v_{i+m})}{4NM} - \overline{f_i^2(u_i, v_i)}} \tag{17.44}$$

where u and v are two mutually perpendicular directions, which can be x and y directions of the pixels or the directions tangential and normal to the image.

After this, the correlation between the gray levels of points $\mathbf{P}_i$ and $\mathbf{P}_j$, $\text{Cor}(\mathbf{P}_i, \mathbf{P}_j)$, and the similarity of three points are calculated.

$$\text{Cor}(\mathbf{P}_i, \mathbf{P}_j) = \frac{\sum_{n=-N}^{N} \sum_{m=-M}^{M} [f_i(u_{i+n}, v_{i+m}) - \overline{f_i(u_i, v_i)}] \times [f_j(u_{j+n}, v_{j+m}) - \overline{f_j(u_j, v_j)}]}{4NM\sqrt{\sigma^2(f_i) \times \sigma^2(f_j)}} \quad (17.45)$$

$$\text{Similarity } (\mathbf{P}_1, \mathbf{P}_2, \mathbf{P}_3) = \frac{1}{3}\text{Cor}(\mathbf{P}_1, \mathbf{P}_2) + \frac{1}{3}\text{Cor}(\mathbf{P}_1, \mathbf{P}_3) + \frac{1}{3}\text{Cor}(\mathbf{P}_2, \mathbf{P}_3) \quad (17.46)$$

The homologous images are searched in a defined neighborhood of points $\mathbf{P}_2$ and $\mathbf{P}_3$. The points with maximum gray-level similarity are selected as the homologous images.

Based on any two of the homologous images among $\mathbf{P}_1$, $\mathbf{P}_2$, and $\mathbf{P}_3$, a point $\mathbf{P}_{ij}$ in space can be determined. The center of gravity of the triangle formed by $\mathbf{P}_{12}$, $\mathbf{P}_{13}$, and $\mathbf{P}_{23}$ is taken as the position of point $\mathbf{P}$.

17.3.2.4 Accuracy Analysis

The main factors influencing the accuracy of photogrammetric systems are as follows:

Defects of optical systems: Lenses can have various types of aberrations. Aberrations cause ambiguity and distortions in images. The distortion coefficient k takes only part of the distortions into account.

Uncertainties caused by image processing: Image processing is a big problem in all types of vision measurements. The gray level of an image changes gradually. There is a transition zone between the image and the background. To determine the boundary between the image and the background, different types of algorithms for selecting a threshold have been developed. The gray level of the image at a point depends not only on the position of the point but also on other factors such as illumination, shadow, background, and neighboring gray levels. The threshold should be adaptive to these factors and the result of the measurement should be close to that obtained from contact measurement. The measuring uncertainty caused by an obscure image is often the largest contributor to the total error budget.

Subdivision errors: Usually, the pixel size of a CCD element is several tens of micrometers. To obtain higher resolutions, different types of subdivision algorithms are applied. Subdivision also contributes certain error to the total budget.

Quality of the target: A laser spot or laser stripe is often used as the target. The diameter of the spot should be small, and the width of the stripe should be narrow and constant when scanning. They must have distinct boundaries.

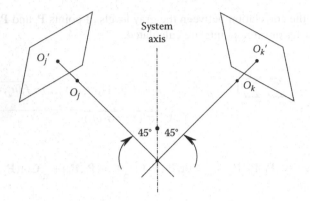

FIGURE 17.15 Optimal arrangement of a photogrammetric system.

The quality of the target has significant influence on the image processing uncertainty. Both the type and intensity of illumination have significant effects on images. The boundary of an image may change with illumination. The most important detail is keeping the illumination stable.

Arrangement of the cameras: All the aforementioned discussions deal with the measuring uncertainty of an individual camera. However, the final measuring uncertainty to a large degree depends on the arrangement of cameras. It has been proved theoretically that the best result is obtained when all the optical axes of cameras, that is, the lines connecting the perspective centers O_j, O_k and the corresponding image centers O_j', O_k', form 45°angles with the system axis, as shown in Figure 17.15 (Zhang 2002). The accuracy of the system also depends on the number of cameras used. Theoretically, two cameras are enough to determine the positions of points in space. However, there might be some obstacles, and not all points are seen by two cameras. With the increase in camera number, both reliability and accuracy are enhanced. However, generally, three or four cameras are enough. Systems with two cameras are also often used. Further, increase in camera number makes the system and algorithm for image matching complicated.

Calibration errors: Most system parameters are obtained from calibration. Calibration is essential for ensuring high accuracy. Calibration errors affect the accuracy of measurements.

Thermal errors and stability of camera mounting: Most measurements are subject to thermal errors. Once the system is mounted on the site and calibrated, camera positions should be kept unchanged. All changes in camera positions would introduce errors in measurement.

Photogrammetry is a good means for noncontact measurement. Its typical measuring uncertainty, when the measuring is performed properly, is on the order of a few parts in 10^5.

17.3.3 Light Pen Coordinate Measuring Machine

A light pen CMM works in touch mode. Several CCD cameras capture the images of the light sources or reticles on the light pen and determine the 3D coordinates of the point at which the light pen touches the workpiece.

17.3.3.1 Working Principle

The light pen CMM shown in Figure 17.16 mainly consists of a specially designed light pen, one or more high-resolution CCD cameras, and a laptop computer (Liu et al. 2005). On the light pen, there are at least three point-shaped light sources (light-emitting diodes [LEDs]) or reticles and a touch-trigger probe with a spherical tip. In case three light sources are used, these light sources and the probe tip must be aligned in one line with known positions. Distances AB, BC, and CD are accurately calibrated. During measurement, a touch-trigger probe or a hard probe contacts the object surface to be measured and the images of the light sources on the light pen are captured by the CCD camera. The images are processed by the computer and the 3D coordinates of the tip center can be calculated. In this way, coordinates of all points are sampled in turn. The geometric features such as the dimensions, form, and position errors of the object are evaluated by the software package based on the obtained 3D coordinates.

The position of the measured point is determined in the camera coordinate system. Origin O of the camera coordinate system is located at the perspective center of the CCD camera. The z axis is along the optical axis passing through the perspective center O and the center of the image plane O'. The x and y axes are parallel to the horizontal and vertical pixels of the CCD element, respectively, as shown in Figure 17.17. The camera and light pen should be calibrated in advance and all their intrinsic parameters known.

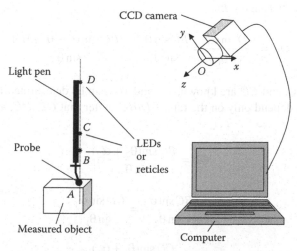

FIGURE 17.16 Light pen coordinate measuring machine.

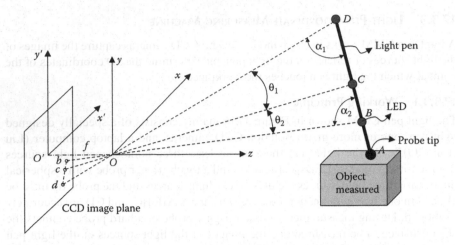

FIGURE 17.17 Model of light pen coordinate measuring machine.

Coordinates of the image points b, c, and d on image plane (X_b, Y_b), (X_c, Y_c), (X_d, Y_d) are obtained from measurement. Their coordinates in the camera system are easy to obtain by subtracting f as their Z coordinate where f is the camera constant.

Angles θ_1 and θ_2 are determined from

$$\theta_1 = \arccos \frac{\mathbf{Oc} \cdot \mathbf{Od}}{|\mathbf{Oc}| \cdot |\mathbf{Od}|} = \arccos \frac{X_c X_d + Y_c Y_d + f^2}{\sqrt{X_c^2 + Y_c^2 + f^2} \sqrt{X_d^2 + Y_d^2 + f^2}} \quad (17.47)$$

$$\theta_2 = \arccos \frac{\mathbf{Ob} \cdot \mathbf{Oc}}{|\mathbf{Ob}| \cdot |\mathbf{Oc}|} = \arccos \frac{X_b X_c + Y_b Y_c + f^2}{\sqrt{X_b^2 + Y_b^2 + f^2} \sqrt{X_c^2 + Y_c^2 + f^2}} \quad (17.48)$$

From triangles OCD and OBC

$$OC = \frac{CD \sin \alpha_1}{\sin \theta_1} = \frac{BC \sin \alpha_2}{\sin \theta_2} = \frac{BC \sin(\alpha_1 + \theta_1 + \theta_2)}{\sin \theta_2} \quad (17.49)$$

Since θ_1, θ_2, BC, and CD are known, α_1 and α_2 can be determined from Equation 17.49 and they depend only on the ratio CD/BC. After that OB, OC, and OD can be determined as

$$OB = \frac{BC \sin(\theta_2 + \alpha_2)}{\sin \theta_2} \quad (17.50)$$

$$OC = \frac{BC \sin \alpha_2}{\sin \theta_2} = \frac{CD \sin \alpha_1}{\sin \theta_1} \quad (17.51)$$

$$OD = \frac{CD \sin(\theta_1 + \alpha_1)}{\sin \theta_1} \quad (17.52)$$

Coordinates of points B, C, and D in the camera system are obtained from OB, OC, and OD by multiplying the corresponding direction cosines.

$$\begin{pmatrix} X_P \\ Y_P \\ Z_P \end{pmatrix} = OP \begin{pmatrix} X_p/\sqrt{X_p^2 + Y_p^2 + f^2} \\ Y_p/\sqrt{X_p^2 + Y_p^2 + f^2} \\ f/\sqrt{X_p^2 + Y_p^2 + f^2} \end{pmatrix} \tag{17.53}$$

where P can be B, C, or D, and p is b, c, or d correspondingly.

The coordinate of the probe tip center A is determined as

$$\begin{pmatrix} X_A \\ Y_A \\ Z_A \end{pmatrix} = \frac{1}{3} \left[\begin{pmatrix} X_B + X_C + X_D \\ Y_B + Y_C + Y_D \\ Z_B + Z_C + Z_D \end{pmatrix} + \begin{pmatrix} X_C - X_D & X_B - X_D & X_B - X_C \\ Y_C - Y_D & Y_B - Y_D & Y_B - Y_C \\ Z_C - Z_D & Z_B - Z_D & Z_B - Z_C \end{pmatrix} \begin{pmatrix} AB/CD \\ AC/BD \\ AD/BC \end{pmatrix} \right] \tag{17.54}$$

The light pen CMM is small, lightweight, and of low cost. A touch-trigger probe can be used to measure some features that are difficult to access by cameras and optical probes. The problems related to the ambiguity of the optical image formed by the object features are also avoided. However, it loses the advantages of noncontact measurement. Unlike other triangulation systems where the target is sensed from two or more reference points, in the light pen CMM three or more target points collinear with the measured point are sensed from a reference point, although multiple reference points are also applicable for enhancing the machine's reliability and accuracy. Commercial systems exist with both one and two cameras.

17.3.3.2 Accuracy Analysis
The main error sources of light pen CMMs are as follows:

Errors related to image extraction and image processing: Since light pen CMMs are based on image extraction and image processing as are photogrammetric systems, all the errors caused by optical systems and image processing, illumination, and subdivision affect the accuracy of light pen CMMs in a similar way, as was stated in Section 17.3.2.4 for photogrammetric systems.

Quality of targets: The targets used in light pen CMMs can be LEDs or reticles. The LEDs give better contrast in images and do not require extra illumination. However, their images may not be round and they generate heat, which causes additional thermal deformation of the pen.

Collinearity error of the targets and tip center: The mathematical model shown in Figure 17.17 is based on the assumption that all the four points A, B, C, and D are lying on one line. An error in measurement is caused when any of these points depart from the common line.

Arrangement of the light pen: To get higher sensitivity in measurements, points B, C, and D in Figure 17.17 should not be close to each other. However, a long pen is difficult to carry and is subject to larger mechanical and thermal deformations, so a compromise must be made. In most cases, the pen is 200–300 mm long. To obtain the highest sensitivity, the pen should be held perpendicular to the z axis. The pen does not work when it is held parallel to the z axis.

Force and thermal deformations: Bending of the pen causes significant error in measurement. The pen should be made of materials with high stiffness-to-mass ratio and it should be held properly. Heat is generated by the operator, LEDs or other illumination, and environmental sources. The pen should be made of materials with low thermal expansion coefficients, and a thermal isolation layer should be put outside the pen so as to minimize thermal effects when the operator handles it.

Calibration errors: Both the camera and the pen should be calibrated well. Pen calibration is still a problem under study. Due to this reason, a light pen CMM offers only about 0.1 mm accuracy and can be used for some shop-floor measurements though U_{95} uncertainties are quoted at 0.025 mm + $L/60,000$ for two-camera systems.* Errors related to image extraction, image processing, and calibration are the main contributors to the total error budget.

17.4 SPHERICAL COORDINATE MEASURING SYSTEMS

Triangulation systems are based on angle measurements. To convert the data obtained from angle measurements to linear data, the system must be calibrated by artifacts to determine its base length. Long calibration rods are difficult to make and use, and they are subject to bending. Initial systems required manual sighting, so they were slow. To overcome these difficulties, spherical coordinate measuring systems were developed (Lau, Hocken, and Haight 1986; Lau and Hocken 1984, 1987; Nakamura et al. 1991). In such systems, the distance L_i from target point $\mathbf{P}_i$ to selected base point O and two angles θ_i and φ_i rotated about the vertical and horizontal axes are measured, as shown in Figure 17.18. Depending on the means used for measuring the distance, spherical coordinate measuring systems are classified into laser trackers, laser radars, and phase difference measuring systems.

17.4.1 LASER TRACKER

A laser tracker consists of a laser interferometer, an angle measuring system, a tracking mechanism and a target. The tracking mechanism makes the laser beam track the target, and the laser interferometer measures the distance from the base point to the target, while the angle measuring system measures the direction of the laser beam. A spherical coordinate system is thereby formed.

---

* As per the manufacturer specification (http://metronor.ewat-cms.com/filesystem/2010/07/product_ brochure_ver_1.3_lr_732.pdf accessed on April 03, 2011).

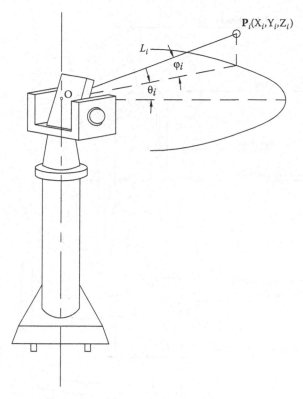

FIGURE 17.18 Spherical CMM.

17.4.1.1 Working Principle

Figure 17.19 shows a schematic construction of a laser tracker. The laser beam is emitted from the laser head through an aperture and it goes to and is reflected by the target. The interferometer inside the case measures the distance between the target and the selected base point. The whole laser interferometer can be rotated about the vertical and horizontal axes by an azimuth motor and an elevation motor, respectively, to aim the laser beam at the target. The rotated angles are measured by azimuth and elevation encoders. The laser interferometer is an incremental measuring device. A spherically mounted retroreflector (SMR) is used as the target, and at the beginning of measurement it is seated in a reference position (called a *bird bath*) and the reading of the interferometer is set to a precalibrated value. The whole tracker is aligned by leveling feet to make the axes of rotation horizontal and vertical.

The working principle of a laser tracking system is shown in Figure 17.20. Target mirror is the object measured. The target mirror moves along an object to be measured or it moves with a moving object such as a robot. The laser beam emitted from the laser head is split into two beams by the beam splitter. One beam passes through the beam splitter, is then reflected by the reference cube corner, and comes back to the beam splitter. Another beam goes to the tracking mirror, is then reflected by the target, and goes back to the tracking mirror. When the target has a motion in a direction perpendicular to incident beam, the reflected beam has a parallel shift and this shift

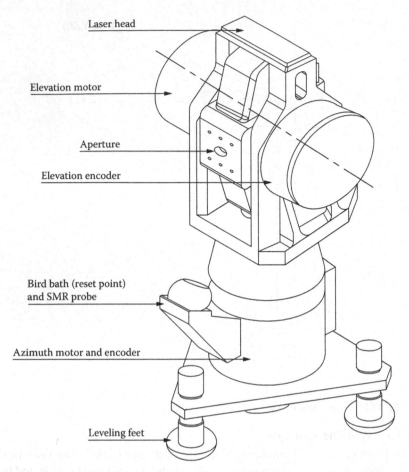

FIGURE 17.19 Laser tracker schematic.

is sensed by the lateral effect photodiode, which receives a part of the returned beam reflected by the beam splitter. It sends a signal to the servomechanism to rotate the tracking mirror until the beam reflected from the tracking mirror goes to the center of the target. Most of the returned beam passes through the beam splitter and goes back to the single beam interferometer. This beam interferes with the beam reflected by reference cube corner, and the fringe counter converts the interference signal to an electric signal and counts the number of interference fringes. In this way, the interferometer always tracks the motion of the target and measures the change of distance between the rotational center of the tracking mirror and the center of the target mirror.

17.4.1.2 Target Component

The target is an important component of a laser tracker. There are basically three types of targets: (1) cube corners, (2) cat's eyes, and (3) plane mirrors. The first two types of targets are used to measure the position of an object and the third one the orientation of an object. The three types of targets are discussed as follows.

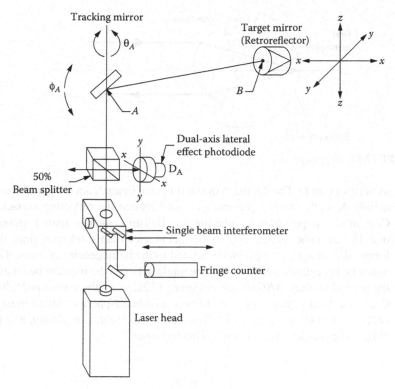

FIGURE 17.20 Working principle of the tracking laser.

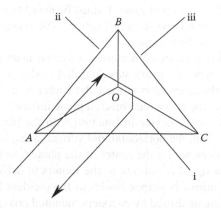

FIGURE 17.21 Cube corner.

1. Cube corner: The cube corner is a corner of a cube with three mutually perpendicular reflecting surfaces i, ii, and iii, as shown in Figure 17.21. Cube corners can be solid or hollow. A solid cube corner is made of a solid glass piece, whereas a hollow one is made of three reflecting plane mirrors glued together. The distinguishing feature of a perfect cube corner is that the reflected beam is always parallel to the incident one and symmetric about

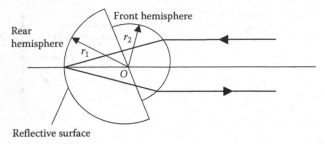

FIGURE 17.22   Cat's eye.

its vertex point O. The deviation of the reflected beam from the parallel one mainly depends on the squareness errors between the reflecting surfaces. Currently, it is possible to make the parallelism error less than 1 arcsecond. Hollow cube corners give higher accuracies than solid ones since the former do not suffer the problems related to the homogeneity of glass. The maximal acceptance angle, that is, the angle between the incident beam and the normal to plane ABC shown in Figure 17.21, of a cube corner is 25.56°.

2. Cat's eye: A cat's eye is composed of two glass hemispheres with a common center, as shown in Figure 17.22 (Takatsuji et al. 1999; Lin, Zhang, and Li 2002). The incident beam is fully reflected when

$$r_2 = \frac{r_1}{n-1} \qquad (17.55)$$

where r_1 and r_2 are radii of the front and rear hemispheres, respectively; and n is the index of refraction of glass. It must be noted that $r_1 = r_2$ when $n = 2$ and the cat's eye then becomes a full sphere. The acceptance angle can be as high as 70° from the normal.

3. Plane mirror: When a cube corner or a cat's eye is used as the reflector, the direction of the reflected beam stays parallel to the incident beam and is independent of the orientation of the cube corner or cat's eye. For sensing the orientation of the measured object, a plane mirror is used as the target. Figure 17.23 shows the device in function with the plane reflector, which can rotate about both the horizontal and vertical axes, is shown. When the reflected beam does not hit the center of the photodetector inside the laser tracking head, a signal is sent out to the motors to drive the plane mirror target until the mirror is perpendicular to the incident beam. The rotated angles φ and β are measured by encoders mounted coaxially with the driving motors, and they indicate the orientation of the measured object.

For measuring the moving object, the target is mounted on the object and moves with it. For measuring a sculptured surface, the target is put on the surface and it moves along the surface. As shown in Figure 17.20, the distance between the center of the target and the corresponding point on a surface is not a constant but depends on the curvature of the surface. For eliminating this error, a target with spherical mount, as shown in Figure 17.24, is used. In this figure, (2) is a cat's eye, which can

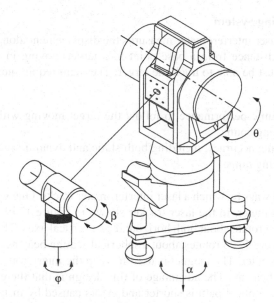

FIGURE 17.23 Target for measuring the orientation of an object.

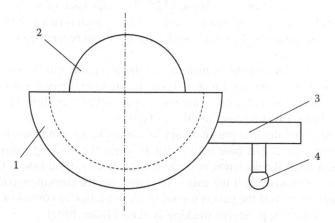

FIGURE 17.24 Target for measuring a sculptured surface.

be replaced by a cube corner; (1) is a wear-proof outer shell with good sphericity, which is fixed concentrically with the cat's eye or cube corner; (3) is a handle; and (4) is an auxiliary ball (only one is seen, but there are two).

In case of a cube corner that is "fixed concentrically," the center of the outer shell is at the vertex point of the cube corner. The outer shell can be kinematically seated on the bird bath, as shown in Figure 17.19. The target can be stably put on the sculptured surface, keeping the distance between the center of cat's eye or cube corner and the measured surface equal to the radius of the outer shell ((1) in Figure 17.24).

17.4.1.3 Tracking System

A conventional laser interferometer measures the displacement along a straight line. To measure the distance from a base point to a target moving in space, the laser interferometer must be able to track the target. The main requirements of the tracking system are

- Good dynamic performance to follow the target moving with high speed without losing count
- High tracking accuracy, including both static and dynamic accuracies
- Wide tracking range

There are several ways by which a laser interferometer tracks. One way is by rotating the laser interferometer. In the laser tracker shown in Figure 17.19, the whole laser interferometer can rotate about both horizontal and vertical axes. The interferometer unit shown in Figure 17.25 rotates about a spherical seating bearing, which assures a stable rotational center. The magnet is used to keep the interferometer unit in close contact with the bearing. The advantage of this design is that there is no reflecting mirror; hence, the optical path is shorter and errors caused by instability of mirror position are eliminated. Its weakness is that the mass of the movable part is relatively large, which causes some negative effects on the dynamic behavior of the system.

The method involves the rotation of one or two tracking mirrors. A design with two rotating mirrors is shown in Figure 17.26. The drawback of such a design is that the internal optical path within the device changes with rotation of the mirrors and no stable base point can be found. Such a design is not recommended for high-accuracy measurements.

In Figure 17.27, a tracking mechanism with a single rotating mirror is shown. The mirror is rotated by a horizontal motor. Then the horizontal motor with mirror is rotated by a vertical motor. Encoders are used for closed loop control. The movable part of vertical motor is still heavy in this mechanism.

It should be noted that the inherent gain of a tracking system changes with the distance between target and base point. That is, when the tracking mirror rotates through a fixed angle the deviation of the beam increases with distance. The tracking system becomes unstable if the gain is too large, but the tracking mirror cannot follow the target motion if the gain is too small. A self-adaptive control of gain was used on some of the first prototype tracking systems (Tullar 1992).

17.4.1.4 System Calibration

A spherical coordinate measuring system measures the vector radius and two angles of the target point. There is no need to determine the scale factor by using a reference bar as in the case of theodolites. However, the laser interferometer is an incremental instrument and the initial distance at zero reading of the interferometer must be calibrated.

The laser tracker can be calibrated by using a conventional laser interferometer, as shown in Figure 17.28; in the figure, (1) is the tracking mirror, (2) is the laser head of the laser tracker, and (5) is the laser head of the conventional laser interferometer.

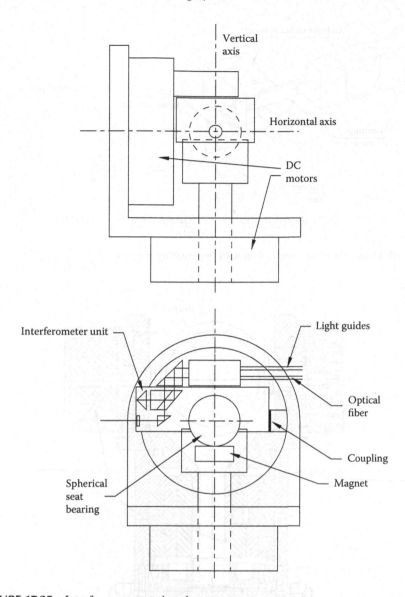

FIGURE 17.25 Interferometer rotating about two axes.

Both cube corner (4) of the conventional laser interferometer and target (6) of the laser tracker are fixed on the ram (3) of a CMM. At the initial position A_0, both the interferometers are set to zero. Then, the ram (3) moves in x direction. Readings from both conventional laser interferometer, b_i, and laser tracker, a_i, are recorded. The following equation can be written:

$$(L_0 + a_i)^2 = L_0^2 + b_i^2 + 2 \cdot L_0 \cdot b_i \cos\theta_0 \qquad (17.56)$$

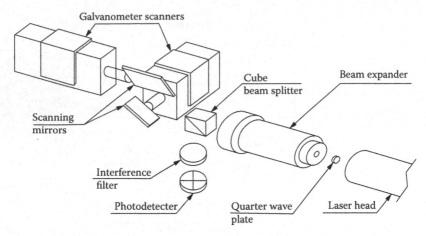

FIGURE 17.26 Tracking mechanism with two rotating mirrors.

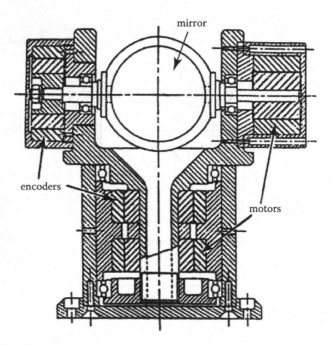

FIGURE 17.27 Tracking mechanism with a mirror rotating about two axes.

where $i = 1$ to n and n is the total number of points measured. The initial distance between the rotational center C of tracking mirror (1) and the initial point A_0 is L_0; θ_0 is the angle formed by CA_0 and the x-axis. It is possible to determine L_0 from

$$\sum \delta_i^2 = \sum [\sqrt{L_0^2 + b_i^2 + 2L_0 \cdot b_i \cdot \cos \theta_0} - (L_0 + a_i)]^2 = \min \qquad (17.57)$$

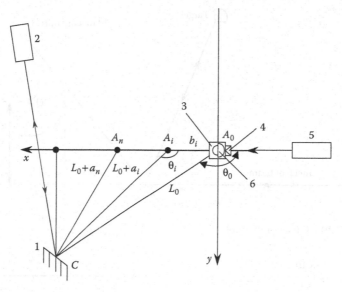

FIGURE 17.28 Laser tracker calibration.

After the target is moved to the bird bath, the reading of the tracking laser interferometer a_b is taken. The absolute distance between point C and bird bath equals $L_0 + a_b$.

17.4.1.5 Accuracy Analysis

The main factors influencing the accuracy of a laser tracker are as follows.

Errors of the laser interferometer: These include variations in refractive index caused by environmental factors.

Errors of the target: Imperfections of the target cause three types of errors. (1) There are many factors that cause changes in the optical path. These factors include flatness error of the reflecting surface of a cube corner, sphericity error, radius errors, eccentricity error, and glue thickness of a cat's eye. (2) The squareness errors of the reflecting surfaces of a cube corner and most imperfections of the cat's eye also cause parallelism errors between incident and reflected beams. In this case, the readings shown by the angle encoders will be different from the real direction of the optical beam, which cause errors in argument angle measurements. (3) The third type of error is the variation of the distance between target center and corresponding point on the measured surface. These errors may be caused by incorrect design of the target mounting mechanism and sphericity error of the outer shell, shown in Figure 17.24.

Tracking errors: Tracking errors may cause errors in both optical path and argument angle measurements. Error motions of the rotational center of the tracking mechanism cause direct errors in the optical path. The optical spot should coincide with the center of rotation. However, the reflecting surface may be located at a distance r from the center of rotation O and the incident beam may have an offset e, as shown in Figure 17.29. Computer simulation

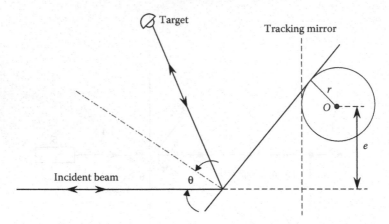

FIGURE 17.29   Offset of the spot.

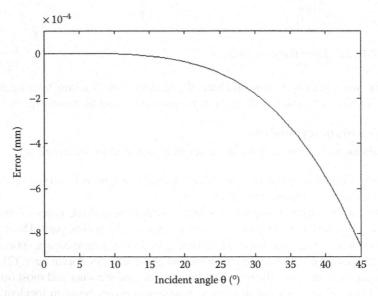

FIGURE 17.30   Error caused by offset of a reflecting mirror.

confirms that the offset e causes only a second-order error in optical path (Zhang et al. 2003). It is negligible when $e < 0.3$ mm. The error caused by r depends on the incident angle θ. The computer simulation result for the case of $r = 0.01$ mm is shown in Figure 17.30. This error is significant as it has direct influence on the overall result. All static and dynamic errors in tracking cause errors in argument angle measurements.

Error of the initial distance (dead path): A laser interferometer is an incremental instrument and the initial distance is obtained through calibration. For the method shown in Figure 17.28, the calibration error of L_0 depends on errors of the displacement measuring interferometer, straightness error motion of the ram, random error of the tracking interferometer, and the number of points sampled n. Nonsystematic errors of the tracking interferometer can be evaluated according to $\sqrt{\left(\sum \delta_i^2\right)/n}$, obtained from Equation 17.57. Computer simulation shows that when all the source errors are on the micrometer level, the calibration error of L_0 can be reduced to submicrometers when $n > 20$. The error of the initial distance is the sum of calibration error of L_0, random error of the tracking interferometer, and locating error of the target on the bird bath.

Errors of angle measuring devices: Generally speaking, errors in measuring the target position in tangential directions are much larger than errors in the radial direction. It is due not only to errors occurring from angle measuring devices but also to other factors including target errors and errors of the tracking mechanism, which influence the errors of angle measurements. When errors of angle measurements are converted to errors in tangential directions, they are multiplied by the distance between the target and the base point. These errors are directly proportional to distance. Also, the distance measured is subject to error. It further increases the total errors in tangential position measurements. The accuracy of current commercially available laser trackers is on the order of 1×10^{-5}. The maximal measuring range is several tens of meters and tracking speed reaches several meters per second.

17.4.2 Systems Based on Absolute Distance Measurement

A laser interferometer is directly traceable to the definition of meter (see also Chapter 2, The International Standard of Length) and gives high-accuracy measurements. However, it is an incremental instrument. The initial distance at its zero reading should be calibrated. Further, when the beam is interrupted or the target moves too fast, the laser interferometer loses its count and the target should be reset on the bird bath before the measurement restarts. Usually, large-scale measurements take a long time. The loss of count can cause many inconveniences in measurements. For solving this problem, systems based on absolute distance measurement were developed, which are now commercially available.

There are two basic types of systems: (1) time-of-flight measuring systems (Section 17.4.2.1) and (2) phase difference measuring systems (Section 17.4.2.2).

17.4.2.1 Time-of-Flight Measuring Systems

In time-of-flight measuring systems, the time needed for a wave to travel from a source to an object and back is measured. The wave can be electromagnetic, such as laser, or some other type of wave. When an electromagnetic wave is used, such a system is usually called *radar*; it is called *lidar* when a laser is used. The time T needed for the wave to travel forth and back depends on the travel speed v and the distance traveled L and

$$T = \frac{2L}{v} \tag{17.58}$$

Both the error in measuring time of flight T and the error of travel speed v cause error in measurement and

$$dL = (v\,dT + T\,dv)/2 \tag{17.59}$$

The speed of light in vacuum equals 299,792,458 m/s. When a laser is used, it is required to measure the time T with an accuracy better than 66.6 ps in order to obtain an accuracy of 1 cm even if there is no error in the speed. This requirement is very strict. An ultrashort pulse laser should be used and the time-measuring system should be very accurate. For this reason, laser radar is mainly used for long-distance range measurements, for example, several kilometers and more, where relatively large measuring uncertainty is allowed. The inaccuracy in the speed data is mainly caused by variations in refractive index of the medium in which the beam is traveling (usually air).

In principle, other waves, such as ultrasonic waves, can also be used for absolute distance measurement. However, ultrasonic waves cannot be used in spherical coordinate measuring systems. The spherical CMS measures the rotated angles of the tracker when the beam aims at the target. The ultrasonic wave has a very large view angle and it cannot provide accurate angle information.

The time-of-flight measuring approach may be used with or without a target. The intensity of the reflected pulse is much higher when using a target than that without a target, and hence higher accuracies can be achieved in the former case. However, there are many cases where targets are difficult to use. In these cases, the return pulses are reflected from the measured object.

17.4.2.2 Phase Difference Measuring System

In this approach, the phase difference between the reflected wave and the initial outgoing wave is measured. The outgoing wave is modulated. The phase difference is measured by comparing the zero crossing points of the outgoing and incoming modulated laser signals. This approach generally provides better resolution and accuracy. However, the phase difference is repeated when the reflected wave changes every period. The measuring range is less than half its wavelength. For expanding the measuring range, waves with two or more different frequencies are used. Two lasers with frequencies γ_1, γ_2 where $\gamma_1 > \gamma_2$ are used. The equivalent wavelength of their beat signal equals $c/(\gamma_1 - \gamma_2) = \lambda_1\lambda_2/(\lambda_2 - \lambda_1)$, where c is the speed of light, and λ_1 and λ_2 are wavelengths of the lasers with frequencies γ_1 and γ_2, respectively. By measuring the phase difference of at least one of the original signals and that of the beat signal, the distance information can be obtained.

The accuracy of measurement mainly depends on the accuracy of measuring the phase difference of the original signal. However, the resulting error in beat signal measurement must be smaller than half the wavelength of the original signal. This approach maintains the high resolution and accuracy of the original signal measurement and expands the measuring range to half the wavelength of the beat signal. For further expanding the measuring range, a third infrared laser can be used. The drawback of multifrequency systems is their complexity.

In the spherical CMS with an absolute distance measuring device, the argument angles are still measured by two optical encoders. The absolute distance measuring devices are often incorporated with theodolites to form spherical coordinate measuring systems. Such systems are called *electronic total stations*.

17.4.3 PERFORMANCE EVALUATION OF LASER-BASED SPHERICAL COORDINATE MEASUREMENT SYSTEMS

Equipment performance should be tested in an acceptable environment comprising acceptable levels of temperature, humidity, electrical power, and vibration. It shall be the responsibility of the user to provide an environment for conducting the performance test, which meets the specified requirements. In certain instances, the environment in which a user intends to use the instrument may not be suitable for bringing out its best performance. In such cases, the results, particularly accuracy, can be significantly worse than specifications. Users need to be aware of the factors that can limit the portability advantage of such systems.

The performance of laser-based spherical coordinate measurement systems, laser trackers, can be evaluated according to the American national standard B89.4.19-2006 (ASME 2006a). This standard specifies broadly which environment should be used for testing and also assigns responsibility for ensuring the environment to the user. A suggested data sheet contains spaces for ranges of temperature, pressure, and humidity, as well as other environmental parameters. Since trackers are sensitive to operator skill as well as numerous environmental factors, the user should plan carefully. It should be noted that vertical temperature gradients, which are common, cause bending of the laser beam, which creates a first-order angular error. Such gradients should be kept at a minimum during testing and instrument use.

Tests are described for two modes of operation: (1) the instrument working in the tracking mode (interferometer) and (2) the instrument working in the absolute distance meter (ADM) mode. Calibrated distances are measured in both modes following the manufacturer's recommended methods. The standard provides recommended positions (to exercise angular motions) and distances that were chosen with the known errors of these instruments in mind. Further, measurements are repeated three times to provide a measure of repeatability. Instruments like a tracker can be used in two positions, the normal position (front sight) and the reverse position (back sight), with the horizontal and vertical angles changed by 180°. This particular procedure is called *two-face measurement* and is to be performed multiple times as it highlights geometric errors of the instrument (primarily squareness [height of standards] and collimation errors). It is recommended that the user of a laser tracker carefully follow the procedures and setup as specified in the standard for evaluating its performance.

17.5 MULTILATERATION SYSTEMS

In a multilateration system several laser trackers are used and the position of the target is determined by the distances of the target from these trackers.

17.5.1 WORKING PRINCIPLE

In the section on single-beam laser trackers (17.4), it was mentioned that one of the most prevalent errors arises from beam bending, which occurs due to temperature

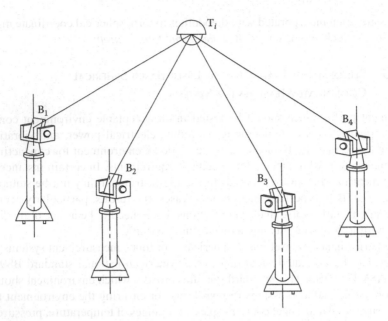

FIGURE 17.31 Multilateration working principle.

gradients in the air through which the beam passes. For the spherical coordinate system utilized by a tracker, this is a first-order error and one that is difficult to quantify and measure in the field. Efforts to get around this problem have led to the technique of multilateration, measuring the length vectors to the target point with beams from at least three trackers simultaneously. The original concept is probably due to Itek company (Greenleaf and Watson 1984; Lin 2002). Since its invention, multiple papers have been published on the subject (Zhang et al. 2003; Schwenke et al. 2005). In this technique, the trackers are used simultaneously or in series (over time) to measure the distance to a target from at least three base points (Hughes, Wilson, and Peggs 2000). In principle, it is impossible to calculate the position of the target if the positions of the base points only are known. Such systems are already commercially available, where the trackers are called "tracers" (www.etalon-ag.com) because they have no angular measuring capability. This method of tracking reduces the errors since beam-bending errors are second order on length measurements. It increases the errors due to movements of base points that are not necessarily wanted or needed. If enough points are taken it is possible to calculate, as in photogrammetry, the base points of the various trackers. Figure 17.31 shows four trackers. Although the detailed mathematics of this method is beyond the scope of this book, it can be found in the works of Takatsuji et al. (1998) and Zhang et al. (2003).

17.6   SUMMARY

Non-Cartesian coordinate measuring systems are a quickly developing area in coordinate measuring techniques. They have wide application prospects in large-scale coordinate measurements and in-situ measurement due to their flexibility and relatively small size and lightweight in comparison with Cartesian CMMs. There are many different approaches to realize coordinate measurements in non-Cartesian coordinate systems. Most of them are based on angular measurements or angular measurements in conjunction with linear measurements. In most cases the coordinates of the measured points cannot be read directly as in Cartesian CMMs. Calibration is essential for non-Cartesian coordinate measuring systems and many of them have measurement accuracy lower than Cartesian ones. It is expected that non-Cartesian coordinate measuring systems will have more and more applications.

relatively small size and lightweight in comparison with Cartesian CMMs. There are many different approaches to realize coordinate measurements in non-Cartesian coordinate systems. Most of them are based on angular measurements or angular measurements in conjunction with linear measurements. In most cases the coordinates of the measured point cannot be read directly as in Cartesian CMMs, so calibration is essential for non-Cartesian coordinate measuring systems, and many of them have in achievement accuracy lower than Cartesian ones. It is expected that non-Cartesian coordinate measuring systems will have more and more applications.

18 Measurement Integration

Robert J. Hocken

CONTENTS

A book on coordinate measuring machines (CMMs) would not be complete without mentioning numerous methods for integrating the measurement function into the manufacturing process. Integration methods are not always independent of measurement techniques, but often evolve as new techniques and instruments are developed. Measurement techniques and instruments are well covered in existing texts (Dotson 2006; Curtis and Farago 2007). This chapter discusses the various ways by which measurements are used for process control in a general sense, concentrating on situations in which CMMs are applicable.

18.1 SELECTION FACTORS

The choice of measurement approach depends on a wide variety of factors including part volume, tolerance level, part complexity and criticality, desired flexibility, measurement purpose, and the like (see also Chapter 15, Application Considerations). Some years ago, a major machine tool company introduced in its advertising a plot that attempted to encapsulate the choices of the type of machining system as a function of part variety and part volume (the exact origin of this plot is unclear, as it has been reproduced numerous times by so many authors and even included in manufacturing textbooks without accreditation; Kalpakjian and Schmid 2009). Figure 18.1 is yet another reproduction of this remarkable encapsulation. Clearly, it oversimplifies the problem, but it does offer some general guidance to manufacturing engineers regarding system choice when faced with the task of producing a machined product. A similar chart has been generated for measurement (Franck 1987). A modified version of the chart is shown in Figure 18.2.

515

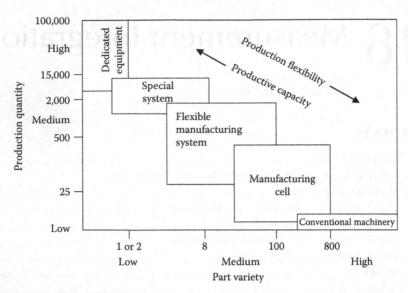

FIGURE 18.1 Manufacturing equipment takes different forms depending on production rates and number of part varieties.

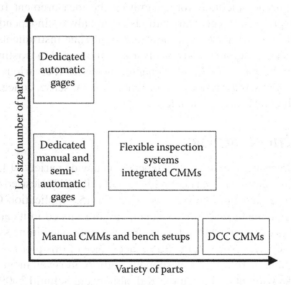

FIGURE 18.2 Measuring equipment options for various production volumes and varieties of parts. (From Franck, G. L., Flexible Inspection Systems for Automated Manufacturing. In *Test, Meas. & Inspect for Quality Control Conference/Exhibition*, 1987. With permission.)

For example, suppose one were faced with the manufacture of hundreds of thousands of simple cylindrical parts whose diameter had to be controlled to 5 μm. Clearly, some form of dedicated automated gaging system might be highly desirable and a CMM would probably not even be considered as an alternative. At the other end of the spectrum, that is, with lot sizes of one or two and a wide variety of parts, a CMM might be very useful, even at relatively low tolerance levels, due to its flexibility. Equally obvious, the CMM or a fully automatic CMM system is highly applicable to those parts that are normally produced by CNC machining centers and flexible manufacturing systems for prismatic parts but not so obvious for those parts that are produced by turning. Because of these considerations, comments in this chapter are confined to measurement methods applicable to prismatic parts produced in medium or small volumes, that is, those parts that can be produced on machining centers and for which a CMM is a likely candidate for the measurement instrument.

18.2 COORDINATE MEASURING MACHINE MEASUREMENTS IN THE MANUFACTURING PROCESS

A scheme for the classification of measurement methods has been introduced by Blaedel (1980) and expanded on by other researchers (Vorburger and Scace 1990; Vorburger et al. 1992, 1994). In this scheme, measurements are classified according to their relationship to the manufacturing process. Particularly, four cases are recognized: (1) preprocess measurement, (2) in-process measurement, (3) process-intermittent measurement, and (4) postprocess measurement. Each of these cases supplies data to the overall factory quality assurance system and has a definite role in a well-run manufacturing enterprise.

Preprocess measurements are those that are made on the manufacturing system itself in order to characterize the machines, arrive at appropriate process plans, and delimit process capability. Calibration and correction of machine tools and measuring machines fall in this category.

In-process measurements are generally of two distinct types. The first type is direct measurement, where a critical feature on the part is measured while the part is being produced, and the dimension, form, or surface finish value obtained is used to alter the process to produce a correct part. The second type of in-process measurement is indirect and is called "deterministic metrology" (Simpson, Hocken, and Albus 1983). The principle here is to monitor the process by measuring variables other than part dimension and to deduce the accuracy of the manufactured part from these parameters.

Process-intermittent measurements, like in-process, are of several different categories. Process-intermittent measurements are performed where the manufacturing process is "momentarily" halted, measurements are made of the part, and those measurements are used to modify process variables. They can be done on or off the machine. For the prismatic parts being discussed here, the most common methods for these types of measurements are the use of on-machine probing using touch-trigger probes or to check the part with a gage, which may, or may not, provide dimensional deviations (differentiating here between go and no-go gages and more complex gages containing measurement sensors). In general, scanning probes (part trace tests), optical tests, and many other techniques are used in different industries either on or off the machine. They will not be discussed here.

Postprocess measurements are the traditional methods of part inspection and can be performed by a tremendous variety of instruments (Curtis and Farago 2007). For the prismatic parts being discussed here, so-called open setups (measurements using height gages and other hand-operated measuring devices on surface plates) and CMMs are the most common. Often such measurements may be augmented with geometry instruments (roundness and cylindricity) when particular part features have tight tolerances on form. Further, such measurements may be open loop (part is passed or rejected) or closed loop, where the part is either reworked (if possible) or data from a measurement are used to correct process parameters on the next part.

18.2.1 PREPROCESS MEASUREMENT AND ANALYSIS

Machine tools and the machining process have been studied systematically for over a century. The major sources of machine tool error that contribute to part errors have been identified. Machine tool errors include geometric errors built into the machine tool by the builder, load-induced errors caused by machine self-loading, part load and cutting forces, thermally induced errors from internal and external heat sources and sinks, spindle error motions, kinematic errors, and vibrations both forced and self-excited. Process errors could be considered to include chucking and fixturing distortions, tool wear, errors in correct machining parameter specification (feeds, speeds, and depth of cut), improper choice of cutting lubricant, poor chip control, and a myriad of other effects (Hocken 1980). Preprocess measurement and careful process planning can reduce these errors and thus constitute a bona fide method for ensuring part conformance to design intent.

One of the most important steps in preprocess metrology is the evaluation of the machine tool. Just as a VDI/VDE test or the ASME B89.4.1 standard is used to evaluate the performance of a CMM (ASME 1997), the ASME B5.54 standard (Methods for the Performance Evaluation of Computer Numerically Controlled Machining Centers) should be used for evaluating the performance of a machine tool (ASME 2005a). This standard includes tests for environmental sensitivity, displacement accuracy, spindle errors, thermal errors, kinematics, and cutting performance (see also Chapter 9, Performance Evaluation). As written, the standard's main purpose is the quantification of these errors, but sophisticated manufacturers are using the tests as diagnostics for correcting the machine errors (Hocken et al. 1977; Donmez et al. 1986; Donmez 1987; Gavin and Yee 1989; Janeczko 1988). Properly characterized, controlled, and corrected machine tools can approach or exceed the accuracy of a measuring machine (Bryan 1979; Estler and Magrab 1985; Patterson 1986).

In many production machines, however, such control and correction are not considered, usually for reasons of past practice although economics is often cited. Whatever the real reason, the vast majority of machines are used in situations of relatively uncontrolled thermal environment and, due to internal heat sources and the other effects mentioned above, produce parts with an accuracy considerably less than the true machine capabilities (Bryan 1990). One classic solution to this problem is manual intervention with a measuring instrument and immediate correction of the cut based on the measured data (this is process-intermittent measurement). Second solution, and one widely practiced when volume permits, is to make a part, measure it, alter the program, measure it again, and repeat this process with careful attention to duty cycles, tooling, and so on, until correct parts are produced, and then keep process parameters as uniform as possible to continue producing conforming parts. This technique is pervasive in the industry. Third method, which is gaining wide acceptance, is to use on-machine probing in which probes are loaded into the spindle with the automatic tool changer and the machine programmed to "measure" the part before and after a cutting cycle. This procedure is discussed more completely in Section 18.2.3.

18.2.2   IN-PROCESS MEASUREMENTS

One of the pervasive aspirations of researchers in machining is the development of sensory-interactive in-process measurement of the part when it is being produced, coupled with feedback to the machine controller for process correction. Laboratory systems have been developed for some operations, but, in general, such systems cannot function in the factory-floor environment (chips and coolant) or cannot access the required part surface due to interference with the tool (Kim, Eman, and Wu 1987). Other drawbacks include system complexity and reliability. Because of a multitude of issues, it is doubtful that such measurements will ever be possible, except on simple parts such as cylinders, without a major breakthrough in sensing technology.

The most viable in-process measurement technique to date has become the indirect sensing approach coupled with the appropriate postprocess measurement (Donmez 1989). Here, the goal is to precalibrate the machine as discussed earlier, monitor on-line appropriate variables, and correct the process according to error models based on a correlation between machining errors and these process variables. Machining errors are quantified by postprocess inspection using traditional metrology instruments such as CMMs. Normally, monitored (besides the traditional machine scales) are temperatures of critical machine elements, cutting forces, and vibration levels, both structural and acoustic. Considerable success has been demonstrated in the laboratory using these approaches, but there have been few efforts to incorporate this technology into production machines (Bryan 1993).

18.2.3   PROCESS-INTERMITTENT MEASUREMENTS

As mentioned earlier, process-intermittent measurements can take many forms. First, it can be performed on or off the machine. If it is done on the machine, it can be performed with an independent measuring instrument or with the machine itself. If it is done off the machine, it is normally done with an independent instrument but even here many variations exist. For example, the part could be measured in the fixture used for machining, in an alternate fixture or in a "free" state. Here the concentration will be on measurements made on the machine, using the machine itself to quickly "measure" the part that was just made. Two approaches are common: using a touch probe in the machine spindle, and an analog probe on higher end machines to trace the surface produced. In machining centers, the touch probe approach dominates. Limited tests designed to evaluate such probing systems are given in two American machine tool standards, one for machining centers (ASME 2005a) and another for turning centers (ASME 1998).

Using touch probes on machine tools for process-intermittent measurement is a complex procedure that requires careful thought. As mentioned earlier, machine tools and the machining process are fraught with error sources, many of which are intrinsic to the machine tool itself. Also machine tools and machining processes are fully deterministic in that they obey the laws of classical mechanics and electro-dynamics within current ability to measure (Hocken, Raja, and Babu 1993). This means that if the scales and geometry of the machine are incorrect when machining, they are still systematically incorrect when measuring. Further, since many of the dominant errors in a machine tool are thermal distortions caused by heating the machine spindles and drives, on-machine probing, since it is done with the spindle off and with a different duty cycle on the drives, and is performed with the machine in an unsteady state. Drifts of the spindle position with respect to the part when the spindle is turned off are often very large (Belforte et al. 1987). These problems can be compounded due to the use of coolant (which may actually be hot), which is also often turned off during on-machine probing causing the part to also be in an unsteady state.

If on-machine probing is applied without a detailed understanding of these and other effects, serious errors can occur. For example, suppose one was milling at 5000 rpm using flood coolant (water based) on an aluminum part in iron fixture and stopped the process for a process-intermittent measurement. The spindle would immediately begin to cool, changing the probe position (and perhaps angular orientation) with respect to time. The part, due to the absence of the flood coolant, also would change dimension rapidly. Finally, the fixtured part might appear to be the correct dimension on the machine, but once removed from the fixture it assumes a very different shape. In the example mentioned, the data obtained on the machine, if used to correct the process, might produce the resultant inaccuracies due to compensation for transients.

Many of these considerations may, however, be taken into account, and on-machine probing does indeed have a valid place in process control as distinct from metrology. For example, properly used probes can give valuable information on tool wear, effects of cutting forces, servo performance, and the like. One of the chief manufacturers of such probes has been careful to address these issues and normally refers to this probing as process "footprinting" or "finger printing." Further, they have developed a unique manufacturing system with well-thought-out procedures to compensate for many of these effects (Somerville 2009). This system is highly automated and uses pallet delivery systems for both tools and workpieces. This high level of automation ensures consistency in process cycle times (and duty cycles), ensuring that even though measurements are made in nonsteady state conditions, these conditions are highly repeatable. In order to address machine tool systematics, the machine tool is used as a comparator rather than a measuring machine per se. Somerville states that

The unique design of the pallet transfer system allowed for storage within the working envelope of a calibrated artifact – replica of the component being machined. Within the machining cycle, component size is established by comparison with this calibrated artifact (replica). Accuracy of measurement is determined by the artifact, not the inherent accuracy of the machine tool. In this way accuracy of measurement tends toward the short-term repeatability of the machine tool.

A secondary artifact, called a calibration block, is also used to calibrate the machine's measurement system on a periodic basis. The control of coolant is not mentioned as well as part distortion due to fixturing, but perhaps they are addressed by appropriate design and control. Note that calibration of both artifacts is performed by traditional methods (CMMs most likely) yielding a measurement system analogous to the master gage system used for decades.

Researchers are also using these techniques, coupled with preprocess machine characterization and correction to improve process accuracy, but commercial application is limited (Bandy 1991; Yee and Gavin 1990; Yee et al. 1992).

In summary, properly utilized process-intermittent measurement can become an important process control tool but must be firmly anchored in process understanding and controlled by independent postprocess measurements of traceable accuracy.

18.2.4 POSTPROCESS MEASUREMENTS

As mentioned in the introduction, the modern metrology laboratory contains a multiplicity of instruments used for the measurement of part attributes including dimension, true position, form, and surface finish at different levels of tolerance. The CMM was never intended to replace all these instruments at all tolerance levels. Rather, such machines can be considered evolving from surface plate measurement technology, where height gages, micrometers, and other measuring instruments are used in conjunction with a nominally flat plate for determining part dimensions and form at moderate accuracy levels.* Such surface plate measurements are still in common use and sometimes called open setup measurements. There are advocates who are convinced that such procedures are inherently more accurate than CMM measurements, and because current tolerance standards (ISO 2004a; ASME 2009) were written with such open setup procedures, problems of "methods divergence" were discussed widely in the past (Pond 1988). Algorithms have advanced this problem (see also Chapter 8, Coordinate Measuring System Algorithms and Filters).

Open setup measurements on a surface plate, even when performed with high-resolution instruments, cannot compete in accuracy, or speed, with a properly operated CMM. The reasons are many. First, and often not considered, is the reliance on the surface plate to provide an accurate datum. Because such plates are never perfectly flat, suffer from thermal distortions and wear, and are often improperly maintained, this assumption of perfect flatness is misleading and erroneous. Second, in open setup measurements, operator skill (and even bias) plays an important role with operator influence through factors such as thermal effects from operator body heat. In the existing standard, operator bias is even recognized, as "rocking" of the part on the datum (plate) is allowed to bring the results within tolerance. Third, in such open setup measurement many other effects such as contact force and proper part fixturing are often poorly controlled and can lead to significant errors. Finally, there is a data integrity issue. It is extremely difficult (currently maybe impossible)

---

* Two schools of thought exist on this issue. The CMM can equally well be thought of as evolving from a machine tool, and in fact, the models produced by Moore Tool and SIP retain this resemblance (SIP 1952; Moore 1970).

to develop completely automated methods for data acquisition and analysis on the variety of instruments and procedures that can be used for this type of measurement. Opportunities for operator error in recording, entering, and analysis of data abound.

Besides surface plate checks, many hand gages are also used on the shop floor. Several of those do not need a surface plate but share similar issues with operator errors and data acquisition. Training and consistent performance also affect the results of manual measurements both with and without surface plates.

Most of these effects can be eliminated or drastically reduced when using a CMM, particularly one under computer control. Fixturing, of course, remains a problem. Datum surfaces may be defined by probing rather than assumed, part form may be computed rapidly from sampled data, and feature-to-feature position relationships come naturally from the CMM's geometry and measurement method. The data management issue becomes straightforward, providing not only a complete record of the measurement but also a documentation of the algorithms used for analysis. Also, the measurements may readily be repeated. Clearly, there are still issues of sampling (Babu, Raja, and Hocken 1993; Hocken, Raja, and Babu 1993; Edgeworth and Wilhelm 1996) and algorithmic choice and correctness (Porta and Wäldele 1986; BSI 1989), which have been discussed in the literature (also refer to Chapter 8, Coordinate Measuring. System Algorithms and Filters); however, even though the sampling density of some CMMs is limited by speed considerations, the actual density is still many times that used in most open setup or shop-floor measurements.

18.3 SUMMARY

The measurements used to ensure quality in a manufacturing system are well established. Although the equipment may vary, optimum process control can be achieved by the following: (1) preprocess metrology to ensure that the manufacturing machines and measuring instruments and machines have the appropriate accuracy, (2) process-intermittent metrology for process control, (3) in-process measurements either direct or indirect where feasible, and finally, (4) postprocess metrology to independently confirm the final results. The system is interconnected, and each portion performs a function that is both independent and interdependent. For example, process-intermittent measurement is an effective means of monitoring the process and reducing the number of times it is necessary to perform independent measurements, but it is not a substitute for the disciplined postprocess measurement. It needs to be pointed out that without the preprocess measurement and analysis, process-intermittent and indirect in-process measurements can compound, rather than correct, quality problems. There is no panacea; proper attention to detail is required in all steps of the manufacturing process, from a machine's acceptance to final part inspection. Only by such careful attention and understanding can manufacturers expect to produce high-quality products that are necessary to maintain global competitiveness.

ACKNOWLEDGMENTS

Special thanks to Paulo H. Pereira from Caterpillar Inc. for kindly reviewing the text and the helpful suggestions.

19 Financial Evaluations

Marion B. (Bill) Grant

CONTENTS

Most projects involving capital investments are analyzed in terms of their projected pay-back to the business. The purchase of coordinate measuring machines* (CMMs) is no exception. Because a CMM is not a production tool, the pay-back analysis tends to be more sophisticated. This chapter points out that a CMM can have an even greater impact on the financial success of a business than a production tool. Today, more than ever before, businesses must produce high-quality products at low cost to survive in the increasingly competitive global market. By keeping processes under control, scrap and rework are minimized, which reduces production costs, and if parts are consistent, then product quality will be superior.

Businesses striving for a strategic advantage in their manufacturing operations must focus on dimensional measurement capability as a key technology toward leveraging their manufacturing prowess. There are many examples to prove that emphasis on measurement enhances the success of manufacturing businesses. Rather than showing detailed formulas, the objective of this chapter is to provide a broad set of factors to consider when evaluating the purchase of a CMM. The point of view of this chapter is that the ultimate economic and, hence, financial value of a CMM is derived from its flexibility and ability to establish statistical process control (SPC) for production and for gage calibration. One of the key characteristics of a CMM, which deals with its strategic value, is its measuring uncertainty relative to the functional and design tolerances of the parts being measured. Given estimates of the cost of obtaining a specific uncertainty and the cost of parts being out of tolerance, a financial evaluation can be carried out comparing the costs of a CMM relative to the cost of bench and automatic dedicated gaging. Example cases show the cost advantages of direct computer-controlled (DCC) CMMs to these other approaches. The actual cost justification techniques will vary with individual business situations and must conform to the financial analysis practices of the particular company.

In keeping with the objective of presenting a broad view of financial evaluations, the chapter will be presented in three sections consisting of "strategic implications," "technical requirements," and "financial analysis." Section 19.1 presents qualitative information pertinent to the choice of a CMM. Section 19.2 presents the quantitative methods of calculation of product quality that affect manufacturing costs. Section 19.3 contains numerical evaluations of some representative examples that demonstrate the financial advantage of CMMs compared with alternative measurement technologies.

19.1 STRATEGIC IMPLICATIONS PERTAINING TO MEASUREMENT

Quality is identified as a key strategic imperative of many manufacturers. Process control is at the heart of quality. It was Taguchi who quantified customer value of quality in the form of the Taguchi Loss Function, postulating that customer-perceived

---

* This chapter discusses examples using a CMM, but it can be similarly applied to other metrology equipment such as articulated arm CMMs and laser trackers considering the proper cost differences (like manual vs. automated operation).

value decreases with the square of the distance from the center or target of the functional tolerance band (Taguchi 1986). Taguchi pointed out that the customer can see whether a producer has process control by the consistent quality that is or is not in the details throughout the producer's product. The customer then buys from the manufacturer that shows in its products that it maintains its processes in control.

19.1.1 MEASUREMENT SERVES DIFFERENT OBJECTIVES IN INDUSTRY

In justifying the purchase of a CMM, its intended use must be clearly identified. For manufacturing industry, there are at least five categories of measurement. These are 100% postprocess inspection, audit inspection, analytic measurement, process control, and gage calibration. A CMM may be used for all five categories with perhaps the exception of performing 100% post process inspection of high-volume production parts.

It is important to understand the thinking process of the managers who will review a proposal to purchase a CMM. There is a principle, which is widely accepted by management that inspection is "nonvalue-added." This principle arises from the truism that "one cannot inspect quality into the product" (Juran and Godfrey 1999), meaning that trying to reach quality targets in product by 100% inspection, sorting conforming product from nonconforming, is not customer led and ultimately fails to protect the customer.

When proposing high-capital-cost equipment such as a CMM, unless it is clear to management that the proposal is in some way adding value to company product, then the proposal might not be given serious consideration, regardless of the financial justification. To prevent this from happening, the value-added use of the CMM must be presented in the request for funds. Process control measurement and audit inspection are value added because they provide critical information to improve the targeting and reduce the overall variance of the process (Kunzmann et al. 2005; Phillips, Baldwin, and Estler 2009). The capability index that measures the combination of targeting and process width is C_{pk} (Grant and Leavenworth 1996). This index has evolved into a quite broadly used metric in the manufacturing sector to maintain high levels of product quality. It will be mathematically defined and evaluated in Section 19.2.2. Analytical measurement is value added because it is part of product engineering or process variance analysis to again improve C_{pk}. 100% postprocess inspection of volume production is rarely justifiable as valueadded.

Postprocess inspection is the measurement, after the process, of certain aspects over 100% of the output, to sort between conforming and nonconforming product. This is considered undesirable, as it is comparable to admitting that mistakes or producing out of tolerance is acceptable and the workpieces must be handled yet another time to sort/rework/scrap them.

Audit inspection is measurement performed postprocess to provide a "check" that parts conform to specification. When circumstances require, it is performed by a customer to check the work of a supplier, when the customer lacks confidence in the supplier's process control. Audit inspection is sometimes utilized when a process is more capable and more stable than is necessary to require SPC to maintain conformity. Then audit inspection can be used to "check" that the process is still performing as previously believed. Audit inspection can also be used in a nonstatistical way to provide retargeting of the process(es).

Analytical measurement is the process of measuring features of output to quantify an effect or condition for the purpose of determining measurands that correlate to product function. This is accomplished by finding a measurand of the product feature that has a deterministic correlation with the acceptable and nonacceptable performance of the product feature. Usually, a sizable quantity of product must be measured in order to establish this correlation. Yet, when this correlation is established, then a functional specification can be set. A capable production process can then be controlled, for example, to have $C_{pk} \geq 1.33$, within this functional specification in order to realize a zero defect status of the product. When vanishing amounts of the product have functional measures outside the functional dimensional tolerance, then vanishing amounts of the product display unsatisfactory functional performance, which is, by definition, a near-zero defect status of the product.

Analytical measurement is also used to determine which aspect(s) of a process is the root cause of product failures or nonconformance to the specifications. Without it, continuous improvement would be impossible. Such analysis is the only way to improve process and product control and gain the resultant quality improvement and added product value.

Process control is measurement that is integral to the production process and maintains the process targeted through statistical methods, adaptive feedback methods, or process metrology. This type of measurement actually improves the capability, C_{pk}, of the process over what the process would exhibit without it because it maintains the process targeting. Value is added by taking a capable process with non-negligible drift that would be making scrap without control, and keeping this process targeted on the center of the product specification. It maintains the process $C_{pk} \geq 1.33$ during long time intervals. This is the express goal of quality engineers and process improvement activities.

Gage and fixture calibration involves high-accuracy measurement made in the central laboratory. With a high-accuracy CMM in the central laboratory, standard masters (e.g., plugs, rings, and thread gages), precision functional or hard gages, and three-dimensional production masters and fixtures can be calibrated to reduce labor-intensive and less accurate manual inspection. As a result, the central laboratory CMM becomes a master traceability path* (ASME 2006b) and is justified in replacing other, less capable devices for precision tooling assessment.

19.1.2 COORDINATE MEASURING MACHINES FOR PROCESS CONTROL

Optimizing the capability of process control implies optimizing all of the aspects of a measurement approach and aspects of its statistical soundness of feedback to control the process that contribute to the overall uncertainty width and targeting of the production process. The uncertainty of the measuring process[†] includes contributions from

---

* Refer also to Chapter 2, The International Standard of Length.
† Refer also to Chapter 14, Measurement Uncertainty for Coordinate Measuring Systems.

- The CMM itself
- The procedure (including the measurement program in software, if applicable)
- The measuring environment (temperature, thermal gradients, humidity, air convection, and vibrations)
- The measuring principle
- The operator
- The calibration and/or setup

The measurement process must be capable of measuring the functional measurand to a level suitable for the range of interest (tolerance or process spread) with sufficient uncertainty and resolution.

Another, even more hidden, uncertainty source comes from lack of standardization. In order to correlate from one measurement to another across locations and makes of CMMs, there must exist standards on the proper measurement method, the proper definition of the data set, the filtering method (what wavelengths of data will be included), the calibration method, and the way to calculate the measurand from the data.

19.1.2.1 Uncertainty Reduction with Standard Computerized Equipment for Statistical Process Control

The uncertainty of measured results can be reduced by using accurate equipment, adhering to standards, and measuring the intended functional measurand, not a derived quantity. The best way to accomplish this is by utilizing commercially available, computerized, programmable equipment such as CMMs, computerized surface metrology instruments, and computerized geometry equipment to provide SPC to the production line. This equipment should be placed in a suitable environment[*] as close as possible to the process.

The bulk of the work on improved process control should be spent improving process capability and placing the best and most widely used inspection technology for statistical process control as close to the process as possible in the best controlled environment possible. Accepted definitions and standards are constantly being developed and improved for top-level, widely used measurement technology, such as contact DCC CMM methods whereby correlation and consistent data integrity are assured. All advanced sensor technologies that are as yet neither widely available nor widely used, and therefore not standardized, must first be correlated to these defined, standardized methods before their use is constructive (also refer to Chapters 7, Multisensor Coordinate Metrology and 9, Performance Evaluation).

19.1.2.2 Optimum Measurement Plan Using Coordinate Measuring Machines

Some sources of errors can be treated when using CMMs to reduce the total uncertainty of the measurement process. Uncertainty as a result of temperature and temperature fluctuation can be reduced by placing the CMM in a temperature-controlled

---

[*] Refer also to Chapter 11, Environmental Control.

enclosure or having a CMM that is self-temperature stabilized or both. Temperature fluctuation in the part can be reduced by allowing the workpiece to soak out sufficiently for thermal stabilization, either on a surface plate in the CMM enclosure or in a temperature-controlled fluid bath. The overall measurement environment must be free from significant air convection and have only slow, relatively low-level temperature drift itself.* The workpiece must be cleaned sufficiently for the measurement to take place in a capable manner to ensure that the actual surface of the part is being measured and not some unknown residue.

However, the measurement must be made such that the delay—the time that the part is machined to the time that the measuring result is available for process control—is compatible with the stability of the process (trend vs. spread). If a process is stable, a longer delay is acceptable. If a process is unstable, either a shorter delay or a redesign of the process is necessary. According to the principles of "zero defect production," processes are required to be capable and kept in control. If this is the case, SPC can be applied, the trend of the process can be predicted, and the acceptable feedback delay can be estimated. These aspects are better optimized, leading to a process in better control if some delay is accepted in exchange for a better accuracy of the measurement rather than having an unreliable measuring result available almost instantly.

These points show that the form of process control with the least uncertainty, the optimum process control, is that which is accomplished through a measurement plan that takes all sources of uncertainty into account and finds the optimum minimization of all of them, or some approximation of this condition. On this basis, it does not make sense to allow all other uncertainties to exist without correction, simply to give the measurement zero time delay. When the process is stable, some delay is acceptable to control all uncertainties in a reasonable, planned way. On this basis, a CMM performing SPC in a controlled environment in parallel with a relatively stable process can be an optimum process-control method, which best treats and minimizes all uncertainty sources. A useful reference on this topic is ASME B89.7.2 (ASME 1999).

19.1.3 STANDARD COORDINATE MEASURING MACHINES FOR PROCESS CONTROL

An advantage of using programmable CMMs is that they are "off-the-shelf." This means lower cost and less risk compared with dedicated gaging because

- The development cost is spread over a higher number of units.
- Specialized workpiece-like pyramid mastering systems (low, high, and mean masters) that are very costly to develop and maintain are eliminated and replaced with standard calibration artifacts such as step gages, ball bars, and ball or hole plates for CMMs.
- DCC inspection equipment such as a CMM can be reprogrammed in software to accommodate design and manufacturing process changes, or reassignment to another product, while dedicated gages need costly hardware changes.

---

* Refer also to Chapter 10, Temperature Fundamentals.

- DCC inspection equipment can be reprogrammed if a product or manufacturing process change renders the current inspection protocol incapacitated. The reprogramming can adjust how many points are used and/or how many data sets are acquired at different locations on the measured feature to bring the overall measurement process back to usable capability.
- Manufacturers have a larger support staff available for standard equipment and are more dedicated to develop their off-the-shelf line than they are for "one-off" dedicated gages.

As seen from the list, standardized, DCC CMMs incorporate a very high degree of versatility and flexibility. The DCC CMM can process acquired data in multiple ways suitable for different processes. A case in point is when the measurand (the characteristic to be measured, i.e., cylindricity) that correlates to component function is not useful for manufacturing. A DCC CMM is able to take the same data set and process it in two different ways to suit both the product engineer and the manufacturing engineer. As an example, where the finished part must meet a cylindricity callout, the CMM could provide the cylindricity results to the product engineer but process the same data set to provide roundness, straightness, and taper results to the manufacturing and quality engineers because those parameters connect more readily to process adjustments they can make.

If standardized, DCC CMMs are used on the shop floor, measuring the same measurand as third level audit/central lab CMMs, and then first level to third level correlation problems will be alleviated. In fact, all the correlation issues caused by the deliberate oversimplification of the measuring task by the dedicated gage will be eliminated and only those caused by insufficient procedures and calibration will remain. This will reduce the cost of inconsistent quality, the improper rejection or acceptance of product, and the engineering and meeting time to resolve these issues.

19.1.4 ALTERNATE MEASUREMENT APPROACHES FOR PROCESS CONTROL

When CMMs are considered for process control, they are frequently compared with two common alternative approaches: on-machine probing and dedicated gaging.

19.1.4.1 On-Machine Probing

The optimum control approach is to place the measurement as close to the process in time and distance as possible. Traditional thinking suggests that this objective can best be accomplished by placing sensors or probes directly on the machine tool to measure the workpiece during the actual process. This approach has fundamental metrological limitations as discussed in Chapter 18, Measurement Integration, and in Section 19.1.2 of this chapter.

19.1.4.2 Dedicated Gaging

Dedicated gaging is known for being a low-cost, shop-hardened, low-cycle time solution. In reality, none of these believed virtues of dedicated gaging are strictly true. Dedicated gaging is not a low-cost approach. Tooling and mastering systems for dedicated gages are normally extremely expensive because they are specialized, tightly toleranced, made-to-order, low-quantity items. Maintenance of dedicated gaging is also often a problem that generally has to be handled by the purchaser, over time, at high cost. Special, part-specific tooling and masters must be maintained, and often even a small derivative type change of the component requires major changes to the gage and its tooling and masters. Operators must tend some forms of dedicated gages during the measurement. For these somewhat manual, dedicated "bench" gages, a certain degree of skill is often required in order to obtain correct and consistent results. DCC CMMs, however, can be installed to run with only engineering support, which can often be obtained from the supplier for a fee, so that relatively unskilled operators can obtain reliable results. The result is that dedicated gages have low-to-medium initial capital outlay with high ongoing costs. CMMs, however, have medium-to-high initial capital outlay and low ongoing costs. Independent of lowest uncertainty considerations, the decision comes down to ongoing expense and disciplined maintenance with dedicated gages versus initial capital outlay with standard CMMs.

19.1.5   Flexible Inspection Systems

Flexibility is critical to competitive manufacturing. Ford Motor Company has generated an internal pamphlet that defines what manufacturing flexibility is and how it must be customer led (Nevins and Winner 1999; Whitney 2000; Wall 2003). Businesses need to be more responsive to changing markets and changing customer needs. This push is not just for small lot sizes, but flexible transfer line technology is used to provide flexibility for intermediate as well as high-production rates. High-speed spindles and very high machine tool slew rates provide single spindle machine tools with increases in productivity that competes with multispindle approaches.

Process control of this technology obviously requires flexibility itself, in order to evaluate the different parts flowing down the line or through the cell, and provide SPC to the different machine tools. The DCC CMM is optimum for these applications, especially if the workpieces contain a mix of cubic and axisymmetric geometries. The DCC CMM can be programmed to measure each part number in the production mix, at various points in their fabrication, and calculate SPC for the different process machines and feedback control signals to retarget the various machines.

Manufacturing engineers need flexible process control rather than the traditional dedicated-gage approach. High-speed CMM technology provides such flexible process control. By utilizing SPC methods, a CMM evaluates production subsets, so that it need not perform at line rates. The workpiece temperature needs to stabilize as it moves into the controlled CMM environment, such as in a soak out chamber, where heat and debris can be removed. The scenario, therefore, usually involves some form of automatic or semiautomatic material handling to transfer workpieces from the process flow onto the CMM. DCC CMMs in these applications have been proven effective.

19.1.6   Acceptance of Coordinate Measuring Machines
for Process Control in Industry

Many manufacturers of high-technology components use this off-the-shelf DCC CMM approach. Caterpillar uses CMM technology with appreciable success wherever possible, rather than dedicated, fixed-point automated gaging. John Deere began using CMM technology in the late 1970s, making it the center of their "Factory-of-the-Future" thrust through the 1980s. The roller-bearing industries in Japan, Europe, and the United States are using CMMs and commercially available surface metrology and geometry (cylindricity) equipment to control their processes. The German automotive industry, including BMW and Daimler Benz and German machine tool suppliers have been using CMMs for some time for process control. These are only a few but representative examples of the many that are available. DCC measuring with CMMs is definitely the trend for process control within industry (see also Chapter 16, Typical Applications).

19.2 TECHNICAL REQUIREMENTS ARE CRITICAL

Today, more than ever, manufacturers need highly technical knowledge of what makes a component perform and what makes it fail. Mating surfaces are more highly loaded, carry higher power densities, and must have several times the life than did components of even the last decade. Mating components with surfaces that rotate or slide against each other must be designed for greater functionality than just the ability to fit together in assembly. It must be verified that stress points do not occur in normal alignment changes or deformations during their operation. Tighter tolerances must be met, and size and location are not the only factors. Geometry (roundness, straightness, flatness, etc.), orientation (squareness, parallelism, and angle), and profile must also be held within tight tolerances. The effects of these aspects are often no longer connected, but geometric form and orientation aspects tend to link to quite different failure modes than do size and location. Size and location can no longer be evaluated by simple, traditional methods, but various types of size and location algorithms are associated more directly to component performance. Finally, tolerances are just tighter; functional and hand gaging methods no longer suffice in many instances. Processes must meet these tighter tolerances and can do so only if they are maintained on target. Variable data are needed to control the process on target, rather than attribute "go/no-go" gage results.

The solution to these problems is the automated capture of digital data by computerized methods. CMM manufacturers have responded to the call by providing machines with dramatically faster slew rates, tighter servo control, higher probing speeds, and greatly increased computer processing speed and capacity. The industry is becoming more standardized so that the data acquired and the processing methods are better defined and more reliable. CMMs are able to evaluate form, orientation, and profile according to geometric dimensioning and tolerancing* (GD&T) standards (ISO 2004a; ASME 2009).

A fact that is gaining broader acceptance is that position, location, and form assessments require higher data densities for more reliable assessment. Much of the demand is a result of the availability of significantly greater computing capability at lower cost. Process control demands greater accuracy than that provided by dial-indicator-based bench or column gages. Failure modes and performance optimization for highly loaded surfaces have become functions of such parameters as the waviness of

---

* Refer also to Chapter 3, Specification of Design Intent: Introduction to Dimensioning and Tolerancing.

the surface. As a result, the waviness content of dimensional data must be discriminated (Bhargava and Grant 1993). Data densities must be sufficiently higher to be able to separate waviness for evaluation relative to assigned tolerances. CMM data rates (analog scanning rates, slew rates, and probing rates) make it possible to assess component features that earlier were only measurable on geometry machines (roundness or cylindricity machines, profiling machines). Even scanning polar coordinate CMMs are being marketed with the potential to increase data rates and reduce uncertainty in the measurement of axisymmetric components and component features.

19.2.1 FUNCTIONAL TOLERANCES REQUIRED FOR ZERO-DEFECT MANUFACTURING

Tolerances are of little value unless they are functional. What is meant by "functional" is that when the component feature aspect is within the functional tolerance, the component feature will perform satisfactorily; when it is outside the tolerance, the component feature will perform anywhere from just less than satisfactory to total failure. In other words, the feature aspect being measured is defined in such a way that the correlation of the dimensional measurand with component feature performance is linear or some other known relationship, and the tolerances of the measurand are set at the bounds of acceptable component feature performance. Feature performance fits categories, such as sealing, wear, elastic load bearing capacity, and so on, which are usually evaluated in engineering testing. Unless this correlation to function is a characteristic of the dimensional measurand and its tolerances that are provided on the print, then the specification is less than completely controlling the quality of the component feature.

If the specification and tolerance set on it are functional, and if the process can be controlled such that $C_{pk} \geq 1.33$, then zero-defect production is possible. If the specification and tolerance are not functional, then even if $C_{pk} \geq 1.33$, the operation will not be defect free because the measurand being inside the tolerance does not guarantee that the feature will not be defective in function.

To determine a functional measurand and its associated functional tolerances, a set of reasonable candidate dimensional measurands must be tested for correlation to component feature performance to determine which one(s) correlate with performance. This is usually accomplished in the form of a designed (Taguchi) experiment, which links dimensional measurands to component performance with the least number of component tests (Bhargava and Grant 1993). Once the measurands that correlate are determined, further testing may be required with more dimensional assessments done until the nature of the correlation is established in detail and the boundaries of acceptable performance are determined with their corresponding dimensional values. These dimensional values then become the tolerance limits.

19.2.1.1 Analytical Measurements with Coordinate Measuring Machines to Establish Functional Tolerances

The testing of the set of candidate measurands is not a trivial task. Sometimes, the functionally correlating measurand is not a standard quantity. The data set for it may need to be processed in a complex, nontraditional manner, so that extensive data often must be acquired to accomplish the task. This is where a CMM is needed for extensive, flexible, analytical measurement to determine what measurands correlate to component function. A special dedicated gage is not suitable for this task. What is needed is a flexible, high-resolution, high-accuracy measurement instrument that is capable of delivering whatever measurement method is needed with whatever type digital data set is needed. The CMM is often the primary device for such an analytical measurement task. Once this laboratory-based or off-line CMM work is done to identify the functional specification, the method for controlling the process may require equally sophisticated data acquisition and computer processing, making a CMM the choice for process control to the functional specification.

19.2.2 PROCESS CONTROL UNCERTAINTY CONTRIBUTION TO OVERALL PROCESS UNCERTAINTY

The method of process control contributes in two ways to the uncertainty of the overall process and to manufacturing products within the specified tolerances. One uncertainty contributor is the repeatability of the gage or process control method, and the other is the targeting or validity of calibration or bias of the gage. When process capability indices such as C_{pk} are determined using the gage that controls the process, the gage targeting or bias and the repeatability of the gage becomes part of the capability of the process (ASME 2005b; Kunzmann et al. 2005; Phillips, Baldwin, and Estler 2009). The process capability indices, C_p and C_{pk}, are given by (Grant and Leavenworth 1996; Montgomery 2009)

$$C_p = (UCL - LCL)/6\sigma \quad \text{and} \quad C_{pk} = \min\{C_{pl}, C_{pu}\}$$

where

$$C_{pl} = (\mu - LCL)/3\sigma \quad \text{and} \quad C_{pu} = (UCL - \mu)/3\sigma$$

and UCL is the upper control limit, LCL is the lower control limit, σ is the process variance, and μ is the process distribution mean. Note that gage targeting affects the C_{pk} of the process more adversely than does the 6σ width of the gage. The sigma of the gage adds to the sigma of the process in the calculation of C_{pk} according to the square root of the sum of the squares.* This is expressed as

$$6\sigma_{(process+gage)} = \sqrt{(6\sigma_{process})^2 + (6\sigma_{gage})^2}$$

However, the gage targeting adds linearly to affect the C_{pk} value more directly. This is expressed as

$$\mu_{(process+gage)} = \mu_{process} + \mu_{gage}$$

---

* See also Chapter 15, Application Considerations.

For an example of this, consider a gage used for SPC for a process that has $C_p = 1.33$ (the 6σ of the process is 75% of the tolerance ignoring targeting). With a perfect gage with no width and perfect targeting, the C_{pk} of this process would be 1.33. If the 6σ of the actual gage consumes 30% of the tolerance but the gage targets with no error, then the apparent process C_{pk} decreases to only 1.24, resulting in approximately 0.01% of the production of the system being outside of the tolerance. But if the actual gage has essentially no width ($\leq$10% of tolerance), but mistargets the process by 30% of the tolerance, then the system C_{pk} drops to 0.53, resulting in 5.5% of production being outside the tolerance (Kunzmann et al. 2005).

This example shows why gage calibration is quite important to effective process control. Using a DCC CMM for SPC provides superior process targeting when compared with traditional process control approaches such as dedicated bench gages and dedicated automatic gages. This superior targeting is because of the standard machine design of CMMs, the standard methods of evaluating their performance, and the standard methods for their calibration.

19.3 FINANCIAL ANALYSIS INVOLVES MANY FACTORS

Before purchasing, a thorough cost analysis needs to be completed when considering alternatives in process control equipment. This analysis can be categorized against the different types of measurement equipment: surface plate inspection, dedicated bench gaging, dedicated automatic gages, manual and DCC CMMs.

19.3.1 COMPARISON OF MEASUREMENT COSTS PER WORKPIECE

The financial analysis undertaken here consists of two pieces. The first is to look at all costs of using different types of equipment to measure a workpiece. The comparative cost per workpiece measured is then calculated. The second is to look at costs and savings, outside of operational costs, in the form of cost avoidances that are present with the DCC CMM approach relative to other methods. The various costs and factors that affect the per piece cost of measurement are listed in Table 19.1 in two categories, fixed costs and variable costs (Duplain 1991; Esswein and Neumann 1991; Callander 1993; Brown and Sharpe 2001).

TABLE 19.1
Fixed and Variable Costs

Fixed Costs	Variable Costs
Engineering cost	Tooling and masters
Cost of equipment	Degree of utilization
Depreciation (years)	Annual maintenance
Annual depreciation	Rate of inspection site costs
Annual interest	Manpower cost of inspection
Floor space costs per year	Labor costs
Initial tooling	Training costs

These costs are briefly reviewed here. Table 19.2 provides a tabulation of the contribution of cost from each of the primary types of measurement for process control: dedicated bench gaging, dedicated automatic gaging, manual CMM, and DCC CMM. Note that this table includes two different degrees of utilization, providing financial estimates for one-shift operation and later the same financial estimates for three-shift operation.

TABLE 19.2
Example of Cost-Efficiency of Different Process Control Measurement Methods

Cost Type	Units or Calculations	Dedicated Bench	Dedicated Automatic	Manual CMM	DCC CMM
Engineering cost	k$	50	150	0	0
Initial cost	k$	126	210	140	280
Initial tooling	k$	102	54	30	30
Depreciation	Years	7	7	7	7
Depreciation per year	k$	40	59	24	44
Interest per year	$0.5 \times cost \times 7\%$	4.5	7.5	5	10
Floor space	k$	0.5	2.5	3	4
Site cost	k$	45	69	32	58
Degree of utilization	*1 Shift (%)*	*10 (each of 3)*	*30*	*80*	*60*
Maintenance per year	(3% to 7%) Cost	6	15	8	8
Tool/masters	30% of Initial	31	16	2	2
Manpower	1 Shift (%, h)	10,600	10,200	80,1600	10,200
Labor cost	k$ (@$25/h)	15	5	40	5
Training	k$ (@$50/h)	3	1	10	1
Total cost per year	k$	100	106	92	74
Cost rate	$/h	167	176	58	62 (1200 h/year)
Measurement time	h	0.01	0.01	0.05	0.03
Measurement rate	$/Part	1.7	1.8	2.9	1.99
Degree of utilization	*3 Shifts (%)*	*10 (each of 3)*	*30 (1800 h/ year)*	*80 (4800 h/year)*	*60 (3600 h/year)*
Maintenance per year	(8% to 15%) Cost	11	21	21	28
Tool/masters	50% of initial	51	27	2	2
Manpower	3 Shifts(%, h)	10,1800	10,600	80,4800	10,600
Labor cost	k$ (@$25/h)	45	15	120	15
Training	k$(@$50/h)	9	3	30	2
Total cost per year	k$	161	135	205	105
Cost rate	$/h utilized	90	75	43	29
Measurement time	h	0.01	0.01	0.05	0.03
Measurement rate	$/Part	0.9	0.8	2.2	0.9

19.3.1.1 Fixed Costs

Initial engineering cost: This is the cost of engineering for dedicated gaging. For CMMs, this initial engineering could be viewed as the cost of initial part programming and special software costs.

Cost of equipment: For surface plate inspection, typically this involves a surface plate, digital height columns, indexing equipment (table or centers), electronic gage heads, gage blocks and masters, and other precision tooling. The cost is approximately $20k to $40k. The cost for dedicated bench gages consists of the gage plus the masters and tooling costs. Initial outlay is similar to surface plate inspection because the gage and tooling are specials. Automatics for high volume are much more expensive, similar to a CMM at approximately $80k to $300k including initial engineering, tooling and setup. Manual CMMs are approximately $20k to $100k. DCC CMMs run from $50k to $500k. CMMs require higher initial capital outlay, but they can replace multiple sets of dedicated equipment, which will be treated later in the cost avoidance calculation.

Initial tooling cost: This is the cost of dedicated masters and tooling for dedicated gaging. For CMMs, this includes costs for special probes and fixtures.

Depreciation: This varies but for capital machines, it is typically 7 or 10 years. Annual depreciation of the capital is the initial capital outlay divided by seven (divided by 10 if 10-year depreciation is used).

Annual interest: This is calculated on half the initial capital outlay due to depreciation.

Floor space costs per year: Floor space for CMMs tends to be more expensive because of the larger total footprint and due to the extra cost of environmentally controlled space. These costs amount to $1000 to $5000 per year.

Total annual fixed costs are the sum of engineering, equipment, depreciation, interest, and floor space.

19.3.1.2 Variable Costs

Degree of utilization: Number of shifts operated multiplied by the time spent in actual use, expressed in hours toward calculating inspection costs. Automatic gaging systems and DCC CMMs tend to have greater utilization.

Annual maintenance: This can be approximated to be a percentage of the initial capital outlay, including calibration and checks.

Tooling and masters: This is the yearly cost of tooling and master replacement and support (indirect costs), which can be estimated based on experience and usage.

The total variable costs are the maintenance plus tooling and masters plus the labor rate times the hours required on the machine to accomplish the process control. Note that the labor rate involvement will be much less for the automatics and DCCs than for the bench and manual CMMs. The total site cost per hour is then the total annual fixed cost plus the total variable costs, which can be compared across machines.

Dividing this total cost by the hours of use provides the cost of the site per hour of each method. Multiplying this site cost per hour by the measurement time per part results in the inspection cost per part with each gaging approach. With these calculations, one can compare the total site cost and the cost per piece of process control using each of the four inspection techniques.

19.3.2 DETERMINING RETURN ON INVESTMENT FOR DIRECT COMPUTER-CONTROLLED COORDINATE MEASURING MACHINES

Return on investment (ROI) is often used to compare capital equipment alternatives. When calculating ROI, other savings can be included in the form of cost avoidances with the selected approach, considering all costs to the company. A list of the major contributors in an ROI calculation is

- Other gaging costs (multiples of other gages to meet capacity or measurement capability)
- Process downtime costs (while pieces are checked to verify initial setup)
- Changeover costs because of part number changes
- Changeover costs because of tolerance reduction
- Engineering support costs of measurement (diagnostics, gage attention)
- Scrap reduction and elimination
- Rework reduction and elimination
- Reinspection reduction and elimination

If the difference in capital outlay results in cost being avoided, then this amount can be included in the ROI calculation. Generally, the versatility of a CMM provides the capability to replace multiple dedicated bench or automatic gages that would have been necessary. Tooling and masters for these multiple gages are also saved. In addition, a CMM can generally perform more tasks than initially envisioned. In these cases, there might be other gage costs avoided. If the CMM also reduces the time for first piece inspection to check initial process setup, this can be estimated by evaluating the improved process productivity.

Tolerance theft is the manner in which control limits must be established inside the tolerance (guard) band to allow for higher uncertainty in gaging (ASME 1999; ASME 2001b; Kunzmann et al. 2005; Phillips, Baldwin, and Estler 2009). To estimate this contribution, the increased cost of running the process to the tighter tolerance must be approximated. Changeover times for part number change or tolerance change is a direct advantage of CMMs over dedicated gaging. The cost and frequency of changing over the dedicated gaging must be evaluated. For the CMM, if there is no change to part fixturing, only a programming change is needed with tolerance or design changes. Training costs are another advantage to the CMM because operator training on dedicated bench gaging is significant, where operator training for the CMM, once fully implemented in production, is minimal. Engineering support costs involve meetings, hours of diagnosis, and problem solving. For CMMs, engineering support is for programming and machine maintenance.

19.3.2.1 Scrap and Rework Reduction

A particularly important aspect of cost avoidance is the elimination of the production of nonconforming product, which reduces process cost, results in customer satisfaction, and increases the prospects for growing market share. The method of process control is important because it affects how much tolerance can be allowed for the process, and it provides the process targeting, both of which directly affect C_{pk}.

Customer protection begins with a sound-tolerancing strategy. Dr. Henrik Nielsen described such a strategy in which lower and upper functional tolerances limits (LFL and UFL) are developed within the component development process by the correlation of measurement with component testing (Nielsen 1992). Yet, these functional tolerance limits are not the lower and upper design limits (LDL and UDL). These design limits are located inside the functional limits by half the uncertainty of the measurement process used in establishing the functional limits, as shown in Figures 19.1 through 19.3. This ensures that when the measurement says the part is within the functional tolerance that it is, in fact, within. When this design goes to manufacturing, whether inside or outside the company, then manufacturing prints should be generated that specify lower and upper control limits (LCL and UCL) inside the design limits by half the uncertainty of the process control methodology to ensure that the product engineers will not reject the product for being outside of the design limits (Figures 19.1 through 19.3). Additionally, lower uncertainty measurement within the component development function (laboratory-based CMM justification) reduces the tolerance theft between the functional limits and the design limits.

The lower uncertainty process control method reduces the tolerance theft between the design limits and the control limits. Therefore, buying high-accuracy, highly traceable process control and laboratory measurement for functional tolerance development does save the money if the company uses a zero-defect process control technique. (Another more complete discussion of how to establish a sound-tolerancing strategy is given in ASME B89.7.3.1–2001; ASME 2001b).

To compare measurement methods, scrap and rework can be estimated from the improved targeting of the process supplied by the process control approach, and from the amount of tolerance theft due to the gage capability. From the equations

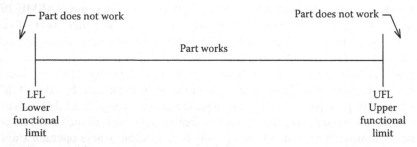

FIGURE 19.1 Functional limits of a tolerance are the limits determined experimentally or by other R&D activities. The feature must lie within these limits for the part to perform satisfactorily.

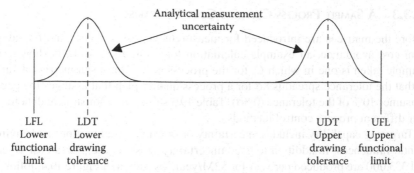

FIGURE 19.2 Drawing (or design) tolerances equal the functional limits interval with half the customer uncertainty subtracted at each end. The functional range is therefore reduced by this measurement uncertainty to obtain the tolerance.

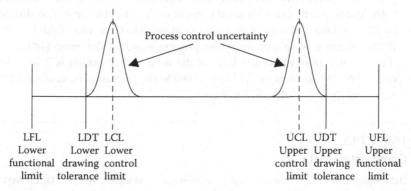

FIGURE 19.3 Control limits are established to avoid having parts rejected and are the limits for the manufacturing process inspection. These limits will ensure proving that all parts are within drawing tolerances.

for the capability indices C_p and C_{pk} given in Section 19.2.2, it can be seen that for a process having a six-standard deviation (6σ) spread that is 75% of the tolerance, C_p is 1.33. For this condition, for C_{pk} to be 1.33, the mean position of the distribution must be centered symmetrically between the UCL and the LCL. If the distance from the distribution mean to one of the control limits is less than 3σ, then C_{pk} is less than one, and the process is producing scrap because a portion of the process distribution is outside of the tolerance.

The amount that it is beyond the control limits can be estimated using a tabulation of the cumulative distribution function (the integral of the normal probability function; Grant and Leavenworth 1996). This mean position in standard deviations of the process distribution from the control limit is estimated and becomes the "z" value in the tabulation. The resultant number is the percentage of the production that must be scrapped or reworked. To determine the scrap cost, calculate the process cost of all the components made in a year and multiply by the scrap percentage. The resulting relative cost savings can then be included in the ROI estimate.

19.3.3 A Sample Process Control Cost Comparison

Before the matrices are introduced for measurement cost per piece and for savings from cost avoidance, an example calculation for scrap and rework is shown. The example taken is one in which C_p for the process is 1.33 for a symmetric tolerance, so that the tolerance spread is 8σ for a process/quality plan that assumes the gaging consumes 10% of the tolerance (0.8σ). Table 19.3 shows the comparative data for the four different process control methods.

Targeting capability includes uncertainty or incomplete calibrations and environmental influences, in addition to gage uncertainty. Assume the pieces cost \$6 each and 333,000 are produced per year for \$2M/year. As shown in Figure 19.4, following are the results for each scenario.

> *Bench gage*: The gage has 30% capability compared with 10% in the quality plan. The control limits must be moved in another 30%/2 – 10%/2 = 10%, or 0.8σ. The targeting error allows the mean of the distribution to run shifted by 20%, or 1.6σ. The total movement of the mean to the control limit is 2.4σ. If the tolerance were four times the process standard deviation (4σ), then the "z" value from the center line for the normal distribution is $2.4\sigma - 4\sigma$ or -1.6σ. The portion of the curve outside the tolerance represents 5.5% scrap and the cost is then \$110k/year.

TABLE 19.3
Comparative Data

Data/Equipment	Bench Gage	Automatic	Manual CMM	DCC CMM
Initial cost	\$40k	\$210k	\$140k	\$280k
Targeting capability (%)	±20	±20	±10	±5
Gage capability (%)	30	20	20	10

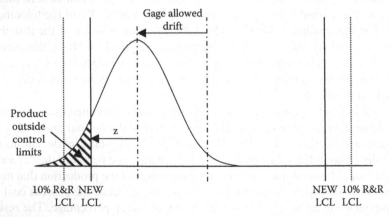

FIGURE 19.4 Scrap produced outside of control limits with less capable gaging and poor calibration.

Automatic gage: The control limits move in by 5%, or 0.4σ, and the targeting moves by 20%, or 1.6σ. The total tolerance loss is 2.0σ. "z" is 2σ – 4σ = –2σ and the table says there is 2.3% outside the control limits with a cost of $46k/year.

Manual CMM: The control limits move in 5% and the targeting moves 10%, so that the total tolerance loss is 1.2σ. "z" is 1.2σ – 4σ = –2.8σ. The table then says that "z" of 2.8 gives 0.26% out of tolerance, with a cost of $5k/year.

DCC CMM: The control limits move in by 0%, and the targeting moves by 5%, or 0.4σ. The total tolerance loss is 0.4σ, so that the table value is 3.6, for 0.02% out of tolerance. This amounts to significantly less that $1k/year out of tolerance.

Notice that these calculations do not include the cost of 100% inspection to sort nonconforming from conforming parts, or the intangible cost of lost sales resulting from poor quality. If the cost of 100% inspection runs as in Table 19.2 as $1/part, then a third of a year's product requires 100% inspection adding $111k to these costs. If rework is involved at $2/part for 50% of the nonconforming product, then this adds another $8k to $20k. Then the total for the top two scenarios above approaches $300k. This is a typical annual cost of quality for an incapable process. The intangible lost sales usually costs considerably more than the typical costs of quality if the product is delivered to the customer with problems. These data are included in the ROI calculations contained in Table 19.4.

If the same assumptions are made for a typical process as above, then all costs can be evaluated, as shown in Tables 19.2 and 19.4. Note that the result of Table 19.2 is that DCC CMM inspection is more cost effective than dedicated bench gaging and dedicated automatic gaging. Also note that manual CMMs are not cost effective.

TABLE 19.4
Savings of DCC CMMs for Process Control over Dedicated Bench Gaging and Dedicated Automatic Gaging

Savings Type (DCC CMM cost)	Cost of Dedicated Bench Gage (k$)	Cost of Dedicated Automatic Gage (k$)
Engineering support ($10k)	8	8
Changeover to new P/N ($10k)	102	54
C/O for process change ($10k)	102	54
Scrap loss ($1k)	110	46
Reinspection	111	111
Rework	20	8
Process downtime cost ($10k)	20	30
More gaging ($10k)	100	100
Total costs ($55k)	573	411
Savings of CMM vs. dedicated	518	386
Depreciation	44	44
Years of amortization	280/(518 + 44) = 0.5 years	280/(386 + 44) = 0.7 years

Manual machines are cost effective only in situations where the only other alternative is surface plate inspection where similar labor cost is required. In Table 19.4, an example of the savings of standard DCC CMMs relative to dedicated bench gaging and dedicated automatic gaging is given.

A major conclusion of this analysis is that, contrary to many managers' expectations, DCC CMMs can be superior to manual CMMs and at least as cost effective as dedicated bench and automatic gaging. When changeover costs are included, the DCC CMM is, once again, clearly superior.

19.3.4 SAMPLE GAGE AND FIXTURE CALIBRATION COST COMPARISON

For gage and fixture calibration, there are four methods in general use. These are surface plate inspection, comparators that compare with gage blocks (ID/OD* comparators and gage block comparators), linear gage-checking stations, and CMMs. CMMs are particularly suited to the calibration of multidimensional gages; that is functional gages, masters or fixtures that have features that must be checked relative to one-another in two or three dimensions. In this case, the only two methods in use are surface plate inspection and CMMs.

For these two cases, the same cost categories as listed in Section 19.3.1 of this chapter are applicable with minor modifications. Table 19.5 shows the cost comparison for the two cases. For the annual savings estimate, as in Section 19.3.2, only two major contributions are considered. These are the savings in inspection costs, and the alleviation of costs because of correlation problems. Correlation problems resulting from poor calibration cause mistargeted processes and a nonconforming product that can nearly shut down the operation or the receiving plant. Costs can escalate as 100% inspection must be used to sort the product to keep production running. These costs include the inspection itself, the scrapping of unrecoverable product, the expedited shipping of conforming product, and engineering time to solve the problem. Table 19.6 shows the savings of the DCC CMM over surface plate inspection used for gage and fixture calibration.

19.3.5 NET PRESENT VALUE CALCULATIONS

The current accounting craze for justifying capital seems to be net present value (NPV) calculations. According to Investopedia (www.investopedia.com, accessed on 28/Mar/2011), NPV is the calculation of the difference between the present value of cash inflows and the present value of cash outflows, used in capital budgeting to analyze the profitability of an investment or project. NPV analysis is sensitive to the reliability of future cash inflows that a capital project will yield (Groppelli and Nikbakht 2006). NPV is calculated by the following equation:

$$NPV = \sum_{t=1}^{T} \frac{C_t}{(1+r)^t} - C_o,$$

---

* ID = internal diameter. OD = outside diameter.

TABLE 19.5
Example of a Cost-Efficiency Audit for Different Gage Calibration Methods

Cost Type	Units and Calculations	Surface Plate	DCC CMM
Engineering cost	k$	0	0
Initial cost	k$	40	280
Initial tooling	k$	30	30
Depreciation	Years	7	7
Depreciation per year	k$	10	44
Interest per year	$0.5 \times$ cost $\times 7\%$	2.5	10
Floor space	k$	4	4
Total site cost	k$	16.5	58
Degree of utilization	1 Shift (%)	100	60
Maintenance per year	(3% to 7%) Cost	2	8
Tool/masters	30% of Initial	9	2
Manpower required	1 Shift (%, h)	100,1600	10,160
Labor cost	k$ (@$25/h)	40	4
Training	k$ (@$50/h)	5	1
Total cost per year	k$	72.5	73
Cost rate	$/h	45	61 (1200 h/year)
Measurement time	h	2	0.5
Measurement rate	$/Part	91	30
Degree of utilization	3 Shifts (%)	100	60 (3600 h/year)
Maintenance per year	(8% to 15%) Cost	5	28
Tool/masters	50% of initial	15	2
Manpower required	3 Shifts (%, h)	100,4800	10,480
Labor cost	k$ (@$25/h)	120	12
Training	k$ (@$50/h)	15	6
Total cost per year	k$	171.5	106
Cost rate	$/h utilized	36	29
Measurement time	h	2	0.5
Measurement rate	$/Part	72	15

where T is the time of duration of the calculation, t is the year that an estimated income or savings C_t comes in, r is the rate of money value depreciation per year over the course of the interval T, and C_o is the initial capital outlay.

NPV compares the value of a dollar today to the value of that same dollar in the future, taking inflation and returns into account. If the NPV of a prospective project is positive, it should be accepted. However, if NPV is negative, the project should probably be rejected because cash flows will also be negative.

Most NPV users note that such a calculation is only as reliable as the validity of assumptions of the income or savings that will come from the capital investment. For this reason, having a metrologist or a quality engineer (rank accounting amateurs at best) provide his own NPV calculation to an MBA business manager or accountant would be viewed with rampant skepticism. For this reason, it is strongly suggested that the metrologist/quality engineer provide the data for his

TABLE 19.6

Savings of DCC CMMs for Gage Calibration over Surface Plate Calibration

Savings Type (DCC CMM cost)	Cost of Surface Plate Gage Cal. (k$)
Normal engineering support ($10k)	10
Calibration cost ($24k)	70
Special shipping	100
Manpower savings ($1k)	100
100% reinspection	30
Scrap	100
Rework	20
Process downtime cost ($10k)	20
Total costs ($45k)	450
Savings of CMM vs. dedicated	405
Depreciation	44
Years of amortization	280/(405 + 44) = 0.62 years

appropriation request, as is exemplified in the tables above, for his specific application, to an allied accountant collaborator to do a professional NPV calculation that will be received with more credibility by the approving business manager. If the allied accountant will help make a case with the NPV to the business manager, then it would further increase the probability of having the request for appropriation approved.

19.4   SUMMARY

This chapter makes clear that an understanding of basic metrology and the need for adequate quality assurance equipment represents an essential part of the production process. The financial evaluation of CMMs needs to take into consideration their potential for providing strategic advantages to manufacturing operations. In many businesses, the core capability that provides a competitive advantage resides in the understanding of product performance and process control technology. Measurement is fundamental to both product performance and process control. Conclusions can be summarized as follows:

- The advanced dimensional capability provided by CMMs offers manufacturing businesses an opportunity to leverage their operations for competitive advantage and improved profitability.
- CMMs used for SPC can provide the customer with visibly consistent quality.
- For SPC, DCC CMMs can provide a strategically and financially viable alternative to dedicated bench and automatic gaging.

- CMMs can provide functional measurement methods that correlate to component function. With such measurement methods, an engineer can set functional tolerances to control component performance.
- A zero-defect condition can be obtained by using DCC CMMs to perform SPC in controlling the manufacturing process to a functional tolerance to $C_{pk} \geq 1.33$. DCC CMMs calibrated with standard methods provide superior process targeting, which is vital for reaching a high level of process performance.
- Using CMMs for process control in comparison to dedicated bench or automatic gaging is (1) comparable in cost when considering the day-to-day cost of measuring workpieces, and (2) lower in cost when considering change-over of processes to new part numbers, new designs of the same part, or new tolerances for the same part.
- Using CMMs for gage and fixture calibration is lower in cost than manual calibration methods.

ACKNOWLEDGMENTS

Appreciation is extended to Dr. Henrik Nielsen and my other industrial metrology colleagues for their many helpful suggestions and contributions. Other such colleagues included Dr. Mark Malburg, Samir Bhargava, Weibo Weng, Patrick Nugent, and Dr. James Salsbury. Our many discussions pertaining to achieving quality through precision measurements were most useful in preparing this chapter. Gratitude is extended to Dr. Paulo Pereira of Caterpillar Inc. for his patient assistance in generating the second edition of this chapter.

References

AIAG. 2002. *Measurement System Analysis*. 3rd ed. Southfield, MI: Automotive Industry Action Group.

Allan, A. L. 1988. The principles of theodolite intersection systems. *Surv Rev* 29(117):226.

Althin, T. K. W. 1948. *C. E. Johansson, 1864–1943: The Master of Measurement*. Stockholm: Ab. C.E. Johansson Corporation.

ANSI/ASQ Z1.4-2003. 2003a. Sampling procedures and tables for inspection by attributes. New York: American National Standards Institute.

ANSI/ASQ Z1.9-2003. 2003b. Sampling procedures and tables for inspection by variables for percent nonconforming. New York: American National Standards Institute.

ANSI/IEEE 268-1992. 1992. Metric practice. New York: American National Standards Institute.

ANSI/NCSL Z540.1-1994* (R2002). 1994. Calibration laboratories and measuring and test equipment—general requirements. New York: American National Standards Institute.

ANSI/NCSL Z540.2-1997 (R2007). 1997. U.S. guide to the expression of uncertainty in measurement. Boulder, CO: National Conference of Standards Laboratories.

ANSI/NCSL Z540.3-2006. 2006. Requirements for the calibration of measuring and test equipment. Boulder, CO: National Conference of Standards Laboratories.

ASA B48.1-1933. 1933. Inch-millimeter conversion for industrial use. New York: American National Standards Association, ASA.

ASHRAE. 1993. ASHRAE Fundamental handbook. New York: American Society of Heating, Refrigerating, and Air-Conditioning Engineers.

ASME B5.54-2005. 2005a. Methods for performance evaluation of computer numerically controlled machining centers. New York: American Society of Mechanical Engineers.

ASME B5.57-1998. 1998. Methods for the performance evaluation of computer numerically controlled lathes and turning centers. New York: American Society of Mechanical Engineers.

ASME B89.1.9-2002. 2002a. Gage blocks. New York: American Society of Mechanical Engineers.

ASME B89.1.13-2001. 2001a. Micrometers. New York: American Society of Mechanical Engineers.

ASME B89.3.4-2010. 2010. Axes of rotation: Methods for specifying and testing. New York: American Society of Mechanical Engineers.

ASME B89.4.1-1997. 1997. Methods for performance evaluation of coordinate measuring machines. New York: American Society of Mechanical Engineers.

ASME B89.4.10-2000. 2000. Methods for performance evaluation of coordinate measuring system software. New York: American Society of Mechanical Engineers.

ASME B89.4.19-2006. 2006a. Performance evaluation of laser-based spherical coordinate measurement systems. New York: American Society of Mechanical Engineers.

ASME B89.4.22-2004. 2004. Methods for performance evaluation of articulated arm coordinate measuring machines. New York: American Society of Mechanical Engineers.

ASME B89.6.2-1973 (R2003). 1973. Temperature and humidity environment for dimensional measurement. New York: American Society of Mechanical Engineers.

ASME B89.7.2-1999. 1999. Dimensional measurement planning. New York: American Society of Mechanical Engineers.

---

* This standard was withdrawn in July 2007 in favor of ISO/IEC 17025-2005 for part 1 and ANSI/NCSL Z540.3-2006 for part 2, as per NCSLi http://www.ncsli.org/NCSLIORG/Store/Core/Orders/product.aspx?catid=84&prodid=647 (Accessed March 18, 2011).

ASME B89.7.3.1-2001. 2001b. Guidelines for decision rules: Considering measurement uncertainty in determining conformance to specifications. New York: American Society of Mechanical Engineers.

ASME B89.7.3.3-2002. 2002b. Guidelines for assessing the reliability of dimensional measurement uncertainty statements in determining conformance to specifications. New York: American Society of Mechanical Engineers.

ASME TR B89.4.10360.2-2008. 2008. Acceptance test and reverification test for coordinate measuring machines (CMMs)—Part 2: CMMs used for measuring linear dimensions. New York: American Society of Mechanical Engineers.

ASME TR B89.7.3.2-2007. 2007. Guidelines for the evaluation of dimensional measurement uncertainty. New York: American Society of Mechanical Engineers.

ASME TR B89.7.4.1-2005. 2005b. Measurement uncertainty and conformance testing: Risk analysis. New York: American Society of Mechanical Engineers.

ASME TR B89.7.5-2006. 2006b. Metrological traceability of dimensional measurements to the SI unit of length. New York: American Society of Mechanical Engineers.

ASME Y14.5-1994. 1994a. Dimensioning and tolerancing. New York: American Society of Mechanical Engineers.

ASME Y14.5-2009. 2009. Dimensioning and tolerancing. New York: American Society of Mechanical Engineers.

ASME Y14.5.1M-1994. 1994b. Mathematical definition of dimensioning and tolerancing principles. New York: American Society of Mechanical Engineers.

Astin, A. V. 1959. *Refinement of Values for the Yard and the Pound.* U.S. Federal Register, July 1, 1959. Washington, DC: U.S. National Bureau of Standards.

Attia, M. H., and L. Kops. 1979. Calculation of thermal deformation of machine tools, in transient state, with the effect of structural joints taken into account. *Ann CIRP* 28/1:241.

Auerbach, F. 1904. *The Zeiss Works and the Carl Zeiss Stiftung in Jena.* London: Marshall, Brookes & Chalkey, Ltd.

Babu, U., J. Raja, and R. J. Hocken. 1993. Sampling methods and substitute geometry algorithms for measuring cylinders in coordinate measuring machines. *Proc Ann Mtg ASPE.* Seattle, WA: ASPE

Bachmann, J. et al. 2004. Aide in decision-making: Contribution to uncertainties in three-dimensional measurement. *Precis Eng* 28:78.

Baldwin, J. M. et al. 2007. Application of simulation software to coordinate measurement uncertainty evaluations. *Measure* 2(4):40.

Balsamo, A., D. Marques, and S. Sartori. 1990. A method for thermal-deformation corrections of CMMs. *Ann CIRP* 39/1:557.

Balsamo, A. et al. 1996. Towards instrument oriented calibration of CMMs. *Ann CIRP* 45/1:479.

Bandy, H. T. 1991. Process-intermittent error compensation. In *NISTIR 4536, Progress Report of the Quality in Automation Project for FY90,* ed. M. A. Donmez, pp. 41–50. Gaithersburg, MD: NIST.

Bartscher, M. et al. 2007. Enhancement and proof of accuracy of industrial computed tomography measurements. *Ann CIRP* 56/1:495.

Battison, A. 1976. *Muskets to Mass Production.* Windsor, VT: The American Precision Museum.

Beers, J. S., and W. B. Penzes. 1992. NIST length scale interferometer measurement assurance, *NISTIR-4998.* Gaithersburg, MD: National Institute of Standards and Technology.

Belforte, G. et al. 1987. Coordinate measuring machines and machine tools self-calibration and error correction. *Ann CIRP* 36/1:359.

Bhargava, S., and M. B. Grant. 1993. Taguchi methods and tolerance assignment: The wavelength aspects of surface metrology. In *Proceedings of the 1993 International Forum on Dimensional Tolerancing and Metrology,* ed. V. Srinivasan and H. B. Voelcker, 27:177. Dearborn, MI: ASME.

BIPM. 1960. New definition of the meter: The wavelength of Krypton-86. In *11th General Council of Weights and Measures*. Paris: Bureau International des Poids et Mesures.

BIPM. 1984. Documents concerning the new definition of the metre. *Metrologia* 19:163. Paris: Bureau International des Poids et Mesures.

BIPM. 2006. *The International System of Units (SI)*. 8th ed. Paris: Bureau International des Poids et Mesures.

Bickersteth, R. 1929. Temperature of adjustment for industrial standards of length. *Machinery.*

Blaedel, K. L. 1980. Error reduction. In *Report of the Machine Tool Task Force, Volume 5: Machine Tool Accuracy,* UCRL-52960-S, ed. R. Hocken. Livermore, CA: University of CA.

Blaedel, K. 1993a. Some comments on changing the reference temperature. *Precis Eng* 15/1:4.

Blaedel, K. 1993b. Thermal effects tutorial. In *Proceedings of the ASPE.* Raleigh, NC: American Society for Precision Engineering.

Bobo, D. 1999. Selecting the right probe for CMMs, Precision Metrology with CMMs, *SME,* Minneapolis, MN, June 8–9.

Bosse, H. et al. 2003. Final report on CCL-S3 supplementary line scale comparison Nano3. *Metrologia* 40.

Braudaway, D. W. 1990. *Standards Laboratory Environments, SAND-90-1962.* Albuquerque, NM: Sandia National Laboratories.

Breyer, K. H., and H. G. Pressel. 1991. Paving the way to thermally stable coordinate measuring machines. In *Proceedings of the IPES 6, Progress in Precision Engineering,* ed. P. Seyfried et al. 56. Braunschweig, New York: Springer-Verlag.

Brown & Sharpe. 2001. Justifying the CMM, *P/N 80-80059-2, Brown & Sharpe Measuring Systems Division.* Precision Park, North Kingstown, RI.

Bryan, J. 1979. Design and construction of an 84 inch diameter diamond turning machine. *Precis Eng* 1:13.

Bryan, J. 1990. International status of thermal error research (1990). *Ann CIRP* 39/2:645.

Bryan, J. 1993. The deterministic approach in metrology and manufacturing. In *Proceeding of the 1993 International Forum on Dimensional Tolerancing and Metrology,* pp. 85–95, ISBN 0-7918-0697-9. New York: ASME.

Bryan, J. B. 1968. International status of thermal error research. *Ann CIRP* 16/2:203.

Bryan, J. B., and D. L. Carter. 1989a. How straight is "straight"? *Amer Mach* (Dec 1, 1989 Issue): 61.

Bryan, J. B., and D. L. Carter. 1989b. Straightness metrology applied to a 100-inch travel creed feed grinder. In *5th International Precision Engineering Sem.,* Monterey, CA (LLNL MISC-4848, October).

Bryan, J. et al. 1966. Thermal effects in dimensional metrology. *ASME 65, Prod. 13.* New York: American Society for Mechanical Engineers.

Bryan, J. et al. 1972. A practical solution to the thermal stability problem in machine tools. *SME MR72–138* and *UCRL 73577.*

Bryan, J. et al. 1973. Reduction of machine tool spindle growth. In *Proceedings of the SME North American Metal Working Research Conference.* McMasters University and UCRL 74672.

Bryan, J. et al. 1982. An order of magnitude improvement in thermal stability using a liquid shower on a measuring machine. In *Proceedings of the SME Workshop on Precision Mach.,* St. Paul, and UCRL 87591: SME.

BSI. 1989. *Assessment of Position, Size and Departure from Nominal Force of Geometric Features.* London: British Standard Institution, BS 7172.

Callander, T. 1993. Justifying MicroVal personal CMMs, P/N 80-80054-3. *Brown & Sharpe Measuring Systems Division.* Precision Park, North Kingstown, RI: Brown & Sharpe.

Chakravarthy, B., H. Cherukuri, and R. G. Wilhelm. 2002. Prediction of thermal soakout time using analytical models. *Precis Eng* 26/1:15.

Charlton, T. 2003. Sensor fusion errors in multi-sensor coordinate measurement. *International Dimensional Workshop*. Nashville, TN.

Chen, Y. H., Y. Z. Wang, and Z. Y. Yang. 2004. Towards a haptic virtual coordinate measuring machine. *Int J Mach Tools Manufac* 44:1009.

Chou, C., and D. B. DeBra. 1990. Liquid temperature control for precision tools. *Ann CIRP* 39/1:535.

Christoph, R. 1989. Bestimmung von geometrischen Größen mit Fotoempfängeranordnungen. Post-doctoral thesis (in German), Jena: Friedrich Schiller University.

Christoph, R., and H. J. Neumann. 2007. *Multi-Sensor Coordinate Metrology*. 2nd ed., ISBN 3-937889-03-5. Landsberg, Germany: Verlag Moderne Industrie.

Christoph, R., E. Trapet, and H. Schwenke. 1997. Verfahren und anordnung zur messung von strukturen eines objects. German Patent DE 198 05 892 A1 (in German).

CIPM. 1931. *Procès-Verbaux des Séances Deuxième Série Tom XIV Session De 1931 Libraire du Bureau Des Longitude*. Comité International des Poids et Mesures, de L'École Polytechnique 55, Quai des Grands-Augustine, 55 Paris.

Clarke, T. A. et al. 2001. Performance verification for large volume metrology systems. In *Proceedings of the Laser Metrology and Machine Performance V*, ed. G. N. Peggs, 105. Birmingham, U.K.: Wessex Institute of Technology; National Physical Laboratory.

Clément, A., P. Bourdet, and R. Weill. 1981. Commande adaptive dimensionelle des machines à mesure 3D. *Ann CIRP* 30/1:429.

CMMA. 1989. *Accuracy Specification for Coordinate Measuring Machines*. London: Coordinate Measuring Machine Manufacturers Association.

Cobleigh, R. 1909. *Handy Farm Devices and How to Make Them*. New York: Orange Judd Co. (Reprinted by Lyons and Burford, 1996).

Cochrane, R. C. 1966. *Measures for Progress: A History of the National Bureau of Standards*. Washington, DC: NBS, U.S. Department of Commerce.

Concheri, G. et al. 2001. Geometric dimensioning and tolerancing (GD&T) versus Geometrical product specification (GPS). In *Proceedings of the XII ADM International Conference on Design Tools and Methods in Industrial Engineering*. Italy. http://adm.ing.unibo.it/ADM%20Rimini/Papers/D1/5-ADM01-D1-024.pdf—as of March 2011.

Curtis, M. A., and F. T. Farago. 2007. *Handbook of Dimensional Metrology*. 4th ed. New York: Industrial Press Inc.

Dang, Q. C., S. Yoo, and S. W. Kim. 2006. Complete 3-D self-calibration of coordinate measuring machines. *Ann CIRP* 55/1:527.

DeBra, D. B. 1992. Vibration isolation of precision machine tools and instruments. *Ann CIRP* 41/2:711.

DeBra, D. B. 1998. Active vibration isolation. Tutorial notes, St. Louis, MO: ASPE Annual Meeting.

DeBra, D. B., R. A. Victor, and J. Bryan. 1986. Shower and high pressure oil temperature control. *Ann CIRP* 35/1:359.

DeFelice, J. D. 1970. Predicting "soak time" before measurement. *J Qual Technol* 2(2):67–71.

Department of Defense. 1988. MIL-STD-45662A,* *Calibration Systems Requirements*. Washington, DC: Department of Defense.

Department of Defense. 1995. MIL-STD-45662A Notice 2, *Calibration Systems Requirements—Notice of Cancellation*. Washington, DC: Department of Defense.

Department of Defense. 1998. AFMAN 32-1094, *Criteria for Air Force Precision Measurement Equipment Laboratory Design and Construction*. Washington, DC: Department of Defense.

Dobosz, M., and A. Wozniak. 2003. Metrological feasibilities of CMM touch-trigger probes. Part II: Experimental verification of the 3D theoretical model of probe pretravel. *Measurement* 34/4:287.

---

* Officially cancelled in 1995 (Department of Defense 1995) in favor of civilian documents ISO 10012 (ISO 2003a) and ANSI/NCSL Z540.1 (ANSI/NCSL, 1994).

Doiron, T. 2004. The history and current status of traceability. In *Summer Topical Meeting, Uncertainty Analysis in Measurement and Design*, ed. W. T. Estler and E. R. Marsh, 28. State College, PA: ASPE.

Doiron, T. 2007. 20°C—A short history of the standard reference temperature for industrial dimensional measurements. *J Res NIST* 112:23.

Doiron, T., A. Schneider, and D. McLaughlin. 2007. Use of air showers to reduce soaking time for high precision dimensional measurements. *NCSL International Annual Workshop & Symposium*. St. Paul, MN: NCSLI.

Doiron, T., and J. Stoup. 1997. Uncertainty and dimensional calibrations. *J Res NIST* 102:647.

Donaldson, R. R. 1972. A simple method for separating spindle error from test ball roundness error. *Ann CIRP* 21/1:125.

Donaldson, R., and S. Patterson. 1983. Design and construction of a large vertical axis diamond turning machine. In *SPIE 27th Annual Technical Symposium*, pp. 62–7.

Donmez, M. A. 1987. Improving quality through deterministic real-time control. In *Quality: Design, Planning, and Control. ASME Winter Annual Meeting*. New York: ASME.

Donmez, M. A. 1989. A real-time control system for CNC machine tool based on deterministic metrology. In *Statistical Process Control in Automated Manufacturing*, ed. J. B. Keats and N. F. Hubels. New York: Marcel Dekker, Inc.

Donmez, M. A. et al. 1986. A general methodology for machine tool accuracy enhancement by error compensation. *Precis Eng* 8:187.

Dotson, C. L. 2006. *Fundamentals of Dimensional Metrology*. 5th ed. Albany, NY: Delmar Cengage Learning.

Drescher, J. 2004. Characterization of a turbine airfoil cooling hole check standard for 5-axis CMMs—uncertainty reduction by redundancy and error reversal. In *Summer Topical Meeting, Uncertainty Analysis in Measurement and Design*, ed. W. T. Estler and E. R. Marsh 107. State College, PA: ASPE.

Duplain, R. J. 1991. Justifying the CMM, *Giddings & Lewis Measurement Systems*. Dayton, OH: Giddings & Lewis Measurement Systems.

Edgeworth, R., and R. G. Wilhelm. 1996. Uncertainty management for CMM probe sampling of complex surfaces. *ASME Manuf Sci Eng* MED-4:511–8.

Edgeworth, R., and R. G. Wilhelm. 1999a. Adaptive sampling for coordinate metrology. *Precis Eng* 23(3):144.

Edgeworth, R., and R. G. Wilhelm. 1999b. Sampling and measurement uncertainty in coordinate metrology. *14th Ann Mtg ASPE*, vol. 20, 389.

Esswein, T., and H. J. Neumann. 1991. Decision-making analysis for the use of CMMs, 60-25-0081/I-e. Oberkochen, FRG: Carl Zeiss, GmbH.

Estler, W. T. 1986. This reversal was first derived and described at the 1986 *SME Precision Machining Workshop*. Cambridge, MA: SME.

Estler, W. T., and E. B. Magrab. 1985. Validation metrology of the large optics diamond turning machine. *NBSIR 85-3182(R)*, Gaithersburg, MD.

Estler, W. T. et al. 1996. Error compensation for CMM touch trigger probes. *Precis Eng* 19:85.

Estler, W. T. et al. 2002. Large-scale metrology—an update. *Ann CIRP* 51/2:587.

Evans, C. 1989. *Precision Engineering: An Evolutionary View*. Bedford, UK: Cranfield Press.

Evans, C. J., R. J. Hocken, and W. T. Estler. 1996. Self-calibration: Reversal, redundancy, error separation, and 'absolute testing. *Ann CIRP* 45/2:617.

Flack, D. 2001a. *Measurement Good Practice Guide No. 41—CMM Measurement Strategies*. NPL, ISSN 1368-6550. London, UK: National Physical Laboratory.

Flack, D. 2001b. *Measurement Good Practice Guide No. 42—CMM Verification*. NPL, ISSN 1368-6550. London, UK: National Physical Laboratory.

Flack, D. 2001c. *Measurement Good Practice Guide No. 43—CMM Probing*. NPL, ISSN 1368-6650: London, UK: National Physical Laboratory.

Franck, G. L. 1987. Flexible inspection systems for automated manufacturing. In *Test, Meas. & Inspect for Quality Control Conference/Exhibition.*

Fu, J. 2000. *Illumination Model and Plate Calibration Method for Vision-Based Coordinate Measuring Machines.* PhD diss., Charlotte, NC: UNC Charlotte.

Gavin, R. J., and K. W. Yee. 1989. Implementing error compensation on machine tools. In *Proceedings of the Southern Manufacturing Technology Conference* (Sponsored by NMTBA, The Association for Manufacturing Technology). Charlotte, NC.

General Service Administration. GSA 1973. GGG-P-463c, Federal specification plate, surface (granite), GSA, Washington, DC.

General Service Administration. GSA 1992. FED-STD-209E, Airborne particulate cleanliness classes in clean rooms and clean zones, Federal Standard 209E, GSA, Washington, DC.*

General Service Administration. 2001. FED-STD-209E, Notice 1, Airborne particulate cleanliness classes in clean rooms and clean zones, Notice of cancellation, GSA, Washington, DC.

Grant, E. L., and R. S. Leavenworth. 1996. *Statistical Quality Control.* New York: McGraw-Hill Publishing Company, Inc.

Greenleaf, A. H., and J. T. Watson. 1984. Self calibrating contour measuring system using fringe counting interferometers. U.S. Patent No. 4,457,625.

Groppelli, A. A., and E. Nikbakht. 2006. *Finance,* Business Review Books. 5th ed. Hauppauge, NY: Barron's Educational Series, Inc.

Halverson, P. G. et al. 2004. Progress towards picometer accuracy laser metrology for the space interferometry mission—update for ICSO 2004. In *Proceedings of the 5th International Conference on Space Optics* (ICSO 2004), 30 March–2 April 2004. Toulouse, France: ESA Publications Division, ISBN 92-9092-865-4.

Hamon, J., P. Giacomo, and P. Carré. 1987. International comparison of measurements of linescales (1976–1984). *Metrologia* 24:187.

Hansen, H. N., and L. De Chiffre. 1997. A combined optical and mechanical reference artefact for coordinate measuring machines. *Ann CIRP* 46/1:467.

Harrington, N. T. 1922. Clearance-testing instrument. U.S. Patent No. 1,437,053.

Hemmelgarn, T. L. et al. 1997. Coordinate measuring machine. U.S. Patent 6,058,618.

Hocken, R. J., ed. 1980. *Technology of Machine Tools, Volume 5: Machine Tool Accuracy.* Report of the Machine Tool Task Force, UCRL-52960-S. Livermore, CA: University of California.

Hocken, R., and B. Borchardt. 1979. On characterizing measuring machine geometry, NBSIR 79-1752. Washington, DC: National Bureau of Standards.

Hocken, R. J., N. Chakraborty, and C. Brown. 2005. Optical metrology of surfaces. *Ann CIRP* 54/2:705.

Hocken, R., J. Raja, and U. Babu. 1993. Sampling issues in coordinate metrology. *Manuf Rev* 6:282.

Hocken, R. et al. 1977. Three dimensional metrology. *Ann CIRP* 26/2:403.

Hopp, T. 1993. Computational metrology. *Manuf Rev* 6:295.

Hughes, E. B., A. Wilson, and G. N. Peggs. 2000. Design of a high-accuracy CMM based on multi-lateration techniques. *Ann CIRP* 49/1:391.

Hume, J. 1953. Linear measurement. In *Engineering Metrology,* vol. 125. London: MacDonald & Co.

Instrument Society of America. 1975. ISA-RP52.1, Recommended environments for standards laboratories. Research Triangle Park, NC: ISA.

ISO 1:2002. 2002a. Geometrical Product Specifications (GPS)—Standard reference temperature for geometrical product specification and verification. Geneva, Switzerland: International Organization for Standardization.

ISO 230-3:2007. 2007. Test code for machine tools—Part 3: Determination of thermal effects. Geneva, Switzerland: International Organization for Standardization.

---

* Cancelled in 2001 (General Service Administration 2001) in favor of ISO 14644 (ISO 1999a).

ISO 230-6:2002. 2002b. Test code for machine tools—Part 6: Determination of positioning accuracy on body and face diagonals (Diagonal displacement tests). Geneva, Switzerland: International Organization for Standardization.

ISO 1101:2004. 2004a. Geometrical Product Specifications (GPS)—Geometrical tolerancing—tolerances of form, orientation, location and run-out. Geneva, Switzerland: International Organization for Standardization.

ISO 9000:2005. 2005a. Quality management systems—Fundamentals and vocabulary. Geneva, Switzerland: Organization for Standardization.

ISO 9001:2008. 2008a. Quality management systems—Requirements. Geneva, Switzerland: International Organization for Standardization.

ISO 10012:2003. 2003a. Measurement management systems—Requirements for measurement processes and measuring equipment. Geneva, Switzerland: International Organization for Standardization.

ISO 10360-1:2000. 2000a. Geometrical Product Specifications (GPS)—Acceptance and reverification tests for coordinate measuring machines (CMM)—Part 1: Vocabulary. Geneva, Switzerland: International Organization for Standardization.

ISO 10360-2:2001.* 2001a. Geometrical Product Specifications (GPS)—Acceptance and reverification tests for coordinate measuring machines (CMM)—Part 2: CMMs used for measuring size. Geneva, Switzerland: International Organization for Standardization.

ISO 10360-2:2009. 2009a. Geometrical Product Specifications (GPS)—Acceptance and reverification tests for coordinate measuring machines (CMM)—Part 2: CMMs used for measuring size. Geneva, Switzerland: International Organization for Standardization.

ISO 10360-3:2000. 2000b. Geometrical Product Specifications (GPS)—Acceptance and reverification tests for coordinate measuring machines (CMM)—Part 3: CMMs with the axis of a rotary table as the fourth axis. Geneva, Switzerland: International Organization for Standardization.

ISO 10360-4:2000. 2000c. Geometrical Product Specifications (GPS)—Acceptance and reverification tests for coordinate measuring machines (CMM)—Part 4: CMMs used in scanning measuring mode. Geneva, Switzerland: International Organization for Standardization.

ISO 10360-5:2010. 2010a. Geometrical Product Specifications (GPS)—Acceptance and reverification tests for coordinate measuring machines (CMM)—Part 5: CMMs using single and multiple-stylus contacting probing systems. Geneva, Switzerland: International Organization for Standardization.

ISO 10360-6:2001. 2001b. Geometrical Product Specifications (GPS)—Acceptance and reverification tests for coordinate measuring machines (CMM)—Part 6: Estimation of errors in computing Gaussian associated features. Geneva, Switzerland: International Organization for Standardization.

ISO 13485:2003. 2003b. Medical devices—Quality management systems—Requirements for regulatory purposes. Geneva, Switzerland: International Organization for Standardization.

ISO 14253-1:1998. 1998. Geometrical Product Specifications (GPS)—Inspection by measurement of workpieces and measuring instruments—Part 1: Decision rules for proving conformance or non-conformance with specification. Geneva, Switzerland: International Organization for Standardization.

ISO 14406:2010. 2010c. Geometrical Product Specification (GPS)—Data extraction. Geneva, Switzerland: International Organization for Standardization.

ISO 14644-1:1999. 1999a. Clean rooms and associated controlled environments—Part 1: Classification of air cleanliness. Geneva, Switzerland: International Organization for Standardization.

---

* Replaced by the 2009 version (ISO 2009a).

ISO 16610 Series. 2006, 2009, 2010. Geometrical Product Specifications (GPS)—Filtration. Geneva, Switzerland: International Organization for Standardization.

ISO 22093:2003. 2003c. Industrial automation systems and integration—Physical device control—Dimensional Measuring Interface Standard (DMIS). Geneva, Switzerland: International Organization for Standardization.

ISO/IEC 17025:2005. 2005. General requirements for the competence of testing and calibration laboratories. Geneva, Switzerland: International Organization for Standardization.

ISO/FDIS 10360-7:2011. 2011. Geometrical Product Specifications (GPS)—Acceptance and reverification tests for coordinate measuring machines (CMM)—Part 7: CMMs equipped with imaging probing systems. Geneva, Switzerland: International Organization for Standardization.

ISO/TR 16015:2003. 2003d. Geometrical Product Specifications (GPS)—Systematic errors and contributions to measurement uncertainty of length measurement due to thermal influences. Geneva: International Organization for Standardization.

ISO/TS 14253-2:1999. 1999b. Geometrical Product Specifications (GPS)—Inspection by measurement of workpieces and measuring equipment—Part 2: Guide to the estimation of uncertainty in GPS measurement, in calibration of measuring equipment and in product verification. Geneva, Switzerland: International Organization for Standardization.

ISO/TS 14253-4:2010. 2010b. Geometrical Product Specifications (GPS)—Inspection by measurement of workpieces and measuring equipment—Part 4: Background on functional limits and specifications limits in decision rules. Geneva, Switzerland: International Organization for Standardization.

ISO/TS 15530-3:2004. 2004b. Geometrical Product Specifications (GPS)—Coordinate Measuring Machines (CMM): Techniques for determining the uncertainty of measurement—Part 3: Use of calibrated workpieces or standards. Geneva, Switzerland: International Organization for Standardization.

ISO/TS 15530-4:2008. 2008b. Geometrical Product Specifications (GPS)—Coordinate Measuring Machines (CMM): Techniques for determining the uncertainty of measurement—Part 4: Evaluating task-specific uncertainty using simulation. Geneva, Switzerland: International Organization for Standardization.

ISO/TS 17450-1:2005. 2005b. Geometrical Product Specifications (GPS)—General concepts—Part 1: Model for geometrical specification and verification. Geneva, Switzerland: International Organization for Standardization.

ISO/TS 23165:2006. 2006. Geometrical Product Specifications (GPS)—Guidelines for the evaluation of coordinate measuring machine (CMM) test uncertainty. Geneva, Switzerland: International Organization for Standardization.

Jalkio, J. A. 1999. The use of optical probes on coordinate measuring machines - strengths and weaknesses. In Precision Metrology with CMMs—SME. Minneapolis, MN: SME.

Janeczko, J. 1998. Machine tool thermal distortion compensation. In Proceedings of the NMTBA 4th Biennial International Manufacturing Tool Technology Conference. McLean, VA: NMTBA

JCGM 100:2008. 2008a. Evaluation of measurement data—Guide to the expression of uncertainty in measurement (GUM). Paris, France: Joint Committee for Guides in Metrology.

JCGM 101:2008. 2008b. Evaluation of measurement data—Supplement 1 to the Guide to the expression of uncertainty in measurement—Propagation of distributions using a Monte Carlo method. Paris, France: Joint Committee for Guides in Metrology.

JCGM 200:2008. 2008c. International vocabulary of metrology—Basic and general concepts and associated terms (VIM). Paris, France: Joint Committee for Guides in Metrology.

Jouy, F., and A. Clément. 1986. Theoretical modelisation and experimental identification of the geometrical parameters of coordinate-machines by measuring a multi-directed bar. Ann CIRP 35/1:393.

Juran, J. M., and A. B. Godfrey. 1999. *Juran's Quality Handbook*. 5th ed. New York: McGraw-Hill, ISBN 0-07-034003-X.

Jusko, O., J. G. Salsbury, and H. Kunzmann. 1999. Results of the CIRP-form intercomparison 1996–1998. *Ann CIRP* 48/1:413.

Kak, A., and A. Slaney. 2001. *Principles of Computerized Tomographic Imaging*. Philadelphia, PA: Society for Industrial Mathematics, ISBN-10:089871494X.

Kalpakjian, S., and S. R. Schmid. 2009. *Manufacturing Engineering and Technology*. 6th ed. Upper Saddle River, NJ: Prentice Hall.

Kim, K., K. Eman, and S. M. Wu. 1987. In-process control of cylindricity in boring operations. *ASME Trans. J Engrg Ind*, 109(4):385–91.

Kim, S. W., and P. A. McKeown. 1996. Measurement uncertainty limit of a video probe in coordinate metrology. *Ann CIRP* 45/1:493.

Knapp, W., U. Tschudi, and A. Bucher. 1991. Comparison of different artifacts for interim coordinate measuring machine checking: A report from the Swiss Standards Committee. *Precis Eng* 13:277.

Krejci, J. V. 1991. Direct computer control coordinate measurement systems: The impact of programmer defined variables on accuracy and repeatability. In *Precision Metrology with Coordinate Measurement Systems—SME*. Southfield, MI: SME.

Krejci, J. V. 1992. Factors to consider in specifying and purchasing a coordinate measuring machine. In *Precision Metrology with Coordinate Measurement Machines—SME*. Livonia, MI: SME.

Křen, P. 2007. Linearisation of counting interferometers with 0.1 nm precision. *Int J Nanotechnol* 4:702.

Kroll, J. J. 2003. *Six Degree of Freedom Optical Sensor for Dynamic Measurement of Linear Axes*. PhD thesis, Charlotte, NC: The University of North Carolina.

Kruth, J.-P., C. Van den Bergh, and P. VanHerck. 2001. Correcting steady-state temperature influences on coordinate measuring machines. *J Manuf Syst* 19/6:365.

Kruth, J.-P., P. VanHerck, and L. D. Jonge. 1994. Self calibration method and software error correction for three-dimensional coordinate measuring using artifact measuring. *Measurement* 14:155.

Küng, A., and F. Meli. 2007. Comparison of three independent calibration methods applied to an ultra-precision µ-CMM. In *Proceedings of 7th International Conference of the European Society for Precision Engineering and Nanotechnology (EUSPEN)*, vol. 1, 230. Bremen, Germany: euspen.

Küng, A., F. Meli, and R. Thalmann. 2007. Ultraprecision micro-CMM using a low force 3D touch probe. *Meas Sci Technol* 18/2:319.

Kunzmann, H. 1989. Today's limits of accuracy in dimensional measurement. In *Proceedings of the IMEKO International Symposium on Metrology and Quality Control*. Beijing, China: IMEKO.

Kunzmann, H., T. Pfeifer, and J. Flügge. 1993. Scales vs. laser interferometers, performance and comparison of two measuring systems, Keynote Paper. *Ann CIRP* 42/2:753.

Kunzmann, H., E. Trapet, and F. Wäldele. 1990. A uniform concept for calibration, acceptance test, and periodic inspection of coordinate measuring machines using reference objects. *Ann CIRP* 39/1:561.

Kunzmann, H., F. Wäldele, and E. Salje. 1983. On testing coordinate measuring machines (CMM) with kinematic reference standards (KRS). *Ann CIRP* 32/1:465.

Kunzmann, H. et al. 1989. Method of measuring rotary-table deviations. U.S. Patent No. 4,819,339.

Kunzmann, H. et al. 2005. Productive metrology—adding value to manufacture. *Ann CIRP* 54/2:691.

Lau, K. C., and R. J. Hocken. 1984. A survey of current robot metrology methods. *Ann CIRP* 33/2:485.

Lau, K. C., and R. J. Hocken. 1987. Three and five axis laser tracking systems. U.S. Patent 4,714,339. Washington, DC.

Lau, K., R. Hocken, and W. C. Haight. 1986. Automatic laser tracking interferometer system for robot metrology. *Precis Eng* 8:3.

Lawall, J., and E. Kessler. 2000. Michelson interferometry with 10 pm accuracy. *Rev Sci Instrum* 71:2669.

Lin, Y. B. 2002. *Optimal Design of Four-Beam Laser Tracking Interferometric Flexible Coordinate Measuring System.* PhD thesis, Tianjin, China: Tianjin University.

Lin, Y. B., G. X. Zhang, and Z. Li. 2002. Design and optimization of a cat's eye retroreflector. *Acta Optica Sinica* 22:1245. (in Chinese).

Lingard, P. S. et al. 1991. Temperature perturbation effects in a high precision CMM. *Precis Eng* 13:41.

Liu, S. G. et al. 2005. A portable 3D vision coordinate measuring system using a light pen. *Key Eng Mater* 295–296:331.

Loewen, E. 1978. Air shower thermal stability. In *Proceedings of the SME Precision Mach. Workshop.* Williamsburg, VA: SME.

Lotze, W. 1994. Multidimensional measuring probe head improves accuracy and functionality of coordinate measuring machines. *Measurement* 13:91.

Lu, E., J. Ni, and S. M. Wu. 1992. An integrated lattice filter adaptive control system for time-varying CMM structural vibration control, Part I: Theory and simulation; Part II: Experimental implementation. In *Proceedings of the ASME Winter Annual Meeting,* 127–42. NY: ASME.

Luttrell, D. 2007. Fundamentals of thermal effects: Precision design principles and measurement and control of temperature. Tutorial notes, Dallas, TX: ASPE Annual Meeting.

Marsh, E. R. 2009. *Precision Spindle Metrology.* 2nd ed. Lancaster, PA: Destech Publications, Inc. ISBN-13:978-1605950037.

Maxwell, J. C. 1890. General considerations concerning scientific apparatus. In *The Scientific Papers of James Clerk Maxwell,* ed. W. D. Niven, pp. 505–22. Cambridge, U.K.: The University Press.

McClure, R. 1969. *Manufacturing Accuracy Through the Control of Thermal Effects.* Dr. Engineering thesis. Berkeley, CA: University of California Berkeley and LLNL.

McMurtry, D. 1987. Footprinting. In *Proceedings of the SME Technical Conference.* Los Angeles, CA: SME.

McMurtry, D. 2003. The development of sensors for CMMs. In *Proceedings of the Laser Metrology and Machine Performance VI,* 205. LAMDAMAP, WIT Press.

Meli, F. et al. 2003. Novel 3D analogue probe with a small sphere and low measurement force. In *Proceedings of the Coordinate Measuring Machines,* 69. Charlotte, NC: ASPE.

Mendenhall, T. C. 1893. Fundamental standards of length and mass. *U.S. Coast and Geodetic Survey Report for 1893.* Washington, DC: US GPO.

Meredith, D. R. 1996. Technical feature: Improving the vibration resistance of CMMs. *Mfg Brown Sharpe Publ Precis Manuf* 3:16.

Michelson, A. A., and E. W. Morley. 1887. On the relative motion of the earth and the luminiferous ether. *Am J Sci* 34:333.

Mikhail, E. M., J. S. Bethel, and J. C. McGlone. 2001. *Introduction to Modern Photogrammetry.* New York: John Wiley & Sons.

Minoru, I., and I. Akira. 1986. Three-view stereo analysis. *IEEE Trans Pattern Anal Mach Intell* PAMI-8/4:524.

Montgomery, D. C. 2009. *Introduction to Statistical Quality Control.* 6th ed. New York: John Wiley.

Moore, W. R. 1970. *Foundations of Mechanical Accuracy.* Bridgeport, CT: The Moore Special Tool Company.

Moroni, G., W. Polini, and M. Rasella. 2003. Manufacturing signatures and CMM sampling strategies. In *Summer Topical Meeting, Coordinate Measuring Machines,* 57. Charlotte, NC: ASPE.

Morse, L. C., and D. L. Babcock. 2009. *Managing Engineering and Technology.* 5th ed. Upper Saddle River, NJ: Prentice Hall.

Morse, E., and H. Voelcker. 1996. Technical note: A tale of two tails. *Mfg Brown Sharpe Publ Precis Manuf* 3:46.

Muralikrishnan, B., J. A. Stone, and J. R. Stoup. 2006. Fiber deflection probe for small hole metrology. *Precis Eng* 30:154.

Muralikrishnan, B., J. A. Stone, and J. R. Stoup. 2008. Area measurement of knife-edge and cylindrical apertures using ultra-low force contact fibre probe on a CMM. *Metrologia* 45:281.

Nakamura, O. et al. 1991. Development of a coordinate measuring system with tracking laser interferometer. *Ann CIRP* 40/1:523.

National Council for Standards Laboratories 2000 (NCSL). *Laboratory Design.* Recommended Practice NCSL-RP-7. 3rd ed. Boulder, CO: National Council for Standards Laboratories.

Neumann, H. J. 1993. *Koordinatenmesstechnik.* Renningen-Malmsheim: Expert (Kontakt & Studium 426, in German).

Neumann, H. J. 2000. *Koordinatenmesstechnik im industriellen Einsat.* Landsberg: Moderne industrie (Die Bibliothek der Technik, 203, in German).

Neumann, H. J. 2005. *Präzisionsmesstechnik in der Fertigung mit Koordinatenmessgeräten,* 2. Aufl. Renningen-Malmsheim: Expert (Kontakt & Studium, 646, in German).

Nevins, J. L., and R. I. Winner. 1999. Ford Motor Company's investment efficiency initiative: A case study. IDA Paper P-3311. Alexandria, VA: Institute for Defense Analyses.

Ni, J., and F. Wäldele. 1995. Coordinate measuring machines. In *Coordinate Measuring Machines and Systems,* ed. J. A. Bosch, 39. New York, NY: Marcel Dekker.

Nielsen, H. S. 1992. Uncertainty and dimensional tolerances. *Quality* 31:25.

Nilsson, J. T. 1995. Application considerations. In *Coordinate Measuring Machines and Systems,* ed. J. A. Bosch, 301. New York, NY: Marcel Dekker.

Ogden, H. 1970. Applications of numerical measuring systems. In *Numerical Control User's Handbook,* 377. New York: McGraw Hill Publishing Co., Ltd.

Patterson, S. R. 1986. *Development of Precision Turning Capabilities at Lawrence Livermore National Laboratory, LLNL.* In *3rd Biennial International Machine Tool Technical Conference,* pp. 147–59. Chicago, IL: Lawrence Livermore National Laboratory.

Pereira, P. H. 2005. Influence of measurement uncertainty and sampling size on process control parameters, invited talk at the *Association for Coordinate Metrology Canada 2005 Workshop.* Hamilton, Ontario: McMaster University.

Pereira, P. H., and D. E. Beutel. 2003. Proposed new tests for evaluating CMM performance. In *Summer Topical Meeting, Coordinate Measuring Machines,* 9. Charlotte, NC: ASPE.

Pereira, P. H., and R. J. Hocken. 2007. Characterization and compensation of dynamic errors of a scanning coordinate measuring machine. *Precis Eng* 31:22.

Peters, C., and H. Boyd. 1920. The calibration and dimensional changes of precision gauge blocks. *Am Mach* 53:627.

Phillips, S. D., J. Baldwin, and W. T. Estler. 2009. Economics of measurement uncertainty and tolerances. In *Proceedings of the ASPE Summer Topical Meeting on Economics of Precis Engineering,* 3. Peoria, IL: ASPE.

Phillips, S. D., K. R. Eberhardt, and B. Parry. 1997. Guidelines for expressing the uncertainty of measurement results containing uncorrected bias. *J Res NIST* 102:577.

Phillips, S. D. et al. 1997. The calculation of CMM measurement uncertainty via the method of simulation by constraints. In *Proceedings of ASPE,* vol. 16, 443. Norfolk, VA: ASPE.

Phillips, S. D. et al. 1998. The estimation of measurement uncertainty of small circular features measured by coordinate measuring machines. *Precis Eng* 22:87.

Phillips, S. D. et al. 2003. The validation of CMM task specific measurement uncertainty software. In *Summer Topical Meeting, Coordinate Measuring Machines,* 51. Charlotte, NC: ASPE.

Pond, J. B. 1988. Answers to CMM problems sought after GIDEP "ALERT." *Metal Working News* 15:23.

Porta, C., and F. Wäldele. 1986. Testing of three coordinate measuring machine evaluation algorithms. *PTB Report EUR 10909 EN.* Brussels: Directorate-General Science Research and Development, Commission of the European Communities.

Preston-Thomas, H. 1990. The international temperature scale of 1990 (ITS-90). *Metrologia* 27:3.

Qberg, E., and F. D. Jones. 1920. *Gage Design and Gage-Making.* 1st ed. New York: The Industrial Press.

Quinn, T. J. 2003. Practical realization of the definition of the metre, including recommended radiations of other optical frequency standards (2001). *Metrologia* 40:103.

Radford, G. S. 1922. *The Control of Quality in Manufacturing.* New York: The Ronald Press Company.

Raugh, M. R. 1985. Absolute two-dimensional submicron metrology for electron beam lithography: A theory of calibration with applications. *Precis Eng* 7/1:3.

Raugh, M. R. 1997. Two-dimensional stage self-calibration: Role of symmetry and invariant sets of points. *J Vac Sci Tech B* 15/6:2139.

Rea, M. S., ed. 2000. *IESNA Lighting Handbook.* 9th ed. New York, NY: Illuminating Engineering Society of North America, ISBN-10 0879951508.

Reeve, C. P. 1974. A method of calibrating two-dimensional reference plates. *NBSIR 74-532.* Washington, DC: National Bureau of Standards.

Reznik, L., and K. P. Dabke. 2004. Measurement models: Application of intelligent methods. *Measurement* 35:47.

Riehle, F. et al. Recommended values of standard frequencies for applications including the practical realization of the metre and secondary representations of the second. *Metrologia* (To be published).

Rivin, E. I. 1995. Vibration isolation of precision equipment. *Precis Eng* 17:41.

Rivin, E. I. 2003. *Passive Vibration Isolation.* New York: ASME. ISBN-10 079180187X.

Roblee, J. 1985. Precision temperature control for optics manufacturing. In *Proceedings of the International Symposium on Optimal and Electro-Optical Applied Science and Engineering.* Cannes, France. UCRL 89756.

Roe, J. W. 1916. *English and American Tool Builders.* New Haven, CT: Yale University Press.

Rolt, F. H. 1929. *Gauges and Fine Measurements.* vol. I. London: MacMillan and Co., Ltd.

Ruck, O. 1998. Coordinate measuring apparatus having a control which drives the probe head of the apparatus in accordance with desired data. U.S. Patent 5,737,244.

Ruijl, T. A. M., and J. van Eijk. 2003. A novel ultra precision CMM based on fundamental design principles. In *Proceedings of the Coordinate Measuring Machines,* 33. Charlotte, NC: ASPE.

Salsbury, J. G. 1995. A simplified methodology for the uncertainty analysis of CMM measurements. Technical Paper, *Conference on Precision Metrology/Applying Imaging and Sensoring, SME,* Indianapolis, IN, IQ95-155, 1.

Salsbury, J. G. 2003a. Calibrating and testing CMMs in a world of uncertainty and accreditation. In *Summer Topical Meeting, Coordinate Measuring Machines,* 63. Charlotte, NC: ASPE.

Salsbury, J. G. 2003b. Implementation of the Estler face motion reversal technique. *Precis Eng* 27:189.

Salsbury, J. G., and M. Inloes. 2006. In-situ calibration of temperature compensation systems using differential length measurements. In *Proceedings of the ASPE 2006 Annual Meeting.* Monterey, CA: ASPE.

Santolaria, J. et al. 2008. Kinematic parameter estimation technique for calibration and repeatability improvement of articulated arm coordinate measuring machines. *Precis Eng* 32:251.

Santolaria, J. et al. 2009. Calibration-based thermal error model for articulated arm coordinate measuring machines. *Precis Eng* 33:476.

Sartori, S., and G. X. Zhang. 1995. Geometric error measurement and compensation of machines. *Ann CIRP* 44/2:599.

Sartori, S. et al. 1989. A method for the identification and correction of thermal deformations in a three coordinate measuring machine. *VDI Ber* 761.

Savio, E., and L. De Chiffre. 2001. Performance verification of CMMs for free form measurements. In *Proceedings of the 2nd EUSPEN International Conference*, 362. Torino, Italy: euspen.

Savio, E., and L. De Chiffre. 2002. An artifact for traceable free form measurements on coordinate measuring machines. *Precis Eng* 26:58.

Schwenke, H. et al. 1994. Experience with the error assessment of coordinate measurements by simulation. In *Proceedings of the 3rd International Conference on Ultraprecision in Manufacturing Engineering*, 370. Aachen, Germany.

Schwenke, H. et al. 2005. Error mapping of CMMs and machine tools by a single tracking interferometer. *Ann CIRP* 54/1:475.

Shakarji, C. M. 1998. Least squares fitting algorithms of the NIST algorithm testing system. *J Res NIST* 103(6):633.

Shakarji, C. M. 2002. Evaluation of one- and two-sided geometric fitting algorithms in industrial software. In *Proceedings of the 17th Annual Meeting*, 100. St. Louis, MO: ASPE.

Shakarji, C. M., and A. Clement. 2004. Reference algorithms for Chebyshev and one-sided data fitting for coordinate metrology. *Ann CIRP* 53/1:439.

Simpson, J., R. Hocken, and J. Albus. 1983. The automated manufacturing research facility of the National Bureau of Standards. *J Manuf Syst* 1/1:17.

SIP. 1952. SIP (Société Génévoise d'Instruments de Physique, or Société d'Instruments de Précision SA), *As ninety years went by...*, Geneva, Switzerland: SIP.

Smith, K. 2002. Metrology: Should you buy or outsource? *Qual Digest*.

Slocum, A. H. 1992a. *Precision Machine Design*. Dearborn, MI: Society of Manufacturing Engineers. ISBN-10 0872634922.

Slocum, A. H. 1992b. Design of three groove kinematic couplings. *Precis Eng* 14:67.

Slocum, A. H., and A. Donmez. 1988. Kinematic couplings for precision fixturing. *Precis Eng* 10:115.

Somerville, L. 2009. The evolution of the Renishaw productivity system. In *Proceedings of the ASPE Summer Topical Meeting on The Economies of Precision Engineering*. Peoria, IL: ASPE.

Spitz, S. N. 1999. *Dimensional Inspection Planning for Coordinate Measuring Machines*. PhD diss. Los Angeles, CA: University of Southern California.

Sprauel, J. M. et al. 2003. Uncertainties in CMM measurements, control of ISO specifications. *Ann CIRP* 52/1:423.

Srinivasan, V. 2004. *Theory of Dimensioning*. New York: Marcel Dekker.

Srinivasan, V. 2005. Elements of computational metrology. In Geometric and Algorithmic Aspects of Computer-Aided Design and Manufacturing *DIMACS Book Series*, ed. R. Janardan, M. Smid, and D. Dutta, vol. 67, 79. Providence, RI: American Mathematical Society.

Srinivasan, S., S. B. Acharya, and S. Anand. 2004. A factor analysis approach for robust inspection of circular features with lobing errors. *Trans No Am Manufac Res Inst SME* 32:135.

Srivastava, S., and J. A. Fessler. 2005. Simplified statistical image reconstruction algorithm for polyenergetic X-ray CT. *IEEE Nucl Sci Symp Conf Rec*.

Stone, J. A. et al. 2009. Advice from the CCL on the use of unstabilized lasers as standards of wavelength: The helium-neon laser at 633 nm. *Metrologia* 46:11.

Stroup, C. G., J. L. Overcash, and R. J. Hocken. 2008. The development of an offset-locked frequency stabilized heterodyne laser source for precision metrology. *Proc ASPE* 44.

Taguchi, G. 1986. On-line quality control during production. Tokyo: Japanese Standards Association/American Supplier Institute.

Takatsuji, T. et al. 1998. Restriction on the arrangement of laser trackers in laser trilateration. *Meas Sci Technol* 9/8:1357.

Takatsuji, T. et al. 1999. Whole-viewing-angle cat's-eye retroreflector as a target of laser tracker. *Meas Sci Technol* 10/7:87.

Taylor, B. N., and C. E. Kuyatt. 1994. Guidelines for evaluating and expressing the uncertainty of NIST measurement results. NIST Technical Note 1297. Gaithersburg, MD: National Institute of Standards and Technology. http://physics.nist.gov/Pubs/guidelines/TN1297/tn1297s.pdf (Accessed April 4, 2011).

Taylor, B. N., and A. Thompson. 2008. The international system of units (SI), *NIST Special Publication 330*. Gaithersburg, MD: National Institute of Standards and Technology.

Teague, E. C. 1997. Basic concept for precision instrument design: Designing instruments and machines to have a high degree of repeatability. Tutorial notes, Norfolk, VA: ASPE Annual Meeting.

Thompson, A., and B. N. Taylor. 2008. Guide for the use of the international system of units (SI), *NIST Special Publication 811*. Gaithersburg, MD: National Institute of Standards and Technology.

Towneley, R. 1666. An extract of a letter, written by Mr. Richard Towneley to Dr. Croon, touching the invention of dividing a foot into many thousand parts, for mathematical purposes. *Philos Trans R Soc Lond* 2:457.

Trapet, E., and F. Wäldele. 1989. Coordinate measuring machines in the production line influence of temperature and measuring uncertainties. In *Proceedings of the IV International Congress on Industrial Metrology*. Zaragoza, Spain.

Tsai, R. 1987. A versatile camera calibration technique for high-accuracy 3D machine vision metrology using off-the-shelf TV cameras and lenses. *IEEE J Rob Autom* 3(4):323.

Tullar, P. M. 1992. *Metrology Testing of a Single Beam Laser Tracking Interferometer.* Master's thesis, Charlotte, NC: University of North Carolina.

U.S. Congress. 2007. *America COMPETES Act,* Public Law 110-69, Section 3013.*

van Vliet, W. P., and P. H. Schellekens. 1996. Accuracy limitations of fast mechanical probing. *Ann CIRP* 45/1:483.

van Vliet, W. P., and P. H. J. Schellekens. 1998. Development of a fast mechanical probe for coordinate measuring machines. *Precis Eng* 22:141.

VDI/VDE 2617-1986. 1986. Parts 1–4, Accuracy of coordinate measuring machines, Düsseldorf, Germany. http://www.vdi.de.

VDI/VDE 2617-2003. 2003. Part 6.2, Accuracy of coordinate measuring machines. Characteristics and testing of characteristics. Guideline for the application of DIN EN ISO 10360 to coordinate measuring machines with optical distance sensors, Düsseldorf, Germany.

VDI/VDE 2617-2007. 2007. Part 6.3, Accuracy of coordinate measuring machines. Characteristics their testing. CMM with multiple probing systems, Düsseldorf, Germany.

Vorburger, T. V., and B. Scace, eds. 1990. *Progress Report of the Quality in Automation Project for FY89-4322*. Gaithersburg, MD: NIST.

Vorburger, T. V. et al. 1992. A strategy for the quality control of automated machine tools. *NISTIR 4773*. Gaithersburg, MD: National Institute of Standards and Technology.

Vorburger, T. V. et al. 1994. Strategy for post-process control of a machine tool. *Manuf Rev* 7:252.

Wäldele, F. 2002. Methoden zur Ermittlung der Messunsicherheit von Koordinatenmessgeräten, course manuals of TAE Esslingen.

---

* Replaces the Metric Act of 1866.

Wäldele, F. et al. 1993. Testing of coordinate measuring machine software. *Precis Eng* 15:121.

Walker, R. 1988. *CMM Form Tolerance Algorithm Testing*. GIDEP Alert X1-A-88-01. Washington, DC: Government-Industry Data Exchange Program, DOD.

Wall, M. 2003. Manufacturing flexibility: What constitutes the Holy Grail for some is just business as usual for others—Supplier Business. *Auto Ind.*

Waurzyniak, P. 2008. Masters of Manufacturing: David McMurtry. Manufacturing Engineering. Vol. 141/1.

Waurzyniak, P. 2008. Masters of Manufacturing: David McMurtry. Manufacturing Engineering. Vol. 141/1.

Weck, M. et al. 1995. Reduction and compensation of thermal errors in machine tools. *Ann CIRP* 44/2:589.

Weckenmann, A., and B. Gawande. 1999. *Koordinatenmesstechni.* München: Carl Hanser (in German).

Weckenmann, A., and M. Heinrichowski. 1985. Problems with software for running coordinate measuring machines. *Precis Eng* 7/2:87.

Weckenmann, A., M. Heinrichowski, and H. J. Mordhorst. 1991. Design of gauges and multipoint measuring systems using coordinate measuring machine data and computer simulation. *Precis Eng* 13:203.

Weckenmann, A., M. Knauer, and H. Kunzmann. 1998. The influence of measurement strategy on the uncertainty of CMM measurements. *Ann CIRP* 47/1:451.

Weckenmann, A., G. Peggs, and J. Hoffmann. 2006. Probing systems for dimensional micro and nano metrology. *Meas Sci Technol* 17:504.

Weckenmann, A. et al. 2004. Probing systems in dimensional metrology. *Ann CIRP* 53/2:657.

Weekers, W. G., and P. H. J. Schellekens. 1997. Compensation for dynamic errors of coordinate measuring machines. *Measurement* 20:197.

Wendt, K., and R. Zumbrunn. 1993. Performance test of automated Theodolite measuring systems in production metrology. BCR project Report, Braunschweig, Germany: European Commission.

Whitney, D. E. 2000. Research issues in manufacturing flexibility, Invited Review—*ICRA 2000 Symposium on Flex*, 383. San Francisco, CA: IEEE.

Wilhelm, R. G., R. Hocken, and H. Schwenke. 2001. Task specific uncertainty in coordinate measurement. *Ann CIRP* 50/2:553.

Woo, T. C., and R. Liang. 1993. Dimensional measurement of surfaces and their sampling. *Comput Aided Des* 25/4:233.

Woschni, H.-G., R. Christoph, and A. Reinsch. 1984. Verfahren zur Bestimmung der Lage einer optisch wirksamen Struktur. *Feingerätetechnik* 33:5 (in German).

Wozniak, A., and M. Dobosz. 2003. Metrological feasibilities of CMM touch-trigger probes, Part I: 3D theoretical model of probe pretravel. *Measurement* 34/4:273.

Wren, J. 1991. America at the wheel. *Auto News.*

Wu, Y., S. Liu, and G. Zhang. 2004. Improvement of coordinate measuring machine probing accessibility. *Precis Eng* 28:89.

Ye, J. 1996. *Errors in High-Precision Mask Making and Metrology*. PhD thesis, Stanford Electronics Laboratory, Stanford University, CA: Palo Alto.

Ye, J. et al. 1997. An exact algorithm for self-calibration of two-dimensional precision metrology stage. *Precis Eng* 20:16.

Yee, K. W., and R. J. Gavin. 1990. Implementing fast part probing and error compensation on machine tools, *NISTIR 4447*. Gaithersburg, MD: NIST.

Yee, K. W. et al. 1992. Automated Compensation of part errors determined by in-process gauging, *NISTIR 4854*. Gaithersburg, MD: National Institute of Standards and Technology.

Yuan, H. 2011. http://www.people.wku.edu/haiwang.yuan/China/proverbs/p.html. Accessed March 5, 2010.

Zhang, H. 2002. *A Study on the Binocular Vision Probe for Sculptured Surface Measurement.* PhD thesis, Tianjin, China: Tianjin University.

Zhang, G. et al. 1985. Error compensation of coordinate measuring machines. *Ann CIRP* 34/1:445.

Zhang, G. et al. 1988. A displacement method for machine geometry calibration. *Ann CIRP* 37/1:515.

Zhang, G. X. et al. 1999. *Coordinate Measuring Machines.* Tianjin, China: Tianjin University Press (in Chinese).

Zhang, G. X. et al. 2002. Towards the intelligent CMM. *Ann CIRP* 51/1:437.

Zhang, G. X. et al. 2003. A study on the optimal design of laser-based multilateration system. *Ann CIRP* 52/1:427.

Zheng, L. 2003. *A Study on the Trinocular Vision Inspection Technology.* Master's thesis, Tianjin, China: Tianjin University.

Zhu, J., J. Ni, and A. J. Shih. 2008. Robust machine tool thermal error modeling through thermal mode concept. *Trans ASME—J Manuf Sci Eng* 130:061006-1.

Zurcher, N. 1996. Using a coordinate measuring machine to calibrate step gages at world class levels of uncertainty. Report Y/AMT-327, Oak Ridge Centers for Manufacturing Technology.

Bibliography

Anbari, N., C. Beck, and H. Trumpold. 1990. The influence of surface roughness in dependence of the probe ball radius with measuring the actual size. *Ann CIRP* 39/1:577.

Balsamo, A. et al. 1997. Results of the CIRP-Euromet intercomparison of ball plate-based techniques for determining CMM parametric errors. *Ann CIRP* 46/1:463.

Balsamo, A. et al. 1999. Evaluation of CMM uncertainty through Monte Carlo simulations. *Ann CIRP* 48/1:425.

Barbato, G., R. Levi, and G. Vicario. 2006. Method of determining the uncertainty of a coordinate measuring machine. International Patent Number WO 2006/064352 Al.

Barini, E. M., G. Tosello, and L. De Chiffre. 2010. Uncertainty analysis of point-to-point sampling complex surfaces using touch probe CMMs, DOE for complex surfaces verification with CMMs. *Precis Eng* 34:16.

Beaman, J., and E. Morse. 2010. Experimental evaluation of software estimates of task specific measurement uncertainty for CMMs. *Precis Eng* 34:28.

Bourdet, P., C. Lartigue, and F. Leveaux. 1993. Effects of data point distribution and mathematical model on finding the best-fit sphere to data. *Precis Eng* 15:150.

Choi, W., T. Kurfess, and J. Cagan. 1998. Sampling uncertainty in coordinate measurement data analysis. *Precis Eng* 22:153.

Coorevits, T., J. David, and P. Bourdet. 1991. Elimination of geometrical errors by permutations—application to a rotary table. *Ann CIRP* 40/1:531.

Cox, M. G., A. B. Forbes, and G. N. Peggs. 2001. Simulation techniques for uncertainty estimation in coordinate metrology. NPL Report. London, UK: National Physical Laboratory.

De Chiffre, L., and H. N. Hansen. 1997. Results from an industrial comparison of coordinate measuring machines in Scandinavia. In *Proceedings of the XIV IMEKO World Congress*, vol. VIII, 22. Tampere, Finland: IMEKO.

DeVries, W. R., and C. J. Li. 1985. Algorithms to deconvolve stylus geometry from surface profile measurements. *J Eng Ind* 107:167.

Edgeworth, R., and R. G. Wilhelm. 1999. Measurement uncertainty due to workpiece error interaction with sampling period. In *Machining Impossible Shapes*, ed. G. J. Olling, B. K. Choi, and R. B. Jerard, 196–200. Norwell, MA: Kluwer Academic Publishers.

Estler, W. T. et al. 1997. Practical aspects of touch-trigger probe error compensation. *Precis Eng* 21/1:1.

Forbes, A. B., and P. M. Harris. 2000. *Simulated Instruments and Uncertainty Estimation*. NPL No. CMSC 01/00. London, UK: National Physical Laboratory.

Haitjema, H. 1992. Uncertainty propagation in surface plate measurements. In *Proceedings of the 4th International Symposium on Dimensional Metrology (ISMQC)*, 304. Tampere, Finland: IMEKO.

Haitjema, H. 1996. Iterative solution of least-squares problems applied to flatness and grid measurements. In *Advanced Mathematical and Computational Tools in Metrology II*, ed. P. Ciarlini et al., pp. 160–70. Singapore: World Scientific Publishing Co.

Haitjema, H., and M. Morel. 2000. The concept of a virtual roughness tester. In *Proceedings of the X International Colloquium on Surfaces*, ed. M. Dietzch and H. Trumpold, 239. Aachen, Germany: Shaker Verlag.

Hall, B. D. 2008. Evaluating methods of calculating measurement uncertainty. *Metrologia* 45:L5.

Hansen, H. N. 1998. A database system for uncertainty estimation in coordinate metrology. In *Proceedings of the 6th IMEKO Symposium "Metrology for Quality Control in Production,"* pp. 215–20. Vienna, Austria: IMEKO

Hansen, H. N., and L. De Chiffre. 1999. An industrial comparison of coordinate measuring machines in Scandinavia with focus on uncertainty statements. *Precis Eng* 23:185.

Hansen, H. N., L. De Chiffre, and E. Savio. 2001. Traceability in coordinate metrology. In *Proceedings of the PRIME 2001*, pp. 363–68. Sestri Levante, Italy.

Henke, R. P. et al. 1999. Methods for evaluation of systematic geometric deviations in machined parts and their relationships to process variables. *Precis Eng* 23(4):273.

Hopp, T. H. 1994. The sensitivity of three-point circle fitting, NISTIR 5501. Gaithersburg, MD: National Institute of Standards and Technology.

Hopp, T. H., and M. S. Levenson. 1995. Performance measures for geometric fitting in the NIST algorithm testing and evaluation program for coordinate measuring systems. *J Res NIST* 100:563.

ISO/TS 14253-3:2002. 2002. Geometrical Product Specifications (GPS)—Inspection by measurement of workpieces and measuring equipment—Part 3: Guidelines for achieving agreements on measurement uncertainty statements. Geneva, Switzerland: International Organization for Standardization.

ISO/TS 14253-4:2010. 2010. Geometrical Product Specifications (GPS)—Inspection by measurement of workpieces and measuring equipment—Part 4: Background on functional limits and specification limits in decision rules. Geneva, Switzerland: International Organization for Standardization.

Knapp, W., and E. Matthias. 1983. Test of the three-dimensional uncertainty of machine tools and measuring machines and its relation to the machine errors. *Ann CIRP* 32/1:459.

Knauer, M., and A. Weckenmann. 1999. A software system to estimate the uncertainty contribution of the workpiece shape in coordinate metrology. In *15th World Congress of the International Measurement Confederation—IMEKO*. Osaka, Japan: IMEKO.

Kruth, J.-P. et al. 2003. Interaction between workpiece and CMM during geometrical quality control in non-standard thermal conditions. *Precis Eng* 26:93.

Kunzmann, H., E. Trapet, and F. Wäldele. 1993. Concept for the traceability of measurements with coordinate measuring machines. In *Proceedings of the 7th International Precision Engineering Seminar*, 41. Kobe, Japan: Springer Verlag.

Kunzmann, H., E. Trapet, and F. Wäldele. 1995. Results of international comparison of ball plate measurements in CIRP and WECC. *Ann CIRP* 44/1:479.

Lingard, P. S. et al. 1991. Length bar and step-gauge calibration using a laser measurement system with a coordinate measuring machine. *Ann CIRP* 40/1:515.

McClure, E. R. 1967. Significance of thermal effects in manufacturing and metrology. *Ann CIRP* 15:61.

Meneghello, R., L. De Chiffre, and A. Balsamo. 2001. Precision of coordinate measurements in industry: AUDIT ITALIANO. In *Proceedings of the 2nd EUSPEN International Conference*, 346. Torino, Italy: euspen.

Miguel, P. C., T. King, and A. Abackerli. 1998. A review on methods for probe performance verification. *Measurement* 23:15.

Nawara, L., and M. Kowalski. 1984. The investigations on selected dynamical phenomena in the heads of multi-coordinate measuring devices. *Ann CIRP* 33/1:373.

Nawara, L., and M. Kowalski. 1985. Influence of the multicoordinate measuring machine head characteristic on circular profiles measurements. *Ann CIRP* 34/1:449.

Nawara, L., and M. Kowalski. 1987. Analysis of the random component of multicoordinate measuring machines and metrological robots position error. *Ann CIRP* 36/1:373.

Pfeifer, T., and G. Spur. 1994. Task specific gauge for the inspection of coordinate measuring machines. *Ann CIRP* 43/1:465.

Phillips, S. D., B. Borchardt, and G. Caskey. 1993. Measurement uncertainty considerations for coordinate measuring machines, *NISTIR 5170*. Gaithersburg, MD: National Institute of Standards and Technology.

Sammartini, M., and L. De Chiffre. 1998. A task specific gauge for pitch measurement of cylindrical gears. In *Proceedings of the 6th ISMQC.* Vienna, Austria: IMEKO.

Sammartini, M., and L. De Chiffre. 2000. Development and validation of a new reference cylindrical gear for pitch measurement. *Precis Eng* 24:302.

Savio, E., H. N. Hansen, and L. De Chiffre. 2002. Approaches to the calibration of freeform artefacts on coordinate measuring machines. *Ann CIRP* 51/1:433.

Schultschik, R., and E. Matthias. 1977. The components of the volumetric accuracy. *Ann CIRP* 26/1:229.

Schwenke, H. et al. 2000. Assessment of uncertainties in dimensional metrology by Monte Carlo simulation: Proposal of a modular and visual software. *Ann CIRP* 49/1:395.

Soons, H., and P. Schellekens. 1992. On the calibration of CMMs using distance measurements. In *Proceedings of the 4th ISMQC*, 321. Tampere, Finland: IMEKO.

Swyt, D. A. 1993. Issues, concepts, and standard techniques in assessing accuracy of coordinate measuring machines, *NIST TN 1400.* Gaithersburg, MD: National Institute of Standards and Technology.

Swyt, D. A. 1994. Uncertainties in dimensional measurements made at nonstandard temperatures. *J Res NIST* 99:31.

Swyt, D. A. 2001. NIST-ASME Workshop on uncertainty in dimensional measurements. *J Res NIST* 106:867.

Tanaka, H., M. Shimojo, and T. Sato. 2004. Evaluation of the geometrical uncertainty of helix deviation measurements using the Monte Carlo simulation. In *Summer Topical Meeting, Uncertainty Analysis in Measurement and Design*, ed. W. T. Estler and E. R. Marsh, 134. State College, PA: ASPE.

Trapet, E., and F. Wäldele. 1991. A reference artefact based method to determine the parametric error components of coordinate measuring machines and machine tools. *Measurement* 9:17.

Trapet, E., and F. Wäldele. 1996. The virtual CMM concept. In *Advanced Mathematical Tools in Metrology II*, 238. Singapore: World Scientific.

Trapet, E. et al. 1999. Traceability of coordinate measurements according to the method of the virtual measuring machine. Final Project Report MAT1-CT94-0076, PTB-report F-35, Part 1 and 2, ISBN 3-89701-330-4.

van Dorp, B. et al. 2001a. Virtual CMM using Monte Carlo methods based on frequency content of the error signal. In *Proceedings of the SPIE 4401 Recent Developments in Traceable Dimensional Measurements*, ed. J. E. Decker and N. Brown, 158. Bellingham, WA: SPIE.

van Dorp, B. et al. 2001b. Calculation of measurement uncertainty for multi-dimensional machines, using the method of surrogate data. In *Advanced Mathematical and Computational Tools in Metrology V*, ed. P. Ciarlini, M. G. Cox, E. Filipe, F. Pavese, and D. Richter. pp. 344–51. Singapore: World Scientific Publishing Co.

Yang, Q. D. et al. 1999. Study of the thermal deformations on a coordinate measuring machine and compensation via a neural network. *Int J Flex Autom Integr Manuf* 7(1&2):129.

Index

Printed in the United States
by Baker & Taylor Publisher Services

Printed in the United States
by Baker & Taylor Publisher Services